Measures and Differential Equations
in Infinite-Dimensional Space

Mathematics and Its Applications (*Soviet Series*)

Volume 76

Measures and Differential Equations in Infinite-Dimensional Space

by

Yu. L. Dalecky
Polytechnic Institute, Kiev, U.S.S.R.

and

S. V. Fomin†

SPRINGER SCIENCE+BUSINESS MEDIA, B.V.

Library of Congress Cataloging-in-Publication Data
available from the Publisher

ISBN 978-94-010-5148-4 ISBN 978-94-011-2600-7 (eBook)
DOI 10.1007/978-94-011-2600-7

This book has been typeset using AMS-LaTeX

SERIES EDITOR'S PREFACE

'Et moi, ..., si j'avait su comment en revenir, je n'y serais point allé.'

Jules Verne

The series is divergent; therefore we may be able to do something with it.

O. Heaviside

One service mathematics has rendered the human race. It has put common sense back where it belongs, on the topmost shelf next to the dusty canister labelled 'discarded non-sense'.

Eric T. Bell

Mathematics is a tool for thought. A highly necessary tool in a world where both feedback and non-linearities abound. Similarly, all kinds of parts of mathematics serve as tools for other parts and for other sciences.

Applying a simple rewriting rule to the quote on the right above one finds such statements as: 'One service topology has rendered mathematical physics ...'; 'One service logic has rendered computer science ...'; 'One service category theory has rendered mathematics ...'. All arguably true. And all statements obtainable this way form part of the raison d'être of this series.

This series, *Mathematics and Its Applications*, started in 1977. Now that over one hundred volumes have appeared it seems opportune to reexamine its scope. At the time I wrote

> "Growing specialization and diversification have brought a host of monographs and textbooks on increasingly specialized topics. However, the 'tree' of knowledge of mathematics and related fields does not grow only by putting forth new branches. It also happens, quite often in fact, that branches which were thought to be completely disparate are suddenly seen to be related. Further, the kind and level of sophistication of mathematics applied in various sciences has changed drastically in recent years: measure theory is used (non-trivially) in regional and theoretical economics; algebraic geometry interacts with physics; the Minkowsky lemma, coding theory and the structure of water meet one another in packing and covering theory; quantum fields, crystal defects and mathematical programming profit from homotopy theory; Lie algebras are relevant to filtering; and prediction and electrical engineering can use Stein spaces. And in addition to this there are such new emerging subdisciplines as 'experimental mathematics', 'CFD', 'completely integrable systems', 'chaos, synergetics and large-scale order', which are almost impossible to fit into the existing classification schemes. They draw upon widely different sections of mathematics."

By and large, all this still applies today. It is still true that at first sight mathematics seems rather fragmented and that to find, see, and exploit the deeper underlying interrelations more effort is needed and so are books that can help mathematicians and scientists do so. Accordingly MIA will continue to try to make such books available.

If anything, the description I gave in 1977 is now an understatement. To the examples of interaction areas one should add string theory where Riemann surfaces, algebraic geometry, modular functions, knots, quantum field theory, Kac-Moody algebras, monstrous moonshine (and more) all come together. And to the examples of things which can be usefully applied let me add the topic 'finite geometry'; a combination of words which sounds like it might not even exist, let alone be applicable. And yet it is being applied: to statistics via designs, to radar/sonar detection arrays (via finite projective planes), and to bus connections of VLSI chips (via difference sets). There seems to be no part of (so-called pure) mathematics that is not in immediate danger of being applied. And, accordingly, the applied mathematician needs to be aware of much more. Besides analysis and numerics, the traditional workhorses, he may need all kinds of combinatorics, algebra, probability, and so on.

In addition, the applied scientist needs to cope increasingly with the nonlinear world and the extra mathematical sophistication that this requires. For that is where the rewards are. Linear models are honest and a bit sad and depressing: proportional efforts and results. It is in the non-linear world that infinitesimal inputs may result in macroscopic outputs (or vice versa). To appreciate what I am hinting at: if electronics were linear we would have no fun with transistors and computers; we would have no TV; in fact you would not be reading these lines.

There is also no safety in ignoring such outlandish things as nonstandard analysis, superspace and anticommuting integration, p-adic and ultrametric space. All three have applications in both electrical engineering and physics. Once, complex numbers were equally outlandish, but they frequently proved the shortest path between 'real' results. Similarly, the first two topics named have already provided a number of 'wormhole' paths. There is no telling where all this is leading - fortunately.

Thus the original scope of the series, which for various (sound) reasons now comprises five subseries: white (Japan), yellow (China), red (USSR), blue (Eastern Europe), and green (everything else), still applies. It has been enlarged a bit to include books treating of the tools from one subdiscipline which are used in others. Thus the series still aims at books dealing with:

- a central concept which plays an important role in several different mathematical and/or scientific specialization areas;
- new applications of the results and ideas from one area of scientific endeavour into another;
- influences which the results, problems and concepts of one field of enquiry have, and have had, on the development of another.

Infinite dimensions tend to make people a bit uneasy: there is so much that can go wrong; and, in fact, when approached naively, things do go wrong. After all our basic intuition is geared to three-dimensional space (if one is a bit optimistic). This applies in particular to measure theory which even in finite dimensions contains plenty of pitfalls for the unwary.

A great deal of thought and research has gone into understanding infinite-dimensional measure theory, and, when looked at in the right way things are elegant and beautiful (as so often in mathematics). But, as explained most clearly in this volume, to reach that state of enlightenment, one has to throw away a number of preconceptions (such as the desirability of rigid motion invariant measures).

There are but very few systematic expositions of the subject. But there is one, by two authors who have contributed substantially, and who have had to use the theory developed. After all, for instance in probability, representation theory, and physics, it is hardly possible to avoid dealing with infinite-dimensional measure theory.

The shortest path between two truths in the real domain passes through the complex domain.

J. Hadamard

La physique ne nous donne pas seulement l'occasion de résoudre des problèmes ... elle nous fait pressentir la solution.

H. Poincaré

Never lend books, for no one ever returns them; the only books I have in my library are books that other folk have lent me.

Anatole France

The function of an expert is not to be more right than other people, but to be wrong for more sophisticated reasons.

David Butler

Bussum, September 1991

Michiel Hazewinkel

Contents

Preface

This preface was written after Sergei Vasilyevich Fomin died. In the summer of 1975 we decided to give two coordinated lecture courses on infinite-dimensional analysis at the Far-Eastern School of Mathematics. One August evening, in a little house on the shore of the Japanese Sea, having just finished a chess play, we were looking through the plan of Vladivostok's environments in view of tomorrow's excursion. A sudden heart attack caused Sergei Vasilyevich's instant death.

Sergei Vasilyevich was an excellent mathematician of wide range and also a remarkable man, these features being strongly related in him: his quiet goodwill greatly facilitated scientific contacts with him and stimulated teamwork.

Though he was often ill in his last years, he did not wish to give up, even for a day, his versatile activities. In the morning of that very day he had given a lecture at the School, and had been listening to lectures of colleagues.

Infinite-dimensional analysis was only one branch of his range of interests. Our collaboration in this field begun in 1963, during the preparation of our joint report to the All-Union Conference on Probability Theory in Tbilisi (see Dalecky–Fomin [1]). Until early 1970 we had presented the results in a number of review articles (Averbukh–Smolyanov–Fomin [1,2], Dalecky [5], Dalecky [8,9]), which form the basis of our work. In the process of preparing these we had given special lecture courses on measure theory in infinite-dimensional spaces at the universities of Moscow and Kiev, respectively, and had obtained a number of new results, both together (Dalecky–Fomin [2]) as well as independently. Some of these results, and related ideas, were reported in our joint report at the First Vilnius Conference on Probability Theory in 1973 (Dalecky–Fomin [3]).

By this time we had made a thorough plan for a book, had continued gathering material according to this plan, and had started to order this material. However, at the height of the work I was left alone, and either had to abandon it all or continue it with double effort and responsibility. Having chosen the second option, I hope that this book will be my best homage to the lasting memory of Sergei Vasilyevich.

Naturally, continuation of the work required more time than anticipated. I have added results obtained with the participation of Sergei Vasilyevich but published posthumously (Dalecky–Fomin [4], Smolyanov–Fomin [1]; I am indebted to Dr. O.G. Smolyanov for his collaboration and for submitting additional material in a form suitable for inclusion in this book). Moreover, I have taken into account results obtained after 1975. Some of these were stimulated by elaborating the manuscript. However, I have conscientiously tried to keep to the original plan of the book.

The result is the present book, which I hope to be completely in the spirit and ideas we have thought out in common.

Of course, I alone am responsible for deficiencies in the realization of our common intention, as well as for the character of the historical comments.

Kiev Yu.L. Dalecky

Preface to the English edition

In the years following the publication of this book in the Russian language, the ideas and results in it have been further developed in various directions. The new branch of 'stochastic analysis' has emerged, and these words are nowadays often encountered in the titles of workshops and conferences. The scope of 'stochastic analysis' and the scope of the present book are close. Nonetheless, I believe that this book has not become obsolete, and wish to express my gratitude to Prof. M. Hazewinkel, whose initiative stimulated the appearance of the English translation.

It was decided to be useful to supplement the Russian edition with new material. This material was prepared in collaboration with my pupils. The supplement to chapters IV–V, 'Cauchy problem for the simplest parabolic equation with essentially infinite-dimensional elliptic operator' (written together with Yu.V. Bogdansky), fills, to a certain extent, a gap mentioned in the comments to chapter IV, by presenting results related to the theory of Laplace–Lévy type differential operators. The supplement 'Stochastic regularization of the ill-posed abstract parabolic problem' (written in collaboration with N.Yu. Goncharuk), is a supplement to chapter VII, and concerns stochastic equations.

The most essential addition is contained in the newly written chapter VIII, 'Smooth measures' (written in collaboration with V.R. Steblovskaya). It concerns smooth measures in infinite-dimensional space. Roughly, these are the measures allowing integration by parts. Various results in chapter II related to absolute continuity and integral transformations are transferred to this case.

Note that in the meantime my book with Ya.I. Belopolskaya, 'Stochastic equations and differential geometry', has appeared, both in Russian and in English translation. This book aims to tranfer certain aspects of the present book to infinite-dimensional smooth manifolds.

Yu.L. Dalecky

Introduction

This book is concerned with certain aspects of analysis in infinite-dimensional spaces. We intend to generalize classical analytic concepts to functions of an infinite-dimensional argument and to related objects.

The natural need for constructing such a mathematical discipline was felt long ago, caused by a development in mathematical physics (variational calculus, in particular). Apparently, the necessity of creating it was first pointed out by J. Hadamard and V. Volterra (see Shilov [1] and the preface to Lévy [1]), and the first steps in this direction were taken by Volterra, R. Gâteaux, P. Lévy, and M. Fréchet.

Up to a certain stage, concepts in analysis can be transferred without complications to the infinite-dimensional case. This is true, in particular, for differential calculus and the simplest problems in the theory of differential equations, including the Cauchy problem for first-order equations (see Lévy [1], Averbukh–Smolyanov [1,2], Dieudonne [1], Dalecky–Kukharchuk [1]). One encounters serious difficulties when turning to integration and the theory of differential equations which, similarly to the classical equations in mathematical physics, cannot be studied without involving integration. Indeed, the theory of measure and integral in an arbitrary measurable space was elaborated rather long ago (see, e.g., Halmos [1]). In the given situation, however, we have to deal with integration theory as related to other structures of the space, since a substantial harmonic analysis is impossible without such a relation.

The first investigations in the theory of means on function spaces and in the corresponding theory of the Laplace operator (the Laplace–Lévy operator) were gathered in Lévy [1]. These studies were conducted by a number of Soviet mathematicians (the initiative by E.M. Polishchuk played an important role; see the references in the comments to chapt. IV). These directions are very interesting, and are, up till now, weakly related to other directions in analysis and its applications (of course, for us such relations must exist).

Another direction of integration theory, related to the development of the theory of stochastic processes and beginning with the work of N. Wiener and A.N. Kolmogorov, turned out to be more fruitful. The theory of Wiener integrals, elaborated in numerous papers by R. Cameron and W. Martin, clarified a number of features properly belonging to integration in function spaces and later studied in more detail for more general situations.

Infinite-dimensional differential equations also naturally arise in the theory of stochastic processes. First, they follow from semigroup actions generated by Markov stochastic processes in infinite-dimensional space. In this connection we note that the theory of stochastic equations, developed by K. Itô, I.I. Gikhman, and A.V. Skorokhod, turned out to be excellently adapted to infinite-dimensional generalizations (see Dalecky [8]). Secondly, they arise when studying so-called statistical solutions of evolution equations in classical mathematical physics (beginning with the paper Hopf [1]; see also the review Vishik–Komech–Fursikov [1]).

Probably, this probabilistic influence on infinite-dimensional analysis can be developed in an independent manner (see Simon [1] for a picturesque admission of the underestimation of probability theory in analysis). However, historically the way was not like this.

In the 1940s, powerful additional incentives to develop infinite-dimensional analysis emerged. In the first place these were related with R. Feynman's work on quantum mechanics and electrodynamics. He introduced and exploited the now famous 'Feynman integral' (though without proper mathematical justification, as is usual in physics; this has not prevented physicists to work with it for some time).

This work, which was related to the Schrödinger equation, has in return influenced the theory of stochastic processes, and has led to the rigorous mathematical results of M. Kac on diffusion equations.

On the other hand, in J. Schwinger's work on quantum electrodynamics there arose equations involving functional derivatives. To study them, an integration procedure in function spaces was needed (see the review Gelfand–Yaglom [1], as well as the historical comments and references in it). Also, the above-mentioned paper by Hopf on statistical hydromechanics, as well as investigations in statistical physics (see Bogolyubov [1]), were a source of equations in functional derivatives.

All these directions were a strong impetus for a mathematical justification. To us the critical moment appears to be 1956, when simultaneously there appeared the review Gelfand–Yaglom [1], the paper Prokhorov [1], and Segal [1], which was the first of a series of papers. Together with the subsequent book by Gelfand–Vilenkin [1], these papers had a considerable effect on further progress in the theory. This progress went in various directions, and we do not pretend to describe it here in full completeness. (A certain reflection of it can be found in the historical comments to the individual chapters in this book).

In any case, now, a quarter of a century after the described events, infinite-dimensional analysis is a completely independent discipline, tightly related to other more classical branches of mathematics and having important physical applications. For these reasons it is worthy of a coherent and consequent presentation. However, at this moment this aim seems too courageous, and the authors did not have it in mind when planning the book.

Instead, we intended to write a relatively short book, in which we would present the basic notions related with differential equations with respect to functions of an infinite-dimensional argument, as well as some methods for studying these. This would necessarily include the basics of measure and integration theory in function spaces.

However, measure theory plays yet another special role in infinite-dimensional analysis. Since there is no standard measure of Lebesgue type in infinite-dimensional space, one cannot canonically identify measures with generalized functions, and canonically couple smooth and generalized functions, a procedure which is so fruitful in finite-dimensional mathematical physics. Thus there arises the necessity to construct an analysis of measures parallel to the analysis of functions. Here, objects dual to smooth measures are generalized functions, and objects dual to smooth functions are generalized measures (distributions). Both constructions are naturally related by Fourier transformation (see Fomin [1,2]).

Already the field of differential equations includes a sufficient number of most varied results. When selecting the material, which involved our personal tastes and interests, we have mainly chosen material related to the study of evolution equations.

The approximate contents of the book is as follows.

In the first chapter we introduce the basic concepts related to measures and more general objects–quasimeasures (which arise when generalizing the notion of finite-dimensional distribution system of a random function to set functions of variable sign). Some conditions for the σ-additivity of cylindrical measures are given, including a sufficiently general version of Kolmogorov's classical theorem on quasimeasures.

The second chapter is devoted to the study of Gaussian measures in Hilbert space. These play a special role in infinite-dimensional analysis for several reasons. First, they naturally arise in problems of physics. Secondly, Gaussian measures possess a quasi-invariance property, allowing a rather consistent harmonic analysis. Finally, certain computations associated with them can be performed to full satisfaction. An abstract analogue of the Feynman measure—the Gaussian measure with complex parameter—is considered at the end of this chapter.

In the third chapter we turn to arbitrary Radon measures in linear topological spaces. Here, conditions for σ-additivity of the type of the Minlos–Sazonov theorem are presented, together with conditions for weak compactness of a family of measures. These conditions are related to results of Yu.V. Prokhorov.

These three chapters form the main part of the book, and are the base for the subsequent exposition. At certain points this part overlaps Bourbaki [2], Gikhman and Skorokhod [3], Kuo [1], and Skorokhod [4], but some questions are treated in a new manner.

The remaining chapters are more special, and are more closely related to the interests and results of the authors.

In the fourth chapter we present a differential calculus of measures and distributions in infinite-dimensional space, introduce differential operators on functions and distributions in infinite-dimensional space, and also consider some questions in harmonic analysis related to integral representations of generalized kernels.

In the fifth chapter we consider linear evolution differential equations with constant and variable coefficients, for functions and distributions in an infinite-dimensional space. The presentation is partly given in the spirit of L. Schwartz classical theory of distributions (see Gelfand–Shilov [1]). For equations with variable coefficients we give a method based on projection on finite-dimensional spaces, with subsequent limit transition.

In the sixth chapter we consider the relations between evolution equations and path (functional) integrals; in particular, representations for solutions of the type of the Feynman–Kac formula. We do not restrict ourselves to linear equations, and will obtain representations of solutions of nonlinear equations as integrals in a branching path space.

In chapter seven we approach certain questions considered in chapters five and six from another viewpoint, based on probabilistic considerations. Here the main tool is an infinite-dimensional version of the theory of Itô stochastic equations and operator multiplicative functionals on Markov stochastic processes. Comparison of the results shows that when this machinery is applicable (i.e., parabolic equations and diffusion type systems), it proves to be more effective than that used in the previous chapters.

To give a description of the material not included in this book is much more difficult, for a number of reasons. In the historical comments we have indicated certain results and trends similar to our exposition. These comments, as well as the references, do not pretend to be complete, and are, as usual, rather subjective. In any case, books and review articles are given for other directions, and additional references can be found in them.

A more serious study of modern mathematical physics should encompass certain other subjects not touched upon here, even not in the references. Such are: integration on infinite-dimensional nonlinear manifolds, infinite-dimensional Lie groups, and the branch which is sometimes called 'noncommutative probability theory'.

To understand the subject matter of the book, the reader is not required to have knowledge beyond that given in basic university courses on functional analysis, measure theory, probability theory, and the elementary theory of partial differential equations. For the reader's convenience we will recall at appropriate places the necessary additional notions from the theories of linear topological spaces and stochastic processes.

CHAPTER I

Measures and quasimeasures. Integration

In this chapter we introduce the basic concepts which will be systematically used throughout this book. We consider realvalued measures on measurable spaces, and give in the purely measurable situation an analogue of Kolmogorov's classical theorem on σ-additivity in product spaces. Further we consider more general objects on the algebra of cylinder sets, namely quasimeasures (in particular, cylindrical measures), and present some elementary notions concerning integration with respect to quasimeasures.

The reader is assumed to be familiar with the basic concepts of general measure and integration theory (see Halmos [1], Neveu [1]). Besides this, an elementary knowledge of general topology and the theory of linear operators is needed for understanding the subject matter of this chapter. For the reader's convenience, the supplement at the end of this chapter briefly reviews some information about topological linear spaces. This will partly be used in the present chapter and in chapter III.

1. Realvalued measures on algebras of sets

1.1. Premeasures. In this section we present some facts related to realvalued additive set functions, in a form convenient for our subsequent exposition.

We call a pair $(X,\mathcal{U})$, where X is a set and $\mathcal{U}$ is an algebra of subsets of it, a *measurable space*. If $\mathcal{U}$ is a σ-algebra, $(X,\mathcal{U})$ is called a *σ-measurable space*.

A *premeasure* on $(X,\mathcal{U})$ taking values in a linear space T is a mapping $\mu\colon \mathcal{U} \to T$ that has the additivity property:

$$\mu(A \cup B) = \mu(A) + \mu(B) \qquad (A, B \in \mathcal{U},\ A \cap B = \emptyset).$$

If T is a topological linear space, the premeasure μ will be called a *measure* if it has the σ-additivity property:

$$\mu\left(\cup_{k=1}^{\infty} A_k\right) = \sum_{k=1}^{\infty} \mu(A_k)$$

$$(A_k \in \mathcal{U},\ k = 1,2,\ldots, \quad \cup_{k=1}^{\infty} A_k \in \mathcal{U}, \quad A_k \cap A_j = \emptyset,\ k \neq j).$$

For a bounded premeasure, σ-additivity is equivalent to continuity: $\mu(A_k) \to 0$ if $A_k \downarrow \emptyset$ $(n \to \infty)$. Throughout this chapter, unless otherwise stated, we consider bounded premeasures with values in $\mathbf{R}^1$. If, in addition, $\mu(A) \geq 0$ $(A \in \mathcal{U})$, then μ is called *positive*.

For a realvalued premeasure μ we define the positive premeasures μ^+ and μ^-, on the same measurable space, by

$$\mu^+(E) = \sup_{A \subset E} \mu(A), \qquad \mu^-(E) = -\inf_{A \subset E} \mu(A),$$

so that

$$\mu(E) = \mu^+(E) - \mu^-(E).$$

The positive premeasure $|\mu|$: $|\mu|(E) = \mu^+(E) + \mu^-(E)$ is called the *total variation* of μ. If μ is a measure, then μ^+, μ^-, and $|\mu|$ are also measures. If $\mathcal{U}$ is a σ-algebra, then there is an $X_0 \in \mathcal{U}$ such that

$$\mu^+(E) = \mu(E \cap X_0), \qquad \mu^-(E) = -\mu(E \cap X \setminus X_0) \qquad (E \in \mathcal{U}).$$

A complexvalued premeasure μ can be naturally expressed in terms of realvalued premeasures μ_1, μ_2:

$$\mu = \mu_1 + i\mu_2, \qquad \text{where } \mu_1 = \Re\mu,\ \mu_2 = \Im\mu.$$

In this case the total variation can be defined as

$$|\mu|(E) = \sup \sum_{j=1}^{n} |\mu(E_j)|,$$

where $E = \cup_{j=1}^{n} E_j$ is a partition into finitely many pairwise disjoint measurable sets, and the supremum is taken over all such partitions. This definition can be transferred to measures taking values in Banach spaces.

For an algebra $\mathcal{U}_0$ we denote by $\mathcal{U} = \sigma(\mathcal{U}_0)$ the minimal σ-algebra containing it. Each measure on $\mathcal{U}_0$ can be uniquely extended to a measure on $\mathcal{U}$.

Let $(X, \mathcal{U})$, $(\tilde{X}, \tilde{\mathcal{U}})$ be a pair of measurable spaces, and let $f\colon X \to \tilde{X}$ be a measurable map: $f^{-1}(\tilde{\mathcal{U}}) = \{A = f^{-1}(\tilde{A})\colon \tilde{A} \in \tilde{\mathcal{U}}\} \subset \mathcal{U}$. Then an arbitrary premeasure μ on $(X, \mathcal{U})$ induces a premeasure $\tilde{\mu} = \mu_f$ on $(\tilde{X}, \tilde{\mathcal{U}})$: $\mu_f(A) = \mu(f^{-1}(\tilde{A}))$ $(\tilde{A} \in \tilde{\mathcal{U}})$, taking values in the same space as μ does.

1.2. Some tests for σ-additivity of premeasures.

DEFINITION 1.1. Let C be a set. A class $\mathcal{K}$ of subsets of it is called *compact* if for any sequence K_n $(n = 1, 2, \dots)$ of elements of $\mathcal{K}$ satisfying $\cap_{n=1}^{\infty} K_n = \emptyset$ there is an $n_0 \in \mathbf{N}$ such that also $\cap_{n=1}^{n_0} K_n = \emptyset$.

PROPOSITION 1.1. *If a class $\mathcal{K}$ of subsets of a set C is compact, then the minimal class $\mathcal{K}_\Delta$ of sets containing all sets of $\mathcal{K}$ and closed under taking finite unions is also compact.*

PROOF. For each $n = 1, 2, \dots$, let $\mathcal{D}_n = \cup_{m=1}^{m_n} K_n^m$, where $K_n^m \in \mathcal{K}$ for all m, n under consideration, and suppose that $\cap_{n=1}^{p} \mathcal{D}_n' \neq \emptyset$ for all $p \in \mathbf{N}$. If we show that $\cap_{n=1}^{\infty} \mathcal{D}_n \neq \emptyset$ as well, the proposition will be proved.

Let $\mathcal{I} = \prod_{n=1}^{\infty} \{1, \dots, m_n\}$ be the space of all sequences $\{j_n\}$ of natural numbers such that, for any n, $j_n \leq m_n$. For each p, let $\mathcal{I}_p = \{\{j_n\} \in \mathcal{I}\colon \cap_{n=1}^{p} K_n^{j_n} \neq \emptyset\}$. Since $\cap_{n=1}^{\infty} \mathcal{D}_n = \cup_{\mathcal{I}} \cap_{n=1}^{p} K_n^{j_n}$ for all p, and since $\cap_{n=1}^{p} \mathcal{D}_n \neq \emptyset$ by requirement, we find that $\mathcal{I}_p \neq \emptyset$ for all p. Moreover, if $p_1 < p_2$, then $\mathcal{I}_{p_1} \supset \mathcal{I}_{p_2}$. Since the space $\mathcal{I}$, being the product of a countable family of topological spaces with the discrete topology, is compact, it follows that $\cap_{p=1}^{\infty} \mathcal{I}_p \neq \emptyset$. Let $\{j_n\} \in \cap_{p=1}^{\infty} \mathcal{I}_p$. Then $\cap_{n=1}^{p} K_n^{j_n} \neq \emptyset$ for all p, and since $\mathcal{K}$ is a compact class, $\cap_{n=1}^{\infty} K_n^{j_n} \neq \emptyset$. However, the latter set is part of the set $\cap_{n=1}^{\infty} \mathcal{D}_n$, which thus turns out to be nonempty. □

DEFINITION 1.2. Let μ be a premeasure on an algebra (semi-algebra) $\mathcal{U}$ of subsets of a set X. We says that a class $\mathcal{K} \subset \mathcal{U}$ *approximates* μ *from below* if for each $A \in \mathcal{U}$ and $\epsilon > 0$ there is a $K \in \mathcal{K}$ such that $K \subset A$ and $|\mu|(A \setminus K) < \epsilon$.

THEOREM 1.1. *Let $\mathcal{U}$ be an algebra (semi-algebra) of subsets of a space X and μ a premeasure on $\mathcal{U}$. For μ to be a measure it suffices that $\mathcal{U}$ contains a compact class $\mathcal{K}$ approximating μ from below.*

PROOF. By proposition 1.1, the case when $\mathcal{U}$ is a semi-algebra reduces to the case when it is an algebra.

Thus, let $\mathcal{U}$ be an algebra. To prove the countable additivity of μ it suffices to establish that if for each natural number n, $A_n \in \mathcal{U}$, $A_1 \supset A_2 \supset \dots$, and $\cap_{n=1}^{\infty} A_n = \emptyset$, then $|\mu|(A_n) \to 0$ $(n \to \infty)$. Let $\epsilon > 0$. For each $n \in \mathbf{N}$, choose a set $K_n \in \mathcal{K}$ such that $K_n \subset A_n$ and $|\mu|(A_n \setminus K_n) \leq \epsilon/2^n$. Since $\cap_{n=1}^{\infty} K_n \subset \cap_{n=1}^{\infty} A_n = \emptyset$, there is a p such that $\cap_{n=1}^{p} K_n = \emptyset$. So,

$$A_p = A_p \setminus \cap_{n=1}^{p} K_n = \cup_{n=1}^{p} (A_p \setminus K_n) \subset \cup_{n=1}^{p} (A_n \setminus K_n),$$

and hence

$$|\mu|(A_p) \leq \sum_{n=1}^{p} |\mu|(A_n \setminus K_n) \leq \epsilon. \quad \square$$

COROLLARY 1.1. *Suppose that a class $\mathcal{K}$ approximating a premeasure μ from below is closed under countable intersection, and let the compactness condition in the theorem be replaced by the weaker requirement that for each $\epsilon > 0$ there is a $K_\epsilon \subset X$ such that the class $\{K \cap K_\epsilon \colon K \in \mathcal{K}\}$ is compact, and such that for each set $F \in \mathcal{K}$ not intersecting with K_ϵ we have $|\mu|(F) < \epsilon$.*

Then the premeasure μ is σ-additive, i.e. it is a measure.

PROOF. In fact, using the same notations, $\cap_{n=1}^{\infty}(K_n \cap K_\epsilon) = \emptyset$, and hence, for some p,

$$K_\epsilon \cap (\cap_{n=1}^{p} K_n) = \cap_{n=1}^{p} (K_n \cap K_\epsilon) = \emptyset.$$

This implies that $|\mu|(\cap_{n=1}^{p} K_n) < \epsilon$, and also

$$|\mu|(A_p) \leq |\mu| \left(\cap_{n=1}^{p} K_n\right) + |\mu| \left(\cup_{n=1}^{p} (A_n \setminus K_n)\right) \leq 2\epsilon. \quad \square$$

An important example of a measurable space is provided by a pair $(X, \mathcal{B})$, where X is a topological space and $\mathcal{B}$ is the σ-algebra of its Borel subsets. In this space the class $\mathcal{K} = \mathcal{K}_X$ of all countably compact subspaces is compact.

DEFINITION 1.3. A premeasure μ on $(X, \mathcal{B})$ is called a *Radon premeasure* if it is approximated from below by the class $\mathcal{K}_X$.

A σ-additive Radon premeasure is called a *Radon measure*.

PROPOSITION 1.2. *Every Radon premeasure on $(X, \mathcal{B})$ is σ-additive.*

This follows directly from theorem 1.1.

Corollary 1.1 can be interpreted as follows.

PROPOSITION 1.3. *Let μ be a premeasure on an algebra $\mathcal{U}$ of subsets of a topological space X, and let $\mathcal{F} \subset \mathcal{U}$ be a class which is closed under intersection, which approximates μ from below, and consisting of closed sets in X. If for every $\epsilon > 0$ there is a compact set K_ϵ such that for every $F \in \mathcal{F}$ it follows from $F \cap K_\epsilon = \emptyset$ that $|\mu|(F) < \epsilon$, then the premeasure μ is σ-additive.*

Now we examine to what extent these sufficient conditions for σ-additivity are also necessary.

PROPOSITION 1.4. *The class $\mathcal{F}$ of closed subsets of a separable metric space X approximates from below every measure on $(X, \mathcal{B})$.*

PROOF. Let $\mathcal{U}_0$ be the subclass of $\mathcal{B}$ consisting of the sets on which μ is approximated from below by $\mathcal{F}$ and closed under taking complements. It is readily verified that $\mathcal{U}_0$ is an algebra, as well as a monotone class. So $\mathcal{U}_0$ is a σ-algebra. Since each closed set in X is the intersection of a sequence of open sets, $\mathcal{U}_0$ clearly contains the class $\mathcal{F}$, which generates $\mathcal{B}$. Therefore $\mathcal{U}_0 = \mathcal{B}$. □

PROPOSITION 1.5. *Let X be a separable complete metric space, $\mathcal{B}$ its σ-algebra of Borel subsets, and μ a positive measure on X. Then there is for every $\epsilon > 0$ a compact set K in X such that $\mu(X \setminus K) < \epsilon$.*

PROOF. Since X is separable, it can be covered by countably many closed balls of radius 1. Choose finitely many balls $S_1^{(1)}, \dots S_{K_1}^{(1)}$ from this cover such that $\mu(\cup_{j=1}^{K_1} S_j^{(1)}) > 1 - \epsilon/2$. Consider now countably many balls of radius 1/2 covering X, and choose a finite subfamily $S_1^{(2)}, \dots S_{K_2}^{(2)}$ from this cover such that $\mu(\cup_{j=1}^{K_2} S_j^{(2)}) > 1 - \epsilon/4$, etc. In the same manner, at the nth step we construct a finite family $S_1^{(n)}, \dots S_{K_n}^{(n)}$ of closed balls of radius $1/n$ and such that $\mu(\cup_{j=1}^{K_n} S_j^{(n)}) > 1 - \epsilon/2^n$. The set $K = \cap_{n=1}^{\infty} \cup_{j=1}^{K_n} S_j^{(n)}$ is obviously closed and totally bounded, and, since X is complete, thus K is a compactum. Moreover, $\mu(X \subset K) \le \epsilon/2 + \dots + \epsilon/2^n + \dots = \epsilon$. □

1.3. Measurable and topological Radon spaces. In connection with the questions under consideration it is useful to introduce the following notions.

DEFINITION 1.4. A measurable space $(X, \mathcal{U})$ is called a *Radon space* if it contains a compact subclass $\mathcal{K}$ approximating from below every measure on $(X, \mathcal{U})$. We will also say that $(X, \mathcal{U}, \mathcal{K})$ is a (measurable) Radon space.

We call a topological space X a *(topological) Radon space* if $(X, \mathcal{B}, \mathcal{K}_X)$ is a Radon space.

In other words, X is a topological Radon space if every σ-additive realvalued measure on $(X, \mathcal{B})$ is a Radon measure.

THEOREM 1.2. *A separable complete metric space is a topological Radon space.*

PROOF. By proposition 1.4, for each $\epsilon > 0$ and Borel set A there is a closed set $F \subset A$ for which $|\mu|(A \setminus F) < \epsilon/2$. The set F is, together with X, a separable complete metric space, and by proposition 1.5 there is a compact set K in F such that $|\mu|(F \setminus K) < \epsilon/2$. Then $|\mu|(A \setminus K) < \epsilon$. □

The class of measurable Radon spaces is rather large, since it is closed under a number of natural operations.

PROPOSITION 1.6. *Let* $(X,\mathcal{U},\mathcal{K})$ *be a measurable Radon space,* $f\colon Y \to X$, $\mathcal{U}_f = f^{-1}(\mathcal{U}) = \{B \subset Y\colon f(B) \in \mathcal{U}\}$. *If the class* $\mathcal{K}_f = f^{-1}(\mathcal{K})$ *is compact in* Y, *then* $(Y,\mathcal{U}_f,\mathcal{K}_f)$ *is a measurable Radon space. This is true, in particular, if* f *is a surjective map.*

PROOF. Let μ be a measure on $(Y,\mathcal{U}_f)$. Then $\mu_f(A) = \mu(f^{-1}(A))$ is a measure on $(X,\mathcal{U})$, and, by requirement, for each $A \in \mathcal{U}$ and $\epsilon > 0$ there is a $K \in \mathcal{K}$ such that $|\mu_f|(A \setminus K) < \epsilon$. So, for each $A_1 \in \mathcal{U}$, $A_1 \subset A \setminus K$, we have $|\mu_f(A_1)| < \epsilon$. Then also $|\mu(f^{-1}(A_1))| < \epsilon$. Since $f^{-1}(A_1)$ is an arbitrary $\mathcal{U}_f$-measurable set satisfying $f^{-1}(A_1) \subset f^{-1}(A) \setminus f^{-1}(K)$, we have $|\mu|(f^{-1}(A) \setminus f^{-1}(K)) < \epsilon$.

If f is also surjective, the assertion follows since $\cap_j f^{-1}(K_j) = \emptyset$ implies $\cap_j K_j = \emptyset$. □

COROLLARY 1.2. *Let* $Y \subset X$, $\mathcal{U}_Y = \{A \cap Y\colon A \in \mathcal{U}\}$, *and suppose that the class* $\mathcal{K}_Y = \{K \cap Y\colon K \in \mathcal{K}\}$ *is compact. Then* $(Y,\mathcal{U}_Y,\mathcal{K}_Y)$ *is a measurable Radon space. In particular, a closed subspace of a topological Radon space is a topological Radon space.*

DEFINITION 1.5. Let $\mathcal{U}_\lambda$ $(\lambda \in \Lambda)$ be a family of algebras of subsets of a space X such that for each pair $\lambda_1, \lambda_2 \in \Lambda$ there is a $\lambda \in \Lambda$ with $\mathcal{U}_{\lambda_1} \cup \mathcal{U}_{\lambda_2} \subset \mathcal{U}_\lambda$. Then the family $\mathcal{U}_\lambda$ is called *directed*, and the algebra $\mathcal{U} = \cup_\Lambda \mathcal{U}_\lambda = \lim_\Lambda \mathcal{U}_\lambda$ is said to be its *limit*.

Also, the measurable space $(X,\mathcal{U})$ will be called the *limit* of the directed family of measurable spaces $(X,\mathcal{U}_\lambda)$ $(\lambda \in \Lambda)$, and $(X,\mathcal{U})$ will be said to have *limit structure*.

PROPOSITION 1.7. *Let* $\mathcal{U}_\lambda$ $(\lambda \in \Lambda)$ *be a directed family of algebras of subsets of a space* X, $\mathcal{U} = \lim_\Lambda \mathcal{U}_\lambda$, $\mathcal{K} \subset \mathcal{U}$ *a compact class, and* $\mathcal{K}_\lambda = \mathcal{K} \cap \mathcal{U}_\lambda$. *If every* $(X,\mathcal{U}_\lambda,\mathcal{K}_\lambda)$ *is a measurable Radon space, then* $(X,\mathcal{U},\mathcal{K})$ *is a measurable Radon space.*

PROOF. Let μ be a measure on $(X,\mathcal{U})$. The restriction $|\mu|_\lambda = |\mu|_{\mathcal{U}_\lambda}$ is a measure on $(X,\mathcal{U}_\lambda)$, which is approximated from below by the compact class $\mathcal{K}_\lambda$, by requirement. Since each $A \in \mathcal{U}$ lies in some $\mathcal{U}_\lambda$, on it the measure μ is approximated from below by the class $\mathcal{K}_\lambda$, and hence also by the class $\mathcal{K} \subset \mathcal{K}_\lambda$. □

Consider a family of measurable spaces $(X_\lambda,\mathcal{U}_\lambda)$ $(\lambda \in \Lambda)$. Take finitely many indices $\lambda_1,\dots,\lambda_n$, and consider the product $X_{\lambda_1,\dots,\lambda_n} = (X_{\lambda_1} \times \dots \times X_{\lambda_n}, \mathcal{U}_{\lambda_1,\dots,\lambda_n})$, where the algebra $\mathcal{U}_{\lambda_1,\dots,\lambda_n}$ is generated by the products $A_{\lambda_1} \times \dots \times A_{\lambda_n}$ of sets $A_{\lambda_j} \in \mathcal{U}_{\lambda_j}$ $(j = 1,\dots,n)$. Let now $X = \prod_\Lambda X_\lambda$ be the space consisting of all sets $\phi = \{\phi_\lambda \in X_\lambda\colon \lambda \in \Lambda\}$. For each set of indices $\lambda_1,\dots,\lambda_n$ the following projection is defined:

$$\pi_{\lambda_1,\dots,\lambda_n}\colon X \to X_{\lambda_1,\dots,\lambda_n}, \quad \pi_{\lambda_1,\dots,\lambda_n}\phi = (\phi_{\lambda_1},\dots,\phi_{\lambda_n}).$$

Consider the following algebras in X:

$$\tilde{\mathcal{U}}_{\lambda_1,\dots,\lambda_n} = \pi^{-1}_{\lambda_1,\dots,\lambda_n}\mathcal{U}_{\lambda_1,\dots,\lambda_n} \quad (\lambda_1,\dots,\lambda_n \in \Lambda).$$

These algebras form a directed family, whose limit,

$$\mathcal{U} = \cup_{n=1}^{\infty} \cup_{(\lambda_1,\dots,\lambda_n)} \tilde{\mathcal{U}}_{\lambda_1,\dots,\lambda_n},$$

is called the *algebra of cylinder sets* of the product $X = \prod_\Lambda X_\lambda$ and is denoted by $\mathcal{U} = \prod_\Lambda \mathcal{U}_\lambda$.

DEFINITION 1.6. The measurable space $(\prod_\Lambda X_\lambda, \prod_\Lambda \mathcal{U}_\lambda)$ will be called the *product* of the family of measurable spaces $(X_\lambda, \mathcal{U}_\lambda)$ $(\lambda \in \Lambda)$.

Hereafter we will denote cylinder sets as follows:

$$\sigma_{\lambda_1,\dots,\lambda_n}(B) = \{\phi \in X : \pi_{\lambda_1,\dots,\lambda_n}\phi \in B\},$$
$$(B \in \mathcal{U}_{\lambda_1,\dots,\lambda_n}, \quad \lambda_1, \dots, \lambda_n \in \Lambda).$$

PROPOSITION 1.8. *Let $\mathcal{K}_\lambda$ be a compact class in $(X_\lambda, \mathcal{U}_\lambda)$ $(\lambda \in \Lambda)$. Then the class $\mathcal{K} = \cup_n \cup_{(\lambda_1,\dots,\lambda_n)} \tilde{\mathcal{K}}_{\lambda_1,\dots,\lambda_n}$, where $\tilde{\mathcal{K}}_{\lambda_1,\dots,\lambda_n}$ is the class consisting of finite unions of the cylinder sets*

$$\sigma_{\lambda_1,\dots,\lambda_n}(K_{\lambda_1} \times \dots \times K_{\lambda_n}) \quad (K_{\lambda_j} \in \mathcal{K}_{\lambda_j}, \quad j = 1, \dots, n), \tag{1.1}$$

is compact in the product $(\prod_\Lambda X_\lambda, \prod_\Lambda \mathcal{U}_\lambda)$.

PROOF. Each set of the form (1.1) can be presented as $\sigma_{\lambda_1,\dots,\lambda_n}(K_{\lambda_1} \times \dots \times K_{\lambda_n}) = \prod_\Lambda K_\lambda$, where

$$K_\lambda = \begin{cases} K_{\lambda_j} & (\lambda = \lambda_j, \quad j = 1, \dots, n), \\ X_\lambda & (\lambda \neq \lambda_1, \dots, \lambda_n). \end{cases}$$

Let $K^j = \prod_\Lambda K_\lambda^j \in \mathcal{K}$ $(j = 1, 2, \dots)$ and $\cap_{j=1}^\infty K^j = \emptyset$. Then there is a $\lambda \in \Lambda$ such that $\cap_{j=1}^\infty K_\lambda^j = \emptyset$ and, since $\mathcal{K}_\lambda$ is compact, for some n_λ we have $\cap_{j=1}^{n_\lambda} \mathcal{K}_\lambda^j = \emptyset$. Clearly, also $\cap_{j=1}^{n_\lambda} K^j = \emptyset$. It remains to apply proposition 1.1. □

PROPOSITION 1.9. *Let $(X_\lambda, \mathcal{U}_\lambda, \mathcal{K}_\lambda)$ be a family of measurable Radon spaces, and $\mathcal{K}$ the class of sets (1.1). Then the product $(\prod X_\lambda, \prod \mathcal{U}_\lambda, \mathcal{K})$ is a measurable Radon space.*

PROOF. By proposition 1.7 it suffices to show that each $(X, \tilde{\mathcal{U}}_{\lambda_1,\dots,\lambda_n}, \tilde{\mathcal{K}}_{\lambda_1,\dots,\lambda_n})$ is a measurable Radon space. Since the map $\pi_{\lambda_1,\dots,\lambda_n}$ is surjective, this reduces to proving that the finite product $(X_{\lambda_1,\dots,\lambda_n}, \mathcal{U}_{\lambda_1,\dots,\lambda_n}, \mathcal{K}_{\lambda_1,\dots,\lambda_n})$ is Radon.

Let μ be a σ-additive measure on $X_{\lambda_1,\dots,\lambda_n}$. It can be assumed to be positive (by taking its variation, if necessary). Let μ_j be its image under the map $\pi_j : X_{\lambda_1,\dots,\lambda_n} \to X_{\lambda_j}$. Since $\mathcal{U}_{\lambda_1,\dots,\lambda_n}$ consists of finite unions of products $A = \prod_{k=1}^n A_k$, it suffices to approximate from below the measure μ on every such product. Let $K_j \subset A_{\lambda_j}$, $K_j \in \mathcal{K}_{\lambda_j}$, and $\mu_{\lambda_j}(A_{\lambda_j} \setminus K_j) < \epsilon/2^j$. Then $A \setminus \prod_{j=1}^n K_{\lambda_j} \subset \cup_{j=1}^n A_{\lambda_1} \times \dots \times A_{\lambda_{j-1}} \times (A_{\lambda_j} \setminus K_{\lambda_j}) \times A_{\lambda_{j+1}} \times \dots \times A_{\lambda_n}$, and $\mu(A \setminus \prod_{j=1}^n K_{\lambda_j}) \leq \sum_j \mu_{\lambda_j}(A_{\lambda_j} \setminus K_{\lambda_j}) < \epsilon$. □

REMARK 1.1. It does not yet follow from proposition 1.9 that a product of topological Radon spaces is a topological Radon space, although this product has the structure of a measurable Radon space (determined by cylinder sets). However, this can be established under additional assumptions.

Let $X = \prod_{n=1}^{\infty} X_n$ be a countable product of topological Radon spaces X_n whose compact subsets are metrizable. Then X is a topological Radon space.

In fact, a product $K = \prod_{n=1}^{\infty} K_n$ of compact sets $K_n \subset X_n$ for which $\mu(X \setminus K) < \epsilon$ can be constructed in a manner analogous to the above. Since K is a metrizable compactum, the restriction of μ to it is a Radon measure, and this implies the assertion. As a consequence of this assertion and corollary 1.2 we find that the projective limit of countably many topological Radon spaces having the property that compact subsets are metrizable, is a topological Radon space.

PROPOSITION 1.10. *Let $(X_j, \mathcal{U}_j, \mathcal{K}_j)$ $(j = 1, 2, \dots)$ be a sequence of measurable Radon spaces satisfying the conditions*

(1) $X_1 \subset X_2 \subset \cdots \subset X_n \subset \dots$;

(2) $\mathcal{U}_n = \{A = \tilde{A} \cap X_n\colon \tilde{A} \in \mathcal{U}_{n+1}\}$.

Put $X = \cup_{n=1}^{\infty} X_n$; $\mathcal{U} = \{A\colon A \cap X_n \in \mathcal{U}_n;\ n = 1, 2, \dots\}$, $\mathcal{K} = \cup_{n=1}^{\infty} \mathcal{K}_n$. Then $(X, \mathcal{U}, \mathcal{K})$ is a measurable Radon space.

PROOF. Let μ be a σ-additive measure on $(X, \mathcal{U})$. Since $X = \lim_{n\to\infty} X_n$, for each $\epsilon > 0$ there is an n_0 such that $|\mu|(X \setminus X_n) < \epsilon$ $(n \geq n_0)$. Let $A \in \mathcal{U}$. Consider $A \cap X_n \in \mathcal{U}_n$. Then there is a $K_n \subset A \cap X_n$ for which $|\mu|((A \cap X_n) \setminus K_n) < \epsilon$, and $|\mu|(A \setminus K_n) < 2\epsilon$. □

COROLLARY 1.3. *Let X be the topological sum $X = \cup_{n=1}^{\infty} X_n$ of a monotone increasing sequence of topological Radon spaces, the topology in X_n being consistent with the one induced by X_{n+1}, and let X_n be closed in X_{n+1}.*

Then X is a topological Radon space. In particular, the strict inductive limit of an increasing sequence of topological, locally convex, linear Radon spaces is a topological Radon space.

PROPOSITION 1.11. *Let $E_1 \subset E_2 \subset \dots$ be an increasing sequence of Banach spaces with completely continuous imbeddings. Then the space $E = \lim\operatorname{ind} E_n$ is a topological Radon space.*

PROOF. Let μ be a countably additive measure on the σ-algebra $\mathcal{B}_E$ of Borel subsets of E, $A \in \mathcal{B}_E$, and $\epsilon > 0$. Let S_n be the unit ball in E_n. Then its closure $\overline{S_n}^{n+1}$ in E_{n+1} is compact in this space. Since for every n the imbedding $E_n \to E$ is continuous, the set $\overline{S_n}^{n+1}$ is compact in E as well. Moreover, it is metrizable (the E_n are Banach spaces). Therefore, by Suslin's theorem, S_n is a Borel subset in E; since $E_n = \cup_{k=1}^{\infty} kS_n$, E_n is also a Borel set in E. The countable additivity of μ implies that $\mu(E \setminus E_n) < \epsilon/2$ for some n, and E_n being Radon implies that there is a compact set K in E_n such that $\mu((A \cap E_n) \setminus K) < \epsilon/2$. Thus $\mu(A \setminus K) < \epsilon$. □

EXAMPLE 1.1. If G is a compact subset of an n-dimensional space, then $\mathcal{D}'_G$, the Schwarz space of distributions, is a Radon space (since it belongs to the class of spaces considered in the previous section). Therefore the space $\mathcal{D}'_{\mathbf{R}^n}$ is a Radon space as well, since $\mathcal{D}' = \lim_j \operatorname{proj} \mathcal{D}'_{G_j}$, where $G_1 \subset G_2 \subset \dots$ and $\cup_{n=1}^{\infty} G_n = \mathbf{R}^n$.

1.4. Cylindrical measures. We will now consider an important and widespread construction, enabling us to examine the σ-additivity of a premeasure. Let a measurable space $(X,\mathcal{U})$ be the limit of a directed family of measurable spaces $(X_\lambda,\mathcal{U}_\lambda)$ $(\lambda \in \Lambda)$.

DEFINITION 1.7. A premeasure μ on $(X,\mathcal{U})$ is called a (bounded) *quasimeasure* if its restrictions $\mu_\lambda = \mu|_{\mathcal{U}_\lambda}$ are σ-additive (i.e. are measures), for all $\lambda \in \Lambda$.

Bounded quasimeasures are sometimes called *cylindrical measures*.

PROPOSITION 1.12. *The total variation $|\mu|$ of a bounded quasimeasure μ is also a bounded quasimeasure.*

PROOF. Since the total variation of a premeasure is a premeasure, we only have to prove that the restrictions $|\mu|_\lambda$ are σ-additive. Let $B^1, B^2, \dots, B^n, \dots$ be a sequence of sets from $\mathcal{U}_\lambda$, $B^j \cap B^k = \emptyset$ $(j \neq k)$, and $B = \cup_{j=1}^\infty B^j$. The additivity of $|\mu|_\lambda$ implies:

$$\sum_{j=1}^\infty |\mu|_\lambda(B^j) \le |\mu|_\lambda(B),$$

and it remains to prove the opposite inequality.

Let $F^r \in \mathcal{U}$ $(r = 1, \dots, n)$ be a finite collection of sets having the properties $B \supset F^r$, $F^{r_1} \cap F^{r_2} = \emptyset$ $(r_1 \neq r_2)$. Then

$$\mu(F^r) = \mu\left(\cup_{j=1}^\infty (F^r \cap B^j)\right) = \sum_{j=1}^\infty \mu(F^r \cap B^j) \quad (r = 1, \dots, n),$$

because there exist algebras $\mathcal{U}_{\bar\lambda}$, containing F^r and $\mathcal{U}_\lambda$, on which μ is σ-additive.

Therefore,

$$\sum_{r=1}^n |\mu(F^r)| \le \sum_{r=1}^n \sum_{j=1}^\infty |\mu(F^r \cap B^j)| = \sum_{j=1}^\infty \sum_{r=1}^n |\mu(F^r \cap B^j)| \le$$
$$\sum_{j=1}^\infty |\mu|(B^j) = \sum_{j=1}^\infty |\mu|_\lambda(B^j). \quad \square$$

THEOREM 1.3. *Let $(X,\mathcal{U}) = \lim_\Lambda (X,\mathcal{U}_\lambda)$, and suppose there is a compact class $\mathcal{K} \subset \mathcal{U}$ such that every $(X,\mathcal{U}_\lambda,\mathcal{K}_\lambda)$, where $\mathcal{K}_\lambda = \mathcal{K} \cap \mathcal{U}_\lambda$, is a measurable Radon space. Then every bounded quasimeasure on $(X,\mathcal{U})$ is a measure.*

PROOF. Let μ be a bounded quasimeasure on $(X,\mathcal{U})$. Then $|\mu|$ is also a bounded quasimeasure, and thus μ may be assumed to be positive. Each set $A \in \mathcal{U}$ belongs to some $\mathcal{U}_\lambda$, on which the restriction of μ is σ-additive and, by requirement, approximated from below by the class $\mathcal{K}_\lambda$ and consequently by $\mathcal{K}$. It remains to apply theorem 1.1. $\square$

THEOREM 1.4. *(Kolmogorov's theorem.) Let $(X,\mathcal{U}) = \prod_\Lambda (X_\lambda,\mathcal{U}_\lambda)$ be the product of measurable Radon spaces $(X_\lambda,\mathcal{U}_\lambda) = (X_\lambda,\mathcal{U}_\lambda,\mathcal{K}_\lambda)$. Then every bounded quasimeasure on $(X,\mathcal{U})$ is a measure.*

PROOF. By the definition of product, $(X,\mathcal{U})$ is the limit of a directed family: $(X,\mathcal{U}) = \lim_{(\lambda_1,\ldots,\lambda_n)}(X,\mathcal{U}_{\lambda_1,\ldots,\lambda_n})$.

It follows from proposition 1.8 that there is a compact class (1.1) for which the conditions of theorem 1.3 hold. □

REMARK 1.2. It follows from the above that the conditions of theorem 1.4 hold if every X_λ is a separable complete metric space with the σ-algebra $\mathcal{U}_\lambda = \mathcal{B}_\lambda$ of Borel sets, or a space obtained from such spaces by means of an arbitrary sequence of passage to the projective limit, taking the topological sum, or transition to a subspace.

2. Cylinder sets and cylindrical functions

2.1. General definition of cylinder set. On a set X, measurable space structure can be constructed by mapping X into other measurable spaces. We have already come across this trick in the construction of a product of measurable spaces. This trick will be used and developed in this section for a linear space X in particular; such X will be our main object in the sequel.

Let $(E_\lambda, \mathcal{B}_\lambda)$ $(\lambda \in \Lambda)$ be a collection of σ-measurable spaces. In this situation it will be termed the standard collection.

Let $F = \{f_\lambda\colon\ \lambda \in \Lambda\}$ be a family of surjective (this clearly does not restrict the generality) maps

$$f_\lambda\colon X \to E_\lambda.$$

This family is said to be *consistent* if for each pair $\lambda_1, \lambda_2 \in \Lambda$ there are a $\lambda = \lambda(\lambda_1,\lambda_2) \in \Lambda$ and a pair of surjective maps

$$\phi_{\lambda_j\lambda}\colon (E_\lambda, \mathcal{B}_\lambda) \to (E_{\lambda_j}, \mathcal{B}_{\lambda_j}) \quad (j = 1,2)$$

for which the diagram

$$\begin{array}{ccc} & f_{\lambda_1} \nearrow & E_{\lambda_1} \\ & & \uparrow \phi_{\lambda_1\lambda} \\ X & \xrightarrow{f_\lambda} & E_\lambda \\ & & \downarrow \phi_{\lambda_2\lambda} \\ & f_{\lambda_2} \searrow & E_{\lambda_2} \end{array} \tag{2.1}$$

is commutative. The maps $\phi_{\lambda_j\lambda}$ are called *connecting maps* for this consistent family. The analogous compatibility condition clearly holds for any finite set $\lambda_1, \ldots, \lambda_n \in \Lambda$: there are a $\lambda \in \Lambda$ and measurable maps $\phi_{\lambda_j\lambda}$ $(j = 1, \ldots, n)$ such that

$$f_{\lambda_k} = \phi_{\lambda_k\lambda} \circ f_\lambda \quad (k = 1, \ldots, n). \tag{2.2}$$

Every map f_λ determines a σ-algebra in X. The family thus obtained is easily seen to be directed: $\mathcal{U}_{\lambda(\lambda_1,\lambda_2)} \supset \mathcal{U}_{\lambda_1} \cup \mathcal{U}_{\lambda_2}$.

Let $\mathcal{U} = \lim_\Lambda \mathcal{U}_\lambda$. Elements of the algebra $\mathcal{U}$ (which is not, in general, a σ-algebra) will be called *F-cylinder sets*, and elements of its σ-subalgebra $\mathcal{U}_\lambda$ will be called *λ-cylinder sets*. Accordingly, $\mathcal{U}$ will be called the *algebra of F-cylinder sets* of X.

We will write $\lambda_1 > \lambda_2$ if $\mathcal{U}_{\lambda_1} \supset \mathcal{U}_{\lambda_2}$ (although, unlike for σ-algebras $\mathcal{U}_\lambda$, here the index set is not a directed sequence). Put $C(\lambda, B) = \{x \in X\colon\ f_\lambda(x) \in B\} = f_\lambda^{-1}(B)$ $(B \in \mathcal{B}_\lambda)$. B is called the *base*, f_λ the *directing map*, and E_λ the *support* of the cylinder set $C(\lambda, B)$. For a cylinder set $C(\lambda, B)$ these three characteristics are clearly not uniquely determined: under the conditions of the diagram (2.1) we have

$$C(\lambda_1, B_1) = C(\lambda, \phi^{-1}_{\lambda_1\lambda}(B_1)) \quad (B_1 \in \mathcal{B}_\lambda, \quad \lambda > \lambda_1). \tag{2.3}$$

We will now consider maps between measurable spaces with the above defined structure. Let $(X,\mathcal{U})$, $(\tilde{X},\tilde{\mathcal{U}})$, $\mathcal{U} = \lim_\Lambda \mathcal{U}_\lambda$, $\tilde{U} = \lim_{\tilde{\Lambda}} \tilde{\mathcal{U}}_{\tilde{\lambda}}$, be a pair of such spaces. A map $f\colon X \to \tilde{X}$ is called *measurable* (*$(\Lambda, \tilde{\Lambda})$-measurable*) if for each $\tilde{\lambda} \in \tilde{\Lambda}$ there is a $\lambda \in \Lambda$ for which f is $(\mathcal{U}_\lambda, \tilde{\mathcal{U}}_{\tilde{\lambda}})$-measurable, i.e.

$$f^{-1}(A_{\tilde{\lambda}}) \in \mathcal{U}_\lambda \quad (A_{\tilde{\lambda}} \in \tilde{\mathcal{U}}_{\tilde{\lambda}}).$$

A $(\Lambda, \tilde{\Lambda})$-measurable map is clearly $(\mathcal{U},\tilde{\mathcal{U}})$-measurable, but the converse is not true, in general.

If X is a space and $(\tilde{X},\tilde{\mathcal{U}})$ is a measurable space with limit structure, then a map $f\colon X \to \tilde{X}$ automatically determines the limit structure on $(X, f^{-1}(\tilde{\mathcal{U}}))$, where $f^{-1}(\tilde{\mathcal{U}}) = \lim_{\tilde{\Lambda}} f^{-1}(\tilde{\mathcal{U}}_{\tilde{\lambda}})$, for which it is $(\Lambda, \tilde{\Lambda})$-measurable.

If $\tilde{\mathcal{U}}$ is an algebra of F-cylinder sets, then $\mathcal{U} = f^{-1}(\tilde{\mathcal{U}})$ is an algebra of $(F \circ f)$-cylinder sets, determined by the family of maps

$$F \circ f = \{f_{\tilde{\lambda}} \circ f : \tilde{\lambda} \in \tilde{\Lambda}\}.$$

Consider two measurable limit structures $(X,\mathcal{U} = \lim_\Lambda \mathcal{U}_\lambda)$ and $(X,\tilde{\mathcal{U}} = \lim_{\tilde{\Lambda}} \tilde{\mathcal{U}}_{\tilde{\lambda}})$ on the same space X. We will say that $\mathcal{U}$ *majorizes* (or *is stronger than*) $\tilde{\mathcal{U}}$, denoted by $\tilde{\mathcal{U}} < \mathcal{U}$, if the identity map id: $X \to X$ is $(\Lambda, \tilde{\Lambda})$-measurable, i.e. for each $\tilde{\lambda}$ there is a λ such that $\tilde{\mathcal{U}}_{\tilde{\lambda}} \subset \mathcal{U}_\lambda$. Clearly, also $\tilde{\mathcal{U}} \subset \mathcal{U}$.

Two structures $(X,\mathcal{U} = \lim_\Lambda \mathcal{U}_\lambda)$ and $(X,\tilde{\mathcal{U}} = \lim_{\tilde{\Lambda}} \tilde{\mathcal{U}}_{\tilde{\lambda}})$ are called *equivalent* if $\tilde{\mathcal{U}} < \mathcal{U}$ and $\mathcal{U} < \tilde{\mathcal{U}}$. Of course, in this case the measurable sets are the same: $\tilde{\mathcal{U}} = \mathcal{U}$.

A collection of maps F is said to be *stronger than* (or *equivalent to*) a collection of maps $\tilde{F}$, $\tilde{F} < F$ ($\tilde{F} = F$) if the measurable structures of the cylinder sets determined by them satisfy the same relation. Clearly, $\tilde{F} \subset F \Rightarrow \tilde{F} < F$.

Nevertheless, there are simple situations in which extension of a family of maps does not make the measurable structure stronger.

PROPOSITION 2.1. *Let F and $\tilde{F}$ be two consistent families of maps of a space X, and let for each $f \in F$ $(f\colon X \to (E, \mathcal{B}))$ there be a collection $\tilde{\lambda}_1, \ldots, \tilde{\lambda}_n \in \tilde{\Lambda}$ and a measurable map*

$$\Psi\colon \prod_{k=1}^{n} (E_{\tilde{\lambda}_k}, \mathcal{B}_{\tilde{\lambda}_k}) \to (E, \mathcal{B})$$

such that

$$f(x) = \Psi(f_{\tilde{\lambda}_1}(x) \ldots, f_{\tilde{\lambda}_n}(x)). \tag{2.4}$$

Then $F < \tilde{F}$.

PROOF. By (2.2) the general case can be reduced to the case $n = 1$, i.e. $f(x) = \Psi(f_{\tilde{\lambda}}(x))$.

This relation implies

$$f^{-1}(B) \subset f^{-1}_{\tilde{\lambda}}(B_{\tilde{\lambda}}) \subset \mathcal{U}_{\tilde{\lambda}}. \quad \square$$

COROLLARY 2.1. *Upon enlarging the system $\tilde{F}$ by maps (2.4) which are superpositions of a measurable map and finitely many maps in $\tilde{F}$, the measurable structure of $(X,\mathcal{U})$ is replaced by an equivalent one.*

REMARK 2.1. Note that this property remains true if instead of finitely many maps from $\tilde{F}$ we take countably many, provided (2.2) is not violated.

In general, enlargement by functions of the form

$$f(x) = \Psi(f_{\tilde{\lambda}_1}(x), \dots, f_{\tilde{\lambda}_n}(x), \dots),$$

with Ψ a measurable map

$$\Psi\colon \prod_{k=1}^{\infty} E_{\tilde{\lambda}_k} \to E$$

and countable sets $\tilde{\lambda}_k$ $(k = 1, 2 \dots)$, leads to a stronger collection of maps than $\tilde{F}$, and hence to enlargement of the algebra $\tilde{\mathcal{U}}$. However, this procedure does not change the minimal σ-algebra $\sigma(\tilde{\mathcal{U}})$ containing $\tilde{\mathcal{U}}$, as can be readily shown.

EXAMPLE 2.1. Consider the product $(\prod_\alpha X_\alpha, \prod_\alpha \mathcal{U}_\alpha)$ of a family of measurable spaces. Let Λ be a set of finite tuples $\lambda = (\alpha_1, \dots, \alpha_n)$, and let $(E_\lambda, \mathcal{B}_\lambda) = \prod_{k=1}^{n}(X_{\alpha_k}, \mathcal{U}_{\alpha_k})$, the map $f_\lambda\colon X \to E_\lambda$ being natural projection. The set $F = \{f_\lambda\}$ is a consistent family, and the relation $\lambda < \lambda_1$ is equivalent to $\lambda \subset \lambda_1$.

2.2. Cylinder sets in a linear space X. In this case it is natural to take finite-dimensional linear subspaces and the Borel σ-algebras in them as the standard collection of spaces $(E_\lambda, \mathcal{B}_\lambda)$, linear maps as the maps f_λ forming a consistent set F, and functions $\phi_{\lambda\lambda_1}$ as the connecting maps (see below).

These functions $\phi_{\lambda\lambda_1}$ can be constructed as follows. Let Y be another linear space, and let the pair (X, Y) be in duality, with duality relation $\langle x, y\rangle$.

Let $\tilde{Y}$ be a subset of Y, not necessarily linear, and let $\Lambda = \Lambda_{\tilde{Y}}$ be a set of ordered finite tuples $\lambda = (y_1, \dots, y_n)$ of linearly independent elements y_i in $\tilde{Y}$.

Each $\lambda \in \Lambda$ gives rise to a map

$$f_\lambda\colon x \to (\langle x, y_1\rangle, \dots, \langle x, y_n\rangle) \tag{2.5}$$

from X onto $\mathbf{R}^n$. If $\tilde{\lambda} = (z_1, \dots, z_m) \in \Lambda$ is another tuple, linearly related to λ by

$$y_j = \sum_{k=1}^{m} a_{jk} z_k \quad (j = 1, \dots, n),$$

then the formula

$$\langle x, y\rangle = \sum_{k=1}^{n} a_{jk}\langle x, z_k\rangle$$

defines a map

$$\phi_{\lambda\tilde{\lambda}}\colon (\langle x, z_1\rangle, \dots, \langle x, z_m\rangle) \to (\langle x, y_1\rangle, \dots, \langle x, y_n\rangle)$$

which clearly satisfies

$$f_\lambda = \phi_{\lambda\tilde{\lambda}} \circ f_{\tilde{\lambda}}.$$

Hence, for each pair of maps f_{λ_1}, f_{λ_2} there is always a map f_λ having the property (2.2); to construct it it suffices to take a maximal linearly independent subset $\lambda \subset \lambda_1 \cup \lambda_2$.

An algebra of cylinder sets of X generated by a set $\tilde{Y}$ will be called an *algebra of $\tilde{Y}$-cylinder sets*; it will be denoted by $\mathcal{U}_{\tilde{Y}}$.

It follows from proposition 2.1 that $\mathcal{U}_{\tilde{Y}} = \mathcal{U}_{\mathcal{L}(\tilde{Y})}$, where $\mathcal{L}(Y)$ is the linear hull of Y. So, without loss of generality, we can take $\tilde{Y}$ to be linear from the very beginning.

Below, as a rule, X will be a separable locally convex topological linear space. In fact, this assumption is not a restriction, since it is always possible to endow X with the weak topology $\sigma(X, Y)$, i.e. the weakest topology in which all linear functionals $\langle x, y\rangle$ ($y \in Y$) are continuous. Then the linear maps (2.5) are also continuous. Therefore, in the definition of algebra of cylinder sets all linear maps can be taken continuous from $(X, \sigma(X, Y))$ to the euclidean space $\mathbf{R}^n$.

As can be readily seen, this leads to a description of the algebra $\mathcal{U}_Y$ (of cylinder sets in the linear space X constructed from the elements of a dual space Y) in terms of the family F of canonical maps $f_\lambda\colon X \to E_\lambda$, where $E_\lambda = X/X_\lambda$ runs through all quotient spaces with respect to $\sigma(X, Y)$-closed subspaces $X_\lambda \subset X$ of finite codimension.

PROPOSITION 2.2. *Let X and Y be linear spaces in duality. Then the algebra $\mathcal{U}_Y$ of cylinder sets in X is determined by the collection $F = \{R_\lambda\colon \lambda \in \Lambda\}$ of projection operators onto finite-dimensional subspaces $\mathcal{L}_\lambda \subset X$ that are continuous in the topology $\sigma(X, Y)$.*

PROOF. Since in a separable space X each finite-dimensional subspace is isomorphic to a euclidean space, projection operators onto such subspaces are the properties required (on the basis of the considerations given above). It remains to show that the whole algebra of cylinder sets is obtained in this way.

In fact, let X_0 be a closed linear subspace in $(X, \sigma(X, Y))$ of finite codimension, P_0 the canonical map from X onto X/X_0, $\tilde{B}$ a Borel subset of X/X_0, and $A = P_0^{-1}B$. Let X_1 be the linear cofactor of X_0 in X; it is a linear subspace of X. Then the restriction of P_0 to X_1 (denoted by P_0') is a bijective continuous map with continuous inverse (since X/X_0 is separated). Thus P_0' is an isomorphism from X_1 onto X/X_0. Put $P = (P_0')^{-1}P_0$ and $B = (P_0')^{-1}\tilde{B}$. Then P is a continuous projection operator into $(X, \sigma(X, Y))$ mapping X onto X_1, B is a Borel subset of X_1, and $A = P^{-1}B$. □

PROPOSITION 2.3. *Let $Y_1 \subset Y$, and suppose that every element of Y is the limit in $(Y, \sigma(X, Y))$ of a sequence of elements of Y_1. Then $\sigma(\mathcal{U}_Y) = \sigma(\mathcal{U}_X)$.*

This proposition follows from remark 2.1.

EXAMPLE 2.2. The description above is rendered explicit for a rigged Hilbert space $Y = \mathcal{H}_+ \subset \mathcal{H} \subset \mathcal{H} + - = X^*$.[1]

If P is a finite-dimensional orthoprojector on $\mathcal{H}$ with range $\mathcal{L}_P$ in Y, then it is continuous in the topology induced in $\mathcal{H}$ by $\sigma(X, Y)$. Indeed, if $(y_1, \ldots, y_n)$ is a basis

[1] See the supplement.

in $\mathcal{L}_P$, orthonormal in $\mathcal{H}$, then for every $h \in \mathcal{H}$,

$$Ph = \sum_{j=1}^{n}(Ph, y_j)_{\mathcal{H}} y_j = \sum_{j=1}^{n}(h, Py_j)_{\mathcal{H}} y_j = \sum_{j=1}^{n} h(y_j) y_j,$$

where for every $y \in Y$ the map $h \to h(y)\colon \mathcal{H} \to \mathbf{R}$ is continuous with respect to the topology $\sigma(X, Y)$. Therefore P can be continuously extended to the whole of $X = \mathcal{H}_-$; this extension is also denoted by P. Let Q be the set of maps $X \to Y$ that can be obtained in this manner. As before, elements of Q will be called finite-dimensional orthoprojectors.

The algebra of cylinder sets of $X = \mathcal{H}_-$ is taken to be the algebra $\mathcal{U}_Q$ generated by the system Q of orthoprojectors on $\mathcal{H}$ acting from $\mathcal{H}_-$ to $\mathcal{H}_+$. It is easily shown to coincide with $\mathcal{U}_Y = \mathcal{U}_{\mathcal{H}_+}$.

2.3. Measurable linear space. Let X be a linear space, Y a sufficient set of linear functionals on X, and $\mathcal{U}_Y$ the set of Y-cylinder subsets of X.

The pair $(X, \mathcal{U}_Y)$ is called a *measurable linear space.*

A linear map $f\colon X_1 \to X_2$ between linear spaces is, as usual, called a *measurable linear map* between measurable linear spaces, $f\colon (X_1, \mathcal{U}_{Y_1}) \to (X_2, \mathcal{U}_{Y_2})$, if the inverse image $f^{-1}(C)$ of each Y_2-cylinder set $C \subset X_2$ is an Y_1-cylinder set in X_1.

Note that a map f has this property if (but not necessarily only if) f is continuous when endowing X_1 and X_2 with the topologies $\sigma(X_1, Y_1)$ and $\sigma(X_2, Y_2)$, respectively.

In fact, in this case, for a continuous linear map P from $(X_2, \sigma(X_2, Y_2))$ to some finite-dimensional euclidean space the map $P_1 = P \circ f$ is continuous, and $f^{-1}(\mathcal{U}^P) = \mathcal{U}^{P_1}$.

A measurable linear space $(X_2, \mathcal{U}_{Y_2})$ is called a *linear extension* of a measurable linear space $(X_1, \mathcal{U}_{Y_1})$ if there is an injective linear map $f\colon X_1 \to X_2$ for which the map $C \to f^{-1}(C)$ $(C \in \mathcal{U}_{Y_2})$ is an isomorphism between $\mathcal{U}_{Y_2}$ and $\mathcal{U}_{Y_1}$.

EXAMPLE 2.3. If X is a linear function space on a set S whose points determine a sufficient set of linear functionals $Y = S$ (by $s(x) = x(s)$), then $(X, \mathcal{U}_S)$ is a measurable linear space. If $X' \supset X$ is another linear function space on S, then $(X', \mathcal{U}_S)$ is a linear extension of $(X, \mathcal{U}_S)$, and f is a natural imbedding.

We will now introduce the notion of *product of two measurable linear spaces* $(X_1, \mathcal{U}_{Y_1})$ and $(X_2, \mathcal{U}_{Y_2})$. It is the measurable linear space $(X, \mathcal{U}_Y)$ for which $X = X_1 \times X_2$, $Y = Y_1 \times Y_2$, and for each $x = (x_1, x_2) \in X$, $y = (y_1, y_2) \in Y$ we define $y(x) = y_1(x_1) + y_2(x_2)$. The algebra of cylinder sets is generated by the cylinder sets of the form

$$C_1 \times C_2 \quad (C_1 \in \mathcal{U}_{Y_1}, \quad C_2 \in \mathcal{U}_{Y_2})$$

(this means that each element of $\mathcal{U}_Y$ is a finite union of such products).

2.4. Cylindrical functions. Let $(X, \mathcal{U} = \lim_\Lambda \mathcal{U}_\lambda)$ be a measurable space with limit structure, and $\mathcal{U}_\lambda$ $(\lambda \in \Lambda)$ a σ-algebra.

Taking a more general point of view than in the above, elements of $\mathcal{U}$ will be called *cylinder sets* also in this case.

DEFINITION 2.1. A function $f\colon X \to Y$, where $(Y, \mathcal{B})$ is a measurable space, is called a *cylindrical function* if there is a $\lambda \in \Lambda$ for which it is measurable as a map

$$f\colon (X, \mathcal{U}_\lambda) \to (Y, \mathcal{B}). \tag{2.6}$$

This collection of cylindrical functions is denoted by $\Phi(X, \mathcal{U}, Y)$.

We note the following properties of cylindrical functions.

Clearly, for any finite collection of cylindrical functions there is a unique $\lambda \in \Lambda$ such that (2.6) holds for all these functions.

Therefore, if Y is a linear space, then $\Phi(X, \mathcal{U}, Y)$ is also a linear space.

More generally, if $f_j \in \Phi(X, \mathcal{U}, Y)$ $(j = 1, \dots, n)$, $\phi\colon \prod_{j=1}^n (Y_j, \mathcal{B}_j) \to (Y, \mathcal{B})$, then $f(x) = \phi(f_1(x), \dots, f_n(x)) \in \Phi(X, \mathcal{U}, Y)$.

If h is a $(\Lambda, \tilde{\Lambda})$-measurable map $h\colon (X, \mathcal{U}) \to (\tilde{X}, \tilde{\mathcal{U}})$, then for each cylindrical function $f \in \Phi(\tilde{X}, \tilde{\mathcal{U}}, Y)$ the cylindrical function $f \circ h \in \Phi(X, \mathcal{U}, Y)$ exists, giving rise to a mapping from $\Phi(\tilde{X}, \tilde{\mathcal{U}}, Y)$ to $\Phi(X, \mathcal{U}, Y)$:

$$f \to f \circ h. \tag{2.7}$$

This map is linear if Y is linear.

If h is invertible and h^{-1} is $(\Lambda, \tilde{\Lambda})$-measurable, then the map (2.7) is an isomorphism (between linear spaces if Y is linear).

Suppose that the algebra $\mathcal{U} = \mathcal{U}_F$ is determined by a consistent (in the sense of (2.1)) system of maps $F = \{f_\lambda\colon \lambda \in \Lambda\}$. Then the functions f_λ as well as superpositions of them of the form

$$\phi(f_{\lambda_1}(x), \dots, f_{\lambda_n}(x)), \tag{2.8}$$

where

$$\phi\colon \left(\prod E_{\lambda_k}, \prod \mathcal{B}_{\lambda_k}\right) \to (Y, \mathcal{B}),$$

are cylindrical.

Since every function measurable with respect to a σ-algebra determined by finitely many functions is known to be presentable in the form (2.8), all cylindrical functions can be obtained in this manner.

EXAMPLE 2.4. Let $(X, \mathcal{U}) = \prod_\Lambda (X_\lambda, \mathcal{U}_\lambda)$ be a product of σ-measurable spaces. It follows from example 2.1 that the cylindrical functions in this case are the functions of the form

$$\phi\colon \{x_\lambda\}_{\lambda \in \Lambda} \to \phi(x_{\alpha_1}, \dots, x_{\alpha_n}) \in Y,$$

for all ordered tuples $\lambda = (\alpha_1, \dots, \alpha_n)$. The set λ will be called a *support* of the cylindrical function ϕ. Clearly, every $\tilde{\lambda} \supset \lambda$ is a support of ϕ. Enlargement of support means addition of variables which are irrelevant to the function.

We now consider in more detail the linear space Φ, under the conditions of proposition 2.2.

Let $\mathcal{P} = (P)$ be a set of projectors $P\colon X \to X_P$ onto finite-dimensional subspaces $X_P \subset X$ which are continuous with respect to the topology $\sigma(X, Y)$. We assume that the following conditions hold:

a) It is possible to approximate with operators from $\mathcal{P}$ the identity map $I = \mathrm{id}_X$ to any degree of accuracy, in the topology $\sigma(X, Y)$ of elementwise convergence;
b) Each two $P_1, P_2 \in \mathcal{P}$ have a common extension P for which

$$P_j P = P_j \quad (j = 1, 2).$$

The space X_P should contain $X_{P_1} + X_{P_2}$.

Such a system $\mathcal{P}$ determines the algebra $\mathcal{U}_{\mathcal{P}}$ of cylinder sets in X.

The cylindrical functions in this case are the functions of the form

$$f_P(x) = f(Px), \tag{2.9}$$

where f is a measurable map from $(X_P, \mathcal{B}_P)$ to a corresponding space $(Y, \mathcal{B})$.

Note that formula (2.9) assigns to a measurable map

$$f\colon (X, \sigma(\mathcal{U})) \to (Y, \mathcal{B})$$

the collection $\{f_P\}_{P \in \mathcal{P}}$ of cylindrical functions, which is called the *collection of finite-dimensional projections* of f.

If X is a Hilbert space, we will assume that $\mathcal{P}$ consists of orthoprojectors, and for a rigged Hilbert space X of orthoprojectors in the sense of example 2.2.

3. Quasimeasures. Integration

3.1. Quasimeasures. The reader is assumed to be familiar with the general theory of integral with respect to a σ-additive measure on a measurable space. In this section more general set functions, quasimeasures, will be considered.

DEFINITION 3.1. Let $(X, \mathcal{U})$ be a measurable space with algebra $\mathcal{U} = \lim_\Lambda \mathcal{U}_\lambda$ which is the limit of a directed family of σ-algebras. An additive (not necessarily bounded) set function μ on $\mathcal{U}$ taking values in a locally convex topological linear space T is called a *quasimeasure* if for each $\lambda \in \Lambda$ its restriction $\mu_\lambda = \mu|_{\mathcal{U}_\lambda}$ is a measure.

The triplet $(X, \mathcal{U}, \mu)$ is called a *quasimeasure space*.

Note that additivity of μ follows automatically from additivity of the restrictions μ_λ, since Λ is directed.

Throughout this chapter, unless otherwise stipulated, a quasimeasure is assumed to take real values. Recall that under certain additional assumptions (theorems 1.3 and 1.4), a bounded quasimeasure is a measure, i.e. is also σ-additive on $\mathcal{U}$.

However, unbounded quasimeasures are also of interest.

Let $\mathcal{U}$ be the algebra of F-cylinder sets determined by the maps $f_\lambda\colon X \to E_\lambda$ $(f_\lambda \in F,\ \lambda \in \Lambda)$.

These maps relate the quasimeasure μ to the totality of measures $\tilde{\mu}_\lambda$ on the spaces $(E_\lambda, \mathcal{B}_\lambda)$:

$$\tilde{\mu}_\lambda(B_\lambda) = \mu_\lambda(f_\lambda^{-1}(B)) = \mu(C(\lambda, B)) \quad (B \in \mathcal{B}_\lambda),$$

which is conventionally (since the E_λ need not be finite-dimensional) called a *system of finite-dimensional distributions* of the quasimeasure μ.

A system of finite-dimensional distributions of a quasimeasure has to satisfy the natural compatibility conditions following from (2.3):

$$\tilde{\mu}_{\lambda_1}(B_{\lambda_1}) = \tilde{\mu}_\lambda(\phi^{-1}_{\lambda_1\lambda}(B_{\lambda_1})) \quad (B_{\lambda_1} \in \mathcal{B}_{\lambda_1}; \quad \lambda > \lambda_1). \tag{3.1}$$

On the other hand, suppose that for every $\lambda \in \Lambda$ there is defined a measure $\tilde{\mu}_\lambda$ on $(E_\lambda, \mathcal{B}_\lambda)$, condition (3.1) being satisfied for all pairs $\lambda > \lambda_1$ of indices of Λ. Then there is a quasimeasure μ on $(X, \mathcal{U})$ with $\{\tilde{\mu}_\lambda\}_\Lambda$ as system of finite-dimensional distributions.

In fact, if $A \in \mathcal{U}$, then there is, by definition, a $\lambda_1 \in \Lambda$ such that $A \in \mathcal{U}_{\lambda_1}$, i.e. $A = f^{-1}(B_{\lambda_1})$, where $B_{\lambda_1} \in \mathcal{B}_{\lambda_1}$ is uniquely determined (since under a surjective map a set is uniquely determined by its inverse image).

We may then put

$$\mu(A) = \tilde{\mu}_{\lambda_1}(B_{\lambda_1}).$$

This definition makes sense, since if also $A = f^{-1}(B_{\lambda_2}) \in \mathcal{U}_{\lambda_2}$ for another $\lambda_2 \in \Lambda$, then there is a $\lambda > \lambda_k$ $(k = 1, 2)$ such that, by (2.1) and (3.1),

$$\tilde{\mu}_{\lambda_1}(B_{\lambda_1}) = \tilde{\mu}_\lambda(\phi^{-1}_{\lambda_1\lambda}(B_{\lambda_1})) = \tilde{\mu}_\lambda(\phi^{-1}_{\lambda_2\lambda}(B_{\lambda_2})) = \tilde{\mu}_{\lambda_2}(B_{\lambda_2}).$$

Since we are dealing with surjective maps, operations on sets $A = f^{-1}(B) \in \mathcal{U}_\lambda$ correspond to similar operations on corresponding sets $B = f(A) \in \mathcal{B}_\lambda$. This and the σ-additivity of $\tilde{\mu}_\lambda$ imply the σ-additivity of the restriction $\mu_\lambda = \mu|_{\mathcal{U}_\lambda}$.

Thus, the following assertion holds.

PROPOSITION 3.1. *Let $(X, \mathcal{U}_F)$ be a measurable space, with $\mathcal{U}_F$ the σ-algebra of F-cylinder sets and $F = \{f_\lambda \colon \lambda \in \Lambda\}$ a consistent family of surjective maps $f_\lambda \colon X \to E_\lambda$.*

Each quasimeasure on $(X, \mathcal{U}_F)$ determines a consistent (in the sense of (3.1)) system of finite-dimensional distributions $\{\tilde{\mu}_\lambda \colon \lambda \in \Lambda\}$ in the spaces $(E_\lambda, \mathcal{B}_\lambda)$, and, vice versa, any such system uniquely determines a quasimeasure.

We now consider mappings between quasimeasure spaces. Let $(X, \mathcal{U} = \lim_\Lambda \mathcal{U}_\lambda, \mu)$ be a quasimeasure space. Consider a surjective map $f \colon Y \to X$. There are natural σ-algebras $\mathcal{U}^f_\lambda = f^{-1}(\mathcal{U}_\lambda)$ $(\lambda \in \Lambda)$ in Y, giving the algebra $\mathcal{U}^f = \lim_\Lambda \mathcal{U}^f_\lambda$. The quasimeasure μ uniquely determines a set function μ^f on $\mathcal{U}^f$:

$$\mu^f(f^{-1}(A)) = \mu(A). \tag{3.2}$$

This definition is correct, since under a surjective map inverse images and sets uniquely determine each other. The function μ^f is clearly σ-additive on every $\mathcal{U}^f_\lambda$, hence is a quasimeasure.

On the other hand, let

$$f \colon \left(X, \mathcal{U} = \lim_\Lambda \mathcal{U}_\lambda\right) \to (\tilde{X}, \tilde{\mathcal{U}})$$

be a measurable map between measurable spaces. Then each quasimeasure μ on $(X, \mathcal{U})$ determines, in the usual manner, an additive function on $(\tilde{X}, \tilde{\mathcal{U}})$:

$$\mu_f(\tilde{A}) = \mu(f^{-1}(\tilde{A})).$$

This function is not, in general, a quasimeasure, since the limit structure of an algebra of sets is not transferred to $\tilde{\mathcal{U}}$. However, if this structure is already present ($\tilde{\mathcal{U}} =$

$\lim_{\tilde\Lambda}\tilde{\mathcal{U}}_{\tilde\lambda}$) and f is $(\Lambda,\tilde\Lambda)$-measurable, the μ_f is a quasimeasure (its restrictions $(\mu_f)_{\tilde\lambda} = (\mu_\lambda)_f$ are defined and σ-additive). Note that in this case (3.2) implies

$$(\mu^f)_f = f.$$

3.2. Integral with respect to a quasimeasure. Consider the space $\Phi = \Phi_b(X,\mathcal{U},\mathbf{R})$ of realvalued bounded cylindrical functions on $(X,\mathcal{U} = \lim_\Lambda \mathcal{U}_\lambda)$. If $f \in \Phi$, then there is a $\lambda \in \Lambda$ for which this function is $\mathcal{U}_\lambda$-measurable. Therefore the following integral makes sense:

$$\int_X f(x)\mu(dx) = \int_X f(x)\mu_\lambda(dx). \tag{3.3}$$

This definition of integral of a cylindrical function with respect to a quasimeasure is correct: it follows from the compatibility condition (3.1) that the righthand side does not change upon replacing λ by a λ_1 for which $\lambda_1 > \lambda$.

Clearly, (3.3) has the usual properties:

a) $\int_X[\alpha f(x) + \beta g(x)]\mu(dx) = \alpha\int_X f(x)\mu(dx) + \beta\int_X g(x)\mu(dx)$ $(\alpha,\beta \in \mathbf{R},\ f,g \in \Phi)$;

b) the conditions $f(x) \geq 0$, $\mu \geq 0$ imply nonnegativity of the integral:
$$\int_X f(x)\mu(dx) \geq 0;$$

c) if a sequence $\{f_n\} \in \Phi$ has a common support λ and converges with respect to a measure $|\mu_\lambda|$ to a function $f(x)$, then under the usual additional restrictions of Lebesgue's theorem (e.g. majorization of the sequence by a $|\mu_\lambda|$-integrable function) the following relation holds:
$$\int_X f(x)\mu(dx) = \lim_{n\to\infty}\int_X f_n(x)\mu(dx);$$

d) let a pair of spaces $(X,\mathcal{U} = \lim_\Lambda \mathcal{U}_\lambda)$ and $(\tilde X,\tilde{\mathcal{U}} = \lim_{\tilde\Lambda}\tilde{\mathcal{U}}_{\tilde\lambda})$ and a $(\Lambda,\tilde\Lambda)$-measurable map ϕ be given. Then the change of variable formula holds:
$$\int_X f(\phi(x))\mu(dx) = \int_{\tilde X} f(\tilde x)\mu^\phi(d\tilde x). \tag{3.4}$$
In fact, by (3.3), each integral can be reduced to an integral with respect to a finite-dimensional projection μ_λ or $\mu^\phi_{\tilde\lambda}$ of the corresponding quasimeasure, and we are allowed to choose λ and $\tilde\lambda$ such that the map ϕ is $(\mathcal{U}_\lambda,\mathcal{U}_{\tilde\lambda})$-measurable. Next to this, (3.4) reduces to the corresponding formula for maps between measures;

e) let $(X_k,\mathcal{U}_k = \lim_{\Lambda_k}\mathcal{U}_k)$ $(k = 1,2)$ be a pair of quasimeasurable spaces with quasimeasures μ_1,μ_2, respectively. Their product is defined in the usual way, and the limit structure $(X,\mathcal{U}) = (X_1 \times X_2,\mathcal{U}_1 \times \mathcal{U}_2)$,
$$\mathcal{U}_1 \times \mathcal{U}_2 = \lim_{\Lambda_1\times\Lambda_2}\mathcal{U}_{\lambda_1} \times \mathcal{U}_{\lambda_2},$$
is naturally introduced for it.

The additive set function μ, defined on products of cylinder sets by

$$\mu(C_1 \times C_2) = \mu_1(C_1)\mu_2(C_2) \quad (C_k \in \mathcal{U}_k)$$

is a quasimeasure, $\mu = \mu_1 \times \mu_2$ (the product of the quasimeasures μ_1 and μ_2). Its finite-dimensional projections are defined by

$$\mu_{(\lambda_1,\lambda_2)} = \mu_{1\lambda_1} \times \mu_{2\lambda_2}.$$

Clearly, a function $f(x_1, x_2)$ on $(X_1 \times X_2, \mathcal{U}_1 \times \mathcal{U}_2)$ is cylindrical if and only if it is cylindrical with respect to each of its arguments individually (uniformly with respect to the other), i.e. if there is a pair (λ_1, λ_2) such that f is $(\mathcal{U}_{1\lambda_1} \times \mathcal{U}_{2\lambda_2})$-measurable. In the same way as above, Fubini's formula for cylindrical functions,

$$(3.5) \quad \int_{X_1\times X_2} f(x_1,x_2)\mu(dx_1 \times dx_2) = \int_{X_1} \mu_1(dx_1) \int_{X_2} f(x_1,x_2)\mu_2(dx_2),$$

is (under the usual assumption of absolute integrability of the function with respect to to the product measure $|\mu_{1\lambda_1}| \times |\mu_{2\lambda_2}|$) established by passing to finite-dimensional distributions.

3.3. Quasimeasures in a measurable linear space. We turn to the special case when $(X, \mathcal{U} = \mathcal{U}_Y)$ is a measurable linear space, determined by a sufficient set Y of linear functionals on a linear space X.

We will first show that a bounded quasimeasure on $(X, \mathcal{U})$ can always be made into a measure by linearly extending the space.

THEOREM 3.1. *Each measurable linear space $(X, \mathcal{U}_Y)$ has a linear extension such that the image μ_f of every bounded quasimeasure μ on $(X, \mathcal{U}_Y)$ under the imbedding $f\colon (X, \mathcal{U}_Y) \to (\tilde{X}, \tilde{\mathcal{U}}_{\tilde{Y}})$ is σ-additive on $(\tilde{X}, \tilde{\mathcal{U}}_{\tilde{Y}})$.*

PROOF. Let S be a Hamel basis of $\mathcal{L}(Y)$, i.e. a subset $S \subset Y$ consisting of linearly independent elements and algebraically generating $\mathcal{L}(Y)$. The algebras of cylinder sets $\mathcal{U}_S$ and $\mathcal{U}_Y$ coincide, as follows from proposition 2.1. We regard elements of X as functions on S:

$$x(s) = s(x) \quad (x \in X, \quad s \in S).$$

Then $(X, \mathcal{U}_Y) = (X, \mathcal{U}_S)$ is naturally imbedded in $(\tilde{X}, \mathcal{U}_S)$, where $\tilde{X}$ is the set of realvalued functions on S (see example 2.3). Also, the image μ_f of the quasimeasure μ is a quasimeasure (since the imbedding f clearly satisfies the condition of $(\Lambda, \tilde{\Lambda})$-measurability; see subsection 1). By theorem 1.4, every bounded quasimeasure on $(\tilde{X}, \mathcal{U}_S)$ is a measure. □

In the study of quasimeasures in linear spaces, Fourier transformation plays an important role. We will first recall some notions on the Fourier transform of a measure ν on a euclidean space $(\mathbf{R}^n, \mathcal{B})$. The *Fourier transformation* is the map

$$\nu \to \chi_\nu(y) = \int_{\mathbf{R}^n} e^{i(x,y)}\nu(dx) \quad (y \in \mathbf{R}^n)$$

assigning to a measure ν its characteristic functional (the latter is called the *Fourier transform*). Since the variation $|\nu|$ is bounded, the characteristic functional is a bounded uniformly continuous function on $\mathbf{R}^n$. This function uniquely determines ν: if $\prod_{k=1}^n [a_k, b_k]$ is a parallelepiped with boundary of $|\nu|$ measure zero, then

$$\nu\left(\prod_{k=1}^{n}[a_k, b_k]\right) =$$

$$\lim_{T\to\infty} \frac{1}{(2\pi)^n} \int_{-T}^{T} \cdots \int_{-T}^{T} \frac{e^{-ia_k y_k} - e^{-ib_k y_k}}{iy_k} \chi_\nu(y_1, \ldots, y_n)\, dy_1 \ldots dy_n.$$

We return to the situation of a measurable linear quasimeasure space $(X, \mathcal{U}, \mu)$. Without loss of generality we can take Y to be a linear space in duality with X, with duality relation $\langle x, y\rangle$. For each $y \in Y$ the function $e^{i\langle x,y\rangle}$ is a bounded cylindrical function on X. Therefore the following integral makes sense:

$$\chi_\mu(y) = \int_X e^{i\langle x,y\rangle} \mu(dx) = \int_{-\infty}^{\infty} e^{iz} \mu^y(dz), \tag{3.6}$$

where $\mu^y(B) = \mu(x\colon \langle x, y\rangle \in B)$ is a one-dimensional projection of μ.

DEFINITION 3.2. The function $\chi_\mu(y)$ $(y \in Y)$ is called the *characteristic functional* of the quasimeasure μ, while the map $\mu \to \chi_\mu$ is called *Fourier transformation*.

Thus, the characteristic functional of a quasimeasure μ is uniquely determined by the one-dimensional projections μ^y of μ. Consider now the finite-dimensional projections of μ. Let $\mathcal{L} \subset Y$ be a finite-dimensional subspace and $y_1, \ldots, y_n$ a basis in it. Denote by $\mu^{(y_1,\ldots,y_n)}$ the image of μ under the map $X \to (\langle x, y_1\rangle, \ldots, \langle x, y_n\rangle)$ from X to the finite-dimensional space $X/\mathcal{L}^\perp$, where $\mathcal{L}^\perp$ is the kernel of the map.

Then the following formula holds:

$$\chi_\mu\left(\sum_{k=1}^{n} t_k y_k\right) = \int_X e^{i\sum_{k=1}^n t_k\langle x,y_k\rangle} \mu(dx) = \int_{\mathbf{R}^n} e^{i\sum t_k z_k} \mu^{(y_1,\ldots,y_n)}(dz),$$

implying the required.

PROPOSITION 3.2. *The characteristic functional of a quasimeasure uniquely determines the finite-dimensional projections of this quasimeasure: the restriction of* χ_μ *to a finite-dimensional subspace* $\mathcal{L} \subset Y$ *is the characteristic functional of the image of the quasimeasure* μ *under the map* $X \to X/\mathcal{L}^\perp$*, where* $\mathcal{L}^\perp \subset X$ *is the subspace annihilated by the functionals in* $\mathcal{L}$.

EXAMPLE 3.1. Consider, in particular, a rigged Hilbert space $\mathcal{H}_+ \subset \mathcal{H} \subset \mathcal{H}_- = \mathcal{H}$ (see example 2.2).

In this case, the characteristic functional of a quasimeasure μ on $X = \mathcal{H}_-$ is a function on the dual space $Y = \mathcal{H}_+$, which can (by the relation $\langle x, y\rangle = (x, y)_\mathcal{H}$) be written as

$$\chi_\mu(y) = \int_X e^{i(x,y)_\mathcal{H}} \mu(dx). \tag{3.7}$$

Let P be a finite-dimensional orthoprojector on $\mathcal{H}$, acting from $\mathcal{H}_-$ to $\mathcal{H}_+$. In this case the space $\mathcal{L} = \mathcal{L}_P = P\mathcal{H}_-$ can be taken euclidean (using the scalar product in

$\mathcal{H}$). Also, $X/\mathcal{L}^\perp$ is naturally identified with $\mathcal{L}$, and we obtain, in accordance with proposition 3.2, for the image μP of μ:

$$\chi_\mu(Py) = \int_{\mathcal{H}_-} e^{i(x,Py)_\mathcal{H}}\mu(dx) = \int_{\mathcal{H}_-} e^{i(Px,y)_\mathcal{H}}\mu(dx) = \int_{\mathcal{L}_P} e^{i(\xi,y)}\mu^P(d\xi) = \chi_{\mu^P}(Py).$$

We now consider transformations of quasimeasures on measurable linear spaces, and the changes in the characteristic functionals connected with them.

a) Let $(X_1,\mathcal{U}_{Y_1})$ and $(X_2,\mathcal{U}_{Y_2})$ be measurable linear spaces. Introduce in them the weak topologies $\sigma(X_1,Y_1)$, $\sigma(X_2,Y_2)$ determined by the spaces Y_1, Y_2 of linear functionals. Then a continuous linear map $S\colon (X_1,\sigma(X_1,Y_1)) \to (X_2,\sigma(X_2,Y_2))$ is clearly (Λ_1,Λ_2)-measurable. If μ is a quasimeasure on $(X_1,\mathcal{U}_{Y_1})$, then its image μ_S is a quasimeasure on $(X_2,\mathcal{U}_{Y_2})$, and by (3.4),

$$\chi_{\mu_S}(y) = \int_{X_2} e^{i\langle x_2,y\rangle}\mu_S(dx_2) = \int_{X_1} e^{i\langle Sx_1,y\rangle}\mu(dx_1) =$$

(3.8) $$\int_{X_1} e^{i\langle x_1,S^*y\rangle}\mu(dx_1) = \chi_\mu(S^*y) \quad (y\in Y),$$

where $S^*\colon Y_2\to Y_1$ is the adjoint map.
In particular, if S is an imbedding and $Y_1 = Y_2/(SX^\perp)^\perp$, then S^* is the natural projection $Y_2\to Y_1$ and χ_{μ_S} is constant on Y_2-cosets with respect to Y_1.

b) Let μ be a quasimeasure on $(X,\mathcal{U})$ and $h\in X$. Consider the shift

$$X\to X+h.$$

It maps μ to a new quasimeasure in the same measurable linear space:

$$\mu_h(A) = \mu(A-h);$$

also,

(3.9) $$\chi_{\mu_h}(y) = \int_X e^{i\langle x,y\rangle}\mu_h(dx) = \int_X e^{i\langle x+h,y\rangle}\mu(dx) = e^{i\langle h,y\rangle}\chi_\mu(y).$$

c) Consider a pair of measurable linear quasimeasure spaces $(X_j,\mathcal{U}_{Y_j},\mu_j)$ and their product

$$(X_1\times X_2,\mathcal{U}_{Y_1}\times\mathcal{U}_{Y_2},\mu_1\times\mu_2).$$

It can be easily seen that $\mathcal{U}_{Y_1}\times\mathcal{U}_{Y_2} = \mathcal{U}_{Y_1\times Y_2}$, the duality between $X = X_1\times X_2$ and $Y = Y_1\times Y_2$ being determined by $\langle x,y\rangle = \langle x_1,y_1\rangle + \langle x_2,y_2\rangle$, where $x = (x_1,x_2)$, $y = (y_1,y_2)$. So,

$$\chi_{\mu_1\times\mu_2} = \int_{X_1\times X_2} e^{i[\langle x_1,y_1\rangle+\langle x_2,y_2\rangle]}(\mu_1\times\mu_2)(dx) =$$

(3.10) $$\int_{X_1} e^{i\langle x_1,y_1\rangle}\mu_1(dx_1)\int_{X_2} e^{i\langle x_2,y_2\rangle}\mu_2(dx_2) = \chi_{\mu_1}(y_1)\chi_{\mu_2}(y_2).$$

d) Let now $X_1 = X_2$, $Y_1 = Y_2$. Consider the map

(3.11) $$S(x_1,x_2) = x_1+x_2, \quad S\colon X_1\times X_1\to X_1.$$

DEFINITION 3.3. The image of the quasimeasure $\mu_1 \times \mu_2$ under the map (3.11), denoted by $\mu_S = \mu_1 * \mu_2$ and being a quasimeasure on $(X_1, \mathcal{U}_{Y_1})$, is called the *convolution* of the quasimeasures μ_1 and μ_2.

Let $A \in \mathcal{U}_{Y_1}$. The function $\mu_2(A - x)$ is cylindrical and $\mathcal{U}_{Y_1}$-measurable. Therefore the following makes sense:

$$(\mu_1 * \mu_2)(A) = (\mu_1 \times \mu_2)(S^{-1}(A)) = \int_{x_1+x_2 \in A} (\mu_1 \times \mu_2)(dx) =$$

$$\int_{X_1} \mu_2(dx_2) \int_{A-x_2} \mu_1(dx_1) = \int_{X_1} \mu_1(A - x_1)\mu_2(dx_1). \tag{3.12}$$

The definition of convolution clearly implies that it is commutative. Relations (3.8) and (3.10) imply that

$$\chi_{\mu_1 * \mu_2}(y) = \chi_{\mu_1 \times \mu_2}(S^* y) = \chi_{\mu_1}(y)\chi_{\mu_2}(y), \tag{3.13}$$

since the map adjoint to (3.11) is

$$S^*(y) = (y, y).$$

3.4. Positive quasimeasures. There is no description of the characteristic functionals of general quasimeasures on X. However, the situation simplifies for quasimeasures of constant sign.

DEFINITION 3.4. A complexvalued function $\Psi(y)$ on a linear space Y is called *positive definite* if for every finite collection of elements $y_1, \dots, y_n \in Y$ and complex numbers $\alpha_1, \dots, \alpha_n$ the following inequality holds:

$$\sum_{j,k=1}^{n} \Psi(y_j - y_k)\alpha_j \bar{\alpha}_k \geq 0.$$

Clearly, a function $\Psi(y)$ is positive definite if and only if all its restrictions to finite-dimensional subspaces are positive definite.

In the finite-dimensional case the following well-known assertion holds (we state it without proof).

THEOREM 3.2. *(Bochner.) A necessary and sufficient condition for a function $\Psi(y)$, $y \in \mathbf{R}^n$, to be the Fourier transform of a nonnegative measure μ on $(\mathbf{R}^n, \mathcal{B}_n)$,*

$$\Psi(y) = \int_{\mathbf{R}^n} e^{i(x,y)} \mu(dx),$$

is that it be positive definite and continuous at $y = 0$.

Without difficulty, this theorem implies

PROPOSITION 3.3. *Let X, Y be a pair of linear spaces in duality. A necessary and sufficient condition for a complexvalued function $\Psi(y)$ $(y \in Y)$ to be the characteristic functional of a nonnegative quasimeasures μ on $(X, \mathcal{U}_Y)$ is that it be positive definite and that every restriction of it to a finite-dimensional subspace $L \subset Y$ be continuous at $y = 0$.*

PROOF. Necessity follows immediately from Bochner's theorem, since the restriction $\Psi|_L$ is the Fourier transform of the corresponding finite-dimensional projection of μ.

For the converse, let Ψ be positive definite and have finite-dimensional restrictions which are continuous at zero (and hence, by the positive definiteness, on the whole space). Let $\mathcal{M} \subset X$ be a finite-codimensional subspace which is closed in the topology $\sigma(X,Y)$, and let $\mathcal{M}^{\perp} \subset Y$ be the finite-dimensional subspace of functionals annihilating $\mathcal{M}$. By Bochner's theorem there is a unique nonnegative measure $\mu_{\mathcal{M}}$ on the quotient space $X/\mathcal{M}$ for which

$$\Psi(y) = \int_{X/\mathcal{M}} e^{i\langle x,y\rangle} \mu_{\mathcal{M}}(dx) \quad (y \in \mathcal{M}^{\perp}).$$

The set of measures thus constructed in the quotient spaces is consistent, since if $\mathcal{M}_1 \subset \mathcal{M}_2$ and $g\colon X/\mathcal{M}_1 \to X/\mathcal{M}_2$ is the canonical map, then the following relation holds for the measure $\nu = (\mu_{\mathcal{M}_1})g$:

$$\chi_{\nu}(y) = \chi_{\mu_{\mathcal{M}_1}}(y)|_{\mathcal{M}_2^{\perp}} = \Psi(y)|_{\mathcal{M}_1^{\perp}}|_{\mathcal{M}_2^{\perp}} = \Psi(y)|_{\mathcal{M}_2^{\perp}} = \chi_{\mu_{\mathcal{M}_2}}(y) \quad (y \in \mathcal{M}_1^{\perp}),$$

proving that $\nu = \mu_{\mathcal{M}_2}$.

This system of finite-dimensional distributions determines a nonnegative quasimeasure μ on X. □

3.5. Integration of noncylindrical functions. Above it has been shown that integral of a cylindrical function with respect to a quasimeasure can be defined without difficulty by reduction to integrals with respect to finite-dimensional projections, which are σ-additive already.

Our next purpose is to extend this integration procedure to certain classes on noncylindrical functions.

Let $(X, \mathcal{U} = \lim_{\Lambda} \mathcal{U}_{\lambda})$ be a measurable space with limit structure. Consider the linear space F of all $\sigma(\mathcal{U})$-measurable bounded realvalued functions on X.

DEFINITION 3.5. A family S_{λ} $(\lambda \in \Lambda)$ of linear operators on F is called a *system of projections* if it has the following properties:

a) $(S_{\lambda} f)(x)$ is a $\mathcal{U}_{\lambda}$-measurable function on X;
b) the relation

$$S_{\lambda_1} S_{\lambda_2} = S_{\lambda_1} \quad (\lambda_1 < \lambda_2) \tag{3.14}$$

holds;
c) There is a topology σ in F for which $\lim_{\Lambda} S_{\lambda}(f) = f$ $(f \in F)$.

Let μ be a quasimeasure on $(X, \mathcal{U})$, and let a system of projections $\{S_{\lambda}\}$ be given. For each $f \in F$, the measurable map $S_{\lambda}(f)\colon (X, \mathcal{U}_{\lambda}) \to (\mathbf{R}, \mathcal{B})$ determines a σ-additive measure on $(\mathbf{R}, \mathcal{B})$:

$$\theta_{\lambda,f}(B) = \mu_{S_{\lambda} f}(B) = \mu_{\lambda}[(S_{\lambda} f)^{-1}(B)] \quad (B \in \mathcal{B}).$$

DEFINITION 3.6. Let G be a linear space of bounded continuous functions on $\mathbf{R}$. A quasimeasure μ is said to be *of type* (F, G) if for each $f \in F$ the directed family of measures $\theta_{\lambda,f}$ $(\lambda \in \Lambda)$ weakly converges as a family of distributions over the basis space, i.e. for every $g \in G$ the limit

$$\lim_{\Lambda} \int_{-\infty}^{\infty} g(y)\theta_{\lambda,f}(dy) = \lim_{\Lambda} \int_X g(S_\lambda f(x))\mu(dx) \stackrel{def}{=} \int_X g(f(x))\mu(dx) \tag{3.15}$$

exists. By definition, this limit is called the *integral of the function* $g(f(x))$ *with respect to the quasimeasure* μ.

REMARK 3.1. By definition, for a quasimeasure of type (F, G) the integral (3.15) exists for every pair $f \in F$, $g \in G$. If G is also a topological space and if the space G' dual to it is weakly complete, then there is a distribution $\theta_f \in G'$ such that

$$\int_X g(f(x))\mu(dx) = \int_Y g(y)\theta_f(dy).$$

Also, convergence of g_n to g in G implies convergence of the integrals:

$$\int_X g_n(f(x))\mu(dx) = \int_X g(f(x))\mu(dx). \tag{3.16}$$

There are several natural ways of defining a system of projections.

EXAMPLE 3.2. Let ν be a probability measure on $(X, \sigma(\mathcal{U}))$, and let $\mathcal{U}_\lambda \subset \sigma(\mathcal{U})$ be a σ-subalgebra. It is well known (see Doob [1]) that for every ν-integrable function $f(x)$ on $(X, \sigma(\mathcal{U}))$ the conditional expectation $M_\lambda f$ is uniquely determined as the $\mathcal{U}_\lambda$-measurable function satisfying

$$\int_X [M_\lambda f(x)] g(x)\nu(dx) = \int_X f(x)g(x)\nu(dx),$$

for all bounded $\mathcal{U}_\lambda$-measurable realvalued functions $g(x)$ Also, the operators

$$S_\lambda f(x) = M_\lambda f(x)$$

satisfy the conditions for being a system of projections. The latter follows from a well-known theorem of Doob [1], stating that

$$\lim_{\Lambda} M_\lambda f(x) = f(x) \quad (\nu\text{-almost everywhere}).$$

Clearly, also

$$\lim_{\Lambda} \int_X g(S_\lambda f(x))\nu(dx) = \int_X g(f(x))\nu(dx)$$

for all continuous functions $g(y)$ and bounded $\sigma(\mathcal{U})$-measurable $f(x)$, the righthand side having the usual meaning.

Thus there are spaces G and F such that every measure ν is of type (F, G) for an appropriate system of projections.

EXAMPLE 3.3. Let X be a topological linear space, and let a measurable limit structure in it be determined by the system $\mathcal{P} = \{P\}$ of orthoprojectors onto finite-dimensional subspaces X_P satisfying the conditions a) and b) in example 2.4.

Then we can define a system of projections by the formula

$$(S_P f)(x) = f(Px) \quad (P \in \mathcal{P}).$$

Also, for every bounded continuous function $f(x)$,

$$\lim_{\mathcal{P}} (S_P f)(x) = \lim_{\mathcal{P}} f(Px) = f(x) \quad (x \in X).$$

Under these conditions we have for every measure ν on $(X, \sigma(\mathcal{U}))$:

$$\lim_{\mathcal{P}} \int_X (S_P f)(x)\nu(dx) = \int_X f(x)\nu(dx).$$

Consider now a class of functionals which are integrable with respect to a quasimeasure. For the sake of simplicity we restrict ourselves to the case of a rigged Hilbert space, considered above.

Let $X = \mathcal{H}_- \supset \mathcal{H} \supset \mathcal{H}_+$. Consider a quasimeasure μ on $\mathcal{H}_-$ with characteristic functional

$$\chi_\mu(\theta) = \int_{\mathcal{H}_-} e^{i(x,\theta)_{\mathcal{H}}}\mu(dx)$$

which is bounded and continuous on $\mathcal{H}_+$.

Let F be the space of functions on $\mathcal{H}_-$ that are the Fourier transform of a complexvalued measure ν on $\mathcal{H}_+$:

$$f(x) = \int_{\mathcal{H}_+} e^{i(x,\theta)}\nu(d\theta). \tag{3.17}$$

PROPOSITION 3.4. *For any function of the form (3.17) the following integral exists:*

$$\int_X f(x)\mu(dx) = \lim_{\mathcal{P}} \int_X f(Px)\mu(dx) = \lim_{\mathcal{P}} \int_{X_P} f(y)\mu_P(dy); \tag{3.18}$$

here μ is a quasimeasure with characteristic functional which is bounded and continuous on $\mathcal{H}_+$, and $\mathcal{P}$ is the system of finite-dimensional orthoprojectors on $\mathcal{H}$ acting from $\mathcal{H}_-$ to $\mathcal{H}_+$.

Also, the following formula holds:

$$\int_{\mathcal{H}_-} f(x)\mu(dx) = \int_{\mathcal{H}_+} \chi_\mu(\theta)\nu(d\theta). \tag{3.19}$$

PROOF. Applying Fubini's theorem for every $P \in \mathcal{P}$ we obtain

$$\int_X f(Px)\mu(dx) = \int_{X_P}\int_{\mathcal{H}_+} e^{i(Py,\theta)}\nu(d\theta)\mu_P(dy) =$$
$$\int_{\mathcal{H}_+}\int_{X_P} e^{i(y,P\theta)}\mu_P(dy)\nu(d\theta) = \int_{\mathcal{H}_+} \chi_\mu(P\theta)\nu(d\theta).$$

Under the conditions stated, at the righthand side we can pass to the limit as $P \to I$. □

REMARK 3.2. The complexvalued measures on $\mathcal{H}_+$ form the Banach algebra $M(\mathcal{H}_+)$, with convolution $\nu_1 * \nu_2$ as product and with norm $\|\nu\| = |\nu|(\mathcal{H}_+)$. The Fourier transformation (3.17) transfers this Banach algebra structure to the space $\Phi(\mathcal{H}_-)$ of functions with norm $\|f\| = \|\nu\|$ and pointwise product.

If $f \in \Phi(\mathcal{H}_-)$ and $\phi(z)$ is a function holomorphic in the disc $\{|z| \leq \|f\|\}$, then $\phi(f(z)) \in \Phi(\mathcal{H}_-)$. Thus, in this case we can take for G the space of functions analytic in a closed disc.

4. Supplement: Some notions related to the topology of linear spaces

Here we give a brief presentation of the concepts of the topology of linear spaces which are used in this book (especially in chapter III), in a form convenient for us. A more detailed treatment can be found in Bourbaki [1], Schaeffer [1], and Pietch [1].

4.1. Prenorms. Let X be a linear space over a field S (i.e. $\mathbf{C}$ or $\mathbf{R}$). A nonnegative function p having the properties

$$p(x+y) \leq p(x) + p(y); \quad p(\alpha x) = |\alpha| p(x) \quad (x, y \in X; \quad \alpha \in S)$$

is called a *prenorm* (*seminorm*) on X.

Consider the linear subspace

$$N_p = \{x \in X : p(x) = 0\}.$$

If $N_p = \{0\}$, then X is a normed linear space with norm $p(x)$. If N_p is nontrivial, we consider the quotient space $X_p = X/N_p$, and let $Q_p \colon X \to X_p$ be the corresponding canonical map. X_p is a normed space, with norm $\|\xi\|_p = p(Q_p^{-1}\xi)$ $(\xi \in X_p)$. We denote by $\overline{X}_p$ the Banach space obtained by completing X_p. Let i_p be the natural imbedding of X_p in $\overline{X}_p$, and let $\overline{Q}_p = i_p Q_p$.

Let q be another prenorm on X for which

$$p(x) \leq C q(x) \quad (x \in X).$$

Then $N_q \subseteq N_p$, and the continuous linear map $\Psi_{pq} \colon \overline{X}_q \to \overline{X}_p$ is canonically defined:

$$\Psi_{pq}(\xi) = \overline{Q}_p(Q_q^{-1}\xi) \quad (\xi \in i_q(X_q));$$

$$\Psi_{pq}(\xi) = \lim_{n\to\infty} \Psi_{pq}(\xi_n) \quad (\xi = \lim_{n\to\infty} \xi_n; \quad \xi_n \in i_q(X_q); \quad \xi \in \overline{X}_q).$$

If N_p has a direct complement in X, then X_p may be identified with it: $X = X_p \dotplus N_p$, and Q_p is projection onto X_p. If also N_q is complementable in N_p:

$$X = N_q \dotplus N_{pq} \dotplus X_p = N_q \dotplus X_q,$$

then Ψ_{pq} is the closure of the restriction $Q_{pq} = Q_p|_{X_q}$.

A prenorm p is called a *Hilbert prenorm* if $p^2(x) = b(x, x)$, where b is a nonnegative bilinear form on $X \times X$. If p is a Hilbert prenorm, then $\overline{X}_p$ is a Hilbert space.

4.2. Locally convex spaces. Let X be a locally convex topological linear space; its topology (which is consistent with the linear structure) is determined by a family Π of prenorms as the weakest of all topologies in which all these prenorms are continuous. In this case a fundamental system of closed neighborhoods of zero in X is formed by sets of the form

$$\{x: \sup_{1\le j\le n} p_j(x) \le \epsilon,\ \epsilon > 0,\ p_j \in \Pi\}.$$

These sets are *absolutely convex*, i.e. with each pair of points x, y they also contain all points $\alpha x + \beta y$ ($|\alpha| + |\beta| \le 1$). They are also *absorbing*, i.e. for every $x \in X$ they also contain all points $\frac{1}{\lambda}x$ for $|\lambda|$ sufficiently large.

Conversely, each absolutely convex absorbing set $A \subset X$ gives rise to a prenorm:

$$p_A(x) = \inf\left\{\lambda > 0 : \frac{1}{\lambda}x \in A\right\},$$

which is continuous if and only if A is a neighborhood of zero in X.

The topology generated by Π is separable if for every $x \ne 0$ there is a $p \in \Pi$ for which $p(x) \ne 0$.

A set in X is called *bounded* if it is absorbed by some neighborhood of zero. If A is a bounded absolutely convex set, then $p_A(x)$ is a norm (in general, not on all of X but only on the subspace $X_A = \{x \in X\colon \exists\lambda > 0\colon \frac{1}{\lambda}x \in A\}$). This set is called a *Hilbert set* if p_A is a Hilbert norm.

4.3. Duality of linear spaces. Let X and Y be linear spaces. They are said to be in *duality* if there is defined a bilinear functional $\langle x, y\rangle$ ($x \in X$, $y \in Y$) with the properties: for each $y \in Y$ ($y \ne 0$) there is an $x \in X$ such that $\langle x, y\rangle \ne 0$, and for each $x \in X$ ($x \ne 0$) there is a $y \in Y$ such that $\langle x, y\rangle \ne 0$. The functional $\langle x, y\rangle$ is called a *duality relation*, *pairing*, or *coupling*.

A locally convex topological linear space X is in duality with its dual space X' (of all continuous linear functionals on X), with coupling $\langle x, f\rangle = f(x)$ ($f \in X'$).

Let X, Y be linear spaces in duality, and let $\langle\ ,\ \rangle$ be the coupling.

A locally convex topology β in X is said to be *consistent* with this duality if $Y = X'_\beta$ algebraically, where X'_β is the space of all linear functionals on X that are continuous with respect to the topology β. The weakest of all such topologies is the so-called *weak topology* $\sigma(X, Y)$ in X; it is determined by the system of prenorms

$$p_y(x) = |\langle x, y\rangle| \quad (y \in Y).$$

There is a strongest topology consistent with the duality of two locally convex topological linear spaces X and $Y = X'$: the *Mackey topology* $\tau(X, X')$. It is the topology of uniform convergence on all absolutely convex $\sigma(X, X')$-compact subsets of $X' = X'_\sigma$. A space X whose topology coincides with $\tau(X, X')$ is called a *Mackey space*.

Every closed, absorbing, absolutely convex subset of a locally convex topological linear space is called a *barrel*. A locally convex topological linear space is called *barrelled* if every barrel in it is a neighborhood of zero. Separable barrelled spaces are Mackey spaces. Conversely, a complete separable Mackey space is barrelled.

The set of barrelled spaces includes the set of Fréchet spaces (which are complete metrizable locally convex topological linear spaces), which, in turn, includes the set of Banach spaces. Spaces obtained from barrelled spaces by passing to a quotient space or to an inductive limit are barrelled.

4.4. Rigged Hilbert spaces. Let $\mathcal{H}$ be a Hilbert space with scalar product $(\cdot,\cdot)_{\mathcal{H}}$, and let Y be a locally convex topological linear space densely imbedded in $\mathcal{H}$ and with a topology which is stronger than the one induced from $\mathcal{H}$. Each element $h \in \mathcal{H}$ determines a continuous linear functional $h(y) = (y,h)_{\mathcal{H}}$ on Y, and in this manner $\mathcal{H}$ is imbedded in $X' = Y$; this imbedding is dense with respect to the topology $\sigma(X,Y)$. (Here, elements of distinct spaces which coincide under imbedding are denoted by the same symbol.)

Thus, the triple of densely imbedded spaces

$$Y \subset \mathcal{H} \subset X = Y'$$

is obtained, and the duality relation $\langle x,y\rangle$ is an extension of the scalar product. To stress this, we will often use the notation $\langle x,y\rangle = (x,y)_{\mathcal{H}}$ for arbitrary $x \in X$, $y \in Y$. In particular, Y can be a Hilbert space (then so is X), or a projective limit of Hilbert spaces (then X is an inductive limit of Hilbert spaces).

The situation above arises when Y is a *quasiHilbert space*, i.e. a locally convex topological linear space on which there exists a continuous Hilbert norm $p(y)$. Then $\mathcal{H}$ is taken to be the completion of Y with respect to this norm.

If the topology of Y is determined by countably many Hilbert norms, then we may assume that $Y = \cap_{k=0}^{\infty}\mathcal{H}_k$, where $\mathcal{H}_{k+1} \subset \mathcal{H}_k$ and $\|x\|_k \le \|x\|_{k+1}$ $(k = 0,1,\dots)$. Such a space Y is called a *countable-Hilbert space*. In this case, also $X = \cup_{k=0}^{\infty}\mathcal{H}_{-k}$, where $\mathcal{H}_{-k} = \mathcal{H}_k'$ and $\mathcal{H}_k \subset \mathcal{H}_0 \subset \mathcal{H}_{-k}$.

4.5. Polars. Let X be a locally convex topological linear space, and $\mathcal{M} \subset X$. The *polar* of $\mathcal{M}$ is the subset

$$\mathcal{M}^0 = \{y \in X' : |\langle x,y\rangle| \le 1, \ \forall x \in \mathcal{M}\}$$

in the dual space X'. The polar is a weakly closed, absolutely convex set in X'. The polar of each neighborhood of zero in X is a weakly compact subset of X'.

If $F \subset X$ is a linear subspace, then $F^0 \subset X'$ is the linear subspace of elements annihilating F:

$$F^0 = \{y \in X' : |\langle x,y\rangle| = 0, \ \forall x \in F\}.$$

Let L be a closed linear subspace of X'_σ, and X'_σ/L the quotient space with the quotient topology (the strongest topology in which the natural map $P\colon X'_\sigma \to X'_\sigma/L$ is continuous). Then the polar $L^0 \subset X$ can be identified with the dual of the quotient space: $L^0 = (X'_\sigma/L)'$, with natural duality

$$\langle x, \{\xi + L\}\rangle \stackrel{\text{def}}{=} \langle x,\xi\rangle \quad (x \in L^0, \ \xi \in X').$$

If $\mathcal{M}$ is an absolutely convex set in X and $\mathcal{M}^0 \subset X'$ is its polar, then $P\mathcal{M}^0$ is the polar of $\mathcal{M} \cap L^0 \subset L^0$ in X'/L.

Note that the polar of a Hilbert set A is a Hilbert set, with $p_{A^0}(y) = \sup_{x\in A}|\langle x,y\rangle|$.

4.6. Nuclear topology. Let X be a locally convex space with topology β. Consider the totality $R(X, \beta)$ of all Hilbert prenorms in X which are continuous with respect to β. Let $q \in R(X, \beta)$, let S be a positive Hilbert–Schmidt operator in $\overline{X}_q$, and $p(x) = q(S\overline{Q}_a x)$. Let $\mathcal{P}(X, \beta)$ be the system of prenorms thus obtained. This system is not empty, since $R(X, \beta) \supset R(X, \sigma)$, and without loss of generality, the spaces $\overline{X}_q$, corresponding to prenorms determining the weak topology, can be considered to be Hilbert spaces (they are finite-dimensional).

The topology in X generated by the system of prenorms $\mathcal{P}(X)$ is called the *nuclear topology* $\tau_S(X, \beta)$ associated with the topology β.[2]

A locally convex topological linear space X is called *nuclear* if $\tau_S(X, \beta) = \beta$.

Since bounded operators on a finite-dimensional space are Hilbert–Schmidt, the space (X, σ) is nuclear: $\tau_S(X, \sigma) = \sigma$, and hence every nuclear topology in X is stronger than the weak topology $\sigma(X, X')$.

The construction above is equivalent to the following one: $p \in \mathcal{P}(X)$ if and only if $p \in R(X, \beta)$ and there is a majorizing prenorm $q \in R(X, \beta)$ (i.e. $p \leq q$) for which the canonical imbedding $\tau_{pq} : X_q \to X_p$ is a Hilbert–Schmidt operator.

As an example, consider a Hilbert space with the topology of strong convergence. Since every continuous prenorm in a Hilbert space is majorized by the norm, $\mathcal{P}(X, \beta)$ consists of all prenorms of the form $p_S(x) = \|Sx\|$, where S runs over the totality of all nonnegative Hilbert–Schmidt operators.

More generally, let the topology of the locally convex topological linear space X be determined by the totality $R(X, \beta)$ of all Hilbert prenorms. Then $P(X, \beta)$ consists of the prenorms

$$\tilde{p}(x) = \|SQ_p(x)\|_p \quad (p \in R(X, \beta)),$$

where $Q_p \colon X \to \overline{X}_p$ is the canonical map, S is a Hilbert–Schmidt operator on $\overline{X}_p$, and $\|\cdot\|_p$ is the norm in $\overline{X}_p$.

An example of a nuclear space is provided by a countable-Hilbert space whose imbedding operators $\mathcal{I}_k \colon \mathcal{H}_{k+1} \to \mathcal{H}_k$ are Hilbert–Schmidt operators (a countable-Hilbert space with Hilbert–Schmidt imbeddings).

4.7. Compactness. We recall some ideas related to the notion of compactness of a set. A set $\mathcal{M} \subset X$ is called *compact* if each open covering of it contains a finite covering. Each infinite subset $\mathcal{M}_1$ of a compact set $\mathcal{M}$ has a limit point $x \in \mathcal{M}$. The weaker property that the limit point $x \in X$ exists, but not necessarily $x \in \mathcal{M}$, is called *relative compactness* of $\mathcal{M}$. The closure of a subset of a compact set, relative to this compact set, is compact. A set $\mathcal{M}$ is called (relatively) *sequentially compact* if each sequence of elements in it contains a convergent subsequence (converging to $x \in X$, respectively $x \in \mathcal{M}$). For a countablet set, relative sequential compactness implies relative compactness.

A convergent sequence has precisely one limit point. If a sequence is relatively compact and has one limit point, then it converges.

Let X, Y be a dual pair of linear spaces. If $(Y, \sigma(Y, X))$ is separable, then a relatively compact sequence $\{x_n\} \subset (X, \sigma(X, Y))$ is also relatively sequentially compact.

[2]This topology is also called the *Sazonov topology*, after Sazonov, who introduced it for Hilbert spaces (see Sazonov [3] and also Kolmogorov [3]).

It suffices to show that each infinite part of this sequence contains a subsequence with a unique limit point. Such a subsequence can be clearly constructed by the diagonal process so as to converge on the elements of a countable set which is dense in $(Y, \sigma(Y, X))$.

A sequence $\{x_n\}$ is called a *fundamental sequence* if every subsequence $\langle x_n, y\rangle$ is such $(y \in Y)$. The space X is called *weakly complete* if every fundamental sequence (in this sense) in it converges. Since a fundamental sequence has precisely one limit point, relative compactness of the sequence implies convergence.

5. Chapter I: Supplementary remarks and historical comments

Algebras of cylinder sets on products of spaces arise naturally in the theory of random functions. In this case a nonnegative normed quasimeasure is nothing but a consistent system of finite-dimensional distributions of a random function. Kolmogorov [1] was the first to note that this set induces a measure in a product of compact spaces. Kolmogorov's proof has been transferred to general situations (see Dynkin [1]), preserving certain topological requirements. In Neveu [1], topological conditions are replaced by a condition in terms of compact classes of sets. This approach was developed by Smolyanov–Fomin [1], who formulated the notion of measurable Radon space. They proved Kolmogorov's theorem for a product of Radon spaces and a realvalued bounded quasimeasure.

The version of Kolmogorov's theorem for a linear space with a sufficient supply of linear functionals was considered by Getoor [1], Segal [2], Gross [5], McShane [1]. Our presentation of properties of cylinder sets in linear spaces and of the theorem about σ-additivity of the image of a bounded quasimeasure under linear extension of a measurable linear space follows Smolyanov–Fomin [1].

A further generalization consists in passing from products of spaces to spaces with measurable limit structure: theorem I.3 includes a generalization of Kolmogorov's theorem to this case.

Integration with respect to a positive quasimeasure arises already in the theory of random functions, when computing averages of functionals by means of finite-dimensional distributions. In general, the problems arising in this situation are of a technical nature (see Fomin [5]). the situation for bounded quasimeasures of variable sign is similar. Note that a number of theorems from the theory of stochastic processes can be formally transferred to this case.

Much more complicated is the case for unbounded quasimeasures. Integrals with respect to such measures first appeared in the famous papers of Feynman (see Feynman–Hibbs [1]). strictly speaking, the 'Feynman measure' is not a quasimeasure in our sense. Its finite-dimensional projections are not bounded, and become bounded only after special regularization.

There is a vast amount of literature on Feynman integrals and other integrals with respect to unbounded quasimeasures. We will touch again on these questions in chapters II and IV, and now refer only to Gelfand–Yaglom [1], Dalecky [5], Albeverio–Hoegh-Krohn [1], and the references therein. Formula (3.19) for computing integrals was used in Dalecky–Ladokhin [1]. In Albeverio–Hoegh-Krohn [1] it is used to define infinite-dimensional Fresnel integrals, which are then put at the foundation of their treatment of Feynman integrals.

CHAPTER II

Gaussian measures in Hilbert space

In this chapter we consider the important class of Gaussian measures in function spaces. The presentation is carried out for Hilbert spaces: it will be shown in the next chapter that every Gaussian cylindrical measure can be realized on such a space.

We consider measurable, square integrable linear functionals and operators, and construct the well-known orthogonal decomposition of the space of functions that are square integrable with respect to a Gaussian measure into spaces of polynomials. The approach to this problem is based on special formulas for integration by parts. The problem of the absolute continuity of Gaussian measures is studied in detail.

We also consider Fourier transformation with respect to a Gaussian measure. Finally we give an elementary treatment of certain Feynman integrals, those with respect to a Gaussian quasimeasure involving a complex parameter.

For understanding this chapter one needs brief knowledge of chapter I, mainly of its definitions. The reader is also required to be able to deal with some elementary concepts in the geometry of Hilbert space, and in the theory of operators on it.

1. Gaussian measures in finite-dimensional spaces

1.1. Characteristic functional and density. Recall that a Gaussian measure in $\mathbf{R}^1$ is a nonnegative measure with density $p(x) = \exp\{-(x-\alpha)^2/(2\sigma^2)\}/(\sigma\sqrt{2\pi})$, which is absolutely continuous with respect to Lebesgue measure. It is uniquely determined by the real number α (mean) and positive number σ^2 (variance). The characteristic functional of a Gaussian measure ρ in $\mathbf{R}^1$ is easily verified to have the form

$$\chi(\omega) = \int_{-\infty}^{\infty} e^{i\omega x}\rho(x)\,dx = \exp\left\{-\frac{1}{2}\sigma^2\omega^2 + i\alpha\omega\right\}. \tag{1.1}$$

A nonnegative measure μ in $\mathbf{R}^n$ is called *Gaussian* if all its one-dimensional projections μ^y $(y \in \mathbf{R}^n)$, where

$$\mu^y(A) = \mu\{x \in \mathbf{R}^n : \ (x,y) \in A\},$$

are Gaussian. This condition leads to an expression for the characteristic functional χ_μ. In fact, for the characteristic functional of a projection μ^y, equation (I.3.6) implies that $\chi_{\mu^y}(\omega) = \chi_\mu(\omega y)$, which implies

$$u(\omega y) = \ln \chi_\mu(\omega y) = -\frac{1}{2}\sigma^2(y)\omega^2 + i\alpha(y)\omega.$$

Differentiating this expression with respect to ω at $\omega = 0$ leads to the conclusion that $\alpha(y) = (\alpha, y)$ $(\alpha = -iu'(0))$ is a linear functional, and $\sigma^2(y) = (By, y)$ $(B = -u''(0))$ is a quadratic functional.

Thus, the characteristic functional of a Gaussian measure μ in $\mathbf{R}^n$ always has the form

$$\chi_\mu(y) = \exp\left\{-\frac{1}{2}(By, y) + i(\alpha, y)\right\}, \tag{1.2}$$

where $\alpha \in \mathbf{R}^n$ is the mean and B is a positive operator on $\mathbf{R}^n$, called the *correlation operator*.

By straightforward computation and diagonalization of the matrix of B it can be readily verified that for $\det B \neq 0$ the function (1.2) is the Fourier transform of the positive function

$$p(x) = \frac{1}{\sqrt{(2\pi)^n \det B}} \exp\left\{-\frac{1}{2}(B^{-1}(x-a), x-a)\right\},$$

which is the density of a nonnegative measure μ in $\mathbf{R}^n$ with respect to Lebesgue measure.

Thus, for $\det B \neq 0$ formula (1.2) describes all (nondegenerate) Gaussian measures in $\mathbf{R}^n$.

If $\det B = 0$, then there is a maximal orthogonal projection operator P in $\mathbf{R}^n$ such that $B = PB = BP$. Now μ is concentrated on the subspace $P\mathbf{R}^n$, shifted along a vector a, and is nondegenerate there.

1.2. Computation of certain integrals. Let μ_B be a nondegenerate, centered (i.e. $a = 0$) Gaussian measure with correlation operator B.

To compute the moments

$$m_{2k}(\phi_1, \dots, \phi_{2k}) = \int_{\mathbf{R}^n} (x, \phi_1) \dots (x, \phi_{2k})\, \mu(dx)$$

it is suitable to use the relation

$$m_{2k}(\phi_1, \dots, \phi_{2k}) = (-1)^k \chi_\mu^{(2k)}(0)(\phi_1, \dots, \phi_{2k}),$$

where the derivative of order $2k$ along $\phi_1, \dots, \phi_{2k}$ appears at the righthand side. This results in

$$\begin{aligned} m_{2k}(\phi_1, \dots, \phi_{2k}) &= (-1)^k \left[e^{-(By,y)/2}\right]_{y=0}^{(2k)} (\phi_1, \dots, \phi_{2k}) = \\ &\frac{1}{k!2^k}[(By, y)^k]^{(2k)}(\phi_1, \dots, \phi_{2k}) = \\ &\frac{1}{k!2^k} \sum_{(j_1, \dots, j_{2k})} (B\phi_{j_1}, \phi_{j_2}) \dots (B\phi_{j_{2k-1}}, \phi_{j_{2k}}). \end{aligned} \tag{1.3}$$

The sum in (1.3) is over all permutations of the index sequence. In particular, $m_2(\phi_1, \phi_2) = (B\phi_1, \phi_2)$, and

$$m_4(\phi_1, \phi_2, \phi_3, \phi_4) = (B\phi_1, \phi_2)(B\phi_3, \phi_4) + (B\phi_1, \phi_3)(B\phi_2, \phi_4) + (B\phi_1, \phi_4)(B\phi_2, \phi_3).$$

As a consequence we arrive at the formulas

$$\int_{\mathbf{R}^n} (Ax, x)\, \mu_B(dx) = \sum_{j,k=1}^n a_{jk} \int_{\mathbf{R}^n} x_j x_k\, \mu_B(dx) = \sum_{j,k=1}^n a_{jk} b_{jk} = \operatorname{Tr} AB, \tag{1.4}$$

$$\int_{\mathbf{R}^n} (Ax,x)^2\,\mu_B(dx) = \sum_{j,k,j_1,k_1=1} a_{jk}a_{j_1k_1}\int_{\mathbf{R}^n} x_jx_kx_{j_1}x_{k_1}\,\mu_B(dx) =$$

$$\sum_{j,k,j_1,k_1=1} a_{jk}a_{j_1k_1}(b_{jk}b_{j_1k_1} + b_{jj_1}b_{kk_1} + b_{jk_1}b_{k_1j} =$$

$$(\operatorname{Tr} AB)^2 + 2\operatorname{Tr}(AB)^2. \tag{1.5}$$

Hence we can easily derive the following estimates.

LEMMA 1.1. *Let $A > 0$. Then*

$$\mu_B\{x:\ (Ax,x) \ge 1\} \le \operatorname{Tr} AB,$$

and

$$\mu_B\{x:\ |(Ax,x) - \operatorname{Tr} AB| \le c\sqrt{\operatorname{Tr} AB}\} \ge 1 - \frac{2}{c^2}\|AB\|.$$

PROOF. Using the well-known Chebyshev inequality, we obtain

$$\mu_B\{x:\ (Ax,x)\ge 1\} = \int_{\{(Ax,x)\ge 1\}} \mu_B(dx) \le \int_{\mathbf{R}^n} (Ax,x)\,\mu_B(dx) = \operatorname{Tr} AB,$$

and, similarly,

$$\mu_B\{x:\ |(Ax,x) - \operatorname{Tr} AB| \ge c\sqrt{\operatorname{Tr} AB}\} \le$$

$$\int_{\mathbf{R}^n} \frac{[(Ax,x) - \operatorname{Tr} AB]^2}{c^2 \operatorname{Tr} AB}\mu_B(dx) = \frac{2}{c^2}\frac{\operatorname{Tr}(AB)^2}{\operatorname{Tr} AB} \le \frac{2}{c^2}\|AB\|. \quad \square$$

We compute a few more integrals of exponential-type functions. By the change $(x,y) = \theta$ and transference of the measure μ_B to the measure μ_B^y in $\mathbf{R}$ we obtain

$$\int_{\mathbf{R}^n} e^{(x,y)}\mu_B(dx) = \int_{-\infty}^{\infty} e^{\theta}\mu_B^y(d\theta) =$$

$$\frac{1}{\sqrt{2\pi(By,y)}}\int_{\infty}^{\infty} e^{\theta - \frac{\theta^2}{2(By,y)}}\,d\theta = e^{(By,y)/2}. \tag{1.6}$$

The integration of a quadratic exponential is somewhat more complicated.

PROPOSITION 1.1. *Let A be an operator satisfying $2(Ax,x) < (B^{-1}x,x)$ $(x\in\mathbf{R}^n)$ (or, equivalently, $2B^{1/2}AB^{1/2} < I$). Then*

$$\int_{\mathbf{R}^n} e^{(Ax,x)}\mu_B(dx) = \frac{1}{\sqrt{\det(I - 2B^{1/2}AB^{1/2})}} = \frac{1}{\sqrt{\det(I-2AB)}}. \tag{1.7}$$

PROOF. By the change $x = B^{1/2}y$ we obtain

$$\int_{\mathbf{R}^n} e^{(Ax,x)}\mu_B(dx) = \frac{1}{(2\pi)^{n/2}}\int_{\mathbf{R}^n} e^{(B^{1/2}AB^{1/2}y,y) - \|y\|^2/2}\,dy.$$

Let $B^{1/2}AB^{1/2} = S\Lambda S^*$, with Λ a diagonal matrix and S an orthogonal matrix. By the change $y = Sz$ we bring the latter integral to the following form (note that under the conditions imposed, Λ has eigenvalues λ_k satisfying $\lambda_k < 1/2$):

$$\frac{1}{(2\pi)^{n/2}} \int_{\mathbf{R}^n} e^{(\Lambda z,z)-\|z\|^2/2}\, dz = \prod_{k=1}^{n} \frac{1}{\sqrt{2\pi}} \int_{-\infty}^{\infty} e^{(\lambda_k - 1/2)z_k^2}\, dz_k =$$

$$\frac{1}{\sqrt{\prod_{k=1}^{n}(1-2\lambda_k)}} = \frac{1}{\sqrt{\det(I - 2A^{1/2}BA^{1/2})}}. \quad \square$$

1.3. Integration by parts. Here we derive some useful formulas involving integrals with respect to a Gaussian measure. Conventionally, these formulas are called *integration by parts formulas*.

Let $u(x)$ be a function which is differentiable at a point $x = (x_1, \dots, x_n) \in \mathbf{R}^n$. We introduce the notations $u'(x) = (\partial u/\partial x_1, \dots, \partial u/\partial x_n)$ for its gradient (the vector of first derivatives), and $u''(x) = \|\partial^2 u/\partial x_k \partial x_j\|_{k,j=1}^n$ for the Hessian (the matrix of second derivatives).

PROPOSITION 1.2. *Let $u(x)$ be a differentiable function in $\mathbf{R}^n$ satisfying*

$$\int_{\mathbf{R}^n} |u(x)|^2 \mu_B(dx) < \infty, \quad \int_{\mathbf{R}^n} \|u'(x)\|^2 \mu_B(dx) < \infty.$$

Then the following formula holds:

$$\int_{\mathbf{R}^n} u(x)(x,\psi)\mu_B(dx) = \int_{\mathbf{R}^n} (u'(x), B\psi)\mu_B(dx) \quad (\psi \in \mathbf{R}^n). \tag{1.8}$$

PROOF. Every function $u(x)$ satisfying the conditions stated, can be approximated in the mean square with respect to μ_B by polynomials, together with its derivative $(u'(x), B\psi)$ along a given direction $B\psi$. It is also easily seen that limit transition is allowed in formula (1.8), and thus it suffices to verify this formula for polynomials, and by linearity it suffices consider monomials $u(x) = (x,\phi_1) \times \cdots \times (x,\phi_s)$. Such monomials render both sides of (1.8) symmetric functionals with respect to the ϕ_k $(k = 1, \dots, s)$, so it suffices to verify (1.8) for $\phi_1 = \cdots = \phi_s = \phi$ and $s = 2p-1$ (if s is even, both sides are easily seen to vanish). In this case (1.3) implies that

$$\int_{\mathbf{R}^n} u(x)(x,\psi)\mu_B(dx) = \int_{\mathbf{R}^n} (x,\phi)^{2p-1}(x,\psi)\mu_B(dx) = \frac{(2p)!}{p!2^p}(B\phi,\phi)^{p-1}(B\psi,\psi)$$

and

$$\int_{\mathbf{R}^n} (u'(x), B\psi)\mu_B(dx) = (2p-1)\int_{\mathbf{R}^n} (x,\phi)^{2p-2}\mu_B(dx)(B\psi,\psi) =$$

$$\frac{(2p-1)!}{(p-1)!2^{p-1}}(B\phi,\phi)^{p-1}(B\psi,\psi). \quad \square$$

COROLLARY 1.1. *If $v(x)$ is also a differentiable function on $\mathbf{R}^n$, and*

$$\int_{\mathbf{R}^n} |u(x)v(x)|^2 \mu_B(dx) < \infty, \quad \int_{\mathbf{R}^n} \{|u(x)|^2 \|v'(x)\|^2 + |v(x)|^2 \|u'(x)\|^2\} \mu_B(dx) < \infty,$$

then, applying (1.8) to the product $u(x)v(x)$, we obtain

$$\int_{\mathbf{R}^n} \{u(x)(x,\phi) - (u'(x), B\phi)\} v(x) \mu_B(dx) = \int_{\mathbf{R}^n} u(x)(v'(x), B\phi) \mu_B(dx). \tag{1.9}$$

PROPOSITION 1.3. *If $u(x)$ and $v(x)$ both satisfy the conditions of proposition 1.2, then*

$$\int_{\mathbf{R}^n} [u(x)(x,\phi) - (u'(x), B\phi)][v(x)(x,\psi) - (v'(x), B\psi)] \mu_B(dx) =$$
$$(B\phi,\phi) \int_{\mathbf{R}^n} u(x)v(x)\mu_B(dx) + \int_{\mathbf{R}^n} (u'(x), B\psi)(v'(x), B\phi)\mu_B(dx). \tag{1.10}$$

PROOF. First we assume that $v(x)$ is twice differentiable, and bounded together with its derivatives. It follows from (1.9) and (1.8) that

$$\int_{\mathbf{R}^n} [u(x)(x,\phi) - (u'(x), B\phi)] v(x)(x,\psi)\mu_B(dx) =$$
$$\int_{\mathbf{R}^n} u(x)\{(v'(x), B\phi)(x,\psi) + v(x)(\psi, B\phi)\}\mu_B(dx) =$$
$$\int_{\mathbf{R}^n} \{u(x)v(x)(B\phi,\psi) + (u'(x), B\psi)(v'(x), B\phi) +$$
$$u(x)(v''(x)B\phi, B\psi)\}\mu_B(dx).$$

Similarly,

$$\int_{\mathbf{R}^n} [u(x)(x,\phi) - (u'(x), B\phi)](v'(x), B\psi)\mu_B(dx) = \int_{\mathbf{R}^n} u(x)(v''(x)B\phi, B\psi)\mu_B(dx).$$

Subtracting termwise these equalities we arrive at (1.10).

In particular, for $u = v$ and $\phi = \psi$ we have

$$\int_{\mathbf{R}^n} [u(x)(x,\phi) - (u'(x), B\phi)]^2 \mu_B(dx) =$$
$$(B\phi,\phi) \int_{\mathbf{R}^n} u^2(x)\mu_B(dx) + \int_{\mathbf{R}^n} (u'(x), B\phi)\mu_B(dx). \tag{1.11}$$

Approximating in the mean square the function $u(x)$ in proposition 1.2 by smooth bounded functions and passing to the limit, we obtain

$$\int_{\mathbf{R}^n} [u(x)(x,\phi) - (u'(x), B\phi)]^2 \mu_B(dx) < \infty,$$

and the quadratic relation (1.11) with corresponding bilinear formula (1.10). □

COROLLARY 1.2. *Let $u(x)$ and $v(x)$ be vector functions taking values in $\mathbf{R}^n$, and let $h(x)$ be a scalar function, all satisfying the conditions of proposition 1.2. Then:*

$$\int_{\mathbf{R}^n} (u(x), x)\mu_B(dx) = \int_{\mathbf{R}^n} \mathrm{Tr}[u'(x)B]\mu_B(dx), \tag{1.12}$$

$$\int[(u(x),x)-\operatorname{Tr}u'(x)B]h(x)\mu_B(dx)=\int_{\mathbf{R}^n}(u(x),Bh'(x))\mu_B(dx), \tag{1.13}$$

$$\int_{\mathbf{R}^n}[(u(x),x)-\operatorname{Tr}u'(x)B][(v(x),x)-\operatorname{Tr}v'(x)B]\mu_B(dx)=$$
$$\int_{\mathbf{R}^n}(Bu(x),v(x))\mu_B(dx)+\int_{\mathbf{R}^n}\operatorname{Tr}[v'(x)Bu'(x)]\mu_B(dx). \tag{1.14}$$

The proof of each of these equations is based on coordinate representations and corresponding equations for scalar functions. It suffices to note the relations

$$(u(x),x)=\sum_k(u(x),\phi_k)(\phi_k,x),\qquad (v(x),x)=\sum_k(v(x),\phi_k)(\phi_k,x),$$

and $(u_k'(x),B\phi)=(u'(x)B\phi,\phi_k)$, where $u_k(x)=(u(x),\phi_k)$.

1.4. Solution of the Cauchy problem. The Gaussian measure μ_k is closely related to the parabolic partial differential equation

$$\frac{\partial u}{\partial t}=\frac{1}{2}\sum_{j,k=1}^{n}b_{jk}\frac{\partial^2}{\partial x_j\partial x_k}\qquad (x\in\mathbf{R}^n,\quad t\geq 0). \tag{1.15}$$

Consider the corresponding Cauchy problem with initial condition $u(x,0)=u_0(x)$. It follows from (1.15) that for the Fourier transform $v(\omega,t)=\int_{\mathbf{R}^n}e^{i(\omega,x)}u(x,t)\,dx$ we have $v'(\omega,t)=-(B\omega,\omega)v(\omega,t)/2$. This implies

$$v(\omega,t)=\exp\left\{-\frac{1}{2}(B\omega,\omega)t\right\}v_0(\omega).$$

Applying the convolution theorem and using the fact that $e^{-(B\omega,\omega)t/2}$ is the characteristic functional of the Gaussian measure μ_{Bt} with correlation operator Bt, we obtain an expression for the solution of problem (1.15) as an integral of the initial condition with respect to a Gaussian measure:

$$u(x,t)=\int_{\mathbf{R}^n}u_0(x-y)\mu_{Bt}(dy). \tag{1.16}$$

This result can be readily verified to hold also for functions of rather rapid growth (such that the integral (1.16) exists, together with the integrals obtained from it by at most twice differentiation with respect to the parameter x).

2. Gaussian measures in Hilbert space

2.1. σ-additivity for a Gaussian cylindrical measure. Throughout this chapter, X is a separable real Hilbert space, $\mathcal{U}$ is its Borel σ-algebra, $\mathcal{U}^{\mathcal{P}}$ is the algebra of cylinder sets determined by some system of finite-dimensional projectors, described in I.2.2, so that $\mathcal{U}=\sigma(\mathcal{U}^{\mathcal{P}})$ is the minimal σ-algebra containing $\mathcal{U}^{\mathcal{P}}$. Also, $X=\overline{X}_{\mathcal{P}}$, where $X_{\mathcal{P}}=\cup_{P\in\mathcal{P}}X_P$ ($X_P=PX$).

A cylindrical measure μ in X is called *Gaussian* if every finite-dimensional (one-dimensional) projection of it is Gaussian.

Recall that a cylindrical measure μ is determined by its characteristic functional χ_μ, which is the common extension of the compatible system of characteristic functionals

$\{\chi^{(P)}\}$ of the projections μ^P. We will now consider a *centered cylindrical measure*, i.e. a cylindrical measure μ for which all projections μ^P are centered:

$$\int_{X_P} (x, \phi)\mu^P(dx) = 0, \quad (\phi \in X_P, \quad P \in \mathcal{P}).$$

It follows from (1.2) that its characteristic functional has the form

$$\chi_\mu(y) = e^{-b(y,y)/2}, \tag{2.1}$$

where $b(y_1, y_2)$ is a nonnegative bilinear functional on the linear set $X_{\mathcal{P}}$, continuous on any finite-dimensional subspace of X_P ($P \in \mathcal{P}$). This functional is called the *correlation* of μ.

If $b(y_1, y_2) = (By_1, y_2)$ with B a positive operator (possibly unbounded) on $X_{\mathcal{P}}$, then B is called the *correlation operator* of μ (we will frequently denote μ by μ_B). Note that the correlation operator B_P of the finite-dimensional projection μ^P has the form $B_P = PBP$.

On the other hand, if $b(y_1, y_2)$ is a nonnegative bilinear functional on a set $\mathcal{D}$ which is dense in X, then it is possible to construct a Gaussian cylindrical measure μ with (2.1) as characteristic functional (and with $\mathcal{P}$ the set of orthoprojectors on finite-dimensional subspaces in $\mathcal{D}$).

THEOREM 2.1. *A Gaussian cylindrical measure μ in X is σ-additive if and only if its correlation functional has the form $b(y_1, y_2) = (By_1, y_2)$, where B is a nonnegative nuclear operator on X.*

PROOF. Let μ be σ-additive. Then its characteristic functional and correlation functional are continuous on X, hence $b(y, y) = (By, y)$ ($B \in \mathcal{L}(X)$). We show that the bounded operator B is nuclear. Assuming the contrary, we can find a finite-dimensional orthoprojector P having arbitrarily large trace $\operatorname{Tr} PBP = R$. Consider the cylinder set $C(P, \mathcal{M})$, where $\mathcal{M} = \{x \in X_P\colon |\|x\|^2 - R| < \alpha\sqrt{R}\}$ lies outside the ball of radius $R - \alpha\sqrt{R}$. Lemma 1.1 implies an estimate from below:

$$\mu(C(P, \mathcal{M})) \geq 1 - \frac{2\|b\|}{\alpha^2} = \frac{1}{2} \quad \text{for } \alpha = 2\sqrt{\|B\|}.$$

Thus, there are balls in X with center at zero and arbitrarily large radius $R - \alpha\sqrt{R}$, but of measure less than $1/2$. This is impossible, by the continuity of μ, since $\mu(X) = 1$. This proves necessity.

Sufficiency follows from lemma 1.1 and proposition I.1.3. In fact, for any cylinder set $C(P, \mathcal{M})$ lying outside a ball of radius R (which is compact in the weak topology), its base $\mathcal{M}$ lies outside the same ball in X_P, and hence

$$\mu(C(P, \mathcal{M})) = \mu^P(\mathcal{M}) \leq \frac{\operatorname{Tr} PBP}{R^2} \leq \frac{\operatorname{Tr} B}{R^2}$$

can be made arbitrarily small for sufficiently large R. □

2.2. Some transformations of Gaussian measures in X. a) Shift $x \to x + a$. This transformation maps a cylindrical measure μ to the cylindrical measure $\mu^{(a)}$ with characteristic functional

$$\chi_{\mu^{(a)}}(y) = \exp\left\{-\frac{1}{2}b(y,y) + i(a,y)\right\}. \tag{2.2}$$

Conversely, the measure with characteristic functional (2.2) can be obtained from the centered measure by shift over the element $a \in X$; a is called the mean of $\mu^{(a)}$.

Clearly, two cylindrical measures μ and $\mu^{(a)}$ are either both σ-additive or both not σ-additive.

b) Linear transformation $x \to Sx$. Under such a transformation the characteristic functional becomes (see (I.3.8)):

$$\chi_{\mu^S}(y) = \chi_\mu(S^* y) = \exp\left\{-\frac{1}{2}(SBS^* y, y)\right\};$$

consequently, the transformed measure μ^S is Gaussian. It is σ-additive if the operator SBS^* is nuclear on SX.

c) Product of measures. Let μ_1, μ_2 be Gaussian cylindrical measures in Hilbert spaces X_1, X_2, respectively. The product $\mu_1 \times \mu_2$ is the Gaussian cylindrical measure in $X_1 \times X_2$ with characteristic functional

$$\exp\left\{-\frac{1}{2}[b_1(y_1,y_1) + b_2(y_2,y_2)]\right\}.$$

Conversely, let $X = X_P \oplus X_Q$ be a decomposition in orthogonal subspaces X_P and X_Q, and let P, $Q = I - P$ be the corresponding orthoprojectors. If P commutes with B, $PB = BP$, then

$$(By, y) = (B_P y_P, y_P) + (B_Q y_Q, y_Q),$$

and hence the Gaussian cylindrical measure μ_B splits into the product

$$\mu_B = \mu_P \times \mu_Q \tag{2.3}$$

of the Gaussian cylindrical measures $\mu_{B_P} = \mu_B^P$, $\mu_{B_Q} = \mu_B^Q$ in the spaces X_P and X_Q.

d) Convolution of measures. Convolution of cylindrical measures leads to multiplication of their characteristic functionals (I.3.13). Thus, the convolution of two Gaussian cylindrical measures $\mu_{B_1}^{(a_1)}$ and $\mu_{B_2}^{(a_2)}$ is the Gaussian cylindrical measure with correlation operator $B_1 + B_2$ and mean $a_1 + a_2$.

2.3. Computation of integrals. We compute certain integrals with respect to a centered Gaussian measure μ with nuclear correlation operator B on a Hilbert space X.

The simplest examples are brought about by considering functions of finitely many linear functionals,

$$f(x) = F((x,\phi_1), \dots, (x,\phi_n)),$$

where $\phi_1, \dots, \phi_n \in X$ is an orthonormal system and $F(t_1, \dots, t_n)$ is a measurable function in $\mathbf{R}^n$.

Such a function $f(x)$ is cylindrical, and so

$$\int_X f(x)\mu(dx) = \int_{\mathbf{R}^n} F(t_1,\dots,t_n)\mu^{(y_1\dots,y_n)}(dt),$$

where $\mu^{(y_1\dots,y_n)}$ is the measure in $\mathbf{R}^n$ obtained from μ by the map $x \to (y_1,\dots,y_n)$, $y_k = (x,\phi_k)$, and having characteristic functional

$$\chi^{(y_1\dots,y_n)}(t_1,\dots,t_n) = \chi_\mu\left(\sum_{k=1}^n t_k\phi_k\right) = \exp\left\{-\frac{1}{2}\sum_{j,k=1}^n t_k t_j(B\phi_k,\phi_j)\right\}.$$

Thus, the following formula holds:

(2.4)

$$\int_X F((x,\phi_1),\dots(x,\phi_n))\mu_B(dx) = \frac{1}{\sqrt{(2\pi)^n \det\beta}}\int_{\mathbf{R}^n} F(t_1,\dots,t_n)e^{-(\beta^{-1}t,t)/2}\,dt,$$

where β is the matrix $\beta = \|(B\phi_k,\phi_j)\|_{j,k=1}^n$.

In particular, for ϕ_k we will often use the eigenvectors e_k of B. In this case $\beta = \|\lambda_k\delta_{jk}\|_{j,k=1}^n$ is a diagonal matrix, and the computation is facilitated.

As an example we note the following formula (which resembles the finite-dimensional case):

$$\int_X e^{(\phi,x)}\mu_B(dx) = e^{(B\phi,\phi)/2}. \tag{2.5}$$

Note that the reasoning above does not use the nuclearity of B, and so it is true for arbitrary Gaussian quasimeasures.

When computing integrals of noncylindrical functionals it is convenient to use example I.3.3. We consider some examples.

a) Let $f(x) = \|x\|^2$. Denote by P_n the orthoprojector on the linear hull $\mathcal{L}_n =$ l.e. $\{e_1,\dots,e_n\}$ of the first n eigenvectors of B. Then (1.4) implies

$$\int_X \|P_n x\|^2\mu(dx) = \int_{\mathbf{R}^n}\|x\|^2\mu_{P_nBP_n}(dx) = \operatorname{Tr} P_nBP_n.$$

Since the sequence $\|P_n x\|$ tends to $\|x\|$ monotonically, we find

$$\int_X \|x\|^2\mu_B(dx) = \lim_{n\to\infty}\operatorname{Tr} P_nBP_n = \operatorname{Tr} B < \infty.$$

Thus, a σ-additive Gaussian measure μ in a Hilbert space X has finite second moment.

Now we can draw the conclusion that the function (Ax,x) is integrable for every bounded operator A, and derive similarly that

$$\int_X (Ax,x)\mu_B(dx) = \operatorname{Tr} AB. \tag{2.6}$$

Similarly, finiteness of higher moments can be shown, and formulas for them can be derived; e.g.

$$\int_X (Ax,x)^2\mu(dx) = [\operatorname{Tr} AB]^2 + 2\operatorname{Tr}(AB)^2. \tag{2.7}$$

REMARK 2.1. It follows from (2.6) and (2.7) that lemma 1.1 can be transferred without changes to an infinite-dimensional space X.

b) We now consider the function $f(x) = e^{(Ax,x)}$, with A a bounded operator on X satisfying

$$A \le \alpha I \quad (2\alpha B < I). \tag{2.8}$$

First, suppose $A = \alpha I$. Then $\exp\{\alpha \|P_n x\|^2\} \to \exp\{\alpha\|x\|^2\}$ monotonically, and (1.7) gives

$$\int_X e^{\alpha\|x\|^2} \mu(dx) = \lim_{n\to\infty} [\det(I - 2\alpha B P_n)]^{-1/2}.$$

Nuclearity of B implies convergence of the product

$$\det(I - 2\alpha B) = \lim_{n\to\infty}(I - 2\alpha B P_n) = \prod_{k=1}^{\infty}(1 - 2\alpha\lambda_k),$$

and thus

$$\int_X e^{\alpha\|x\|^2}\mu(dx) = [\det(I - 2\alpha B)]^{-1/2}.$$

In particular, the function $e^{\alpha\|x\|^2}$ is integrable and, under the condition (2.8), so is the function $e^{(Ax,x)}$. Moreover, $e^{(Ax,x)} \le e^{\alpha\|x\|^2}$. Therefore limit transition under the integral sign in (1.7) is allowed as $P \to I$, resulting in the formula

$$\int_X e^{(Ax,x)}\mu(dx) = [\det(I - 2B^{1/2}AB^{1/2})]^{-1/2}. \tag{2.9}$$

c) Suppose $(Ax,x) \ge 0$, $\alpha > 0$. Replacing A in (2.9) by $-\lambda A$, multiplying by $e^{-\lambda\alpha}$ the inequality obtained, and integrating termwise gives

$$\int_X \frac{e^{-\lambda_0(Ax,x)}}{\alpha + (Ax,x)}\mu(dx) = \int_{\lambda_0}^{\infty} \frac{e^{-\lambda_0}\, d\lambda}{\sqrt{\det(I + 2\lambda B^{1/2}AB^{1/2})}}, \qquad \lambda_0 \ge 0. \tag{2.10}$$

2.4. Gaussian cylindrical measures with arbitrary correlation operator. Let $\mu = \mu_B$ be the centered Gaussian cylindrical measure in X with correlation operator B. B is not yet assumed to be nuclear. Consider a linear map $S\colon X \to X_1$, where X_1 is also a Hilbert space. It gives rise to the cylindrical measure μ^S in X_1 with characteristic functional

$$\chi_{\mu^S}(y) = \chi_\mu(S^*y) = \exp\left\{-\frac{1}{2}(BS^*y, S^*y)_X\right\} = \exp\left\{-\frac{1}{2}(SBS^*y, y)_{X_1}\right\}.$$

Its correlation operator $B_S = SBS^*$ may already turn out to be nuclear on X_1, and so the cylindrical measure μ^S may be σ-additive (in this case S is called a *Radonifying operator* for μ).

In particular, consider the case when S is a dense imbedding of X into X_1. The restriction of the scalar product $(\cdot,\cdot)_{X_1}$ to X is continuous, hence

$$(Sy, Sy)_{X_1} = (Ty, Ty)_X, \tag{2.11}$$

with T a bounded operator on X. It follows from (2.11) that TS^{-1} acts isometrically from $SX \subset X_1$ into X, and so it has a continuous extension $\tilde{T}$ ($\tilde{T}S = T$), to $X_1 = \overline{SX}$ and with values in X, for which $(\tilde{T}x_1, \tilde{T}y_1)_X = (x_1, y_1)_{X_1}$. Then $(Sx, y_1)_{X_1} =$

$(Tx, \tilde{T}y_1)_X = (x, T^*\tilde{T}y_1)_X$, that is, $S^* = T^*\tilde{T}$. Also, the correlation operator has the form $B_S = SBT^*\tilde{T}$ ($B_S|_{SX}S = BT^*T$); this operator is nonnegative with respect to the scalar product $(\cdot,\cdot)_{X_1}$.

PROPOSITION 2.1. *Let $\mu = \mu_B$ be a Gaussian cylindrical measure in a Hilbert space X. It is always possible to introduce a scalar product $(x,y)_1 = (Tx, Ty)$ in X such that μ is σ-additive in the completion X_1 of X with respect to the norm $\|x\|_1^2 = (x,x)_1$.*

PROOF. The operator T can be chosen such that TBT^* is nuclear on X. Also, B_S turns out to be nuclear on X_1, since

$$\sum_{k=1}^{\infty}(SBT^*\tilde{T}e_k, e_k)_{X_1} = \sum_{k=1}^{\infty}(TBT^* f_k, f_k) < \infty$$

for a certain orthonormal basis $\{e_k\}$ in X_1, and $f_k = \tilde{T}e_k$. It remains to apply theorem 2.1. □

COROLLARY 2.1. *Let B be a bounded operator on X and T a nondegenerate Hilbert–Schmidt operator. Then μ_B has a σ-additive extension to the space X_1 obtained by completing X with respect to the norm $\|x\|_1 = \|Tx\|$. In this case the imbedding operator $S\colon X \to X_1$ is also a Hilbert–Schmidt operator:*

$$\sigma_2^2(S) = \sum_{k=1}^{\infty} \|Sf_k\|_1^2 = \sum_{k=1}^{\infty} \|Tf_k\|^2 = \sigma_2^2(T) < \infty.$$

In particular, corollary 2.1 is true for the cylindrical measure μ_I with the identity I on X as correlation operator.

EXAMPLE 2.1. Let $X = \ell_2$ be the space of sequences $x = (x_1, x_2, \dots)$ with finite norm $\|x\|^2 = \sum_{k=1}^{\infty} |x_k|^2$. The Gaussian cylindrical measure μ_I is defined on cylinder sets

$$C(P_{k_1,\dots,k_n}, \mathcal{M}) = \{x : (x_{k_1}, \dots, x_{k_n}) \in \mathcal{M}\}$$

by the formula

$$\mu_I(C(P_{k_1,\dots,k_n}, \mathcal{M})) = \frac{1}{\sqrt{(2\pi)^n}} \int_{\mathcal{M}} \exp\left\{-\frac{1}{2}\sum_{j=1}^{n} |x_{k_j}|^2\right\} \prod_{j=1}^{n} dx_{k_j}.$$

This measure is σ-additive in the space X_1 of sequences $x = (x_1, x_2, \dots)$ with finite norm $\|x\|_1^2 = \sum_{k=1}^{\infty} \alpha_k^2 |x_k|^2$, where $\sum_{k=1}^{\infty} |\alpha_k|^2 < \infty$.

A Gaussian measure μ_I that is σ-additive in an extension X_1 of a space X with imbedding operator S satisfying $\sigma_2^2(S) < \infty$ (a quasinuclear imbedding) is called a *canonical Gaussian measure*.

PROPOSITION 2.2. *A centered Gaussian measure μ in a Hilbert space X can be represented by a canonical Gaussian measure.*

PROOF. Let $\mu = \mu_B$. Without loss of generality we may assume that the correlation operator B is nondegenerate (otherwise we use the subspace of X on which this holds). We make the domain of definition of the unbounded operator $B^{-1/2}$ into a complete Hilbert space, $\mathcal{H}_0$, with scalar product $(x, y)_0 = (B^{-1/2}x, B^{-1/2}y)$. Let S be the imbedding operator of $\mathcal{H}_0$ into X. If $\{e_k\}$ is an orthonormal basis in $\mathcal{H}_0$ consisting of eigenvectors of B, $Be_k = \lambda_k^2 e_k$, then

$$\sum_k \|Se_k\|^2 = \sum_k \|B^{-1/2}B^{1/2}e_k\|^2 = \sum_k \|B^{1/2}e_k\|_0^2 = \sum_k \lambda_k^2 < \infty.$$

Here $B^{1/2}$ plays the role of T. The cylindrical measure μ in $\mathcal{H}_0$ is easily verified to have the identity as correlation operator. In fact,

$$\int_{\mathcal{H}_0} (x, \phi)_0 (x, \psi)_0 \mu(dx) = \int_X (x, B^{-1}\phi)(x, B^{-1}\psi)\mu(dx) =$$
$$(BB^{-1}\phi, B^{-1}\psi) = (B^{-1/2}\phi, B^{-1/2}\psi) = (\phi, \psi)_0. \quad \square$$

In the sequel we will often use the description of Gaussian measures mentioned in proposition 2.2, putting $X = \mathcal{H}_-$. Then, introducing the Hilbert space structure of $\mathcal{H}_+$ on the set $B^{-1}X$ by means of the scalar product $(x, y)_+ = (B^{-1}x, B^{-1}y) = (T^{-1}x, T^{-1}y)_0$, we obtain the rigged Hilbert space $\mathcal{H}_+ \subset \mathcal{H}_0 \subset \mathcal{H}_-$.

In this situation it is convenient to let the characteristic functional of a measure μ be that of the cylindrical measure μ_I induced by μ in $\mathcal{H}_0$. The latter can be defined by the integral

$$\chi(\theta) = \int_{\mathcal{H}_-} e^{i(x,\theta)_0} \mu(dx) = e^{-\|\theta\|_0^2/2}$$

for $\theta \in \mathcal{H}_+$ only, and it can be continuously extended to the whole of $\mathcal{H}_0$.

REMARK 2.2. The arguments in the prof of (2.3) are valid for any orthogonal projector P on $\mathcal{H}_0$ with values in $\mathcal{H}_+$. Therefore, for any such projector the factorization

$$\mu = \mu_P \times \mu_{I-P} \tag{2.12}$$

holds, where μ_{I-P} is a canonical Gaussian measure in the rigged Hilbert space

$$\mathcal{H}_+ \cap (\mathcal{H}_0 \Theta P\mathcal{H}_0) \subset \mathcal{H}_0 \Theta P\mathcal{H}_0 \subset \mathcal{H}_-^{(P)} = \{x \in \mathcal{H}_- : \ Px = 0\}.$$

3. Measurable linear functionals and operators

3.1. Measurable linear functionals. In this section we consider measurable linear functionals on a Hilbert space X equipped with a Gaussian measure μ. Without loss of generality μ can be assumed to be canonical, $\mu = \mu_I$, in the rigged Hilbert space $X = \mathcal{H}_- \supset \mathcal{H}_0 \supset \mathcal{H}_+$ with quasinuclear imbedding S: $\sigma_2(S) < \infty$.[1] Let $\mathcal{P}$ be the set of finite-dimensional orthogonal projectors P on $\mathcal{H}_0$ such that $\mathcal{H}_P = P\mathcal{H}_- \subset \mathcal{H}_+$.

[1] See corollary 2.1.

PROPOSITION 3.1. *Every μ-integrable function $f(x)$ can be approximated in the mean, with arbitrary degree of accuracy, by cylindrical functions of the form*

$$f_P(x) = f_P(Px) = \int f(Px + (I - P)y)\mu_{I-P}(dy), \tag{3.1}$$

where μ_{I-P} is the measure in the factorization (2.12).

PROOF. Consider the decomposition $\mathcal{H}_- = \mathcal{H}_P \dotplus \mathcal{H}_-^{(P)}$. Fubini's theorem implies that the function (3.1) is defined almost everywhere (mod μ_P) in $\mathcal{H}_P$, and

$$\int_{\mathcal{H}_P} f_P(Px)\mu_P(dPx) = \int_{\mathcal{H}_-} f(x)\mu(dx).$$

Putting $f_P(x) = f_P(Px)$, we extend it as a cylindrical function to $\mathcal{H}_-$, and we have

$$\int_{\mathcal{H}_-} f_P(x)\mu(dx) = \int_{\mathcal{H}_P} f_P(Px)\mu_P(dPx) = \int_{\mathcal{H}_-} f(x)\mu(dx). \tag{3.2}$$

The map $f \to f_P$ is clearly linear. It follows from (3.2) that it is isometric on the cone of nonnegative functions in $\mathcal{L}_1(\mathcal{H}_-, \mu)$. By representing an arbitrary function $f(x)$ as a difference of nonnegative functions, $f(x) = f^+(x) - f^-(x)$, we infer from $f_P(x) = (f^+)_P - (f^-)_P$ the continuity of this map:

$$\int_{\mathcal{H}_-} |f_P(x)|\mu(dx) \le \int_{\mathcal{H}_-} \{(f^+)_P(x) + (f^-)_P(x)\}\mu(dx) =$$
$$\int_{\mathcal{H}_-} \{f^+(x) + f^-(x)\}\mu(dx) = \int_{\mathcal{H}_-} |f(x)|\mu(dx).$$

For a cylindrical function $g(x) = g(Px)$ we have $g_P(x) \equiv g(x)$, and hence

$$\int_{\mathcal{H}_-} |f(x) - f_P(x)|\mu(dx) \le$$
$$\int_{\mathcal{H}_-} |f(x) - g(x)|\mu(dx) + \int_{\mathcal{H}_-} |g_P(x) - f_P(x)|\mu(dx) \le$$
$$2\int_{\mathcal{H}_-} |f(x) - g(x)|\mu(dx).$$

It remains to note that there is always a cylindrical function $g(x)$ making the righthand side sufficiently small. □

LEMMA 3.1. *If a measurable function $f(x)$ is invariant under shifts by elements $y \in \mathcal{H}_+$:*

$$f(x + y) = f(x) \pmod{\mu} \quad (y \in \mathcal{H}_+), \tag{3.3}$$

then it coincides almost everywhere with a constant:

$$f(x) \equiv const \pmod{\mu}. \tag{3.4}$$

PROOF. Replacing if necessary $f(x)$ by $e^{if(x)}$, we may take $f(x)$ bounded, and hence integrable with respect to the Gaussian measure μ. It follows from (3.1) that (3.3) also holds for $f_P(x)$. Since $\mathcal{H}_P \subset \mathcal{H}_+$, (3.4) holds for $f_P(x)$. It remains to use proposition 3.1. □

THEOREM 3.1. *A measurable linear set $\mathcal{E} \subset \mathcal{H}_-$ has measure $\mu(\mathcal{E})$ equal to* 1 *or* 0. *Also, $\mu(\mathcal{E}) > 0$ implies $\mathcal{E} \supset \mathcal{H}_0$.*

PROOF. Suppose $\mu(\mathcal{E}) > 0$. Use corollary 4.1, which says that in this case $\mu(\mathcal{E} + \lambda x) > 0$ for all $\lambda \in \mathbf{R}$ and $x \in \mathcal{H}_0$ (the proof of this result is not based on the theorem under consideration). For $x \notin \mathcal{E}$ we thus obtain a continuum of disjoint sets $\mathcal{E}_\lambda = \mathcal{E} + \lambda x$ of positive measure, which is impossible. Hence $\mathcal{E} \supset \mathcal{H}_0$. Together with $\mathcal{E}$, its indicator function $j_{\mathcal{E}}(x)$ is invariant under shifts by elements of $\mathcal{H}_0$, hence, by lemma 3.1, $j_{\mathcal{E}}(x) \equiv \text{const} \pmod{\mu}$. This constant can be 1 only, so that $\mu(\mathcal{E}) = \int_{\mathcal{H}_-} j_{\mathcal{E}}(x)\mu(dx) = 1$. □

DEFINITION 3.1. A measurable function $f(x)$ on $\mathcal{H}_-$ is called a *measurable linear functional* if it is the limit of a μ-almost everywhere convergent sequence of continuous linear functionals:

$$f(x) = \lim_{n\to\infty} (x, \phi_n)_0 \pmod{\mu}, \quad \phi_n \in \mathcal{H}_+. \tag{3.5}$$

THEOREM 3.2. *A measurable linear functional on $\mathcal{H}_-$ is square integrable with respect to μ, and its restriction to $\mathcal{H}_0$ is continuous. Conversely, a continuous linear functional on $\mathcal{H}_0$ has a measurable linear extension onto $\mathcal{H}_-$.*

In this way a bijective correspondence is set up between the space of ($\pmod{\mu}$ classes of) measurable linear functionals on $\mathcal{H}_-$ and the space $\mathcal{H}_0$: $f(x) \leftrightarrow f$. This correspondence is isometric in the sense that

$$\int_{\mathcal{H}_-} |f(x)|^2 \mu(dx) = \|f\|_0^2.$$

PROOF. Let $\phi_n \in \mathcal{H}_+$ be a sequence of elements as in (3.5), and let $\mathcal{E}$ be a linear set of full measure on which convergence takes place. By theorem 3.1, $\mathcal{E} \supset \mathcal{H}_0$, and consequently ϕ_n converges weakly in $\mathcal{H}_0$ to an element $f \in \mathcal{H}_0$ so that:

$$f(x) = \lim_{n\to\infty} (x, \phi_n)_0 = (x, f)_0 \quad (x \in \mathcal{H}_0).$$

Clearly, f is uniquely determined by $f(x)$. It follows from the relation

$$\int_{\mathcal{H}_0} \exp\{i(x, \phi_n - \phi_m)\}\mu(dx) = \exp\left\{-\frac{1}{2}\|\phi_n - \phi_m\|_0^2\right\}$$

that the sequence ϕ_n also converges strongly in $\mathcal{H}_0$, and hence there is mean square convergence of the functionals:

$$\int_{\mathcal{H}_0} |(x, \phi_n) - (x, \phi_m)|^2 \mu(dx) = \|\phi_n - \phi_m\|_0^2 \to 0 \quad (m, n \to \infty). \tag{3.6}$$

Letting $m \to \infty$ in this formula, we find

$$\int_{\mathcal{H}_0} |(x, \phi_n) - f(x)|^2 \mu(dx) = \|\phi_n - f\|_0^2 \to 0 \quad (n \to \infty),$$

and hence

$$\|f\|^2 = \lim_{n\to\infty} \|\phi_n\|^2 = \lim_{n\to\infty} \int_{\mathcal{H}_0} |(x, \phi_n)|^2 \mu(dx) = \int_{\mathcal{H}_0} |f(x)|^2 \mu(dx).$$

Conversely, let $f \in \mathcal{H}_0$. Then there is a sequence $\phi_n \in \mathcal{H}_+$ such that $\|f - \phi_n\|_0 \to 0$. Also, the sequence of linear functionals (x, ϕ_n), which are continuous on $\mathcal{H}_-$, mean square converges. Choosing an almost everywhere convergent subsequence ϕ_{n_k} from it, we put $f(x) = \lim_{k\to\infty}(x, \phi_{n_k})$ (mod μ). Repeating the considerations above for this linear functional, we arrive at formula (3.6), showing that $f(x)$ is uniquely defined (mod μ). □

REMARK 3.1. The theorem implies that every measurable linear functional $f(x)$ is defined on $\mathcal{H}_0$ and is uniquely determined by its values on this space. Below we will always identify $f(x)$ with the corresponding element $f \in \mathcal{H}_0$, and use the notation

$$f(x) = (x, f)_0 \quad (x \in \mathcal{H}_-, \quad f \in \mathcal{H}_0)$$

also for $x \notin \mathcal{H}_0$.

REMARK 3.2. It follows from the proof that the map $x \to (x, f)_0$ leads to a Gaussian measure μ^f in $\mathbf{R}^1$ with characteristic functional $\exp\{-\|f\|_0^2/2\}$.

It is easily seen that

$$\int_{\mathcal{H}_-} (x, f)_0 (x, g)_0 \mu(dx) = (f, g)_0.$$

REMARK 3.3. It is useful to state the result obtained in terms of a Hilbert space X equipped with a σ-additive Gaussian measure μ_B with nondegenerate correlation operator B. The set of continuous linear functionals on X can be identified with X: $\phi(x) = (x, \phi)$, and also

$$\int_X |(x, \phi)|^2 \mu(dx) = (B\phi, \phi).$$

So, the set of measurable linear functionals, which coincides with the set of square integrable linear functionals, can be identified with the space X_B (the completion of X with respect to the norm

$$\|x\|_B^2 = \|B^{1/2}x\|^2 = (Bx, x).)$$

3.2. Measurable linear operators. For a pair of Banach spaces $\mathcal{B}_1$, $\mathcal{B}_2$, we denote by $\mathcal{L}(\mathcal{B}_1, \mathcal{B}_2)$ the space of bounded operators from $\mathcal{B}_1$ into $\mathcal{B}_2$; if $\mathcal{B}_1$ and $\mathcal{B}_2$ are both Hilbert spaces, $\mathcal{L}_2(\mathcal{B}_1, \mathcal{B}_2)$ denotes the subspace of $\mathcal{L}(\mathcal{B}_1, \mathcal{B}_2)$ consisting of the Hilbert–Schmidt operators. Further, $\mathcal{L}(B) = \mathcal{L}(B, B)$, $\mathcal{L}_2(B) = \mathcal{L}_2(B, B)$.

DEFINITION 3.2. A μ-measurable function $A(x)$ on $\mathcal{H}_-$, being the limit of an almost everywhere convergent sequence of linear continuous functions

$$A(x) = \lim_{n\to\infty} A_n(x) \ (\mathrm{mod}\ \mu), \quad A_n \in \mathcal{L}(\mathcal{H}_-, \mathcal{K}), \tag{3.7}$$

is called a *measurable linear operator* on $\mathcal{H}_-$ (with values in the Hilbert space $\mathcal{K}$).

Each continuous linear operator $A \in \mathcal{L}(\mathcal{H}_-, \mathcal{K})$ gives rise to a measurable linear operator $\tilde{A}(x)$. The restriction $A_0 = AS$ of the latter to $\mathcal{H}_0$ is a Hilbert–Schmidt operator ($A_0 \in \mathcal{L}_2(\mathcal{H}_0, \mathcal{K})$), since the imbedding operator S is Hilbert–Schmidt.

On the other hand, since convergence in (3.7) takes place on a linear set $\mathcal{E}_A$ of positive measure and containing $\mathcal{H}_0$, the operator $A \in \mathcal{L}(\mathcal{H}_0, \mathcal{K})$, $Ax = \lim_{n\to\infty} A_n x$

($x \in \mathcal{H}_0$), is uniquely determined. Thus, a measurable linear operator on $\mathcal{H}_0$ has a continuous restriction to $\mathcal{H}_-$.

Note that for $\phi \in \mathcal{K}$ the function $(\tilde{A}(x), \phi)_{\mathcal{K}}$ is a measurable linear functional on $\mathcal{H}_-$, coinciding on $\mathcal{H}_0$ with $(Ax, \phi)_{\mathcal{K}} = (x, A^*\phi)_0$. Therefore the following formula holds:

$$(\tilde{A}(x), \phi)_{\mathcal{K}} = (x, A^*\phi) \pmod{\mu}. \tag{3.8}$$

THEOREM 3.3. *A square integrable measurable linear operator $\tilde{A}x$ has a continuous restriction Ax to $\mathcal{H}_0$ satisfying the condition $A \in \mathcal{L}_2(\mathcal{H}_0, \mathcal{K})$. Conversely, an operator $A \in \mathcal{L}_2(\mathcal{H}_0, \mathcal{K})$ has a measurable linear extension $\tilde{A}(x)$ to $\mathcal{H}_-$. This bijective correspondence between $\mathcal{L}_2(\mathcal{H}_0, \mathcal{K})$ and the space of measurable linear operators is isometric:*

$$\int_{\mathcal{H}_-} \|\tilde{A}(x)\|^2 \mu(dx) = \sigma_2^2(A) \overset{def}{=} \sum_j \|A\phi_j\|^2$$

($\{\phi_j\}$ is an orthonormal basis in $\mathcal{H}_0$).

PROOF. Let $\{\phi_j\}$ be an orthonormal basis in $\mathcal{K}$. Integrating termwise the equality

$$\|\tilde{A}(x)\|_{\mathcal{K}}^2 = \sum_j |(\tilde{A}(x), \phi_j)_{\mathcal{K}}|^2 = \sum_j |(x, A^*\phi_j)_0|^2 \pmod{\mu},$$

we find that $A \in \mathcal{L}_2(\mathcal{H}_0, \mathcal{K})$:

$$\int_{\mathcal{H}_-} \|\tilde{A}(x)\|_{\mathcal{K}}^2 \mu(dx) = \sum_j \int_{\mathcal{H}_-} |(x, A^*\phi_j)_0|^2 \mu(dx) =$$
$$\sum_j \|A^*\phi_j\|_0^2 = \sigma_2^2(A^*) = \sigma_2^2(A).$$

Conversely, if $A \in \mathcal{L}_2(\mathcal{H}_0, \mathcal{K})$, then for $x \in \mathcal{H}_0$,

$$Ax = \sum_k (Ax, \phi_k)\phi_k = \sum_k (x, A^*\phi_k)\phi_k. \tag{3.9}$$

Every term on the righthand side can be measurably extended to $\mathcal{H}_-$, and since the series converges in $\mathcal{L}_2(\mathcal{H}_-, \mu)$, its sum can be measurably extended to $\mathcal{H}_-$ as well. It is possible to choose an almost everywhere convergent subsequence from the sequence of partial sums of this series, which, by (3.7), determines a measurable linear operator. □

3.3. Integration by parts. We have shown that to each operator $A \in \mathcal{L}_2(\mathcal{H}_0, \mathcal{K})$ there corresponds a measurable function $\tilde{A}x$ on $\mathcal{H}_-$ coinciding on $\mathcal{H}_0$ with Ax (and vice versa). The situation changes, however, if $A = A_x$ depends on x. In this case the expression $\tilde{A}_x x$ need not make sense at all on a set of full (or even positive) measure, since the domain of definition of the measurable linear operator induced by A_x need not contain the element x, except for a set of x's of measure zero.

We will show that the function $\tilde{A}_x x$ is well defined for $A_x \in \mathcal{L}(\mathcal{H}_-, \mathcal{K})$ and can be 'regularized', giving a meaning to the expression obtained also in a more general case.

First we will introduce some notation (see chapter IV for more details).

If $f(x)$ is a function on $\mathcal{H}_-$ taking values in the Hilbert space $\mathcal{K}$, then its *derivative* (along $\mathcal{H}_0$) at a point $x \in \mathcal{H}_-$ is the operator $f'(x) \in \mathcal{L}(\mathcal{H}_0, \mathcal{K})$ satisfying

$$f(x_0 + h) - f(x) = f'(x)h + o(\|h\|_0) \quad (h \in \mathcal{H}_0, \quad x \in \mathcal{H}_-).$$

We denote by $C_{12}(\mathcal{H}_-, \mathcal{K})$ the space of continuous and bounded functions on $\mathcal{H}_-$, taking values in $\mathcal{K}$, and with first derivative $f'(x) \in \mathcal{L}_2(\mathcal{H}_0, \mathcal{K})$, and by $\mathfrak{H}_1(\mathcal{H}_-, \mathcal{K}, \mu)$ the subspace resulting from completing it with respect to the norm

$$\|f\|_1^2 = \int_{\mathcal{H}_-} \{\|f(x)\|_{\mathcal{K}}^2 + \sigma_2^2(f'(x))\}\mu(dx).$$

By $\mathcal{L}^n(\mathcal{H}, \mathcal{K})$ we denote the Banach space of n-linear operators acting from $\mathcal{H}$ to $\mathcal{K}$, and by $\mathcal{L}_2^n(\mathcal{H}, \mathcal{K})$ its subspace of Hilbert-Schmidt operators, i.e. the operators T with finite norm

$$\sigma_2^2(T) = \sum_{j_1,\dots,j_n=1}^{\infty} \|T(e_{j_1}, \dots, e_{j_n})\|_{\mathcal{K}}^2,$$

where $\{e_j\}_{j=1}^{\infty}$ is an (any) orthonormal basis in $\mathcal{H}$. The space $\mathcal{L}_2^n(\mathcal{H}, \mathcal{K})$ is a Hilbert space with the scalar product

$$(T, V)_n = \sum_{j_1,\dots,j_n=1}^{\infty} (T(e_{j_1}, \dots, e_{j_n}), V(e_{j_1}, \dots, e_{j_n}))_{\mathcal{K}}. \tag{3.10}$$

The *trace of a bilinear operator* $A \in \mathcal{L}^2(\mathcal{H}, \mathcal{K})$ is the vector expression

$$\operatorname{Tr} A = \sum_{j=1}^{\infty} A(e_j, e_j), \tag{3.11}$$

which belongs to $\mathcal{K}$ if the series is absolutely convergent. The set of such bilinear operators with trace in $\mathcal{K}$ is denoted by $N(\mathcal{H}, \mathcal{K})$.

It is readily verified that $T \in \mathcal{L}_2^2(\mathcal{H}, \mathcal{K})$ implies that the operator TA, $TA(f, g) = T(Af, g)$, belongs to $N(\mathcal{H}, \mathcal{K})$. In particular, $(f(x)A)' \in N(\mathcal{H}, \mathcal{K})$ if $f \in C_{12}(\mathcal{H}_-, \mathcal{L}_2(\mathcal{H}_0, \mathcal{K}))$.

THEOREM 3.4. *Let $A, B \in \mathcal{L}_2(\mathcal{H}_0, \mathcal{H}_0)$, $f(x), g(x) \in \mathfrak{H}(\mathcal{H}_-, \mathcal{L}_2(\mathcal{H}_0, \mathcal{K}))$. Then the equations*

$$\int_{\mathcal{H}_-} f(x)(\tilde{A}x)\mu(dx) = \int_{\mathcal{H}_-} \operatorname{Tr}(f(x)A)'\mu(dx) \tag{3.12}$$

and

$$\int_{\mathcal{H}_-} (f(x)\tilde{A}x - \operatorname{Tr}(f(x)A)', g(x)\tilde{B}x - \operatorname{Tr}(g(x)B)')_{\mathcal{K}}\mu(dx) =$$
$$\int_{\mathcal{H}_-} \{(f(x)A, g(x)B)_1 + ((f(x)A)', (g(x)B)')_2\}\mu(dx) \tag{3.13}$$

are true.

The proof of this theorem will be carried out right after the consideration of two particular cases, which are of interest in themselves.

The first of these is obtained by taking

$$\mathcal{K} = \mathbf{R}^1, \quad A(x) = (x, \phi), \quad B(x) = (x, \psi); \qquad \phi, \psi \in \mathcal{H}_0; \quad f, g \in \mathfrak{H}_1(\mathcal{H}_-, \mathbf{R}^1, \nu). \tag{3.14}$$

Now theorem 3.4 takes the following form:

PROPOSITION 3.2. *Under the conditions (3.14), the following equations hold:*

$$\int_{\mathcal{H}_-} f(x)(x, \phi)\mu(dx) = \int_{\mathcal{H}_-} (f'(x), \phi)\mu(dx) \tag{3.15}$$

and

$$\int_{\mathcal{H}_-} [f(x)(x, \phi) - (f'(x), \phi)][g(x)(x, \psi) - (g'(x), \psi)]\mu(dx) = (\phi, \psi) \int_{\mathcal{H}_-} f(x)g(x)\mu(dx) + \int_{\mathcal{H}_-} (f'(x), \psi)(g'(x), \phi)\mu(dx). \tag{3.16}$$

PROOF. First, suppose $f, g \in C_1(\mathcal{H}_-, \mathbf{R}^1)$ and $P \in \mathcal{P}$. Proposition 1.3 implies that the formulas (3.15), (3.16) are valid for the functions $f(Px)$ and $g(Px)$. The limit transition $P \to I$ is possible, since the integrands are bounded and there is pointwise convergence. This proves the formulas for $f, g \in C_1(\mathcal{H}_-, \mathbf{R}^1)$.

Setting $g = f$, $\psi = \phi$, (3.16) implies

$$\int_{\mathcal{H}_-} [f(x)(x, \phi) - (f'(x), \phi)]^2 \mu(dx) = \|\phi\|^2 \int_{\mathcal{H}_-} f^2(x)\mu(dx) + \int_{\mathcal{H}_-} (f'(x), \phi)^2 \mu(dx) \leq \|\phi\|^2 \|f\|_1^2,$$

showing that the linear operator

$$l_\phi(f) = f(x)(x, \phi) - (f'(x), \phi),$$

defined on C_1, acts continuously from $\mathfrak{H}_1$ into $\mathfrak{H}$, the space of functions that are square integrable with respect to μ. This operator can be continuously extended to $\mathfrak{H}_1$. Since each integral in (3.15), (3.16) is continuous on $\mathfrak{H}_1$, by limit transition these formulas hold for $f, g \in \mathfrak{H}_1$. □

We now consider the more general case of vector functions $f(x)$, $g(x)$.

PROPOSITION 3.3. *Let $f, g \in \mathfrak{H}_1(\mathcal{H}_-, \mathcal{K}, \nu)$, $A, B \in \mathcal{L}_2(\mathcal{H}_0, \mathcal{K})$. Then the following formulas hold:*

$$\int_{\mathcal{H}_-} (f(x), \tilde{A}x)_{\mathcal{K}} \mu(dx) = \int_{\mathcal{H}_-} \operatorname{Tr}[f'(x)A^*]\mu(dx) \tag{3.17}$$

and

$$\int_{\mathcal{H}_-} [(f(x), \tilde{A}x)_{\mathcal{K}} - \operatorname{Tr}(f'(x)A^*)][(g(x), \tilde{B}x)_{\mathcal{K}} - \operatorname{Tr}(g'(x)B^*)]\mu(dx) = \int_{\mathcal{H}_-} \{(A^* f(x), B^* g(x)) + \operatorname{Tr} A^* f'(x) B^* g'(x)\}\mu(dx). \tag{3.18}$$

PROOF. Let $\{\phi_k\}$ be a basis in $\mathcal{K}$. Integrating termwise the mean square convergent expansion

$$(f(x)\tilde{A}x) = \sum_{k=1}^{\infty}(f(x), \phi_k)(x, A^*\phi_k),$$

we obtain the first formula in the proposition:

$$\int_{\mathcal{H}_-}(f(x), \tilde{A}x)_{\mathcal{K}}\mu(dx) = \sum_k \int_{\mathcal{H}_-}(f(x), \phi_k)(x, A^*\phi_k)\mu(dx) =$$
$$\sum_k \int_{\mathcal{H}_-}(f'(x)A * \phi_k, \phi_k)\mu(dx) = \int_{\mathcal{H}_-}[\operatorname{Tr} f'(x)A^*]\mu(dx).$$

To prove the second formula we consider the expansions

$$(3.19)\quad \begin{aligned}(f(x), Ax) - \operatorname{Tr} f'(x)A^* &= \textstyle\sum_k\{(f(x), \phi_k)(A^*\phi_k, x) - (f'(x)A^*\phi_k, \phi_k)\},\\ (g(x), Bx) - \operatorname{Tr} g'(x)B^* &= \textstyle\sum_k\{(g(x), \phi_k)(B^*\phi_k, x) - (g'(x)B^*\phi_k, \phi_k)\}.\end{aligned}$$

First assume that each of them contains finitely many summands. Multiplying these, integrating termwise, and applying proposition 3.2, we obtain the relation required:

$$\int_{\mathcal{H}_-}[(f(x), Ax) - \operatorname{Tr}(f'(x)A^*)][(g(x), Bx) - \operatorname{Tr}(g'(x)B^*)]\mu(dx) =$$
$$\sum_{k,j}\int_{\mathcal{H}_-}\{(f(x), \phi_k)(x, A^*\phi_k) - (f'(x)A^*\phi_k, \phi_k)\} \times$$
$$\{(g(x), \phi_j)(B^*\phi_j, x) - (g'(x)B^*\phi_j, \phi_j)\}\mu(dx) =$$
$$\sum_{k,j}\int_{\mathcal{H}_-}\{(A^*\phi_k, B^*\phi_j)(f(x), \phi_k)(g(x), \phi_j) +$$
$$(f'(x)B^*\phi_j, \phi_k)(g'(x)A^*\phi_k, \phi_j)\}\mu(dx) =$$

$$(3.20)\quad \int_{\mathcal{H}_-}\{(A^*f(x), B^*g(x)) + \operatorname{Tr} f'(x)B^*g'(x)A^*\}\mu(dx).$$

It remains to show that limit transition is allowed in (3.20) to get rid of the assumption that (3.19) be finite expansions. For $g = f$ and $B = A$ the formula

(3.21)
$$\int_{\mathcal{H}_-}[(f(x), Ax) - \operatorname{Tr}(f'(x)A^*)]^2\mu(dx = \int_{\mathcal{H}_-}\{\|A^*f(x)\|^2 + \operatorname{Tr}[f'(x)A^*]^2\}\mu(dx)$$

follows from (3.20). In this equation both sides are defined for $f \in \mathfrak{H}_1$, $A \in \mathcal{L}_2(\mathcal{H}_0, \mathcal{K})$, and they coincide on the set of finite-dimensional operators A, which is dense in $\mathcal{L}_2(\mathcal{H}_0, \mathcal{K})$. Therefore the expression

$$\ell_A(f)(x) = (f(x), Ax) - \operatorname{Tr} f'(x)A^*$$

is an A-linear operator acting from a dense set in $\mathcal{L}_2(\mathcal{H}_0, \mathcal{K})$ into $\mathfrak{H}$. Its continuity follows from (3.21):

$$\int_{\mathcal{H}_-}|\ell_A(f)(x)|^2\mu(dx) \leq \|A\|^2\|f\|_1^2.$$

Hence it can be extended to all of $\mathcal{L}_2(\mathcal{H}_0, \mathcal{K})$, resulting in the fact that (3.21) holds for $f \in \mathfrak{H}_1$. Now it is not difficult to pass to the limit in the corresponding bilinear form (3.20) also, or to derive it directly from the quadratic version of (3.21). □

PROOF OF THEOREM 3.4. We first derive some auxiliary formulas. Differentiating the equality

$$(f(x)A\psi, \phi)_{\mathcal{K}} = (\psi, A^* f^*(x)\phi)_{\mathcal{H}_0} \quad (\psi \in \mathcal{H}_0, \quad \phi \in \mathcal{K})$$

termwise along $h \in \mathcal{H}_0$, we arrive at

$$(f'(x)(h, A\psi), \phi)_{\mathcal{K}} = (\psi, A^* f^{*\prime}(h, \phi))_{\mathcal{H}_0}.$$

This implies $(A^*[f^*(x)\phi]' e_s, e_j)_{\mathcal{H}_0} = (A^* f^{*\prime}(x)(e_s, \phi), e_j)_{\mathcal{H}_0} = (f'(x)(e_s, Ae_j), \phi)_{\mathcal{K}}$. Summation over $s = j$ leads to

$$\text{(3.22)} \qquad \operatorname{Tr} A^*[f^*(x)\phi]' = \sum_s (f'(x)(e_s, Ae_s), \phi)_{\mathcal{K}} = (\operatorname{Tr}(f(x)A)', \phi)_{\mathcal{K}}.$$

Similarly,

$$\operatorname{Tr} A^*[f^*(x)\phi]' B^*[g^*(x)\psi]' =$$
$$\sum_{s,j} (f'(x)(e_s, Ae_j), \phi)(g'(x)(e_j, Be_s), \psi) = \sum_{s,j} ((f(x), A)'(e_s, e_j), \phi)((g(x)B)'(e_j, e_s), \psi),$$

whence, defining the operator $\check{C} \in \mathcal{L}^2(\mathcal{H}_0, \mathcal{K})$ for $C \in \mathcal{L}_2(\mathcal{H}_0, \mathcal{K})$ by $\check{C}(\phi, \psi) = C(\psi, \phi)$, we obtain

$$\sum_m \operatorname{Tr} A^*[f^*(x)\phi_m]' B^*[g^*(x)\phi_m]' =$$
$$\sum_{s,j} ((f(x)A)'(e_s, e_j), (g(x)B)'(e_j, e_s)) = ((f(x)A)', (g(x)B)')_2.$$

All series are easily checked to be absolutely convergent, and so the considerations above are correct.

We now proceed with the proof of the theorem. Formulas (3.17) and (3.22) imply

$$\int_{\mathcal{H}_-} (f(x)Ax, \phi)\mu(dx) = \int_{\mathcal{H}_-} (Ax, f^*(x)\phi)\mu(dx) =$$
$$\int_{\mathcal{H}_-} \operatorname{Tr} A^*[f^*(x)\phi]' \mu(dx) = \int_{\mathcal{H}_-} (\operatorname{Tr}(f(x)A)'\phi)\mu(dx),$$

leading to (3.12), since ϕ is arbitrary. Finally, using the expansion

$$f(x)Ax - \operatorname{Tr}[f(x)A]' = \sum_m (f(x)Ax - \operatorname{Tr}((f(x)A)', \phi_m)\phi_m =$$
$$\sum_m \{(Ax, f^*(x)\phi_m) - \operatorname{Tr} A^*(f^*(x)\phi_m)'\}\phi_m,$$

and the same expansion for $g(x)$, we infer from (3.18) and (3.22) that

$$\int_{\mathcal{H}_-} (f(x)Ax - \operatorname{Tr}[f'(x)A]', g(x)Bx - \operatorname{Tr}[g(x)B]')\mu(dx) =$$

$$\sum_m \int_{\mathcal{H}_-} \{(A^*f^*(x)\phi_m, B^*g^*(x)\phi_m) + \operatorname{Tr} A^*[f^*(x)\phi_m]' B^*[g^*(x)\phi_m]'\}\mu(dx) =$$

$$\int_{\mathcal{H}_-} \{(A^*f^*(x), B^*g^*(x))_1 + ((\check{f}(x)A)', (\check{g}(x)B)')_2\}\mu(dx). \quad \square$$

Consider now the following expression in more detail:

$$\beta(A, f)(x) = f(x)Ax - \operatorname{Tr}[f(x)A]'. \tag{3.23}$$

It is defined for $f \in \mathfrak{H}_1\{\mathcal{H}_-, \mathcal{L}_2(\mathcal{H}_0, \mathcal{K}), \mu\}$, $A \in \mathcal{L}_2(\mathcal{H}_0, \mathcal{H}_0)$, and is a measurable function on $\mathcal{H}_-$ taking values in $\mathcal{K}$. Theorem 3.4 implies that $\beta(A, f)(x) \in \mathfrak{H}(\mathcal{H}_-, \mathcal{K}, \mu)$, and the following estimate:

$$\|\beta(A, f)(x)\|_{\mathfrak{H}}^2 \leq \int_{\mathcal{H}_-} \{\sigma_2^2[f(x)A] + \sigma_2^2[(f(x)A)']\}\mu(dx). \tag{3.24}$$

The righthand side, unlike the lefthand side, is also defined if $A \in \mathcal{L}(\mathcal{H}_0, \mathcal{H}_0)$ is not a Hilbert–Schmidt operator. Further, the A-linear function $\beta(A, f)$, taking values in $\mathfrak{H}$, is clearly continuous. Note that the righthand side of (3.24) is continuous with respect to A, convergence being assumed strong. Let $A \in \mathcal{L}(\mathcal{H}_0, \mathcal{H}_0)$, $A_n \in \mathcal{L}_2(\mathcal{H}_0, \mathcal{H}_0)$, and $A\phi = \lim_{n\to\infty} A_n\phi$ ($\phi \in \mathcal{H}_0$). The estimate

$$\begin{aligned} &\|\beta(A_k, f) - \beta(A_j, f)\|_{\mathfrak{H}}^2 \leq \\ &\int_{\mathcal{H}_-} \{\sigma_2^2[f(x)(A_k - A_j)] + \sigma_2^2[(f(x)(A_k - A_j))']\}\mu(dx) \end{aligned} \tag{3.25}$$

implies that it is possible to define

$$\beta(A, f) = \lim_{n\to\infty} \beta(A_n, f),$$

the limit being independent of the approximating sequence chosen. These considerations suggest the following result.

THEOREM 3.5. *Let $f \in \mathfrak{H}_1(\mathcal{H}_-, \mathcal{L}_2(\mathcal{H}_0, \mathcal{K}), \mu)$. For $A \in \mathcal{L}(\mathcal{H}_0, \mathcal{H}_0)$ the A- and f-linear differential expression $\beta(A, f)(x)$ is defined. Its restriction to $A \in \mathcal{L}_2(\mathcal{H}_0, \mathcal{H}_0)$ has the form (3.23). It satisfies the relation*

$$\begin{aligned} &\int_{\mathcal{H}_-} (\beta(A, f)(x), \beta(B, g)(x))_{\mathcal{K}}\mu(dx) = \\ &\int_{\mathcal{H}_-} \{(f(x)A, g(x)B)_1 + ((\check{f}(x)A)', (\check{g}(x)B)')_2\}\mu(dx). \end{aligned} \tag{3.26}$$

In particular, for $\beta(f) = \beta(I, f)$ we have

$$\int_{\mathcal{H}_-} \|\beta(f)\|^2\mu(dx) \leq \int_{\mathcal{H}_-} \{\sigma_2^2(f(x)) + \sigma_2^2(f'(x))\}\mu(dx). \tag{3.27}$$

REMARK 3.4. Under the additional assumptions $f \in \mathcal{L}(\mathcal{H}_-, \mathcal{K})$, $f' \in N(\mathcal{H}_0, \mathcal{K})$, (3.23) remains valid also for $A \in \mathcal{L}(\mathcal{H}_0, \mathcal{H}_0)$. In particular, $\beta(f) = f(x)x - \operatorname{Tr} f'(x)$.

REMARK 3.5. Note the version of theorem 3.5 under the conditions of proposition 3.3: The expression $l_I(f) = (f(x), x) - \operatorname{Tr} f'(x)$, defined for $f(x) \in \mathcal{H}_+$, can be extended to $f \in \mathfrak{H}(\mathcal{H}_-, \mathcal{H}_0, \mu)$, and

$$\int_{\mathcal{H}_-} |l_I(f)(x)|^2 \mu(dx) = \int_{\mathcal{H}_-} \|f(x)\|^2 + \operatorname{Tr} f'^2(x)\}\mu(dx) \leq$$

$$\int_{\mathcal{H}_-} \{\|f(x)\|^2 + \sigma_2^2(f'(x))\}\mu(dx). \tag{3.27'}$$

REMARK 3.6. All these results can be stated in terms of a Gaussian measure μ_B with correlation operator B on a Hilbert space X. E.g., (3.21) becomes

$$\int_X |(f(x), Ax) - \operatorname{Tr} A^* f'(x)B|^2 \mu_B(dx) = \int_X \{(ABA^* f(x), f(x)) + \operatorname{Tr}[A^* f'(x)B]^2\}\mu(dx).$$

Also, for a bounded operator $f'(x)$ the differential expression

$$l_A^{(B)}(f)(x) = (f(x), Ax) - \operatorname{Tr} A^* f'(x)B$$

can be extended to operators A (possibly unbounded) such that $ABA^* \in \mathcal{L}(X, X)$, $BA^* \in \mathcal{L}_2(X, X)$.

3.4. Expansion into orthogonal polynomials. In this section we will use the above mentioned integration by parts formulas to give an orthogonal decomposition of the Hilbert space $\mathfrak{H} = L_2(\mathcal{H}_-, \mathcal{K}, \mu)$, of square μ-integrable functions on $\mathcal{H}_-$ with values in $\mathcal{K}$, into subspaces of polynomial functions.

First some notation. As before, $\mathcal{L}_2^n(\mathcal{H}_0, \mathcal{K})$ is the Hilbert space of n-linear operators from $\mathcal{H}_0$ into $\mathcal{K}$ (by definition, $\mathcal{L}_2^0(\mathcal{H}_0, \mathcal{K}) = \mathcal{K}$). Consider the linear operator

$$\mathcal{I}_{n,j}\colon \mathcal{L}_2^n(\mathcal{H}_0, \mathcal{K}) \to \mathcal{L}_2^j(\mathcal{H}_0, \mathcal{L}_2^{n-j}(\mathcal{H}_0, \mathcal{K})) \quad (n = 1, 2 \ldots;\ j = 1, \ldots, n-1)$$

defined by identifying an n-linear operator A, with values in $\mathcal{K}$, with the j-linear ($j < n$) operator $\mathcal{I}_{n,j}A$, with values in $\mathcal{L}_2^j(\mathcal{H}, \mathcal{K})$):

$$[\mathcal{I}_{n,j}(A)(\phi_1, \ldots, \phi_j)](\phi_{j+1}, \ldots, \phi_n) = A(\phi_1, \ldots, \phi_j, \phi_{j+1}, \ldots, \phi_n)$$
$$(\phi_1, \ldots, \phi_n \in \mathcal{K}).$$

The relation

$$\sigma_2^2(\mathcal{I}_{n,j}(A)) = \sum_{k_1, \ldots, k_n} \|\mathcal{I}_{n,j}(A)(e_{k_1}, \ldots, e_{k_j})\|^2_{\mathcal{L}_2^{n-j}} =$$

$$\sum_{k_1, \ldots, k_j} \sum_{s_1, \ldots, s_{n-j}} \|A(e_{k_1}, \ldots, e_{k_j}, e_{s_1}, \ldots, e_{s_{n-j}})\|^2_{\mathcal{K}} == \sigma_2^2(A)$$

shows that the operator $\mathcal{I}_{n,j}$ (which is clearly invertible) is unitary: $\mathcal{I}_{n,j}^* = \mathcal{I}_{n,j}^{-1}$. We will use the notation $\mathcal{I}_{n,0} = I$ in the space $\mathcal{L}_2^n(\mathcal{H}, \mathcal{K})$.

By $G_n(\mathcal{H}_0, \mathcal{K})$ we denote the space of functions $A_n(x) = A(x, \dots, x)$, $A \in \mathcal{L}_2^n(\mathcal{H}_0, \mathcal{K})$ $(x \in \mathcal{H}_0)$. We set $G_0(\mathcal{H}_0, \mathcal{K}) = \mathcal{K}$, and

$$S_n(\mathcal{H}_0, \mathcal{K}) = G_0(\mathcal{H}_0, \mathcal{K}) \dot{+} \cdots \dot{+} G_n(\mathcal{H}_0, \mathcal{K})$$

the space of polynomials $\sum_{k=0}^n \tilde{A}_k(x)$. Further, we denote by $\mathcal{L}_{20}^n$, G_{n0}, S_{n0} the corresponding subspaces of functions that are finite-dimensional with respect to $A \in \mathcal{L}_{20}^n(\mathcal{H}, \mathcal{K})$ if $A(\phi_1, \dots, \phi_n) = A(P_1\phi_1, \dots, P_n\phi_n)$ for certain finite-dimensional orthoprojectors $P_1, \dots, P_n$.

Our purpose is to construct a linear map $\theta_n \colon \mathcal{L}_2^n \to S_n$ associating to a multilinear operator A the polynomial function $\theta_n(A)(x) = \theta_{n,A}(x)$ satisfying the recurrence condition

$$\begin{aligned} &\theta_n(A)(x) = \beta(\theta_{n-1}(\mathcal{I}_{n,n-1}(A)))(x) = \\ &[\theta_{n-1}(\mathcal{I}_{n,n-1}(A))(x)](x) - \operatorname{Tr} \mathcal{I}_{2,1}^*[\theta_{n-1}(\mathcal{I}_{n,n-1}(A))]'(x). \end{aligned} \tag{3.28}$$

We do this by induction, first for finite-dimensional $A \in \mathcal{L}_{20}^n$, and subsequently, by passing to the limit, for the general case. We first verify that formula (3.28) correctly determines θ_n by means of θ_{n-1} for finite-dimensional operators, when all functions under consideration are cylindrical and, thus, defined on $\mathcal{H}_-$, and when there is also no doubt about convergence of the trace of the operator considered. In fact, if $A \in \mathcal{L}_{20}^n(\mathcal{H}_0, \mathcal{K})$, then $\mathcal{I}_{n,n-1}(A) \in \mathcal{L}_{20}^{n-1}(\mathcal{H}_0, \mathcal{L}_{20}^1(\mathcal{H}_0, \mathcal{K}))$. If the map θ_{n-1} is defined on $\mathcal{L}_2^{n-1}(\mathcal{H}_0, \mathcal{L}_{20}^1(\mathcal{H}_0, \mathcal{K}))$, then $\theta_{n-1}(\mathcal{I}_{n,n-1}(A))$ is a polynomial function, with values in $\mathcal{L}_2^1(\mathcal{H}_0, \mathcal{L}_2^1(\mathcal{H}_0, \mathcal{K}))$. Its derivative $[\theta_{n-1}(\mathcal{I}_{n,n-1}(A))]'(x) \in \mathcal{L}_2^1(\mathcal{H}_0, \mathcal{L}_2^1(\mathcal{H}_0, \mathcal{K}))$ is identified (by $\mathcal{I}_{2,1}^*$) with a certain element of $\mathcal{L}_2^2(\mathcal{H}_0, \mathcal{K})$ whose trace belongs to $\mathcal{K}$. Since in this case the first term in (3.28) is also an element of $\mathcal{K}$, $\theta_n(A)$ is uniquely determined, and clearly $\theta_n(A) \in \mathcal{P}_{n0}$.

This allows us to define $\theta_n(A)$ by induction, beginning with θ_0: $\theta_0(f) = f$. Note that, obviously, $\theta_1(A)(x) = Ax$ for $A \in \mathcal{L}_2^1(\mathcal{H}_0, \mathcal{K})$, and so, for $A \in \mathcal{L}_2^2(\mathcal{H}_0, \mathcal{K})$:

$$\theta_1(\mathcal{I}_{21}(A))(x) = \mathcal{I}_{21}(A)(x), \quad [\theta_1(\mathcal{I}_{21}(A))]'(x) = \mathcal{I}_{21}(A).$$

Consequently,

$$\theta_2(A)(x) = [\mathcal{I}_{21}(A)(x)](x) - \operatorname{Tr} \mathcal{I}_{21}^* \mathcal{I}_{21}(A) = A(x, x) - \operatorname{Tr} A.$$

We will now clarify some of the properties of the maps θ_n.

PROPOSITION 3.4. *Let $A \in \mathcal{L}_{20}^n(\mathcal{H}_0, \mathcal{K})$ be symmetric with respect to permutation of arguments. Then*

$$\theta_n(A)'(x) = n\theta_{n-1}(\mathcal{I}_{n,n-1}(A))(x), \tag{3.29}$$

and relation (3.28) can be brought to the form

$$\theta_n(A) = [\theta_{n-1}(\mathcal{I}_{n,n-1}(A))(x)](x) - (n-1) \operatorname{Tr}[\theta_{n-2}(\mathcal{I}_{n,n-2}(A))(x)]. \tag{3.30}$$

PROOF. For $n = 1$, (3.29) becomes the obvious identity $(Ax)' = A$. Assume (3.29) holds for $n \le m - 1$. Then, for $n \le m$, (3.30) follows from the simple relation $\mathcal{I}_{21}^* \cdot \mathcal{I}_{n-1,n-2}\mathcal{I}_{n,n-1} = \mathcal{I}_{n,n-2}$. Differentiation of this formula gives

$$\theta_m(A)'(x)h =$$
$$\theta_{m-1}(\mathcal{I}_{m,m-1}(A))(x)h + \theta_{m-1}(\mathcal{I}_{m,m-1}(A)'(x)(h,x)) - (m-1)\operatorname{Tr}\theta_{m-2}(\mathcal{I}_{m,m-2}(A))'(x)h$$

or, using the symmetry of A,

$$\theta_{m-1}(\mathcal{I}_{m,m-1}(A)'(x)(h,x)) = (m-1)\theta_{m-2}(\mathcal{I}_{m-1,m-2}\mathcal{I}_{m,m-1}(A))(x)(h,h);$$

$$\theta_m(A)'(x) = \theta_{m-1}(\mathcal{I}_{m,m-1}(A))(x) +$$
$$(m-1)\{\theta_{m-1}(\mathcal{I}_{m-1,m-2}\mathcal{I}_{m,m-1}(A))\}(x)(x) - \operatorname{Tr}\mathcal{I}_{21}^*[\theta_{m-2}(\mathcal{I}_{m-1,m-2}\mathcal{I}_{m,m-1}(A))]'(x) =$$
$$m\theta_{m-1}(\mathcal{I}_{m,m-1}(A))(x). \quad \square$$

The following result can be derived from (3.26), (3.28), and (3.29).

PROPOSITION 3.5. *For $A \in \mathcal{L}_{20}^k(\mathcal{H}_0, \mathcal{K})$, $B \in \mathcal{L}_{20}^j(\mathcal{H}_0, \mathcal{K})$ the following formula is true:*

$$\int_{\mathcal{H}_-} (\theta_k(A)(x), \theta_j(B)(x))\mu(dx) =$$
$$\int_{\mathcal{H}_-} \{(\theta_{k-1}(\mathcal{I}_{k,k-1}(A))(x), \theta_{j-1}(\mathcal{I}_{j,j-1}(B))(x))_1 +$$
$$(k-1)(j-1)(\theta_{k-2}(\mathcal{I}_{k,k-2}(A))(x), \theta_{j-2}(\mathcal{I}_{j,j-2}(B))(x))_2\}\mu(dx). \tag{3.31}$$

COROLLARY 3.1. *Under the same conditions there is the following orthogonality relation:*

$$\int_{\mathcal{H}_-} (\theta_k(A)(x), \theta_j(B)(x))\mu(dx) = \delta_{kj}k!(A,B)_k. \tag{3.32}$$

In fact, (3.31) is obvious for $j = k = 0$, and follows immediately from (3.12) for $j = 0$ and arbitrary $k > 0$. Assuming it holds for all $j < m$ and $k < r$, we get

$$\int_{\mathcal{H}_-} (\theta_{r-1}(\mathcal{I}_{r,r-1}(A))(x), \theta_{m-1}(\mathcal{I}_{m,m-1}(B)(x)))_1\mu(dx) =$$
$$\delta_{mr}(r-1)!(\mathcal{I}_{r,r-1}(A), \mathcal{I}_{m,m-1}(B)) = \delta_{mr}(r-1)!(A,B)_r$$

and, similarly,

$$(m-1)(r-1)\int_{\mathcal{H}_-} (\theta_{r-2}(\mathcal{I}_{r,r-2}(A))(x), \theta_{m-2}(\mathcal{I}_{m,m-2}(B)(x)))_2\mu(dx) =$$
$$\delta_{mr}(r-1)(m-1)(r-2)!(A,B)_r.$$

Adding these expressions we find the required result.

Now we are ready to remove the condition of finite-dimensionality in the determination of θ_n.

THEOREM 3.6. *There is a unique map*

$$\theta_n : \mathcal{L}_2^n(\mathcal{H}, \mathcal{K}) \to \mathcal{L}_2(\mathcal{H}_-, \mathcal{K}, \mu)$$

associating to a multilinear operator $A \in \mathcal{L}_2^n(\mathcal{H}, \mathcal{K})$ the measurable polynomial $\theta_n(A)(x)$ defined by the relations

$$\begin{aligned} \theta_1(A)(x) &= \hat{A}x; \\ \theta_n(A)(x) &= \beta(\theta_{n-1}(\mathcal{I}_{n,n-1}(A)))(x) \quad (n > 1). \end{aligned} \tag{3.33}$$

The following orthogonality relation also holds:

$$\int_{\mathcal{H}_-} (\theta_k(A)(x), \theta_j(B)(x))\mu(dx) = \delta_{kj} k!(A, B)_k; \tag{3.34}$$

and for a symmetric A the following differentiation formula holds:

$$\theta_n(A)'(x) = n\theta_{n-1}(\mathcal{I}_{n,n-1}(A))(x).$$

PROOF. A map possessing the required properties has been defined already on the dense set $\mathcal{L}_{20}^n(\mathcal{H}, \mathcal{K})$ in $\mathcal{L}_{20}^n(\mathcal{H}, \mathcal{K})$ consisting of the finite-dimensional operators. The formula

$$\int_{-\mathcal{H}_-} \|\theta_n(A)(x)\|_{\mathcal{K}}^2 \mu(dx) = n!\sigma_2^2(A), \tag{3.35}$$

which follows from (3.34), shows that θ_n is isometric. This allows us to extend the map to the whole of $\mathcal{L}_2^n(\mathcal{H}, \mathcal{K})$ while preserving the properties mentioned. Uniqueness follows from the recurrence formula (3.33). □

COROLLARY 3.2. *The space $\mathfrak{H} = L_2(\mathcal{H}_-, \mathcal{K}, \mu)$ of square μ-integrable functions on $\mathcal{H}_-$ with values in $\mathcal{K}$ can be decomposed as an orthogonal sum $\mathfrak{H} = \otimes \sum_{n=1}^{\infty} \Lambda_n$ of subspaces $\Lambda_n = \theta_n(\mathcal{L}_2^n(\mathcal{H}, \mathcal{K}))$ of measurable polynomials.*

The orthogonality follows from (3.34), while equality follows from the denseness in $\mathfrak{H}$ of the set of polynomials.

4. Absolute continuity of Gaussian measures

4.1. Equivalence of measures in a product space. Suppose we are given two measures μ_1 and μ_2 in a measurable space $(X, \mathcal{U})$. Recall that μ_2 is said to be *absolutely continuous* with respect to μ_1 ($\mu_2 < \mu_1$) if $\mu_1(A) = 0$ ($A \in \mathcal{U}$) implies $\mu_2(A) = 0$. The measures μ_1 and μ_2 are called *equivalent* ($\mu_1 \sim \mu_2$) if $\mu_2 < \mu_1$ and $\mu_1 < \mu_2$.

In this section we deal with finite nonnegative measures only. In this case, $\mu_2 < \mu_1$ if and only if

$$\mu_2(A) = \int_A p(x)\mu_1(dx), \tag{4.1}$$

where $p(x)$ is a nonnegative μ_1-integrable function (the *density* or *Radon–Nikodým derivative*), or equivalently, if and only if

$$\int_X f(x)\mu_2(dx) = \int_X f(x)p(x)\mu_1(dx) \tag{4.2}$$

for every bounded measurable function $f(x)$.

An opposite situation arises when the measures are concentrated on disjoint sets: there is an $A \in \mathcal{U}$ such that $\mu_1(A) = \mu_1(X)$, $\mu_2(A) = 0$. The measures μ_1 and μ_2 are then said to be *mutually singular* (or *orthogonal*, $\mu_1 \perp \mu_2$).

Although intermediate situations are possible in general, we now consider the important case when either one of the measures is absolutely continuous with respect to the other, or the measures are orthogonal.

Let $X = \prod_{k=1}^{\infty} X_k$ be an infinite product of measurable spaces, and suppose that on each of them two nonnegative normalized measures $\mu_k < \nu_k$ are given, with density $p_k(x) = \mu_k(dx)/\nu_k(dx)$. Consider the infinite product measures $\mu = \prod_{k=1}^{\infty} \mu_k$, $\nu = \prod_{k=1}^{\infty} \nu_k$.

THEOREM 4.1. *(Kakutani [1].) Depending on whether the infinite product*

$$\prod_{k=1}^{\infty} \alpha_k, \quad \alpha_k = \int_{X_k} \sqrt{p_k(x_k)} \nu_k(dx_k), \tag{4.3}$$

is convergent or divergent to zero, the measure μ is absolutely continuous or orthogonal with respect to ν, respectively. In the first case the sequence

$$r_n(x) = \prod_{k=1}^{n} p_k(x_k) \tag{4.4}$$

converges in the mean to the integrable function $r(x) = \prod_{k=1}^{\infty} p_k(x_k)$. This function $r(x)$ is a density: $r(x) = \mu(dx)/\nu(dx)$.

PROOF. Since $\alpha^2 = |\int_{X_k} \sqrt{p(x_k)} \nu_k(dx_k)|^2 \leq \int_{X_k} p_k(x_k) \nu_k(dx_k) = 1$, the product (4.3) cannot diverge to infinity. If it diverges to zero, then there is a sequence $\beta_s = \prod_{k=n_s}^{m_s} \alpha_k$ for which the series $\sum_{s=1}^{\infty} \beta_s$ converges. Consider the sequence of cylinder sets $A_s = \{x \colon \prod_{k=n_s}^{m_s} p_k(x_k) \geq 1\}$. The estimate

$$\nu(A_s) = \int_{A_s} \nu(dx) \leq \int_{A_s} \sqrt{\prod_{k=n_s}^{m_s} p_k(x_k) \nu(dx)} = \prod_{k=n_s}^{m_s} \int_{X_k} \sqrt{p_k(x_k)} \nu_k(dx_k) = \beta_s$$

implies, since $\sum_{s=1}^{\infty} \beta_s$ converges, that

$$\nu(A) = 0 \quad (A = \limsup_{s \to \infty} A_s).$$

On the other hand, if $B_s = X \setminus A_s$, then

$$\mu(B_s) = \int_{B_s} \mu(dx) \leq \int_{B_s} \left\{ \prod_{k=n_s}^{m_s} p_k(x_k) \right\}^{-1/2} \mu(dx) =$$

$$\int_{B_s} \sqrt{\prod_{k=n_s}^{m_s} p_k(x_k) \nu(dx)} \leq \prod_{k=n_s}^{m_s} \int_{X_k} \sqrt{p_k(x_k)} \nu_k(dx_k) = \beta_s,$$

and hence $\mu(\limsup_{s \to \infty} B_s) = 0$. Therefore $\mu(A) \geq \mu(\liminf A_s) = 1$, i.e. $\mu(A) = 1$, which implies $\mu \perp \nu$.

Suppose now that the product (4.3) is convergent. Consider the sequence of cylindrical functions $\phi_n(x) = \sqrt{\prod_{k=1}^n p_k(x_k)}$. The relation

$$\int_X |\phi_{n+p}(x) - \phi_n(x)|^2 \nu(dx) = \int_X \prod_{k=1}^n p_k(x_k) \left| \sqrt{\prod_{k=n+1}^{n+p} p_k(x_k)} - 1 \right|^2 \nu(dx) =$$
$$\int_{X_{n+1}} \cdots \int_{X_{n+p}} \left| \sqrt{\prod_{k=n+1}^{n+p} p_k(x_k)} - 1 \right|^2 \prod_{k=n+1}^{n+p} \nu_k(dx_k) = 2\left(1 - \prod_{k=n+1}^{n+p} \alpha_k\right)$$

implies that the sequence ϕ_n is fundamental in $L_2(X, \nu)$. Also, the sequence $r_n = \phi_n^2$ converges in the mean, since

$$\int_X |r_{n+p}(x) - r_n(x)| \nu(dx) \le$$
$$\left\{\int_X |\phi_{n+p}(x) - \phi_n(x)|^2 \nu(dx)\right\}^{1/2} \left\{\int_X [|\phi_{n+p}(x)| + |\phi_n(x)|]^2 \nu(dx)\right\}^{1/2} \le$$
$$2\left\{\int_X |\phi_{n+p}(x) - \phi_n(x)|^2 \nu(dx)\right\}^{1/2}.$$

Let $r(x) = \lim_{n\to\infty} r_n(x)$. For any bounded cylindrical function $f(x)$ we have

$$\int_X f(x)\mu(dx) = \int_X f(x_1, \dots, x_n) \prod_{k=1}^n \mu_k(dx_k) =$$
$$\int_X f(x_1, \dots, x_n) r_n(x) \prod_{k=1}^n \nu_k(dx_k) = \int_X f(x) r_n(x) \nu(dx) =$$
$$\lim_{n\to\infty} \int_X f(x) r_n(x) \nu(dx) = \int_X f(x) r(x) \nu(dx).$$

Approximating a bounded measurable function by cylindrical functions, we obtain (4.2) for any such function. □

Now we apply this theorem to the problem of absolute continuity of Gaussian measures on the product of real lines. Let $X = \mathbf{R}^\infty = \prod_{k=1}^\infty X_k$ ($X_k = \mathbf{R}^1$),

$$\frac{\nu_k(dx_k)}{dx_k} = \frac{1}{\sigma_k\sqrt{2\pi}} e^{-(x_k-\gamma_k)^2/(2\sigma_k^2)}, \quad \frac{\mu_k(dx_k)}{dx_k} = \frac{1}{\lambda_k\sqrt{2\pi}} e^{-(x_k-\beta_k)^2/(2\lambda_k^2)}.$$

In this case,

$$p_k(x_k) = \frac{\sigma_k}{\lambda_k} \exp\left\{-\frac{1}{2\sigma_k^2\lambda_k^2}[(x_k-\beta_k)^2\sigma_k^2 - (x_k-\gamma_k)^2\lambda_k^2]\right\} \tag{4.5}$$

and an easy computation gives

$$\alpha_k = \int_{-\infty}^{\infty} \sqrt{p_k(x_k)}\nu(dx_k) =$$

$$\frac{1}{\sqrt{2\pi\sigma_k\lambda_k}} \int_{-\infty}^{\infty} \exp\left\{-\frac{1}{4\sigma_k^2\lambda_k^2}[(x_k-\beta_k)^2\sigma_k^2 + (x_k-\gamma_k)^2\lambda_k^2]\right\} dx_k =$$

$$\sqrt{\frac{2\lambda_k\sigma_k}{\lambda_k^2-\sigma_k^2}} \exp\left\{-\frac{(\beta_k-\gamma_k)^2}{4(\lambda_k^2+\sigma_k^2)}\right\} \tag{4.6}$$

Now we will consider some special cases.

PROPOSITION 4.1. *Let $\lambda_k = \sigma_k$ ($k = 1, 2 \dots$). A necessary and sufficient condition for the equivalence of two Gaussian measures μ, ν which are different only as regards their means, is that*

$$\sum_{k=1}^{\infty} \frac{(\beta_k-\gamma_k)^2}{\sigma_k^2} < \infty. \tag{4.7}$$

In this case the density $r(x) = \mu(dx)/\nu(dx)$ has the representation

$$r(x) = \lim_{n\to\infty} \exp\left\{\sum_{k=1}^{n} \frac{x_k(\beta_k-\gamma_k)}{\sigma_k^2} - \frac{\beta_k^2-\gamma_k^2}{2\sigma_k^2}\right\}. \tag{4.8}$$

If (4.7) is not true, then μ and ν are orthogonal.

PROOF. In the case under consideration, $\alpha_k = \exp\{-(\beta_k-\gamma_k)^2/(8\sigma_k^2)\}$, and convergence of the infinite product (4.3) is equivalent to convergence of the series (4.7). Formula (4.8) follows from (4.4) and (4.5). Since the measures μ and ν are interchangeable in this situation, their absolute continuity implies their equivalence. □

PROPOSITION 4.2. *Let $\beta_k = \gamma_k = 0$. Then two Gaussian measures with zero means are equivalent if and only if*

$$\sum_{k=1}^{\infty} \frac{(\lambda_k-\sigma_k)^2}{\lambda_k\sigma_k} < \infty, \tag{4.9}$$

and in this case

$$r(x) = \lim_{n\to\infty} \prod_{k=1}^{n} \frac{\sigma_k}{\lambda_k} \exp\left\{-\frac{x_k^2}{2}\left(\frac{\sigma_k^2-\lambda_k^2}{\sigma_k^2\lambda_k^2}\right)\right\}. \tag{4.10}$$

If (4.10) is not satisfied, μ and ν are orthogonal.

PROOF. In this case $\alpha_k = \sqrt{2\sigma_k\lambda_k/(\sigma_k^2+\lambda_k^2)}$, and convergence of the product (4.3) is clearly equivalent to convergence of the product

$$\prod_{k=1}^{\infty} \alpha_k^2 = \prod_{k=1}^{\infty} \left(1 + \frac{(\sigma_k-\lambda_k)^2}{2\sigma_k^2\lambda_k^2}\right),$$

and hence to that of the series (4.9). Now (4.5) also implies (4.10). □

4.2. Equivalence of Gaussian measures which differ by their means. We return to the consideration of Gaussian measures in a Hilbert space X. Let $\nu = \mu_B$ be a centered Gaussian measure with nuclear correlation operator B on X, and let $\mu = \mu_{B,a}$ be the measure obtained from μ_B by shift over the element $a \in X$. To study absolute continuity of these measures, it is convenient to represent them as products of one-dimensional measures.

Let $\{e_k\}$ be an orthonormal basis of eigenvectors of B in X:

$$Be_k = \lambda_k e_k, \quad \lambda_k > 0 \quad (k = 1, 2, \dots), \quad \sum_{k=1}^{\infty} \lambda_k < \infty.$$

The operator $Ux = (x_1, x_2, \dots)$, $x_n = (x, e_n)$, unitarily identifies X and ℓ_2, regarded as part of its linear extension $\mathbf{R}^\infty$. The projection operator $P_n x = \sum_{k=1}^{n} (x, e_k) e_k$ becomes $\tilde{P}_n x = (x_1, \dots, x_n)$.

The image of the Gaussian measure μ_B is also a Gaussian measure, $\mu_{\tilde{B}}$, with correlation operator $\tilde{B} = U^* B U$ which is represented on ℓ_2 by a diagonal matrix $\|\lambda_j, \delta_{jk}\|_{j,k=1}^{\infty}$. Therefore the finite-dimensional projection $\mu_{\tilde{P}_n B \tilde{P}_n}$ of this measure has density

$$p_n(x) = \prod_{k=1}^{n} \frac{1}{\sqrt{2\pi\lambda_k}} \exp\left\{-\frac{x_k^2}{2\lambda_k}\right\}$$

with respect to Lebesgue measure in $\mathbf{R}^n$.

In the same manner, $\mu_{B,a}$ is related to the measure $\mu_{\tilde{B},\tilde{a}}$ with finite-dimensional projections having densities

$$q_n(x) = \prod_{k=1}^{n} \frac{1}{\sqrt{2\pi\lambda_k}} \exp\left\{-\frac{(x_k - a_k)^2}{2\lambda_k}\right\}; a_k = (a, e_k), \quad \sum_{k=1}^{\infty} |a_k|^2 < \infty.$$

Both these measures, with supports in ℓ_2, may be regarded in $\mathbf{R}^\infty$ as infinite products of one-dimensional measures. Neither absolute continuity nor orthogonality is affected by this transformation, and

$$\frac{\mu_{\tilde{B},\tilde{a}}(U\,dx)}{\mu_{\tilde{B}}(U\,dx)} = \frac{\mu_{B,a}(dx)}{\mu_B(dx)}. \tag{4.11}$$

THEOREM 4.2. *The Gaussian measure $\mu_{b,a}$ obtained from a centered Gaussian measure μ_B by shift over an element $a \in X$ is equivalent to μ_B if and only if*

$$a \in B^{1/2} X. \tag{4.12}$$

In this case the density has the form

$$r(x) = \frac{\mu_{B,a}(dx)}{\mu_B(dx)} = \exp\left\{(B^{-1}a, x) - \frac{1}{2}\|B^{-1/2}a\|^2\right\}. \tag{4.13}$$

If (4.12) does not hold, $\mu_{B,a}$ and μ_B are orthogonal.

PROOF. For $a = \sum_{k=1}^{\infty} a_k e_k$, (4.12) can be written as

$$\sum_{k=1}^{\infty} \frac{a_k^2}{\lambda_k} < \infty, \tag{4.14}$$

and coincides with (4.7). Now proposition 4.1 implies the assertion about equivalence and orthogonality. To prove (4.13) it remains to note that the expression

$$\lim_{k\to\infty}\sum_{k=1}^{n}\frac{x_k a_k}{\lambda_k}=(B^{-1}a,x)$$

is a measurable linear functional on X, and therefore (4.8) may be rewritten as (4.13). □

REMARK 4.1. If $B^{-1}a\in X$, i.e. $\sum_{k=1}^{\infty}a_k^2/\lambda_k^2<\infty$, then the derivative $r(x)$ is continuous on X.

REMARK 4.2. It is useful to rewrite the result obtained in other terms, taking $X=\mathcal{H}_-\supset\mathcal{H}_0\supset\mathcal{H}_+$ and μ a canonical Gaussian measure: $B=I_{\mathcal{H}_0}$. Condition (4.14) then means that $a\in\mathcal{H}_0$. Then (a,x) is a measurable linear functional on X, and

$$r(x)=\exp\left\{(x,a)-\frac{1}{2}\|a\|^2\right\}.$$

Note that $r(x)\in L_p(X,\mu)$ for all $p\geq 1$:

$$\int_{\mathcal{H}_-}|r(x)|^p\mu(dx)=e^{-p\|a\|^2/2}\int_{\mathcal{H}_-}e^{p(x,a)}\mu(dx)=\exp\left\{\frac{p(p-1)}{2}\|a\|^2\right\}.$$

COROLLARY 4.1. *Let $\mathcal{E}$ be a measurable set of positive measure, $\mu(\mathcal{E})>0$, in $\mathcal{H}_-$, and let $a\in\mathcal{H}_0$. Then the shifted set $\mathcal{E}_a=\mathcal{E}+a$ also has positive measure:*

$$\mu(\mathcal{E}+a)=\int_{\mathcal{E}}\mu_a(dx)=\int_{\mathcal{E}}\exp\left\{(x,a)-\frac{1}{2}\|a\|^2\right\}\mu(dx)>0.$$

PROOF. The integrand is everywhere positive. □

4.3. Equivalence of Gaussian measures with distinct correlation operators. In the sequel we will need some properties of so-called *regularized determinants.* These are briefly presented below; more details can be found in Gokhberg–Krein [1].

Consider on X an operator of the form $A=I+S$, and let, first, S be a finite-dimensional operator with nonzero eigenvalues $\lambda_1,\dots,\lambda_n$ (listed taking into account multiplicity)[2]. Put

$$\Delta(A)=\det A=\prod_{j=1}^{n}(1+\lambda_j),$$

$$\tilde{\Delta}(A)=\det A\exp(-\operatorname{Tr}S)=\prod_{j=1}^{n}(1+\lambda_j)e^{-\lambda_j}.$$

By limit transition, $\Delta(A)$ can be defined for all $A=I+S$ with nuclear S, and $\tilde{\Delta}(A)$ can be defined for all $A=I+S$ with Hilbert–Schmidt operator S. Also,

a) the function $\Delta(I+S)$ is bounded and uniformly continuous on bounded sets of the space N of nuclear operators; there is the following product formula:

$$\Delta(I+S_1)\Delta(I+S_2)=\Delta[(I+S_1)(I+S_2)];$$

[2] More precisely, not X but its complexification should be considered.

b) the function $\tilde{\Delta}(I+S)$ is bounded and uniformly continuous on bounded sets of the space $\mathcal{L}_2^1$ of Hilbert–Schmidt operators; there is the following product formula:

$$\tilde{\Delta}(I+S_1)\tilde{\Delta}(I+S_2)\exp(-\operatorname{Tr} S_1S_2) = \tilde{\Delta}[(I+S_1)(I+S_2)].$$

LEMMA 4.1. *Two centered Gaussian measures μ_{B_1}, μ_{B_2} in a Hilbert space X are equivalent if their sets X_{B_1}, X_{B_2} of linear measurable functionals coincide. If this condition does not hold, then $\mu_{B_1} \perp \mu_{B_2}$.*

PROOF. The first part of the lemma readily follows since transition to an equivalent measure does not affect almost everywhere convergence.

Let now $X_{B_1} \neq X_{B_2}$; hence the norms $\|X_{B_j}\| = (B_jx, x)$ $(j=1,2)$ are nonequivalent in X. Therefore there is a sequence of elements $\{x_n\} \subset X$ for which

$$\|x_n\|_{B_1} \to 0 \quad (n \to \infty), \qquad \|x_n\|_{B_2} = 1.$$

Consider the sequence of continuous linear functionals (x_n, x) on X. Since $\int_X |(x_n,x)|^2 \mu_{B_1}(dx) = \|x_n\|_{B_1}^2 \to 0$, it is possible to choose a subsequence (also denoted by x_n) which converges to zero μ_{B_1}-almost everywhere. The linear set $\mathcal{L} = \{x\colon (x_n, x) \to 0\}$ is clearly measurable with respect to each measure. Also, $\mu_{B_1}(\mathcal{L}) = 1$ but $\mu_{B_2}(\mathcal{L}) = 0$, since otherwise $\mu_{B_2}(\mathcal{L}) = 1$ by theorem 3.1, contradicting the relation

$$\int_X |(x_n, x)|^2 \mu_{B_2}(dx) = \|x_n\|_{B_2}^2 = 1. \quad \square$$

Thus, in subsequent considerations, excluding the orthogonal case, it is possible to put $X_{B_1} = X_{B_2}$, i.e. the operators $B_1^{1/2}B_2^{-1/2}$ and $B_2^{1/2}B_1^{-1/2}$ are bounded on X. In other words, $B_2^{1/2} = SB_1^{1/2} = B_1^{1/2}S^*$, where S is a bounded and invertible operator on X.

We now apply the transformation mapping μ_{B_1} to a canonical Gaussian measure μ. In other words, we take $X = \mathcal{H}_-$, and $\mathcal{H}_0 = B_1^{1/2}X$ with scalar product $(\phi,\psi)_0 = (B_1^{-1/2}\phi, B_1^{-1/2}\psi)$. Then the measure μ_{B_2} has correlation operator $C = B_2B_1^{-1} = B_1^{1/2}S^*SB_1^{-1/2}$ (this is a bounded and invertible operator on $\mathcal{H}_0$, since $B_1^{-1/2}$ is unitary from X to $\mathcal{H}_0$):

$$\int_X |(x,\phi)_0|^2 \mu_{B_2}(dx) = \int_X |(x, B_1^{-1}\phi)|^2 \mu_{B_2}(dx) = (B_1^{-1}B_2B_1^{-1}\phi, \phi) = (B_2B_1^{-1}\phi,\phi)_0.$$

In this way we arrive at a pair $\mu = \mu_I$, μ_C $(C \in \mathcal{L}(\mathcal{H},\mathcal{H}))$ of Gaussian measures with supports in $\mathcal{H}_-$.

PROPOSITION 4.3. *Let $C = I + K$, $K \in \mathcal{L}_2(\mathcal{H},\mathcal{H})$, be an invertible operator on $\mathcal{H}_0$. Then the measure μ_C is equivalent to the canonical measure $\mu = \mu_I$, and*

$$r(x) = \frac{\mu_C(dx)}{\mu(dx)} = \frac{1}{\sqrt{\tilde{\Delta}(I+K)}} \exp\{\theta_2(K)(x) - \|KC^{-1/2}\|^2\}. \tag{4.15}$$

PROOF. Let $\{e_k\}$ be an orthonormal basis of eigenvectors of K in $\mathcal{H}_0$. Consider the unitary map

$$U\colon \mathcal{H}_0 \to \ell_2\colon Ux = (x_1, x_2, \dots), \quad x_k = (x, e_k) \quad (k = 1, 2 \dots).$$

Here, the space $\mathcal{H}_-$ is mapped to $\mathbf{R}^\infty$, and the measures μ, μ_C become products of one-dimensional Gaussian measures with densities

$$\frac{1}{\sqrt{2\pi}} e^{-x_k^2/2}, \quad \frac{1}{\sqrt{2\pi(1+\lambda_k)}} e^{-x_k^2/(2(1+\lambda_k))} \quad (Ke_j = \lambda_j e_j).$$

Condition (4.9) takes the form $\sum(\sqrt{1+\lambda_j} - 1)^2/\sqrt{1+\lambda_j} < \infty$, and is easily seen to be equivalent to the condition $\sum_j |\lambda_j|^2 < \infty$. Therefore $\tilde{\mu} \sim \tilde{\mu}_C$, and hence $\mu \sim \mu_C$.

In this case the expression (4.10) takes the form

$$r(x) = \lim_{n\to\infty} r_n(x) = \lim_{n\to\infty} \prod_{k=1}^{n} (1+\lambda_k)^{-1/2} \exp\left\{ -\frac{x_k^2}{2} \frac{1-(1+\lambda_k)}{1+\lambda_k} \right\} =$$
$$\lim_{n\to\infty} \prod_{k=1}^{n} [(1+\lambda_k)e^{-\lambda_k}]^{-1/2} \exp\left\{ \frac{\lambda_k x_k^2}{2} - \frac{\lambda_k}{2} - \frac{\lambda_k^2 x_k^2}{2(1+\lambda_k)} \right\}.$$

The relations

$$\prod_{k=1}^{n} (1+\lambda_k)e^{-\lambda_k} = \tilde{\Delta}(P_n C), \quad \sum_{k=1}^{n} \frac{\lambda_k^2 x_k^2}{1+\lambda_k} = \|P_n K(I+K)^{-1/2} x\|^2,$$

where $P_n x = \sum_{k=1}^{n} (x, e_k) e_k$, are easily checked, and then

$$\sum_{k=1}^{n} (\lambda_k x_k^2 - \lambda_k) = \theta_2(KP_n)(x).$$

Thus,

$$r_n(x) = [\tilde{\Delta}_n(I + P_n K)]^{-1/2} \exp\left\{ \frac{1}{2}\theta_2(KP_n)(x) - \frac{1}{2}\|P_n K(I+K)^{-1/2} x\}^2 \right\},$$

and the proof is finished by passing to the limit as $n \to \infty$. Note that thereby $\tilde{\Delta}(I+P_nK) \to \tilde{\Delta}(I+K)$ and the expression under the exp sign converges in $L_2(\mathcal{H}_-, \mu)$ to $(\theta_2(K)(x) - \|K(I+K)^{-1/2}x\|^2)/2$. So there is a subsequence converging almost everywhere, and since $\{r_n(x)\}$ is a priori known to be convergent in $L_1(\mathcal{H}_-, \mu)$, we obtain (4.15). □

COROLLARY 4.2. *Let μ_{C_1}, μ_{C_2} be two Gaussian measures in $\mathcal{H}_-$ having bounded and invertible correlation operators C_1, C_2 on $\mathcal{H}_-$. If $K = C_1 - C_2 \in \mathcal{L}_2(\mathcal{H}_0, \mathcal{H}_0)$, then $\mu_{C_1} \sim \mu_{C_2}$.*

In fact, it is possible to change variables so that the measure μ_{C_1} becomes μ_I after renormalization of the space $\mathcal{H}_0$, μ_{C_2} becomes $\mu_{I+KC_1^{-1}}$, while $KC_1^{-1} \in \mathcal{L}_2(\mathcal{H}_0, \mathcal{H}_0)$.

Now we are able to proof our final result.

THEOREM 4.3. *Let $C = C^* \geq 0$ be a bounded operator on $\mathcal{H}_0$ with bounded inverse C^{-1}. The Gaussian measure μ_C with support in $\mathcal{H}_-$ is equivalent to the canonical measure μ_I if and only if its correlation operator is representable in the form $C = I + K$ with $K \in \mathcal{L}_2(\mathcal{H}_0, \mathcal{H}_0)$.*

If this condition does not hold, then μ_C and μ_I are orthogonal.

PROOF. Sufficiency follows from proposition 4.3.

To prove necessity we replace C by an operator $C + K$ ($K \in \mathcal{L}_2(\mathcal{H}_0, \mathcal{H}_0)$) having a complete system of eigenvectors (such a K exists by a well-known theorem of von Neumann). Then $\mu_{C+K} \sim \mu_C$. Now it is possible, using the method above, to turn to sequence spaces in which the corresponding measures $\tilde{\mu}_{C+K}$ and $\tilde{\mu}_I$ have diagonal correlation matrices with eigenvalues $\lambda_k^2, \sigma_k^2 = 1$, respectively. By proposition 4.2, these measures are either equivalent or orthogonal, and so are μ_{C+K} and μ_C. The equivalence condition takes the form $\sum_{k=1}^{\infty}(\lambda_k - 1)^2/\lambda_k < \infty$, meaning $[(C+K)^{1/2} - I](C+K)^{-1/4} = K_1 \in \mathcal{L}_2(\mathcal{H}_0, \mathcal{H}_0)$. This implies the representation $C = I + K_2$, where

$$K_2 = K_1[(C+K)^{-1/4}(I + (C+K)^{1/2}) - I] \in \mathcal{L}_2(\mathcal{H}_0, \mathcal{H}_0). \quad \square$$

REMARK 4.3. For applications, other expressions of the density $\mu_C(dx)/\mu(dx)$ are useful.

Consider first the operator $\hat{K} = C^{-1} - I$, belonging together with K to $\mathcal{L}_2(\mathcal{H}_0, \mathcal{H}_0)$. Since $K\hat{K}$ is nuclear, the obvious relation $K + \hat{K} + K\hat{K} = 0$ implies that

$$\theta_2(K)(x) - \|KC^{-1/2}x\|^2 = -\theta_2(\hat{K})(x) - \theta_2(K\hat{K})(x) - (K^2(I + \hat{K})x, x) =$$
$$-\theta_2(\hat{K})(x) + \operatorname{Tr} K\hat{K} - (K(\hat{K} + K + \hat{K}K)x, x) = -\theta_2(\hat{K})(x) + \operatorname{Tr} K\hat{K}.$$

Using formula (4.15) we obtain the expression

$$r(x) = \sqrt{\tilde{\Delta}(C^{-1})} \exp\left\{\frac{1}{2}\theta_2(C^{-1} - I)(x)\right\}. \tag{4.16}$$

Let now $S \in \mathcal{L}_2(\mathcal{H}_0, \mathcal{H}_0)$ satisfy the relation $C^{-1} = (I + S^*)(I + S)$. By a simple transformation and the use of the relations

$$\tilde{\Delta}(I + S) = \tilde{\Delta}(I + S^*), \quad \operatorname{Tr} S = \operatorname{Tr} S^*, \quad \theta_2(S) = \theta_2(S^*),$$

(4.16) implies the following formula:

$$r(x) = \tilde{\Delta}(I + S) \exp\left\{-\theta_2(S)(x) - \frac{1}{2}\|Sx\|^2\right\}. \tag{4.17}$$

REMARK 4.4. We restate these results in terms of the initial Hilbert space X in which the centered Gaussian measures μ_{B_1}, μ_{B_2} with nuclear correlation operators B_1 and B_2 are supported.

Let the operator $B_1^{-1/2} B_2 B_1^{-1/2}$ be bounded and invertible. If

$$B_1^{-1/2} B_2 B_1^{-1/2} - I$$

is a Hilbert–Schmidt operator, then the measures μ_{B_1}, μ_{B_2} are equivalent. Otherwise they are orthogonal. In the case of equivalence, the density can be expressed as

$$\frac{\mu_{B_2}(dx)}{\mu_{B_1}(dx)} =$$

$$\frac{1}{\sqrt{\tilde{\Delta}(B_2B_1^{-1})}} \exp\frac{1}{2}\left\{\theta_2(B_2B_1^{-1}-I)(x) - (B_2B_1^{-1})((B_2B_1^{-1}-I)^{-1/2}x,x)\right\} =$$

$$\sqrt{\tilde{\Delta}(B_1B_2^{-1})}\exp\left\{-\frac{1}{2}\theta_2(B_2B_1^{-1}-I)(x)\right\} =$$

$$\tilde{\Delta}((B_2B_1^{-1})^{-1/2})\exp\left\{-\theta_2((B_2B_1^{-1})^{-1/2}-I)(x) - \frac{1}{2}\|[B_2B_1^{-1})^{-1/2}-I]x\|^2\right\}.$$

REMARK 4.5. If S is nuclear, (4.17) takes the simpler form

$$r(x) = \Delta(I+S)\exp\left\{-(Sx,x) - \frac{1}{2}\|Sx\|^2\right\}.$$

4.4. Absolute continuity of measures obtained from Gaussian measures by certain transformations of space. Let $y = \Phi(x)$ be a function, with values in a space X equipped with a measure μ, defined in some neighborhood U_{x_0} of a point $x_0 \in X$, and invertible in a neighborhood of the point $y_0 = \Phi(x_0)$. Then the measure $\mu^{\Phi^{-1}}$ is defined in U_{x_0} by

$$\mu^{\Phi^{-1}}(A) = \mu(\Phi(A)).$$

If f is a bounded continuous function with support in U_{y_0}, then

$$\int_X f(\Phi(x))\mu^{\Phi^{-1}}(dx) = \int_X f(y)\mu(dy).$$

Let $\mu^{\Phi^{-1}} < \mu$ and $\mathcal{J}(\Phi;x) = \mu^{\Phi^{-1}}(dx)/\mu(dx)$. Then

$$\int_X f(\Phi(x))\mathcal{J}(\Phi;x)\mu(dx) = \int_X f(y)\mu(dy), \tag{4.18}$$

where the integration at the righthand side is, in fact, restricted to U_{y_0}.

We consider some particular cases.

a) Linear transformation. Let $X = \mathcal{H}_-$, $\mu = \mu_I$ a canonical Gaussian measure, and $\Phi(x) = x + Sx$, where S is a measurable linear operator on $\mathcal{H}_-$ of class $\mathcal{L}_2(\mathcal{H}_0,\mathcal{H}_0)$. In this case the measure $\mu^{\Phi^{-1}}$ is also a Gaussian measure, with correlation operator $(I+S)^{-1}(I+S^*)^{-1}$.

It follows from theorem 4.3 and remark 4.1 that $\mu^{\Phi^{-1}} \sim \mu$, and

$$\mathcal{J}(\Phi;x) = \tilde{\Delta}(I+S)\exp\left\{-\theta_2(S)(x) - \frac{1}{2}\|Sx\|^2\right\}. \tag{4.19}$$

Let now $\Phi(x) = (I+S)x + a = \Phi_1 x + a$ ($a \in \mathcal{H}_0$). The measure $\mu^{\Phi_1^{-1}}$ has the same correlation operator as $\mu^{\Phi^{-1}}$, and its mean is $(I+S)^{-1}a$. Therefore (see theorem 4.2),

$$\frac{\mu^{\Phi^{-1}}(dx)}{\mu^{\Phi_1^{-1}}(dx)} = \exp\left\{-((I+S)x,a) - \frac{1}{2}\|a\|^2\right\}.$$

Combining this expression with (4.19) we obtain the formula

$$\mathcal{J}(\Phi;x)=\tilde{\Delta}(I+S)\exp\left\{-\theta(S+a)(x)-\frac{1}{2}\|Sx+a\|^2\right\},$$

where

$$\theta(S+a)(x)=\theta_2(S)(x)+\theta_0(a)(x)=\theta_2(S)(x)+(a,x). \tag{4.20}$$

b) Nonlinear transformation on a finite-dimensional space. Let $X=\mathbf{R}^n$, Φ a map which is differentiable in a neighborhood of x_0, and $\det\Phi'(x_0)\neq 0$. Introduce the auxiliary linear map $z=C(x)=y_0+\Phi'(x_0)(x-x_0)$, and the nonlinear map $y=C_1(z)=\Phi(x_0+[\Phi'(x_0)]^{-1}(z-y_0))$. Then $\Phi=C_1C$, and

$$\left.\frac{\mu^{\Phi^{-1}}(dx)}{\mu(dx)}\right|_{x=x_0}=\left.\frac{\mu^{C^{-1}C_1-1}(dx)\mu^{C^{-1}}(dx)}{\mu^{C^{-1}}(dx)\mu(dx)}\right|_{x=x_0}=\left.\frac{\mu^{C^{-1}}(dx)}{\mu(dx)}\right|_{x=x_0},$$

since $C_1'(Cx_0)=I$, and hence $(\mu^{C^{-1}C_1^{-1}}(dx)/\mu^{C^{-1}}(dx))|_{x=x_0}=1$.

Using (4.20) we obtain an expression for the density:

$$\mathcal{J}(\Phi;x)=\Delta(\Phi'(x))\exp\left\{-(C(x)-x,x)-\frac{1}{2}\|C(x)-x\|^2\right\}=$$

$$\tilde{\Delta}(\Phi'(x))\exp\left\{-l(\Phi-I)(x)-\frac{1}{2}\|\Phi(x)-x\|^2\right\}, \tag{4.21}$$

first for $x=x_0$, and, since the point x_0 may be replaced by an arbitrary point $x\in U_{x_0}$, also for $x\in U_{x_0}$. Recall that $l(f)=(f(x),x)-\operatorname{Tr}f'(x)$ (see remark 3.5).

c) Nonlinear transformation on an infinite-dimensional space. Let again $X=\mathcal{H}_-$, $\mu=\nu$ a canonical Gaussian measure, and suppose the map $y=\Phi(x)$ has the form $\Phi(x)=x+S(x)$ in a neighborhood of $x_0\in\mathcal{H}_-$, where the function $S(x)$ satisfies the following conditions:

A_1) S is a bounded continuous map from U_{x_0} to $\mathcal{H}_0$;

A_2) for every point $x\in U_{x_0}$ the $\mathcal{H}_0$-derivative $S'(x)\in\mathcal{L}_2(\mathcal{H}_0,\mathcal{H}_0)$ exists, and S' is a bounded continuous map from U_{x_0} to $\mathcal{L}_2(\mathcal{H}_0,\mathcal{H}_0)$;

A_3) for $x\in U_{x_0}$ the linear operator $[I+S'(x)]^{-1}$ is bounded on $\mathcal{H}$.

To study the absolute continuity of $\mu^\Phi<\mu$ we use finite-dimensional approximations.

Let P be a finite-dimensional orthoprojector on $\mathcal{H}$ taking values in $\mathcal{H}$ ($P\in\mathcal{P}$ and $Q=I-P$). Consider the map

$$\Phi^{(P)}(x)=x+PS(x_0+P(x-x_0)).$$

Represent the measure μ as a product, $\mu=\mu_P\times\mu_Q$, corresponding to the decomposition $\mathcal{H}_-=P\mathcal{H}_0\dot{+}\mathcal{H}_-^{(P)}$ (see (2.12)). Shrinking the neighborhood U_{x_0}, if necessary, we may regard it as a product of neighborhoods of the projections $Px_0\in P\mathcal{H}_0$ and $Qx_0\in\mathcal{H}_-^{(P)}$ of x_0: $U_{x_0}=U_{Px_0}\times U_{Qx_0}$. On this product the map $\Phi_{(x)}^{(P)}$ is easily seen to be equivalent to a pair of transformations, $\Phi^{(P)} = (\Phi_P^{(P)}, \Phi_Q^{(P)})$:

$$\Phi_P^{(P)}(Px) = Px + PS(x_0 + P(x - x_0)),$$
$$\Phi_Q^{(P)}(Qx) = Qx.$$

The first of these is finite-dimensional, while the second is the identity. So the study of $\Phi^{(P)}$ reduces to that of the transformation $\Phi_P^{(P)}$ on $P\mathcal{H}_0$.

First of all, note that $\Phi_P^{(P)\prime}(Px)|_{x=x_0} = P + PS'(x_0)P$. This operator is invertible on $P\mathcal{H}_0$ if the operator $I + PS'(x_0)P$ is invertible on $\mathcal{H}_0$. It follows from A_2-A_3 that this is true if P approximates I with sufficient accuracy, i.e. if $\sigma_2(PS'(x_0)P - S'(x_0)) < \epsilon$. Then (4.21) yields

$$\mathcal{J}(\Phi^{(P)}; x_0) = \mathcal{J}(\Phi_P^{(P)}; Px_0) =$$
$$\tilde{\Delta}(I + PS'(Px_0)P) \exp\left\{-l(PS)(x_0) - \frac{1}{2}\|PS(x_0)\|^2\right\}.$$

This formula holds in a neighborhood of x_0 in which the operator $I + S'(x_0)$ is invertible. For functions f having support in this neighborhood, (4.21) takes the form

$$\int_{\mathcal{H}_-} f(x + PS(x))\tilde{\Delta}(I + PS'(Px)P) \exp\left\{-l(PS)(x) - \frac{1}{2}\|PS(x)\|^2\right\} \mu(dx) =$$
$$\int_{\mathcal{H}_-} f(x)\mu(dx). \tag{4.22}$$

The next step is to carry out the limit transition $P \to I$ in (4.22). It follows from A_1 and A_2 and the estimate (3.27) that

$$l(PS)(x) \to l(S)(x)$$

(mean square convergence). So there is a sequence P_n such that the integrand at the lefthand side of (4.22) converges almost everywhere in U to $f(\Phi(x))\mathcal{J}(\Phi; x)$, where

$$\mathcal{J}(\Phi; x) = \tilde{\Delta}(I + S'(x)) \exp\left\{-l(S)(x) - \frac{1}{2}\|Sx\|^2\right\}.$$

It remains to verify any condition enabling one to carry out limit transition under the integral sign.

PROPOSITION 4.4. *The following estimate holds:*

$$\int_{U_{x_0}} |\mathcal{J}(I + S; x)|^2 \mu(dx) \le C \int_{U_{x_0}} \mathcal{J}(I + 2S; x)\mu(dx),$$

where

$$C = \sup_{U_{x_0}} \{\Delta[(I + S'^2(x))(I + 2S'(x))^{-1}] e^{\|Sx\|^2}\}.$$

PROOF. Using the properties of determinants and the fact that $S'^2(I+2S')^{-1} \in N$, we obtain by straightforward calculation:

$$\frac{|\mathcal{J}(I+S:x)|^2}{\mathcal{J}(I+2S;x)} = \frac{\tilde{\Delta}^2(I+S)}{\tilde{\Delta}(I+2S')}e^{\|S(x)\|^2} =$$

$$\Delta[(I+S'^2(x))(I+2S'(X))^{-1})]e^{\|Sx\|^2}, \tag{4.23}$$

which proves the proposition. □

THEOREM 4.4. *Suppose the function $\Phi(x) = x + S(x)$ satisfies conditions A_1-A_3 in a neighborhood U_{x_0} of a point $x_0 \in \mathcal{H}_-$. Then in this neighborhood the measures $\mu^{\Phi^{-1}}$ and μ are equivalent, and the density $\mathcal{J}(\Phi;x) = \mu^{\Phi^{-1}}(dx)/\mu(dx)$ is square integrable on U_{x_0} and can be represented as*

$$\mathcal{J}(\Phi;x) = \tilde{\Delta}(I+S'(x))\exp\left\{-l(S)(x) - \frac{1}{2}\|Sx\|^2\right\}. \tag{4.24}$$

PROOF. This assertion has already been proved for the maps $\Phi^{(P_n)}(x) = x + P_nS(P_nx)$. It follows from (4.22) for $f \equiv 1$ that $\int_{U_{x_0}} \mathcal{J}(\Phi^{(P_n)};x)\mu(dx) \le 1$, hence, because of the estimate obtained in proposition 4.4, $\int_{U_{x_0}} |J(\Phi^{(P_n)};x)|^2\mu(dx) \le \text{const}$ $(n = 1,2,\dots)$. This is sufficient in order that limit transition as $n \to \infty$ be possible in (4.22). The square integrability of $\mathcal{J}(\Phi;x)$ follows from Fatou's lemma. □

COROLLARY 4.3. *If, in addition to the requirements of theorem 4.4, $S'(x) \in N(\mathcal{H}_0,\mathcal{H}_0)$, then*

$$\mathcal{J}(\Phi;x) = \Delta(I+S'(x))\exp\left\{-(S(x),x) - \frac{1}{2}\|Sx\|^2\right\}. \tag{4.25}$$

REMARK 4.6. We rephrase these results in terms of a Gaussian measure μ_B with nondegenerate nuclear correlation operator B on a Hilbert space.

Let $\Phi(x) = x + S(x)$ be a map defined in a neighborhood U_{x_0} of a point $x_0 \in X$ and satisfying the following conditions:

A_1) the function $\Lambda(x) = B^{-1/2}S(x)$ is a bounded continuous map from U_{x_0} to X;

A_2) the function $\Lambda(x+B^{1/2}y)$ is differentiable along X with respect to y for each $x \in X$ and $C(x) = \Lambda'(x)B^{1/2}$ is a bounded continuous map from U_{x_0} to $\mathcal{L}_2(X,X)$.

A_3) for $x \in U_{x_0}$ the operator $[I+C(x)]^{-1}$ is bounded on X.

Then $\mu^{\Phi^{-1}} < \mu$ in U_{x_0}, and

$$\mathcal{J}(\Phi;x) = \tilde{\Delta}(I+C(x))\exp\left\{-l^{(B)}_{B^{-1/2}}(\Lambda)(x) - \frac{1}{2}\|\Lambda x\|^2\right\}, \tag{4.26}$$

where $l^{(B)}_{B^{-1/2}}(\Lambda)(x)$ is a measurable linear functional (see remark 3.6) that is an extension of the functional $(\Lambda(x), B^{-1/2}x) - \operatorname{Tr} C(x)$ defined for $x \in \mathcal{D}_{B^{-1/2}}$.

If, moreover, the function $B^{-1}S(x) = B^{-1/2}\Lambda(x)$ is continuous and $C(x) \in N$, then

$$\mathcal{J}(\Phi;x) = \Delta(I+C(x))\exp\left\{-(B^{-1/2}(\Lambda)(x),x) - \frac{1}{2}\|\Lambda x\|^2\right\}, \tag{4.27}$$

5. Fourier–Wiener transformation

5.1. Fourier transformation with respect to a Gaussian measure. Consider the Fourier transform

$$\tilde{f}(\omega) = \int_{\mathcal{H}_-} f(x) e^{i(\omega,x)} \mu(dx) \tag{5.1}$$

of a function $f \in \mathfrak{H} = L_2(\mathcal{H}_-, \mathbf{R}^1, \mu)$ with respect to the canonical Gaussian measure μ. It is defined for $\omega \in \mathcal{H}_0$, since then (ω, x) is a measurable linear functional on $\mathcal{H}_0$. An easy computation shows that

$$|\tilde{f}(\omega)| \leq \|f\|_{\mathfrak{H}}, \quad |\tilde{f}(\omega_2) - \tilde{f}(\omega_1)| \leq \|f\|_{\mathfrak{H}} \|\omega_2 - \omega_1\|_0. \tag{5.2}$$

DEFINITION 5.1. Let $f \in \mathfrak{H}$. The *Fourier–Wiener transform* of f is the function $\hat{f}(\omega) = \tilde{f}(\omega) e^{\|\omega\|^2/2}$, defined on $\mathcal{H}_0$.

PROPOSITION 5.1. *If $f(x) = f(Px)$ $(P \in \mathcal{P})$ is a cylindrical function, then $\hat{f}$ is also a cylindrical function:*

$$\hat{f}(\omega) = \hat{f}(\omega) = e^{\|P\omega\|^2/2} \tilde{f}(P\omega). \tag{5.3}$$

Hence it can be extended to $\mathcal{H}_-$ as a cylindrical function.

PROOF. Using (2.12) we find

$$\tilde{f}(\omega) = \int_{\mathcal{H}_-} f(Px) \exp\{i(Px,\omega) + i((I-P)x,\omega)\} \mu(dx) =$$

$$\int_{P\mathcal{H}} f(Px) e^{i(Px,\omega)} \mu^{(P)}(dx) \int_{\mathcal{H}_-^{(P)}} e^{i((I-P)x,\omega)} \mu^{(I-P)}(dx) = e^{-\|(I-P)\omega\|^2/2} \hat{f}(P\omega),$$

implying (5.3). □

PROPOSITION 5.2. *Let $f \in \mathfrak{H}$, $g_P(\omega) = \hat{f}(P\omega)$. The following inversion formula for the Fourier–Wiener transformation holds:*

$$f(x) = \lim_{P \to I} \hat{g}_P(-Px).$$

PROOF. By (2.12),

$$g_P(\omega) = e^{\|P\omega\|^2/2} \int_{\mathcal{H}_-} f(x) e^{i(x,P\omega)} \mu(dx) =$$

$$e^{\|P\omega\|^2/2} \int_{P\mathcal{H}} \left\{ \int_{\mathcal{H}_-} f(Px + (I-P)x) \mu^{(I-P)}(dx) \right\} e^{i(Px,P\omega)} \frac{1}{\sqrt{(2\pi)^n}} e^{-\|Px\|^2/2} \, d(Px).$$

This shows that the Fourier transform of the function $f^{(P)}(x) e^{-\|Px\|^2/2}$, where $f^{(P)}(x) = \int_{\mathcal{H}_-^{(P)}} f(Px + (I-P)y) \mu^{(I-P)}(dy)$, is $g_P(\omega) e^{-\|P\omega\|^2/2}$. Inverting the Fourier transformation over the finite-dimensional space $P\mathcal{H}$, we find

$$f^{(P)} = \frac{1}{\sqrt{(2\pi)^n}} e^{\|Px\|^2/2} \int_{P\mathcal{H}} g_P(\omega) e^{-\|P\omega\|^2/2} e^{-i(Px,P\omega)}\, d(P\omega) =$$

$$e^{\|Px\|^2/2} \int_{P\mathcal{H}} g_P(\omega) e^{i(Px,P\omega)} \mu^{(P)}(d\omega) = \hat{g}_P(-Px). \tag{5.4}$$

It remains to apply proposition 3.1, which gives

$$f(x) = \lim_{P\to I} f^{(P)}(x) \quad (\text{mod } \mu). \quad \square$$

REMARK 5.1. Suppose $\hat{f}(\omega)$ is extended to $\mathcal{H}_-$ such that limit transition is possible under the integral sign in (5.4). Then

$$f(x) = e^{\|x\|^2/2} \int_{\mathcal{H}_-} \hat{f}(\omega) e^{i(x,\omega)} \mu(d\omega) = \hat{f}(-x) \quad (x \in \mathcal{H}_0). \tag{5.5}$$

If $f(x)$ is uniformly continuous on $\mathcal{H}_-$, then this formula determines it uniquely for all $x \in \mathcal{H}_-$.

In particular, this is true for a cylindrical function f, since then $\hat{f}$ is also cylindrical.

We now introduce a class of functions whose Fourier–Wiener transforms have more natural properties. Let $\mu_{I/2}$ be a Gaussian measure in $\mathcal{H}_-$ with correlation operator $I/2$. Put $\mathfrak{H}_{1/2} = L_2(\mathcal{H}_-, \mathbf{R}^1, \mu_{I/2})$.

Consider the Hilbert space $\tilde{\mathfrak{H}}$ of functions for which

$$\|f\|_{\tilde{\mathfrak{H}}} = \int_{\mathcal{H}_-} |f(x)|^2 \mu(dx) + \int_{\mathcal{H}_-} |f(x)|^2_{\mu_{I/2}}(dx) =$$

$$\int_{\mathcal{H}_{--}} \left\{ |f(x)|^2 + \left| f\left(\frac{x}{\sqrt{2}}\right)\right|^2 \right\} \mu(dx) < \infty.$$

This space contains the class G of Borel functions f that satisfy the estimate

$$|f(x)| \le A e^{\alpha\|x\|^2_-} \quad \left(\alpha < \frac{1}{4}\right).$$

Clearly, the linear set G is dense in $\tilde{\mathfrak{H}}$.

THEOREM 5.1. *The Fourier–Wiener transformation can be extended to $\mathfrak{H}_{1/2}$ as a unitary operator $\hat{f} = Uf$ so that Parseval's equality holds:*

$$\int_{\mathcal{H}_-} \left| f\left(\frac{x}{\sqrt{2}}\right)\right|^2 \mu(dx) = \int_{\mathcal{H}_-} \left| \hat{f}\left(\frac{\omega}{\sqrt{2}}\right)\right|^2 \mu(d\omega). \tag{5.6}$$

The inverse transformation is given by the formula

$$(U^{-1}f)(x) = \hat{f}(-x). \tag{5.7}$$

PROOF. Let first $f(x) = f(Px)$ be a cylindrical function of class G. Then $f^{(P)}(x) = f(x)$ and, as has been shown in the proof of proposition 5.3, the Fourier transform of $f(Px)e^{-\|Px\|^2/2}$ over $P\mathcal{H}$ equals $\hat{f}(\omega)e^{-\|P\omega\|^2/2}$. So, Parseval's equality holds:

$$\int_{P\mathcal{H}} |f(Px)e^{-\|Px\|^2/2}|^2\,dx = \int_{P\mathcal{H}} |\hat{f}(P\omega)e^{-\|P\omega\|^2/2}|^2\,d\omega,$$

and, after change of variables, takes the form

$$\int_{P\mathcal{H}} \left| f\left(\frac{Px}{\sqrt{2}}\right)\right|^2 \mu^{(P)}(dx) = \int_{P\mathcal{H}} \left|\hat{f}\left(\frac{P\omega}{\sqrt{2}}\right)\right|^2 \mu^{(P)}(d\omega).$$

This is equivalent to (5.6), since f and $\hat{f}$ are cylindrical functions. Thus, the operator U, defined on a set dense in $\mathfrak{H}_{1/2}$, is isometric. So it can be extended to all of $\mathfrak{H}_{1/2}$, preserving isometry. In the same manner the operator $Vf(x) = \hat{f}(-x)$ can be shown to have similar properties. It follows from remark 5.1 that on the dense set of cylindrical functions the equality $UV = VU = I$ holds, which can be extended (continuously) to the whole space. □

5.2. Fourier–Wiener transformation of entire functions. Let $\tilde{\mathcal{H}}_-$ be the complex hull of a real Hilbert space $\mathcal{H}_-$, i.e. the orthogonal sum of two copies of $\mathcal{H}_-$, with elements presented in the form $z = x + iy$ $(x, y \in \mathcal{H}_-)$ and with norm $\|x + iy\|_-^2 = \|x\|^2 + \|y\|^2$.

We denote by Γ_α the class of functions $f(z)$ defined on $\tilde{\mathcal{H}}_-$ and having the following properties:

a) $f(x + \lambda y)$ is an entire function of the complex variable λ for $x, y \in \mathcal{H}_-$;
b) for some $A = A_f$ the estimate $|f(x + \lambda y)| \le Ae^{\alpha\|x+\lambda y\|_-^2}$ holds.

Note that if $f \in \Gamma_\alpha$ $(\alpha < 1/4)$, then for a fixed y we have $f_{\lambda y}(x) = f(x + \lambda y) \in G$. We introduce the integral transformation

$$F_f(\lambda, y) = \int_{\mathcal{H}_-} f(x + \lambda y)\mu(dx) \quad (y \in \mathcal{H}_-). \tag{5.8}$$

Clearly, for $f \in \Gamma_\alpha$ $(\alpha < 1/2)$ the function $F_f(\lambda, y)$ is defined, and is of class $\Gamma_{2\alpha|\lambda|}$ for $\alpha < 1/4$ and of class G for $|\lambda| \le 1$, $\alpha < 1/4$. The transformation (5.8) leaves invariant the class $\Gamma_{+0} = \bigcap_{\alpha>0}\Gamma_\alpha$.

PROPOSITION 5.3. *For functions of class Γ_α the transformation (5.8) coincides at $\lambda = i$ with Fourier–Wiener transformation:*

$$F_f(i, y) = \hat{f}(y) \quad (y \in \mathcal{H}_-). \tag{5.9}$$

If $\hat{f} \in \Gamma_\alpha$, then the following inversion formula holds:

$$f(x) = F_{\hat{f}}(-i, y) = \int_{\mathcal{H}_-} \hat{f}(y - ix)\nu(dy). \tag{5.10}$$

PROOF. Consider $F_f(\lambda, y)$ for $y \in \mathcal{H}_0$, and let, first, $\lambda > 0$. By (4.13) we have

$$F_f(\lambda, y) = \int_{\mathcal{H}_-} f(x + \lambda y)\mu(dx) = \int_{\mathcal{H}_-} f(x)\mu_{\lambda y}(dx) = e^{-\lambda^2\|y\|^2/2} \int_{\mathcal{H}_-} f(x)e^{\lambda(x,y)}\mu(dx).$$

The functions figuring in the lefthand and righthand sides of the equality are defined for all complex values of λ, and are analytic. This readily follows from the estimates satisfied by the integrands. They coincide for $\lambda > 0$, and hence for all complex λ. In particular, for $\lambda = i$ we find (5.9) with $y \in \mathcal{H}_0$. If f is a cylindrical function, then both sides of (5.9) are cylindrical, and thus they can be extended to $\mathcal{H}_-$, preserving equality. So, the operators F_f and Uf coincide on the set of cylindrical functions, which is dense in $\mathfrak{H}$. For $f \in \Gamma_\alpha$ we consider the approximating sequence of cylindrical functions $f_n(x) = f(P_n x)$. This sequence has a majorant, and it can be readily checked that $F_{f_n} \to F_f$, $Uf_n \to Uf$ in $\mathfrak{H}_{1/2}$. By limit transition we obtain (5.9) in the general case. Since all considerations above are also applicable to $\hat{f}$, (5.10) follows from (5.7). □

5.3. Connection between the Fourier–Wiener transformation and orthogonal polynomials.

THEOREM 5.2. *Let $A \in \mathcal{L}_2^n(\mathcal{H}, \mathcal{K})$. Then*

$$\widehat{\theta_n(A)}(x) = i^n \tilde{A}(x) \quad (x \in \mathcal{H}_0). \tag{5.11}$$

PROOF. First, let $A \in \mathcal{L}_{20}^n(\mathcal{H}, \mathcal{K})$. Consequently, the function $\tilde{A}(x) = A(x, \dots, x)$ is cylindrical. We will show (5.11) to be true in this case. It suffices to prove that

$$\theta_n(A)(x) = i^n \int_{\mathcal{H}_-} \tilde{A}(y)e^{i(x,y)}\mu(dy)e^{\|x\|^2/2} \quad (x \in \mathcal{H}_0). \tag{5.12}$$

Because of linearity, this formula needs to be verified only for operators $A(x_1, \dots, x_n) = (x_1, \phi_1)\dots(x_n, \phi_n)\psi$ ($\phi_k \in \mathcal{H}_0$, $\psi \in \mathcal{K}$). This can be done by straightforward substitution. So, the righthand side of (5.12) satisfies the recurrence relations (3.27), determining θ_n uniquely. Formula (5.12) is thus established for cylindrical operators, as is the formula

$$e^{\|x\|^2/2} \int_{\mathcal{H}_-} \theta_n(A)(y)e^{i(x,y)}\mu(dy) = i^n A(x, \dots, x), \tag{5.13}$$

obtained from it by inversion.

Let now $A \in \mathcal{L}_2^n(\mathcal{H}, \mathcal{K})$ and $A = \lim_{k\to\infty} A_k$, $A_k \in \mathcal{L}_{20}^n(\mathcal{H}, \mathcal{K})$. It is easily seen that termwise limit transition is allowed in (5.13), which proves (5.11). □

COROLLARY 5.1. *Let $A_{n,j} \in \mathcal{L}_2^n(\mathcal{H}_-, \mathbf{R}^1)$ ($j = 1, 2, \dots$; $n = 0, 1, \dots$) be a double sequence of operators for which $\{\theta_n(A_{n,j})(x)\}$ is a complete orthonormal system in $L_2(\mathcal{H}_-, \mathbf{R}^1, \mu)$.*

It follows from (5.13) that for $\Im\lambda = 0$,

$$\int_{\mathcal{H}_-} \theta_n(A)(x) \exp\left\{i\lambda(x, y) + \frac{\lambda^2}{2}\|y\|^2\right\} \mu(dx) = (i\lambda)^n A(y, \dots, y) \quad (y \in \mathcal{H}_0).$$

Because of analyticity with respect to λ, the lefthand and righthand sides coincide for all λ. Hence, for $\lambda = -i$ we find

$$\int_{\mathcal{H}_-} \theta_n(A)(x) \exp\left\{(x,y) - \frac{1}{2}\|y\|^2\right\} \mu(dx) = A(y,\ldots,y).$$

This implies that for any $y \in \mathcal{H}_0$ the following orthogonal expansion in $L_2(\mathcal{H}_-, \mathbf{R}^1, \mu)$ holds:

$$e^{(x,y)-\|y\|^2/2} = \sum_{n,j} A_{n,j}(y,\ldots,y)\theta_n(A_{n,j})(x).$$

REMARK 5.2. It is easily seen that the polynomials $\theta_n(A)(x\sqrt{2})$ are orthogonal in $\mathfrak{H}_{1/2}$. Proposition 5.3 implies that the Fourier–Wiener transform of a polynomial is a polynomial. Since the Fourier–Wiener transformation is unitary, the orthogonality relation is not affected, and hence $U\theta_n(A)(x\sqrt{2}) = C_n\theta_n(A)(x\sqrt{2})$, where C_n is easily seen to be equal to i^n.

Thus, the Fourier–Wiener transform of a function $g \in \mathfrak{H}_{1/2}$ can be determined as follows: if $g(x/\sqrt{2}) = \sum C_{n,j}\theta_n(A_{n,j})(x)$, then

$$Ug(x) = \sum C_{n,j} i^n \theta_n(A_{n,j})(x\sqrt{2}).$$

6. Complexvalued Gaussian quasimeasures

6.1. Feynman integrals. Let B be a positive ($(Bx,x) > 0$, $x \neq 0$) symmetric operator on a Hilbert space X with dense domain of definition $\mathcal{D}_B$. Consider the function $\chi(z,\theta) = \exp\{-(z/2)(B\theta,\theta)\}$ ($\theta \in \mathcal{D}_B$, $\Re z \geq 0$), depending on a complex parameter z. It is absolutely integrable with respect to θ over any finite-dimensional subspace $PX = X_P$ contained in $\mathcal{D}_B$. Under these conditions the function $\chi(z,\theta)$ determines a quasimeasure μ_{zB} in X for which it is the characteristic functional.

In the sequel we take B bounded ($\mathcal{D}_B = X$).

Proposition I.3.4 implies

PROPOSITION 6.1. *For every function of the form $f(x) = \int_X e^{i(x,\theta)}\nu(d\theta)$, where ν is a complexvalued measure ($\|\nu\| < \infty$), the following integral exists for $\Re z > 0$*

$$\int_X f(x)\mu_z B(dx) = \lim_{P\to I}\int_X f(Px)\mu_{zB}^{(P)}(dx) = \int_X e^{-(z/2)(B\theta,\theta)}\nu(d\theta). \tag{6.1}$$

The case $\Re z = 0$ is of special interest, setting $z = i$.

The quasimeasure ν_{iB} is called the *Feynman quasimeasure*.

The finite-dimensional distribution

$$\frac{\mu_{zB}^{(P)}}{dx} = \frac{1}{\sqrt{(2\pi)^n z^n \det PBP}} \exp\left\{-\frac{1}{2z}((PBP)^{-1}x,x)\right\} \tag{6.2}$$

is only locally integrable on X_P, and hence does not have a characteristic functional in the usual sense. However, the function $\exp\{-(z/2)(BP\theta,\theta)\}$ is the Fourier transform of the distribution (6.2) in the class of distributions (generalized functions). The relation

$$\int_{X_P} f(Px)\mu_{zB}^{(P)}(dx) = \int_X \chi(z,P\theta)\nu(d\theta)$$

remains true for functions $f(x)$ that are integrable over any finite-dimensional subspace X_P. So, a weakened version of proposition 6.1 holds.

PROPOSITION 6.2. *If under the conditions of proposition 6.1 we have*

$$\int_{X_P} |f(Px)|\,dx < \infty \quad (P \in \mathcal{P}),$$

then formula (6.1) holds for $\Re z = 0$ *as well.*

Another approach is based on changing the meaning of 'integral with respect to the quasimeasure μ_{iB}. By definition, we set

$$\widetilde{\int_X} f(x)\mu_{iB}(dx) = \lim_{z \to i} \int_X f(x)\mu_{zB}(dx), \tag{6.3}$$

if this limit exists.

Since under the conditions of proposition 6.1 the function at the righthand side of (6.3) is analytic inside the right halfplane and is continuous on the boundary, the limit transition for $z \to i$ is possible. Therefore, under the conditions stated, the integral (6.3) exists, and the following formula holds:

$$\widetilde{\int_X} f(x)\mu_{iB}(dx) = \int_X e^{-(i/2)(B\theta,\theta)}\nu(d\theta). \tag{6.4}$$

6.2. Integration of analytic functionals. We now consider another class of functions for which the integral (6.3) exists. Let $\tilde{X} = X + iX$ be the complexification of the space X introduced in section 5. Consider a function $f(z)$ defined on elements $z = \lambda x$, $\lambda \in \bar{V}$, $V = \{\lambda = \rho e^{i\gamma}\colon\ 0 < \gamma < \pi/4\}$, and satisfying

B$_1$) $f(\lambda x)$ is continuous with respect to $x \in X$ for $\lambda \in \bar{V}$;

B$_2$) $f(\lambda x)$ is analytic with respect to λ in V for every $x \in X$, and is continuous up to the boundary (except, possibly, at $\lambda = 0$);

B$_3$) the following estimate holds:

$$|f(xe^{i\gamma})| \le Ce^{(Ax,x)\sin^2\gamma + g(x)}\phi(\|x\|) \quad (A \le B^{-1}), \tag{6.5}$$

where $\phi(\|x\|)$ is bounded, and integrable over any finite-dimensional space, and $g(x)/(B^{-1}x,x) \to 0$ as $\|x\| \to \infty$, uniformly over balls in any finite-dimensional space.

THEOREM 6.1. *For a function $f(x)$ satisfying the conditions B_1-B_3 the Feynman integral exists, and the following formula holds:*

$$\widetilde{\int_X} f(x)\mu_{iB}(dx) = \int_X f(e^{\pi i/4}x)\mu_B(dx). \tag{6.6}$$

PROOF. Let $P \in \mathcal{P}$, $B_P = PBP$. Consider the following integrals:

$$\int_{X_P} f(Px)\mu_{\lambda^2 B_P}(dx) = \int_{X_P} f(Px)\frac{1}{\sqrt{(2\pi\lambda^2)^n \det B_P}} \exp\left\{-\frac{1}{2\lambda^2}(B_P^{-1}x,x)\right\}\,dx$$

and

$$\int_{X_P} f(\lambda Px)\mu_{B_P}(dx) = \int_{X_P} f(\lambda Px)\frac{1}{\sqrt{(2\pi)^n \det B_P}} \exp\left\{-\frac{1}{2}(B_P^{-1}x, x)\right\} dx.$$

A simple change of variables shows that they coincide for $\lambda > 0$. For every fixed x the integrands are analytic with respect to λ in V, and continuous in $\bar{V}\backslash\{0\}$. We show that the integrands can be uniformly majorated with respect to λ in $V_m = V \cap \{\Re\lambda^2 \geq 1/m, |\lambda| \geq 1/2\}$ $(m = 1, 2, \dots)$ by integrable functions. Then both integrals are analytic functions in $\cup_m V_m = \tilde{V}$, and their coincidence for $\lambda > 0$ implies coincidence in $\tilde{\bar{V}}$.

For $\lambda \in V_m$ we have

$$\left| f(Px)\frac{1}{\sqrt{(2\pi\lambda^2)^n \det B_P}} \exp\left\{-\frac{1}{2\lambda^2}(B_P^{-1}x, x)\right\}\right| \leq$$

$$C\phi(\|x\|)\frac{\exp\left\{g(x) - \Re\frac{1}{2\lambda^2}(B_P^{-1}x, x)\right\}}{|\lambda|^n(2\pi)^{n/2}\sqrt{\det B_P}} \leq$$

$$C_1\phi(\|x\|)\exp\left\{-\frac{1}{2\lambda^2}(B_P^{-1}x, x)\left(1 - \frac{mg(x)}{(B_P^{-1}x, x)}\right)\right\} \leq C_2\phi(\|x\|),$$

$\|x\|$ being sufficiently large.

On the other hand,

$$\left| f(\lambda Px)\frac{1}{\sqrt{(2\pi)^n \det B_P}} \exp\left\{-\frac{1}{2}(B_P^{-1}x, x)\right\}\right| \leq$$

$$C\phi(\|x\|)\frac{\exp\{(A_Px, x) + 2g(x) - (B_P^{-1}x, x)\}}{(2\pi)^{n/2}\sqrt{\det B_P}} \leq$$

$$C_1\phi(\|x\|)\exp\{(A_Px, x) - (1-\epsilon)(B_P^{-1}x, x)\} \leq C_2\phi(\|x\|)$$

for sufficiently large $x \in X_P$, since on a finite-dimensional space $A_P < B_P^{-1}$ implies $A_P < (1-\epsilon)B_P^{-1}$, $\epsilon > 0$ being sufficiently small.

Thus, for $\lambda \in \tilde{V}$,

$$\int_{X_P} f(Px)\mu_{\lambda^2 B_P}(dx) = \int_{X_P} f(\lambda Px)\mu_{B_P}(dx) = \int_X f(\lambda Px)\mu_B(dx).$$

However, the continuity of f with respect to x and the estimate (6.5) imply that the limit transition for $P \to I$ is possible.

So, for $\lambda \in \tilde{V}$ the following limit exists:

$$\int_X f(x)\mu_{\lambda^2 B}(dx) = \lim_{P\to I}\int_X f(Px)\mu_{\lambda^2 B_P}(dx) = \int_X f(\lambda x)\mu_B(dx).$$

It remains to note that the righthand side is continuous with respect to $\lambda \in \bar{V}$; in particular for $\lambda = i \in \tilde{\bar{V}}$, and hence the following limit exists:

$$\widetilde{\int}_X f(x)\mu_{iB}(dx) = \lim_{\lambda^2\to i}\int_X f(x)\mu_{\lambda^2 B}(dx) = \int_X f(e^{\pi i/4}x)\mu_B(dx).$$

□

6.3. Computation of certain Feynman integrals. a) Let $f(x) = F((x, \phi_1), \ldots, (x, \phi_n))$, $\phi_k \in X$ $(k = 1, \ldots, n)$, where $(x_1, \ldots, x_n)$ is a function that is integrable over $\mathbf{R}^n$ with respect to a Gaussian measure. The map $x \to (x_1, \ldots, x_n)$ $(x_k = (x, \phi_k))$ maps X to $\mathbf{R}^n$, and a quasimeasure with correlation operator $\lambda^2 B_n$ is mapped to a complexvalued Gaussian quasimeasure in $\mathbf{R}^n$ with correlation matrix $\lambda^2 B_n = \lambda^2 \|(B\phi_j, \phi_k)\|$. Therefore,

$$\int_X F((x, \phi_1), \ldots, (x, \phi_n)) \mu_{iB_n}(dx) = \int_{\mathbf{R}^n} F(y_1, \ldots, y_n) \mu_{iB_n}(dy) =$$
$$\lim_{\lambda^2 \to i} \frac{1}{\sqrt{(2\pi i)^n \det B_n}} \int_{\mathbf{R}^n} F(y_1, \ldots, y_n) e^{-(1/2\lambda^2)(B_n y, y)}\, dy.$$

In particular, $\int_X e^{(\alpha,x)} \mu_{\lambda^2 B}(dx) = e^{(\lambda^2/2)(B\alpha,\alpha)}$, and by limit transition as $\lambda^2 \to i$ we obtain

$$\widetilde{\int_X} e^{(\alpha,x)} \mu_{iB}(dx) = e^{(i/2)(B\alpha,\alpha)}.$$

b) Replacing x by λx, the function $(Ax, x)^m$ becomes $\lambda^{2m}(Ax, x)^m$, which satisfies the relevant conditions with $\phi(x) = (Ax, x)^m \exp\{-\sqrt{(B^{-1}x, x)}\}$. So,

$$\widetilde{\int_X} (Ax, x)^m \mu_{iB}(dx) = i^m \int_X (Ax, x)^m \mu_B(dx).$$

In particular,

$$\widetilde{\int_X} (Ax, x) \mu_{iB}(dx) = i \operatorname{Tr} AB;$$

$$\widetilde{\int_X} (Ax, x)^2 \mu_{iB}(dx) = -\{[\operatorname{Tr} AB]^2 + 2[\operatorname{Tr}(AB)^2]\}.$$

c) With $A_1 = A_1^*$, $A_2 \geq 0$, the function $f(x) = \exp\{i(A_1 x, x) - (A_2 x, x)\}$ satisfies the conditions of theorem 6.1. Therefore,

$$\widetilde{\int_X} f(x) \mu_{iB}(dx) = \int_X e^{-(A_1 x, x) - i(A_2 x, x)} \mu_B(dx) =$$
$$\{\det[I + B^{1/2}(A_1 + iA_2)B^{1/2}]\}^{-1/2},$$

if $A_1 > B^{-1/2}$.

7. Chapter II: Supplementary remarks and historical comments

The special Gaussian measure in the space of continuous functions, the Wiener measure, constructed by N. Wiener in the 1920s (Wiener [2]), was historically the first Gaussian measure ever studied. This study was conducted in a series of papers by R. Cameron, B. Martin, et al. An analysis of certain aspects of this work, and a description of it, can be found in Gelfand–Yaglom [1] and in the review by Kovalchik [1].

The general notion of Gaussian measure in an infinite-dimensional (Banach) space was introduced in 1935 by Kolmogorov [2] by means of the characteristic functional, defined in the same paper. A description of Gaussian measures in Hilbert space by means of nuclear operators (theorem 2.1) was obtained by Mourier [1].

Segal [3] considered linear transformations of a canonical Gaussian cylindrical measure in Hilbert space. Continuing these studies, Feldman [1] and, independently, Hajek [1] established that Gaussian measures in Hilbert space are either equivalent or orthogonal, and obtained an equivalence condition (theorem 4.3) and a formula for the density in the case of a correlation operator C satisfying the condition $\operatorname{Tr}(C-I) < \infty$. Consequently, absolutely continuous linear (Seidman [1]) and nonlinear (Gross [5]) transformations of a canonical Gaussian measure in Hilbert space were considered.

In the following years the theory of Hilbert space equipped with a Gaussian measure was developed, elaborated, and generalized by a number of researchers. Here we mention only the most important works, which contain additional references: Gikhman–Skorokhod [2,3], Rozanov [1], Skorokhod [1,4].

Most general results on nonlinear transformations of Gaussian measures were obtained by Skorokhod [4] (who did not restrict himself to Gaussian measures) and Ramer [1]. Theorem 4.4 is a version of Ramer's theorem adapted to the case of a Hilbert space. Our presentation is based on a theorem of Kakutani [1], and uses properties of measurable linear functionals.

Measurable linear and multilinear functionals on infinite-dimensional measure spaces were considered by Vershik [2, 3], Smolyanov [1], Shilov–Fahn DucThihn [1], Rozanov [1], Gikhman–Skorokhod [3], and Skorokhod [4].

Orthogonal decompositions of spaces of functions that are square integrable with respect to a Gaussian measure were constructed for the Wiener measure by various methods by Cameron–Martin [1] and Itô [6]. For general Gaussian measures this was done by Vershik [1].

Our approach to this decomposition is based on formulas for integration by parts with respect to a Gaussian measure. The simplest formula, (3.15), was essentially obtained by Cameron [3], and other versions were given by Donsker [1] and Novikov [2].

The formulas presented here were obtained by Dalecky–Paramonova [1,2], and are connected with the study of stochastic integrals with respect to arbitrary Gaussian random processes (see also Skorokhod [5]).

The Fourier–Wiener transformation, especially for Wiener measures, was introduced by Cameron–Martin [2]. Their work was developed and generalized by Guseinov [1,2], and Gadzhiev [1].

About Feynman integrals see also the comments to chapt. I and chapt. VI. Abstract Gaussian quasimeasures with a complex parameter in Hilbert space were introduced by Dalecky–Stremsky [1]. The method of integration of analytic functionals generalizes a result of Cameron and Martin related to Feynman integrals in a strict sense, and obtained by analytic continuation of Wiener integrals. The theory of integration with respect to an abstract 'Feynman measure' was recently constructed by Albeverio–Höegh-Krohn [1], and, in another way, by Uglanov [5].

A nonnegative Gaussian cylindrical measure can always be extended to a σ-additive measure in a certain Hilbert space. However, one often has to study the more narrow

Banach spaces in which the measure is supported.

Gross [1] has started such a study, introducing the notion of abstract Wiener space. Sato [1] has shown that every Gaussian measure in a separable or reflexive Banach space can be realized as an abstract Wiener measure. Papers devoted to this direction are: Goodman [1], Ramer [1] (who constructed the theory of transformations of Gaussian measures in these terms), Kuo [1], and others. Kuo's book contains a rather complete presentation of certain aspects of this direction.

Another direction of study of Gaussian measures in Banach space, based on notions from correlation theory, was initiated by Vakhaniya [1], and continued by Chobanyan–Tarieladze [1].

Finally we must mention the profound investigations of Vershik and Sudakov on various questions of measure theory in infinite-dimensional spaces. These are, in certain aspects, also related to Gaussian measures (Vershik [2,3], Sudakov [2], Vershik–Sudakov [1]).

CHAPTER III

Measures in linear topological spaces

This chapter is devoted to the consideration of nonnegative cylindrical measure in linear topological spaces. σ-additivity conditions (generalizations of the Minlos–Sazonov theorem), conditions of weak compactness (generalizations of the Lévy and Prokhorov theorems), as well as conditions for a measure space to be weakly complete are presented.

As an application we obtain results in the spectral theory of operators on Hilbert space of the type of Stone's theorem.

The presentation is based on results from chapter I; some information given in the supplement to chapter I is used in an essential way, especially results concerning the theory of linear topological spaces.

1. σ-additivity conditions for nonnegative cylindrical measures in the space X' dual to a locally convex space X

1.1. Sufficient conditions for σ-additivity. Strong regularity. Consider a nonnegative cylindrical measure (a bounded quasimeasure) μ in X'. Recall that it is determined by the consistent collection of nonnegative finite-dimensional distributions $\mu^{(x_1,\dots,x_n)}$ $(x_1,\dots,x_n \in X)$ in the finite-dimensional spaces $\mathcal{L}^{(x_1,\dots,x_n)}$; without loss of generality, these spaces may be regarded as Hilbert spaces.

On the other hand, the cylindrical measure μ is uniquely determined by its characteristic functional

$$\chi_\mu(x) = \int_{X'} e^{i(x,y)}\mu(dy) = \int_{-\infty}^{\infty} e^{iz}\mu^{(x)}(dz),$$

which is a positive definite function on X that is continuous at zero along each finite-dimensional subspace of X.

Without loss of generality we can take $\mu(X') = 1$ and $\chi_\mu(0) = 1$.

In chapter I it was shown that the cylindrical measure μ has a σ-additive extension to a certain linear extension of the space X'. These linear extension consists in fact of a collection of functions on X (not necessarily continuous linear functionals). Here we will give additional conditions for this σ-additive measure to have support in X'.

We first prove some auxiliary results.

PROPOSITION 1.1. *Let μ be a nonnegative measure in $\mathbf{R}^n$, $\chi_\mu(\theta)$ its characteristic functional, and A, B nonnegative operators on $\mathbf{R}^n$. If the inequality*

$$|1 - \Re\chi_\mu(y)| \leq \epsilon$$

holds on the set $\{y \in \mathbf{R}^n\colon (Ay,y) \leq 1\}$, *then*

$$\mu\{x \in \mathbf{R}^n : (Bx,x) > c\} \leq \frac{\sqrt{e}}{\sqrt{e}-1}\left(\epsilon + \frac{2}{c}\operatorname{Tr} AB\right) \quad (c > 0). \tag{1.1}$$

PROOF. Without loss of generality we may assume that $c = 1$. Let ν_B be a Gaussian measure in $\mathbf{R}^n$ for which

$$e^{-(Bx,x)/2} = \int_{\mathbf{R}^n} e^{i(x,y)}\nu_B(dy).$$

Since $(Bx,x) \geq 1$ implies that

$$\frac{\sqrt{e}}{\sqrt{e}-1}(1 - e^{-(Bx,x)/2}) \geq 1,$$

we have the following equalities:

$$\begin{aligned}\mu\{x : (Bx,x) \geq 1\} &= \int_{(Bx,x)\geq 1} \mu(dx) \leq \frac{\sqrt{e}}{\sqrt{e}-1}\int_{\mathbf{R}^n}(1 - e^{-(Bx,x)/2})\mu(dx) = \\ \frac{\sqrt{e}}{\sqrt{e}-1}\Re\int_{\mathbf{R}^n}\nu_B(dy)\int_{\mathbf{R}^n}&(1 - e^{i(x,y)})\mu(dx) = \frac{\sqrt{e}}{\sqrt{e}-1}\int_{\mathbf{R}^n}(1 - \Re\chi_\mu(y))\nu_B(dy).\end{aligned} \tag{1.2}$$

Note that under the conditions stated

$$\int_{(Ay,y)\leq 1}(1 - \Re\chi_\mu(y))\nu_B(dy) \leq \epsilon\int_{(Ay,y)\leq 1}\nu_B(dy) \leq \epsilon;$$

moreover,

$$\int_{(Ay,y)>1}(1 - \Re\chi_\mu(y))\nu_B(dy) \leq 2\int_{(Ay,y)>1}\nu_B(dy) \leq 2\int_{\mathbf{R}^n}(Ay,y)\nu_B(dy) = 2\operatorname{Tr} AB.$$

Adding these relations and taking into account (1.2), we obtain the required proposition. □

PROPOSITION 1.2. *Let X be a finite-dimensional linear space, $Y = X'$ its algebraic dual, p, q Hilbert prenorms in X satisfying the condition $p(x) \leq Cq(x)$ $(x \in X)$, ψ_{pq} the canonical map from X_q to X_p.*

Let μ be a normalized nonnegative measure on the Borel σ-algebra of Y, satisfying the condition

$$|1 - \Re\chi_\mu(x)| \leq \epsilon \text{ for } p(x) \leq 1 \quad (\epsilon < 1).$$

Then for each $r > 0$ the measure of the complement of the polar of the set $S(q,r) = \{x \in X\colon q(x) < r\}$ satisfies the estimate

$$\mu\{Y \setminus S^0(q,r)\} \leq \frac{\sqrt{e}}{\sqrt{e}-1}[\epsilon + 2r^2\sigma_2^2(\psi_{pq})], \tag{1.3}$$

σ_2 being the Hilbert–Schmidt norm.

PROOF. We introduce a scalar product in X as follows. Identify X_q with some direct complement of N_q, and let Q_q be the projection operator on X_q corresponding to the decomposition

$$X = N_q \dot{+} X_q. \tag{1.4}$$

Put

$$(x_1, x_2) = ((I - Q_q)x_1, (I - Q_q)x_2)_0 + (Q_q x_1, Q_q x_2)_1,$$

where $(\cdot,\cdot)_1$ is the scalar product in X_q inducing the prenorm q and $(\cdot,\cdot)_0$ is a scalar product in N_q.

Let $p(x) = \|Ax\|$, where A is an operator on X (without loss of generality: nonnegative). The following decomposition, analogous to (1.4), holds: $X = N_p \dot{+} X_p$, where $N_p = N(A) \supset N_q$ and $X_p = R(A) \perp N_p$. The operator ψ_{pq} is the orthoprojector from X_q onto X_p. If $\{e_j\}_1^m$ is an orthonormal basis in X_q such that $e_j \in X_p$ for $j \geq k$, then

$$\sigma_2^2(\psi_{pq}) = \sum_{j=1}^m p^2(\psi_{pq} e_j) = \sum_{j=1}^m p^2(e_j) = \sum_{j=1}^m \|Ae_j\|^2 = \sigma_2^2(A). \tag{1.5}$$

Consider now the norm q_δ $(\delta > 0)$ in X defined by

$$q_\delta^2(x) = q^2(Q_q x) + \delta^2 \|(I - Q_q)x\|^2 = \|B_\delta x\|^2; \quad B_\delta = Q_q + \delta(I - Q_q).$$

The corresponding space X_{q_δ} coincides with X, since it follows from (1.5) that, as before, $\sigma_2^2(\psi_{pq_\delta}) = \sigma_2^2(A)$. In fact, the above introduced scalar product in X varies only on $N_q \subset N(A)$, and hence the relation $p(x) = \|Ax\|$ is preserved and the operator A remains nonnegative. It is readily verified that as $\delta \to 0$, $S(q_\delta, r) \uparrow S(q, r)$ monotone, and hence $S^0(q_\delta, r) \downarrow S^0(q, r)$, resulting in

$$\mu(S^0(q_\delta, r)) \downarrow \mu(S^0(q, r)).$$

Therefore it suffices to establish (1.3) for the nondegenerate prenorms q_δ. Identify Y with the euclidean space X equipped with the scalar product defined above. Then the polar of the set $S(q_\delta, r) = \{x\colon \|B_\delta x\| < r\}$ is easily seen to be the set $S^0(q_\delta, r) = \{y\colon \|B_\delta^{-1} y\| \leq 1/r\}$. Proposition 1.1 implies that

$$\mu(Y \setminus S^0(q_\delta, r)) = \mu\left\{y : \|B_\delta^{-1} y\| \geq \frac{1}{r^2}\right\} \leq \frac{\sqrt{e}}{\sqrt{e} - 1}(c + 2r^2 \operatorname{Tr} A^2 B_\delta^{-2}),$$

where

$$\operatorname{Tr} A^2 B_\delta^{-2} = \operatorname{Tr} A B_\delta^{-2} A = \sigma_2^2(A B_\delta^{-1}) = \sigma_2^2(A) = \sigma_2^2(\psi_{pq}),$$

since $AB_\delta^{-1} = A\{Q_q + (I - Q_q)/\delta\} = A$. □

We now introduce some important notions which will be used in this chapter in an essential manner.

DEFINITION 1.1. A realvalued cylindrical measure μ in X' is called *X-regular* if for each $\epsilon > 0$ there is a subset K_ϵ, compact in X'_σ and such that for each X-cylinder set B lying outside K_ϵ $(B \cap K_\delta = \emptyset)$ the condition $|\mu(B)| < \epsilon$ holds (and thus, $|\mu|(B) < \epsilon$).

μ is called *strongly X-regular* if K_ϵ can be chosen to be an absolutely convex Hilbert subset.

PROPOSITION 1.3. *Every X-regular realvalued cylindrical measure μ on $(X', \mathcal{U}_X)$ is σ-additive.*

PROOF. Consider the class F of X-cylinder sets that are closed in X'_σ, i.e. the cylinder sets whose bases are closed in the corresponding finite-dimensional supports. Proposition I.1.4 implies that F approximates μ from below. So the conditions of proposition I.1.3 are fulfilled. □

COROLLARY 1.1. *Let X be a reflexive separable Banach space. A realvalued X'-cylindrical measure in X is σ-additive if and only if for each $\epsilon > 0$ there is an $R > 0$ with the property that for any X-cylinder set B lying outside the ball $\{x\colon \|x\| \le R\}$ the estimate $|\mu(B)| < \epsilon$ holds.*

PROOF. Sufficiency follows from proposition 1.3, since $X = (X')'$. Necessity follows from proposition I.1.5, since every compact set is bounded. □

This assertion can be extended to reflexive separable Fréchet spaces if, in the formulation, the ball is replaced by a weakly compact set.

Note that in this case strong X-regularity of a measure μ which is σ-additive on the Borel σ-algebra of X'_σ is equivalent to

$$|\mu|(X' \setminus K_\epsilon) < \epsilon$$

for the set K_ϵ figuring in the definition.

THEOREM 1.1. *Let μ be a nonnegative cylindrical measure on the algebra $\mathcal{U}_X$ of X-cylinder sets of the space X' dual to a locally convex space X.*

If the characteristic functional $\chi_\mu(x)$ is continuous at zero in the nuclear topology $\tau_S(X, \beta)$ corresponding to a topology β on X and compatible with the duality between X and X', then the cylindrical measure μ is strongly X-regular, and hence σ-additive, on $\mathcal{U}_X$.

PROOF. To prove this theorem it suffices to establish the existence of a compact set K_ϵ satisfying the estimate given in the definition of strong X-regularity.

Without loss of generality we may assume that $\beta = \tau(X, X')$ is the Mackey topology, i.e. the strongest topology compatible with the duality, since continuity of functions is preserved by replacing the initial topology by a stronger one. Note that the polar of a neighborhood of zero in the space $(X, \tau(X, X'))$ is compact and absolutely convex in X'_σ; this will be used below when constructing K_ϵ.

The continuity of χ_μ implies that for each $\delta > 0$ there is a Hilbert prenorm[1] $p \in \mathcal{P}(X, \tau(X, X'))$ for which $|1 - \Re\chi_\mu(x)| < \delta$ if $p(x) < 1$. By the definition of nuclear topology there is a Hilbert prenorm q, continuous in the topology $\tau(X, X')$, for which the canonical map $\psi_{pq}\colon \bar{X}_q \to \bar{X}_p$ is a Hilbert–Schmidt map. We show that the polar $S^0(q, r)$ of the set $S(q, r) = \{x \in X\colon q(x) < r\}$ has the required properties for suitable δ, r.

Let B be an X-cylinder set lying outside $S^0(q, r)$, and let $L \subset X'$ be the finite-codimensional subspace defining B. By passing to the quotient space X'/L, we replace B by its base B_0, which has the same measure: $\mu_L(B_0) = \mu(B)$. The canonical image

[1]See the supplement to chapter I.

of the polar $S^0(q,r)$ is the polar of the set $S(q,r) \cap L^0 = \{x \in L^0;\ q(x) < r\}$, which lies in the finite-dimensional space L^0 dual to X'/L.

In this situation we may use proposition 1.2. Note that by passing to the subspace L^0 the map ψ_{pq} is replaced by its restriction, and so the quantity $\sigma_2(\psi_{pq})$ does not increase. Hence the following estimate holds:

$$(1.6)\quad \mu(B) = \mu_L(B_0) \leq \mu_L\{X'/L \setminus (S(q,r) \cap L^0)^0\} \leq \frac{\sqrt{e}}{\sqrt{e}-1}(\delta + 2r^2\sigma_2^2(\psi_{pq})).$$

Choosing $\delta > 0$ sufficiently small, and then $r > 0$ sufficiently small gives the estimate for $K = S^0(q,r)$. □

REMARK 1.1. As follows from the proof, the estimate of the value $\mu(B)$ depends only on the neighborhood determined by the characteristic functional of μ. Strictly speaking, the following assertion is true.

Let p, q be continuous Hilbert prenorms in X such that the canonical imbedding $\psi_{pq}\colon \bar{X}_q \to \bar{X}_p$ satisfies $\sigma_2(\psi_{pq}) < \infty$. Then there is for each $\epsilon > 0$ a $\delta > 0$ and an $r > 0$ such that for every positive normalized cylindrical measure μ on X' satisfying the condition

$$|1 - \Re\chi_\mu(x)| < \delta \quad (p(x) \leq 1),$$

the condition

$$\mu(B) < \epsilon \quad (B \in \mathcal{U}_X;\quad B \cap S^0(q,r) = \emptyset)$$

holds.

1.2. Necessary conditions for σ-additivity. We now state a result which is to some extent opposite to theorem 1.1.

THEOREM 1.2. *Let X be a locally convex space, and μ a realvalued measure on the Borel σ-algebra of the space X'_σ. If μ is strongly X-regular, then its characteristic functional χ_μ is continuous in the nuclear topology τ_S corresponding to the Mackey topology $\tau(X,X')$.*

In particular, if the space (X,β) is barrelled, then χ_μ is continuous in the nuclear topology $\tau_S(X,\beta)$ corresponding to β (and hence in β also).

PROOF. To prove the theorem it suffices to determine for each $\epsilon > 0$ a $\tau(X,X')$-continuous Hilbert prenorm $p_\epsilon(x)$ and a Hilbert–Schmidt operator A_ϵ in $\bar{X}_{p_\epsilon}$ satisfying the estimate

$$|\chi_\mu(x_1) - \chi_\mu(x_2)| < \epsilon \quad \text{if } p_\epsilon(A_\epsilon\bar{Q}_{p_\epsilon}(x_1 - x_2)) < 1.$$

Without loss of generality (passing to the variation, and normalizing) we may assume that μ is a probability measure. It is readily seen that continuity at zero of the characteristic functional $\chi_{|\mu|}$ of the variation $|\mu|$ implies continuity of χ_μ in the whole space:

$$\left|\int_{X'} e^{i(x_1,y)}\mu(dy) - \int_{X'} e^{i(x_2,y)}\mu(dy)\right| \leq \int_{X'} |e^{i(x_1-x_2,y)} - 1||\mu|(dy) =$$
$$\sqrt{2}\int_{X'} |1 - \cos(y, x_1 - x_2)||\mu|(dy) = \sqrt{2}(1 - \Re\chi_\mu(x_1 - x_2)).$$

Let K be a subset as in the strong X-regularity condition and such that $\mu(X'\backslash K) < \epsilon/4$, and let K^0 be its polar in X, which is, as is K, a Hilbert subset. Let p_K, p_{K^0} be Hilbert prenorms in X_K, X_{K^0}, respectively, corresponding to these subsets. The definition of polar implies that

$$p(x) = p_{K^0}(x) = \sup_{y \in K} |(x,y)|. \tag{1.7}$$

Since the Mackey topology is determined by uniform convergence on a kind of subsets including K, the prenorm p is $\tau(X,X')$-continuous. Hence K^0 is a neighborhood of zero in $(X,\tau(X,X'))$. We estimate:

$$\begin{aligned} 1 - \Re\chi_\mu(x) &= \int_{X'} [1 - \cos(x,y)]\mu(dy) \leq \\ &\frac{\epsilon}{2} + \int_K [1 - \cos(x,y)]\mu(dy) \leq \\ &\frac{\epsilon}{2} + \frac{1}{2}\int_K |\langle x,y\rangle|^2 \mu(dy). \end{aligned} \tag{1.8}$$

Since $(x,y) = 0$ $(x \in N_p,\ y \in K)$, the integral at the righthand side may be regarded as a quadratic functional on $\bar{X}_p$:

$$\phi(\tilde{x}) = \int_K |(x,y)|^2 \mu(dy) \quad (\tilde{x} = x + N_p).$$

Let A be a nonnegative operator on $\bar{X}_p$ such that $\phi(\tilde{x}) = p^2(A\tilde{x})$, and let $A_\epsilon = A/\sqrt{\epsilon}$. The inequality $p(A_\delta Q_p x) < 1$ implies that $\phi(Q_p x) < \epsilon$, and by (1.8),

$$|1 - \Re\chi_\mu(x)| < \epsilon.$$

It remains to prove that $\sigma_2(A) < \infty$. By the compactness of K, $\bar{X}_p$ is a separable space, and if $\{e_j\}$ is an orthonormal basis in it, then

$$\begin{aligned} \sum_j p^2(Ae_j) &= \sum_j \phi(e_j) = \sum_j \int_K |(e_j,y)|^2 \mu(dy) = \\ \int_K \sum_j |(e_j,y)|^2 \mu(dy) &= \int_K |p_K(y)|^2 \mu(dy) \leq 1, \end{aligned}$$

because of the above mentioned identification and the inequality $p_K(y) \leq 1$ $(y \in K)$. □

COROLLARY 1.2. *Let μ be a strongly X-regular cylindrical measure on the algebra $\mathcal{U}_X$ of cylinder subsets of X'. By theorem I.4.3 it is σ-additive and admits an extension satisfying the conditions of theorem 1.2. Therefore its characteristic functional χ_μ is continuous in the topology $\tau_S(X,\tau(X,X'))$.*

An important result follows from theorems 1.1 and 1.2.

THEOREM 1.3. *(Minlos.) Let X be a nuclear locally convex space. For a function $\chi(x)$ to be the characteristic functional of a nonnegative Radon measure in X'_σ it is sufficient, and necessary if X is barrelled, that it be positive definite and continuous at zero in the topology of X.*

REMARK 1.2. Let $\mathcal{M}$ be a bounded collection of realvalued measures, uniformly strongly X-regular in the sense that the set K_ϵ in the definition of strong X-regularity does not depend on $\mu \in \mathcal{M}$. It is easily seen that all arguments in the proof of theorem 1.2, except the proof that $\sigma_2(A) < \infty$, may be carried out regardless wether or not

$$\phi(\tilde{x}) = \sup_{\mu \in \mathcal{M}} \int_K |(x,y)|^2 \mu(dy).$$

Under these conditions the collection $\tilde{\mathcal{M}}$ of characteristic functionals of measures from $\mathcal{M}$ appears to be equicontinuous in the topology $\tau(X, X')$, but not in the nuclear topology.

REMARK 1.3. If in theorem 1.2 the strong X-regularity condition is weakened to X-regularity, then it follows from the proof of this theorem that

$$|1 - \Re\chi_\mu(x)| \leq \epsilon \quad \text{if } \phi(x) < \epsilon.$$

The prenorm ϕ is, in general, not Hilbert. Also, the subsequent result still holds (but is weaker than in theorem 1.2).

THEOREM 1.2'. *If μ is a regular σ-additive measure in the space X' dual to a barrelled space X, then its characteristic functional χ_μ is continuous in the topology of X. If $\mathcal{M}$ is a uniformly regular set of such measures, then the set $\tilde{\mathcal{M}}$ of their characteristic functionals is equicontinuous.*

1.3. The Hilbert space case.

EXAMPLE 1.1. Let (X, τ) be a Hilbert space with the strong topology, so that X' can be identified with X. The nuclear (Sazonov) topology $\tau_S(X, \tau)$ is determined by the collection of prenorms $\|x\|_S = \|Sx\|$, with S a Hilbert–Schmidt operator. Since every continuous prenorm is Hilbert, the strong X-regularity condition reduces to X-regularity, and it was shown in chapter I that every σ-additive measure on $\mathcal{B}$ is X-regular. So, theorems 1.1 and 1.2 combine to the following assertion.

THEOREM 1.3'. *(Sazonov.) A nonnegative normalized cylindrical measure on the Borel σ-algebra of a Hilbert space X is σ-additive if and only if for each $\epsilon > 0$ there is a positive Hilbert–Schmidt operator S_ϵ such that*

$$\|S_\epsilon x\| < 1 \Rightarrow |1 - \Re\chi_\mu(x)| < \epsilon.$$

Note that, in particular, this condition is satisfied if the characteristic functional is continuous at zero in the topology determined by the single norm $\|\cdot\|_S$; here $S_\epsilon = S/\delta(\epsilon)$, where $\delta(\epsilon) \to 0$ if $\epsilon \to 0$.

In particular, consider a Gaussian cylindrical measure μ in X with characteristic functional $\chi_B(x) = e^{-(Bx,x)/2}$, where $B \geq 0$. From the considerations above the following result, obtained independently in chapter II, follows immediately: a sufficient and necessary condition for σ-additivity of a Gaussian measure is that its correlation operator be nuclear.

We note that the function (Bx, x) is weakly continuous in X only in the case of a finite-dimensional operator B. Thus, continuity of the characteristic functional of a measure in the weak topology does not follow from σ-additivity of the measure alone.

This show how essential the condition of being barrelled is in theorem 1.3: a Hilbert space with the weak topology does not have this property.

We also note that in theorem 1.1 the condition that χ_μ be continuous in the nuclear topology cannot be replaced by weak sequential continuity. In fact, for a completely continuous operator B the functional $\chi_B(x)$ is not weakly sequentially continuous, but if B is not nuclear, then μ_B is not σ-additive.

EXAMPLE 1.2. Consider a pair of densely imbedded Hilbert spaces $X \subset X_-$, and let the imbedding operator be $J\colon X \to X_-$, $\sigma_2^2(J) = \sum_k \|J(e_k)\|_-^2 < \infty$ ($\{e_k\}$ is an orthonormal basis in X).

Let μ be a nonnegative cylindrical measure in X with continuous characteristic functional $\chi_\mu(x)$ on X. The operator J transforms it to a cylindrical measure $\tilde{\mu}$ on the algebra $\tilde{\mathcal{U}}_X$ of X-cylinder subspaces whose pre-images belong to $\mathcal{U}_X$. Here, $\chi_{\tilde{\mu}}(x) = \chi_\mu(J^*x)$ $(x \in X)$. Since $\sigma_2(J^*) < \infty$ and χ_μ is continuous on X, $\chi_{\tilde{\mu}}$ is continuous on X_- in the nuclear topology, and hence the cylindrical measure $\tilde{\mu}$ extending μ to X_- is σ-additive.

EXAMPLE 1.3. Let the characteristic functional figuring in theorem 1.1 be continuous at zero with respect to a fixed prenorm $p(x) = q(SQ_qx)$, where $\sigma_2(S) < \infty$ in $\bar{X}_q$. Using the formula

$$\chi(x + N_p) = \chi_\mu(x)$$

and subsequent closure, the characteristic functional is seen to be continuous on $\bar{X}_p$. The Hilbert spaces $\bar{X}_q \subset \bar{X}_p$ are densely imbedded, and hence their duals $\bar{X}'_p \subset \bar{X}'_q \subset X'$, being subspaces of X' also have this property. Identifying $\bar{X}'_p$ with $\bar{X}_p$, we are in the situation of the previous example, which implies that the cylindrical measure $\tilde{\mu}$ is supported and σ-additive in $\bar{X}'_q$

In particular, for a countably-Hilbert space $X = \lim_{n\to\infty} X_n$ with Hilbert-Schmidt imbeddings, a cylindrical measure whose characteristic functional is continuous on X_n has a σ-additive extension to $X_{-(n+1)}$.

1.4. Integral representations of the group of unitary operators. We now apply the results obtained to some aspects of the spectral theory of operators. Let X be a locally convex space and $\mathcal{H}$ a Hilbert space. Consider the representation $x \to T_x \in \mathcal{L}(\mathcal{H})$ of X by unitary operators on $\mathcal{H}$:

$$T_{x+y} = T_x \cdot T_y; \quad T_0 = I; \quad T_x^* = T_x^{-1} = T_{-x} \quad (x, y \in X).$$

For each $h \in \mathcal{H}$ the function $\phi_h(x) = (T_xh, h)$ is positive definite:

$$\sum_{j,k} \phi_h(x_j - x_k)c_j\bar{c}_k = \sum_{j,k} c_j\bar{c}_k(T_{x_j}T_{-x_k}h, h) = \|\sum_j c_jT_{x_j}h\|^2 \geq 0.$$

If (T_xh, h) is continuous with respect to x on every finite-dimensional subspace $\mathcal{L} \subset X$, then by (I.3.2) the following representation holds:

$$\phi_h(x) = \int_{X'} e^{i(x,y)}\mu_h(dy), \tag{1.9}$$

where μ_h is the nonnegative X-cylindrical measure in X uniquely determined by h.

From (1.9) we find

$$(T_x h_1, h_2) = \int_{X'} e^{i(x,\theta)} \mu_{h_1,h_2}(d\theta),$$

where

$$\mu_{h_1,h_2} = \frac{1}{4}(\mu_{h_1+h_2} - \mu_{h_1-h_2} + i\mu_{h_1+ih_2} - i\mu_{h_1-ih_2})$$

is a complexvalued cylindrical measure which depends bilinearly on h_1, h_2.

Under these conditions we have for each X-cylinder set A,

$$\mu_{h_1,h_2}(A) = (E(A)h_1, h_2),$$

where $E(A) \subset \mathcal{L}(\mathcal{H})$ is a cylindrical measure with operator values. Formula (1.9) is equivalent to the representation

$$T_x = \int_{X'} e^{i(x,\theta)} E(d\theta).$$

Since T_x is unitary, it follows that

$$E^*(A) = E(A) \quad (A \in \mathcal{U}_X),$$

while the group relation gives

$$\int_{X'} e^{i(x_1,y)} E(dy) \int_{X'} e^{i(x_2,y)} E(dy) = \int_{X'} e^{i(x_1+x_2,y)} E(dy),$$

which means that

$$E(A_1 \cap A_2) = E(A_1)E(A_2); \quad E(\emptyset) = 0.$$

(This can be readily checked.) Thus, the operators $E(A)$ are orthoprojectors on $\mathcal{H}$ mapping disjoint cylinder sets to disjoint projections.

The σ-additivity of the measure

$$\mu_h(A) = (E(A)h, h) = \|E(A)h\|^2$$

for every $h \in \mathcal{H}$ implies weak (equivalently, strong) σ-additivity of the projector-valued measure $E(A)$.

Conditions for σ-additivity of cylindrical measures can be obtained from theorem 1.4, taking into account that the corresponding characteristic functionals are given by

$$\chi_h(x) = \chi_{\mu_h}(x) = \int_{X'} e^{i(x,y)} \mu_h(dy) = (T_x h, h).$$

Thus, the following result holds.

THEOREM 1.4. *Let (X, β) be a locally convex space with a topology that is compatible with the duality between X and X'. Let T_x be a unitary representation of X is a Hilbert space $\mathcal{H}$ such that for all $h \in \mathcal{H}$ the function $x \to (T_x h, h)$ is continuous at zero in the topology $\tau_S(X, \beta)$ (which contains the X-cylinder sets).*

Then there is a σ-additive (in the strong meaning) orthogonal measure $E(A)$, with as values projectors on $\mathcal{H}$ and concentrated on the σ-algebra $\mathcal{U}_X$ of X, such that

$$T_x h = \int_{X'} e^{i(x,y)} E(dy)h. \tag{1.10}$$

REMARK 1.4. Let $\mathcal{H}_- \subset \mathcal{H} \subset \mathcal{H}_+$ be a rigged Hilbert space with Hilbert–Schmidt imbeddings: $J \in \mathcal{L}_2(\mathcal{H}_+, \mathcal{H})$, $J^* \in \mathcal{L}_2(\mathcal{H}, \mathcal{H}_-)$. Let $S\colon \mathcal{H}_+ \to \mathcal{H}_-$ be an isometric operator such that $(x, y)_+ = (x, Sy)_0$.

Let $\{e_k\}_{k=1}^{\infty}$ be an orthonormal basis in $\mathcal{H}_+$. We define a nonnegative measure ν in X' by

$$\nu(\Delta) = \sum_{k=1}^{\infty}(E(\Delta)Je_k, Je_k)_0 = \sum_{k=1}^{\infty} \|E(\Delta)Je_k\|_0^2 \le \sum_{k=1}^{\infty} \|Je_k\|_0^2 = \sigma_2^2(J).$$

For any $f \in \mathcal{H}_+$ the following expansion holds:

$$J^*E(\Delta)Jf = \sum_k (J^*E(\Delta)Jf, e_k)_0 Se_k = \sum_k (Jf, E(\Delta)Je_k)_0 Se_k,$$

which implies

$$\|J^*E(\Delta)Jf\|_-^2 = \sum_k |(Jf, E(\Delta)Je_k)_0|^2 \le \|Jf\|_0^2 \nu(\Delta) \le \|f\|_+^2 \nu(\Delta).$$

Hence it follows that the operator-valued measure $J^*E(\Delta)J \in \mathcal{L}(\mathcal{H}_+, \mathcal{H}_-)$ is absolutely continuous with respect to ν. Hence its Radon–Nikodým derivative $P(\theta) \in \mathcal{L}(\mathcal{H}_+, \mathcal{H}_-)$ exists (almost everywhere with respect to $\theta \in X$), so that

$$J^*E(\Delta)J = \int_\Delta P(y)\nu(dy). \tag{1.11}$$

The representation (1.10) can now be written as

$$(T_xJf, Jg)_0 = \int_{X'} e^{i(x,y)}(J^*E(dy)Jf, g)_0 = \int_{X'} e^{i(x,y)}(P(y)f, g)_0\nu(dy) \quad (f, g \in \mathcal{H}_+). \tag{1.12}$$

In particular, for $x = 0$:

$$(Jf, Jg)_0 = \int_{X'} (P(y)f, g)_0\nu(dy). \tag{1.13}$$

If the group T_x leaves $\mathcal{H}_+$ invariant, i.e. $JT_x^+ f = T_xJf$, where $T_x^+ = T_x|_{\mathcal{H}_+}$, then (1.13) implies

$$(JT_x^+ f, Jg)_0 = \int_{X'} (P(y)T_x^+ f, g)_0\nu(dy),$$

which leads, after comparison with (1.12), to

$$(P(y)T_x^+ f, g)_0 = e^{i(x,y)}(P(y)f, g)_0 \quad \nu\text{-almost everywhere.} \tag{1.14}$$

In the same manner the group invariance of the bilinear functional $(P(y)f, g)$ can be proved:

$$(P(y)T_x^+ f, T_x^+ g)_0 = (P(y)f, g)_0 \quad \nu\text{-almost everywhere.} \tag{1.15}$$

1.5. Continuous cylindrical measures. We consider other conditions for σ-additivity of nonnegative cylindrical measures in the space X' dual to a locally convex space X, not based on properties of the Fourier transformation.

Let μ be a normalized cylindrical measure: $\mu(X) = 1$. Consider the collection of its one-dimensional projections:

$$\mu^{(x)}(\Delta) = \mu\{y \in X' : \langle x, y\rangle \in \Delta\} \quad (\Delta \subset \mathbf{R}, \quad x \in X).$$

This collection is called *continuous* if it is continuous at all points $x \in X$ in the sense of weak convergence of measures in $\mathbf{R}$, i.e. if the function

$$\mu\{f : x\} = \int_{-\infty}^{\infty} f(t)\mu^{(x)}(dt) \tag{1.16}$$

is continuous on X for any function $f(t)$ which is continuous and bounded on $\mathbf{R}$.

PROPOSITION 1.4. *The collection of one-dimensional projections of a nonnegative cylindrical measure μ is continuous if $\mu\{f : x\}$ is continuous on X for all f in the class C_0 of continuous functions on $\mathbf{R}$ of compact support.*

PROOF. We first establish an important auxiliary fact. For any $x_0 \in X$, $\epsilon > 0$ there are an $r > 0$ and a neighborhood of zero U_{ϵ,x_0} in X such that

$$\mu^{(x)}\{t : |t| > r\} < \epsilon \quad (x - x_0 \in U_{\epsilon,x_0}). \tag{1.17}$$

Choose first $r_1 > 0$ from the condition $\mu^{(x_0)}\{t : |t| > r_1\} < \epsilon/2$, and let for some $r > r_1$ $\psi(t)$ be a continuous function:

$$\begin{gathered}\psi(t) = 1 \quad (|t| \le r_1);\\ \psi(t) \in [0,1] \quad (r_1 < |t| < r);\\ \psi(t) = 0 \quad (|t| \ge r).\end{gathered}$$

By the requirement, there is a neighborhood of zero $U \subset X$ such that

$$\left|\int_{-\infty}^{\infty} \psi(t)\mu^{(x)}(dt) - \int_{-\infty}^{\infty} \psi(t)\mu^{(x_0)}(dt)\right| < \frac{\epsilon}{2} \quad (x - x_0 \in U).$$

Then also

$$\begin{gathered}\mu^{(x)}\{t : |t| > r\} = 1 - \mu^{(x)}\{t : |t| \le r\} \le\\ 1 - \int_{-\infty}^{\infty} \psi(t)\mu^{(x)}(dt) \le\\ 1 - \int_{-\infty}^{\infty} \psi(t)\mu^{(x_0)}(dt) + \left|\int_{-\infty}^{\infty} \psi(t)\mu^{(x_0)}(dt) - \int_{-\infty}^{\infty} \psi(t)\mu^{(x)}(dt)\right| \le\\ \left[1 - \mu^{(x_0)}\{t : |t| \le r\}\right] + \frac{\epsilon}{2} < \epsilon \quad (x - x_0 \in U).\end{gathered}$$

Let now $f \in C^1(\mathbf{R})$, and let $\phi \in C_0(\mathbf{R})$ coincide with f on the interval $[-r, r]$ and not exceed it in norm. Then for $x - x_0 \in U$ we have

$$\left|\int_{-\infty}^{\infty} f(t)\mu^{(x)}(dt) - \int_{-\infty}^{\infty} f(t)\mu^{(x_0)}(dt)\right| \leq$$

$$\left|\int_{-\infty}^{\infty} \phi(t)\mu^{(x)}(dt) - \int_{-\infty}^{\infty} \phi(t)\mu^{(x_0)}(dt)\right| + \int_{|t|>r} \{|f(t)| + |\phi(t)|\}[\mu^{(x)}(dt) + \mu^{(x_0)}(dt)] \leq$$

$$4\epsilon\|f\| + \left|\int_{-\infty}^{\infty} \phi(t)\mu^{(x)}(dt) - \int_{-\infty}^{\infty} \phi(t)\mu^{(x_0)}(dt)\right|.$$

Choose, for a given $\delta > 0$ a neighborhood of zero V such that

$$\left|\int_{-\infty}^{\infty} \phi(t)\mu^{(x)}(dt) - \int_{-\infty}^{\infty} \phi(t)\mu^{(x_0)}(dt)\right| < \frac{\delta}{2} \quad (x - x_0 \in V).$$

Then for $\epsilon = \delta/(8\|f\|)$ the required estimate holds:

$$\left|\int_{-\infty}^{\infty} f(t)\mu^{(x)}(dt) - \int_{-\infty}^{\infty} f(t)\mu^{(x_0)}(dt)\right| < \delta \quad (x - x_0 \in U \cap V). \quad \square$$

REMARK 1.5. Since for $x_0 = 0$ the measure $\mu^{(0)}$ is concentrated at zero, it follows from the proof that for $x_0 = 0$ (1.17) is valid for any $r > 0$. Thus, the following assertion holds:

If the collection of one-dimensional projections of a nonnegative cylindrical measure μ is continuous, then, for any $\epsilon > 0$, $r > 0$, there is a neighborhood of zero $U_{\epsilon,r} \subset X$ such that

$$\mu\{y \in X' \; |\langle x, y\rangle| > r\} < \epsilon \quad (x \in U_{\epsilon,r}). \tag{1.18}$$

DEFINITION 1.2. A cylindrical measure μ is called *continuous* if for any $\epsilon > 0$, $r > 0$ (equivalently, for any $\epsilon > 0$ and $r = 1$) there is a neighborhood of zero $U_{\epsilon,r} \subset X$ for which property (1.18) holds.

PROPOSITION 1.5. *Continuity of a nonnegative cylindrical measure μ is equivalent to continuity of its collection of one-dimensional projections.*

PROOF. By the remark above it suffices to derive the continuity of the collection $\{\mu^{(x)}\}$ from that of μ. Let $\phi \in C_0(\mathbf{R})$. Then

$$\left|\int_{-\infty}^{\infty} \phi(t)\mu^{(x)}(dt) - \int_{-\infty}^{\infty} \phi(t)\mu^{(x_0)}(dt)\right| =$$

$$\left|\int_{X'} \{\phi(\langle x, y\rangle) - \phi(\langle x_0, y_0\rangle)\}\mu(dy)\right| \leq$$

$$\left|\int_{|\langle x - x_0, y\rangle| \geq \delta} \{\phi(\langle x_0, y\rangle + \langle x - x_0, y\rangle) - \phi(\langle x_0, y\rangle)\}\mu(dy)\right| +$$

$$\left|\int_{|\langle x - x_0, y\rangle| < \delta} \{\phi(\langle x, y\rangle) - \phi(\langle x_0, y_0\rangle)\}\mu(dy)\right|. \tag{1.19}$$

Using the uniform continuity, choose $\delta > 0$ such that

$$\max|\phi(t+\tau) - \phi(t)| < \frac{\epsilon}{2\mu(X')} \quad (|\tau| < \delta).$$

Then the second integral in (1.19) is less than $\epsilon/2$. The first one does not exceed $2\|\phi\|\mu\{x\colon\ |\langle x - x_0, y\rangle| > \delta\}$, and can be made less than $\epsilon/2$ since the measure is continuous and by choosing a sufficiently small neighborhood of zero V such that $x - x_0 \in V$. $\square$

We now establish a connection between continuity of a cylindrical measure and properties of its Fourier transform.

THEOREM 1.5. *Continuity of a nonnegative cylindrical measure is equivalent to continuity of its characteristic functional.*

PROOF. The formula

$$\chi_\mu(\tau x) = \int_{X'} e^{i\tau\langle x,y\rangle}\mu(dy) = \int_{-\infty}^{\infty} e^{i\tau t}\mu^{(x)}(dt)$$

for $t = 1$ readily implies that continuity of $\mu^{(x)}$ implies continuity of χ_μ, since $e^{it} \in C(\mathbf{R})$.

The same formula gives a connection between the characteristic functional of μ and those of the one-dimensional projections: $\chi_{\mu^{(x)}}(\tau) = \chi_\mu(\tau x)$.

Let U be an absolutely convex neighborhood of zero in X for which

$$|1 - \Re\chi_\mu(x)| < \epsilon \quad (x \in U)$$

for some $\epsilon > 0$ (the existence of such a neighborhood follows from the continuity of χ_μ). Since $tx \in U$ for $x \in U$, $|t| \leq 1$, we have

$$|1 - \Re\chi_{\mu^{(x)}}(t)| < \epsilon \quad (|t| < 1, \quad x \in U).$$

Applying the one-dimensional version of proposition 1.1 we obtain

$$\mu\{y \in X' :\ |\langle x, y\rangle| > c\} = \mu^{(x)}\{t :\ |t| > c\} \leq \frac{\sqrt{e}}{\sqrt{e}-1}\left(\epsilon + \frac{2}{c}\right).$$

For sufficiently large c this leads to the result required (it suffices to verify (1.18) for some $r > 0$). $\square$

The next theorem is a result of combining theorems 1.3 and 1.5.

THEOREM 1.6. *Let X be a locally convex space and μ a nonnegative cylindrical measure in X'. A sufficient condition for (regular) σ-additivity of μ is continuity of μ in the nuclear topology (if X is nuclear, in the initial topology) of X. If X is barrelled, this condition is also necessary.*

The continuity condition (1.18) may be reformulated so that it becomes a sufficient condition for σ-additivity of a realvalued cylindrical measure.

THEOREM 1.7. *Let μ be a realvalued cylindrical measure in X' satisfying the condition:*

for each $\epsilon > 0$ and some (hence, any) $r > 0$ there is a neighborhood of zero U in $(X, \tau_S(X))$ such that for any $A \in \mathcal{U}_X$, $x \in U$, the following estimate holds:

$$|\mu(A \cap \{y \in X' : |\langle x, y\rangle| > r\})| < \epsilon. \tag{1.20}$$

Then μ is regularly σ-additive.

If X is barrelled, this condition is also necessary.

PROOF. Formula (1.20) immediately implies that

$$|\mu|\{y \in X' : |\langle x, y\rangle| > r\} < 2\epsilon,$$

implying the continuity of the nonnegative cylindrical measure $|\mu|$ in the corresponding topology. It remains to apply theorem 1.6. Necessity is proved in the same manner. □

REMARK 1.6. The condition in this theorem can be easily restated so as to apply to measures with values in a locally convex space T. For this, $\epsilon > 0$ (in the definition of regularity and in (1.20)) must be replaced by a neighborhood of zero $V \subset T$.

The corresponding result is proved by passing from μ to the realvalued measures $\langle \mu, \phi\rangle$ ($\phi \in T'$), and uses the equivalence of weak and strong σ-additivity of a vector measure.

A useful sufficient condition for σ-additivity of a cylindrical measure is related to the notion of correlation form of a measure.

DEFINITION 1.3. Let μ be a cylindrical measure with finite second moments,

$$\int_{X'} |\langle y, x\rangle|\mu(dy) = \int_{-\infty}^{\infty} z^2 \mu^{(x)}(dz) < \infty \quad (x \in X). \tag{1.21}$$

The bilinear functional

$$B_\mu(x_1, x_2) = \int_{X'} \langle x_1, y\rangle \langle x_2, y\rangle \mu(dy) = \int_{\mathbf{R}^2} z_1 z_2 \mu^{\langle x_1, x_2\rangle}(dz_1 \times dz_2)$$

is called the *correlation form* of μ.

The correlation form can be easily expressed in terms of the characteristic functional by means of twice differentiation:

$$B_\mu(x, x) = \int_{X'} |(x, y)|^2 \mu(dy) = -\chi''(0)(x, x).$$

Here only the values of χ on the one-dimensional space $\{tx\}$ are used, i.e. the characteristic functional of the corresponding one-dimensional distribution.

On the other hand, if the correlation form exists, an estimate for the characteristic functional easily follows:

$$1 - \Re\chi_\mu(x) = \int_{X'} \{1 - \cos(x, y)\}\mu(dy) \le \frac{1}{2} B_\mu(x, x).$$

This estimate leads by theorem 1.1 and example 1.3, to

THEOREM 1.8. *If the correlation form $B_\mu(x, y)$ of a cylindrical measure μ exists in X' and is continuous in the nuclear topology $\tau_S(X)$, then μ is σ-additive in X'.*

If, moreover, $B_\mu(x, y)$ is continuous with respect to a fixed Hilbert prenorm q and $B_\mu(x, x) = q(Sx)$, where $S \in \mathcal{L}_2(\bar{X}_q)$, then μ is supported in the Hilbert space X'_q.

REMARK 1.7. Let μ be a nonnegative measure in X' satisfying condition (1.21). Consider the Hilbert space $\mathcal{L}_2(X', \mu)$. The map $S\colon x \to \langle x, y\rangle$ imbeds X into $\mathcal{L}_2(X', \mu)$. The closure of the image SX is the Hilbert space $\mathcal{H}$ of measurable linear functionals on X that are square integrable with respect to μ. It is isometric to the completion of X with respect to the norm $B_\mu(x, x)$. The imbedding $S\colon X \to \mathcal{H}$ results in a dual imbedding $\mathcal{H}' \to X'$. Canonically identifying $\mathcal{H}'$ and $\mathcal{H}$, we obtain the rigged Hilbert space

$$X \subset \mathcal{H} \subset X'.$$

The measure μ induces the cylindrical measure in $\mathcal{H}$ whose correlation form coincides with the scalar product. In this case there is a Hilbert extension $\mathcal{H}_1$ of $\mathcal{H}$ in which this cylindrical measure is σ-additive. Note, however, that $\mathcal{H}_1$ is not necessarily contained in X'. All this is applicable to an arbitrary Gaussian cylindrical measure, which has thus support in a Hilbert space.

2. Sequences of Radon measures

2.1. Weak compactness in a space of measures. In this section we do not consider individual measures that can be extended to Radon measures, but collections of them.

First we clarify some facts related to measures in an arbitrary (not necessarily linear) completely regular topological space X.

Let $C(X)$ be the Banach space of bounded realvalued continuous functions on X, with norm $\|f\|_C = \sup_X |f(x)|$; let $\mathfrak{M}(X)$ be the Banach space of realvalued Radon measures on X, with norm $\|\mu\| = |\mu|(X)$. These spaces are in duality:

$$\langle f, \mu\rangle = \int_X f(x)\mu(dx).$$

To a measure $\mu \in \mathfrak{M}(X)$ we associate a continuous linear functional F_μ on $C(X)$ in such a way that $\|F_\mu\| = \|\mu\|$. In this way, $\mathfrak{M}(X)$ is linearly and isometrically imbedded in the dual space $C'(X)$.

PROPOSITION 2.1. *Let X be a completely regular topological space, $\Phi \in C'(X)$, and suppose that for each $\epsilon > 0$ there is a compact set $K_\epsilon \subset X$ such that*

$$|\Phi(f)| \le \epsilon\|f\| \tag{2.1}$$

for all $f \in C(X)$ satisfying $f(x) = 0$ $(x \in K_\epsilon)$.

Then there is exactly one Radon measure μ for which

$$\Phi(f) = \int_X f(x)\mu(dx) \quad (f \in C(X)). \tag{2.2}$$

PROOF. For each element $\Phi \in C'(X)$ there is a bounded additive set function μ on the algebra Σ generated by the closed sets for which (2.2) is true. This measure is *regular* in the following sense: for each $E \in \Sigma$ and $\delta > 0$ there are $F, G \in \Sigma$ such that $\bar{F} \subset E \subset \operatorname{int} G$ and $|\mu|(G \setminus F) < \delta$. Here f may be taken compact.

For normal spaces, a proof of a similar result is given in Dunford–Schwartz [1]. We do not reproduce this proof here, but we note that it remains valid if the normality requirement is replaced by the following, which holds in completely regular spaces: if

F_1, F_2 are disjoint closed sets and at least one of them is compact, then there is a continuous function f such that $0 \le f(x) \le 1$, $f(x) = 0$ $(x \in F_1)$, $f(x) = 1$ $(x \in F_2)$. Here the set F in the definition of regularity given above must be chosen compact.

We prove that the measure μ under consideration is a Radon measure. Let $F \subset G \subset X \setminus K_\epsilon$, with F compact, G open, and $|\mu|(G \setminus F) < \epsilon$. Consider a function $f \in C(X)$ with the properties $0 \le f(x) \le 1$, $f(x) = 1$ $(x \in F)$, $f(x) = 0$ $(x \notin G)$. Then

$$|\mu(F)| = \left|\int_F \mu(dx)\right| = \left|\int_X f(x)\mu(dx) - \int_{G\setminus F} f(x)\mu(dx)\right| \le \Phi(f) + \|f\|\cdot|\mu|(G\setminus F) \le 2\epsilon.$$

By regularity, a similar estimate holds for other sets in Σ and disjoint with K_ϵ. This implies that μ is a Radon measure.

It remains to prove uniqueness. Let μ be a Radon measure for which $\int_X f(x)\mu(dx) = 0$ (for all $f \in C(X)$), but $\mu(B) = \delta \ne 0$ for some Borel set B. There are $K \subset B \subset U$, with K compact, U open, and $|\mu|(U \setminus K) < \delta/3$. Consider the function $f \in C(X)$: $f(x) \in [0,1]$, $f(x) = 1$ $(x \in K)$, $f(x) = 0$ $(x \notin U)$. This implies

$$\left|\int_X f(x)\mu(dx)\right| = \left|\int_U f(x)\mu(dx)\right| \ge \left|\int_K f(x)\mu(dx)\right| - |\mu|(U \setminus K) \ge \frac{2}{3}\delta - \frac{1}{3}\delta > 0,$$

a contradiction. This proves that $\mu = 0$. □

DEFINITION 2.1. A set $\mathcal{M} \subset \mathfrak{M}(X)$ is called *uniformly regular* if for each $\epsilon > 0$ there is a compact set $K_\epsilon \subset X$ such that $|\mu|(X \setminus K_\epsilon) < \epsilon$ for all $\mu \in \mathcal{M}$.

THEOREM 2.1. *(Prokhorov's theorem). A bounded and uniformly regular set $\mathcal{M} \subset \mathfrak{M}(X)$ is relatively compact in $(\mathfrak{M}(X), \sigma(\mathfrak{M}, C))$.*

PROOF. The set $\mathcal{M}$ is bounded in $\mathfrak{M}(X)$, and thus in $C'(X)$. By Alaoglu's theorem, it is relatively compact in $(C'(X), \sigma(C', C))$. It remains to show that the closure $\bar{\mathcal{M}}$ in $C'(X)$ belongs to $\mathfrak{M}(X)$. For an arbitrary $\epsilon > 0$ we choose a corresponding set K_ϵ. Let $\Phi \in \bar{\mathcal{M}}$, and let $f \in C(X)$ be such that $f(x) = 0$ $(x \in K_\epsilon)$. Since every δ-neighborhood of Φ (determined by f) contains elements μ from $\mathcal{M}$, we have

$$|\Phi(f)| \le |\langle f, \Phi - \mu\rangle| + |\langle f, \mu\rangle| < \delta + \epsilon\|f\|,$$

and hence (2.1) holds. The final result follows from proposition 2.1. □

COROLLARY 2.1. *Suppose X is a space with the property that every weakly fundamental sequence in it is uniformly regular. Then $\mathfrak{M}(X)$ is weakly sequentially complete.*

In fact, a weakly fundamental sequence is bounded in $\mathfrak{M}(X)$, and hence relatively compact in $(\mathfrak{M}, \sigma(\mathfrak{M}, C))$. Moreover, it cannot have more than one limit point, hence is convergent.

REMARK 2.1. Suppose that every compact subset of a space X is metrizable. Then $(C(X), \sigma)$ can be shown to be separable. Indeed, for a sequence of compact (hence metric) sets $K_{1/n}$ we have to choose countable dense sets in the separable spaces $C(K_{1/n})$, and extend their elements to all of X with preservation of the norm. We obtain a countable set which is dense in $(C(X), \sigma)$.

Separability of the dual space and relative compactness of a sequence of measures imply relative sequential compactness of the sequence. This leads to the following strengthening of theorem 2.1.

THEOREM 2.1'. *(Le Cam [1]). If all compact sets in a space X are metrizable, then every bounded, uniformly regular set of Radon measures in X is weakly sequentially compact.*

2.2. Weak completeness of spaces of measures. Now we will exhibit a fairly large class of spaces X for which the space $\mathfrak{M}(X)$ of Radon measures is weakly sequentially complete.

DEFINITION 2.2. A collection of measures $\mathcal{M} \subset \mathfrak{M}(X)$ is called *uniformly small on open sets* if for any sequence U_n $(n = 1, 2, \dots)$ of pairwise disjoint open sets in X the following relation holds:

$$|\mu|(U_n) \to 0 \quad \text{(uniformly with respect to } \mu \in \mathcal{M}).$$

A space X is said to have *property R* (or, to be an *R-space*) if every collection of measures in $\mathfrak{M}(X)$ that is uniformly small on open sets is uniformly regular.

THEOREM 2.2. *Every weak fundamental sequence of Radon measures in a completely regular space X is uniformly small on open sets. Therefore, if X is an R-space, then $\mathfrak{M}(X)$ is weakly sequentially complete.*

PROOF. The latter assertion follows from the first one by corollary 2.1, so we begin by proving it.

Let $\{\mu_n\}$ be a weak fundamental sequence of Radon measures in X. Assume that there is a sequence of open sets $\{U_n\}$ $(U_j \cap U_k = \emptyset$ if $j \neq k)$ for which the condition of uniform smallness of $\{\mu_n\}$ is violated, i.e. there is a subsequence μ_{n_s} such that for some $\delta > 0$,

$$|\mu_{n_s}|(U_s) > \delta \quad (s = 1, 2, \dots).$$

Since μ_{n_s} is a Radon measure, we can find a compact set K_s such that $|\mu_{n_s}|(K_s) > \delta$ as well. Now the statement required can be reduced to the following lemma. □

LEMMA 2.1. *Let X be a completely regular space and U_n and $K_n \subset U_n \subset X$ $(n = 1, 2, \dots)$ be sequences of open, respectively compact, sets, with $U_j \cap U_k = \emptyset$ $(j \neq k)$.*

If $\mathcal{M} = \{\mu_n\}$ is a weak fundamental sequence in $\mathfrak{M}(X)$, then $\lim_{n\to\infty} |\mu_n|(K_n) = 0$ uniformly with respect to $\mu \in \mathcal{M}$.

PROOF. Assuming the contrary, choose for some $\epsilon > 0$ a sequence of measures $\mu_{n_j} \in \mathcal{M}$ and compact sets $\tilde{K}_{n_j} \subset K_{n_j}$ $(n_1 < n_2 < \dots)$ so that the inequality $|\mu_{n_j}(\tilde{K}_{n_j})| > \epsilon$ $(j = 1, 2, \dots)$ holds. For each j we construct an open set V_j such that

$$\tilde{K}_{n_j} \subset V_j \subset U_{n_j}, \quad |\mu_{n_j}(V_j \setminus \tilde{K}_{n_j})| < \frac{\epsilon}{4},$$

and a continuous function ψ_j,

$$\psi_j(x) \in [0,1] \quad (x \in X),$$
$$\psi_j(x) = 1 \quad (x \in \tilde{K}_{n_j}),$$
$$\psi_j(x) = 0 \quad (x \in X \setminus V_j).$$

Also,

$$\left| \int_X \psi_j(x) \mu_{n_j}(dx) \right| > \frac{3}{4}\epsilon. \tag{2.3}$$

Let now $c = \{c_j\} \in \ell_\infty$ be a bounded sequence of real numbers, $\|c\| = \sup |c_j|$. The function

$$\psi^c(x) = \sum_{j=1}^{\infty} c_j \psi_j(x) \tag{2.4}$$

is continuous and bounded: $|\psi^c(x)| \leq \|c\|$. So, the fundamentality of $\mathcal{M}$ implies that the numerical sequence

$$\int_X \psi^c(x) \mu_{n_j}(dx) = \sum_k c_k a_j^k \quad (j \to \infty), \tag{2.5}$$

where $a_j^k = \int_X \psi_k(x) \mu_{n_j}(dx)$, converges. Absolute convergence of the series (2.5) clearly follows from the absolute and uniform convergence of (2.4). In particular, for $c_k = 1$ $(k = 1, 2, \dots)$ we obtain $\sum_k |a_j^k| < \infty$.

Thus, the sequence

$$a_j = (a_j^1, a_j^2, \dots) \quad (j = 1, 2, \dots)$$

of elements of ℓ_1 is fundamental in the weak topology $\sigma(\ell_1, \ell_\infty)$, by (2.5). Convergence of a sequence in this topology is known to coincide with convergence in the norm of ℓ_1, so

$$\lim_{n \to \infty} \sum_{k=n}^{\infty} |a_j^k| = 0 \quad \text{(uniformly with respect to } j = 1, 2, \dots).$$

However, this contradicts the estimate $|a_j^j| > 3\epsilon/4 > 0$ following from (2.3). □

PROPOSITION 2.2. *Let $\mathcal{M}$ be a collection of measures that is uniformly small on the open sets of a completely regular space X, $\mathcal{M} \subset \mathfrak{M}(X)$.*

Then the property of being uniformly small on the open sets of a new space Y is possessed by:

a) *the collection $\mathcal{M}_f = \{|\mu|_f : \mu \in \mathcal{M}\}$ of images of the variations of the measures in $\mathcal{M}$ under a continuous map $f: X \to Y$ into a completely regular space Y;*

b) *the collection $\mathcal{M}_Y$ of restrictions of the measures in $\mathcal{M}$ to a closed subspace $Y \subset X$ equipped with the topology inherited from X.*

PROOF. a) Let $\{U_k\}$ be a sequence of disjoint open sets in Y. Then $f^{-1}(U_k)\}$ is a similar sequence in X, and hence $|\mu|_f(U_k) = |\mu|(f^{-1}(U_k)) \to 0$ (uniformly with respect to $\mu \in \mathcal{M}$).

b) Let $\{U_k\}$ be a sequence of disjoint open sets in Y, and let there be a sequence $\{\mu_n\} \subset \mathcal{M}$ such that $|\mu_n(U_n)| > \epsilon$ for some $\epsilon > 0$. For each n, let K_n be a compact subset of U_n such that $|\mu|(K_n) > 3\epsilon/4$. Since the closures of a subset of Y in X and Y coincide, we have $K_n \cap (\cup_{j=n+1}^{\infty} K_j) = \emptyset$. In this case (see lemma 2.2 below) there are pairwise disjoint open sets $V_n \supset K_n$ in X and, by the properties of Radon measures, open sets V_n': $K_n \subset V_n' \subset V_n$ such that $|\mu(V_n')| > \epsilon/2$. This contradicts the uniform smallness of $\mathcal{M}$ on the open subsets of X. □

2.3. Properties of R-spaces. In this section we consider various operations over R-spaces.

THEOREM 2.3. *A product $X = \prod_{n=1}^{\infty} X_n$ of R-spaces X_n is an R-space.*

PROOF. Let $\mathcal{M}$ be a collection of measures in X that is uniformly small on open sets. By proposition 2.2, for each n the collection $\tilde{\mathcal{M}}_n = \{|\mu|_n\colon \mu \in \mathcal{M}\}$ of projections on X_n of the variations of the measures in $\mathcal{M}$ has the property of being uniformly small on open sets. Therefore, for all n the set $\tilde{\mathcal{M}}_n$ is uniformly regular.

It remains to show that the set $\tilde{\mathcal{M}} = \{|\mu|\colon \mu \in \mathcal{M}\}$ of nonnegative measures in X, for which every projection $\tilde{\mathcal{M}}_n$ is uniformly regular, is itself uniformly regular.

In fact, for any $\epsilon > 0$ and $n = 1, 2, \ldots$ there is a compact set $K_n \subset X_n$ such that $|\mu|_n(X_n \setminus K_n) < \epsilon/2^n$. Consider the compact set $K = \prod_{k=1}^{\infty} K_n$ (in X). It satisfies

$$|\mu|(X \setminus K) \leq \sum_{k=1}^{\infty} |\mu|\{\mathrm{Pr}_n^{-1}(X_n \setminus K_n)\} =$$

$$\sum_{k=1}^{\infty} |\mu|_n(X_n \setminus K_n) < \epsilon. \quad \square$$

THEOREM 2.4. *A closed subspace Y of an R-space X is an R-space.*

PROOF. Let $\mathcal{M}$ be a collection of measures in $\mathfrak{M}(Y)$ that is uniformly small on open sets. By proposition 2.2, the images of these measures under the imbedding of Y in X form a collection $\tilde{\mathcal{M}}$ that is also uniformly small on open sets (of X). So, for any $\epsilon > 0$ there is a compact set $\tilde{K}_\epsilon$ in X for which $|\tilde{\mu}|(X \setminus \tilde{K}_\epsilon) < \epsilon$ $(\tilde{\mu} \in \tilde{\mathcal{M}})$ Also, $K_\epsilon = Y \cap \tilde{K}_\epsilon$ is compact in Y and

$$|\mu|(Y \setminus K_\epsilon) < \epsilon \quad (\mu \subset \mathcal{M}). \quad \square$$

COROLLARY 2.2. *The projective limit of a countable sequence of R-spaces is an R-space.*

In fact, the projective limit of a countable sequence of spaces is homeomorphic to a closed subspace of the product of these spaces.

In the sequel we will need the following auxiliary assertion.

LEMMA 2.2. *Let X be a completely regular space, and C_n $(n = 1, 2, \ldots)$ a sequence of pairwise disjoint subsets of X for which the sets $B_n = \cup_{k=n+1}^{\infty} C_k$ are closed.*

Then there is a sequence $U_n \supset C_n$ $(n = 1, 2, \ldots)$ of pairwise disjoint open subsets of X.

PROOF. By the complete regularity of X and the compactness of C_1 there are open sets $U_1 \supset C_1$, $V_1 \supset B_1$ with disjoint closures. Suppose that the open sets $U_j \supset C_j$ $(j = 1, 2, \dots)$, $V_n \supset B_n$ with pairwise disjoint closures have already been chosen. Then construct open sets $U_{n+1} \supset C_{n+1}$, $V_{n+1} \supset B_{n+1}$ with disjoint closures such that $U_{n+1} \subset V_n$, $V_{n+1} \subset V_n$. □

THEOREM 2.5. *Let X be the topological sum $X = \cup_{n=1}^{\infty} E_n$ of a monotone increasing sequence $E_n \subset E_{n+1}$ of closed subspaces, where for each n the topology of E_n coincides with the topology inherited from E_{n+1} and E_n is closed in E_{n+1}.*

If E_n are R-spaces, then so is X.

PROOF. Let $\mathcal{M}$ be a collection of measures from $\mathfrak{M}(X)$ that is uniformly small on open sets of X. By the conditions of the theorem, for each $\epsilon > 0$ there is a number n such that

$$|\mu|(X \setminus E_n) < \frac{\epsilon}{2} \quad (\forall \mu \in \mathcal{M}). \tag{2.6}$$

On the other hand, since all measures in $\mathcal{M}$ are Radon measures, there would exist a sequence of measures $\mu_n \in \mathcal{M}$ and a sequence of compact sets $K_n \subset X$ such that, for some $\delta > 0$,

$$K_n \cap E_n = \emptyset; \quad \mu_n(K_n) > \delta \quad (n = 1, 2, \dots).$$

If we show that the sets $F_i = \cup_{n=i}^{\infty} K_n$ are closed, then by lemma 2.2 the sets K_n can be included in pairwise disjoint open sets U_n, contradicting the uniform smallness of $\mathcal{M}$.

Let $x \notin F_i$, and let j be the smallest index for which $x \in E_j$. Since E_j intersects only finitely many K_n, there is a neighborhood W_j of x in E_j for which $W_j \cap F_i = \emptyset$. Furthermore, there is a neighborhood W_{j+1} in E_{j+1} for which $W_{j+1} \cap E_j = W_j$; $\bar{W}_{j+1} \cap F_i = \emptyset$; etc. Hence $W = \cup_{k=j}^{\infty}$ is also a neighborhood of x in X, disjoint from F_i.

Thus, (2.6) holds. Consider the set $\mathcal{M}_n = \{\mu_n\}$ of restrictions to E_n of measures in $\mathcal{M}$. By proposition 2.2 it is uniformly small on open sets in E_n. So there is a set $K_{n,\epsilon}$, compact in E_n (and hence in X), for which $|\mu_n|(E_n \setminus K_n) < \epsilon/2$. Consequently, $|\mu_n|(X \setminus K_{n,\epsilon}) < \epsilon$ $(\mu \in \mathcal{M})$. □

THEOREM 2.5'. *The strong inductive limit $X = \lim_{n\to\infty} E_n$ of an increasing sequence of locally convex R-spaces, each being closed in its successor, is an R-space.*

This can be proved similarly to the above.

COROLLARY 2.3. *Since a compact space is an R-space, every σ-compact space (the free union of an increasing sequence of compact subspaces) is an R-space as well.*

2.4. Examples of R-spaces.

EXAMPLE 2.1. Every complete metric space X is an R-space.

Let $\mathcal{M} \subset \mathfrak{M}(X)$ be uniformly small on open sets of X. It suffices to show that for each $\epsilon > 0$ there is a compact set K_ϵ such that

$$|\mu|(X \setminus K_\epsilon^\epsilon) < \epsilon \quad (\mu \in \mathcal{M}), \tag{2.7}$$

where $A^\epsilon = \{x\colon \rho(x, A) \leq \epsilon\}$. In fact, the set $C_m = \cap_{n=m}^\infty K_{1/2^n}^{1/2^n}$ is compact, since it is closed and totally bounded. Also, for each measure in $\mathcal{M}$ we have the following estimate:

$$|\mu|(X \setminus C_m) \leq \sum_{n=m}^\infty |\mu|(X \setminus K_{1/2^n}^{1/2^n}) < \sum_{n=m}^\infty \frac{1}{2^n} = \frac{1}{2^{m-1}}.$$

This proves the uniform regularity of $\mathcal{M}$.

Assuming (2.7) to be violated, we can construct a compact set $K_1 \subset X$ and a measure μ_1 with $|\mu_1(K_1)| > \epsilon$; then a compact set $K_2 \subset X \setminus K_1^\epsilon$ and a measure μ_2 with $|\mu_2(K_2)| > \epsilon$. In general, having constructed K_n and μ_n, we can construct a compact set $K_{n+1} \subset X \setminus \cup_{j=1}^\infty K_j^\epsilon$ and a measure μ_{n+1} with $|\mu_{n+1}(K_{n+1})| > \epsilon$. The open sets $V_n = \operatorname{int} K_n^{\epsilon/2}$ are pairwise disjoint, and $|\mu|_n(V_n) > \epsilon$, contradicting the uniform smallness of $\mathcal{M}$.

EXAMPLE 2.2. Let $X = \cap_n E_n$ be a complete countably-normed space with completely continuous imbeddings. Then its dual X' is σ-compact, and thus an R-space.

In fact, the complete continuity of the imbeddings $E_{n+1} \subset E_n$ implies that X is a Montel space (a barrelled space in which all bounded sets are relatively compact). Therefore it is reflexive, and its strong dual is also a Montel space. The reflexivity of X' implies that each bounded set in it is contained in the polar of a neighborhood of zero in X, and X' being Montel implies that the polar U^0 of a neighborhood of zero in $U \subset X$ is compact in X'. Thus, if $\{U_n\}$ is a fundamental system of neighborhoods of zero in X, then $\{U_n^0\}$ is a fundamental system of compact sets in X.

EXAMPLE 2.3. It follows from theorems 2.3–2.5$'$ and corollaries 2.2–2.3 that the operations of taking the topological product and sum or by passing to a subspace (in particular, taking injective or projective limits) the previous examples of R-spaces yield R-spaces.

In particular, the well-known spaces of test functions and distributions $\mathcal{D}$, $\mathcal{D}'$, $\mathcal{D}_K$, $\mathcal{D}'_K$, $\mathcal{S}$, and $\mathcal{S}'$ (K a compact subset of a euclidean space), used in analysis, are R-spaces.

2.5. Weak compactness of a family of measures in a space X'.

We consider conditions for weak compactness of a family of Radon measures in the space X' dual to a locally convex space X.

LEMMA 2.3. *Let $\mathcal{M}$ be a set of positive normalized X-cylindrical measures in X'. If the set $\tilde{\mathcal{M}}$ of characteristic functionals of these measures is equicontinuous at zero in $(X, \tau_S(X, \tau(X, X')))$, then $\mathcal{M}$ is uniformly regular.*

PROOF. By the definition of equicontinuity, for each $\delta > 0$ there is a neighborhood V in the topology $\tau_S(X, \tau(X, X'))$ such that

$$|1 - \Re\chi_\mu(x)| < \delta \quad (\mu \in \mathcal{M}).$$

This neighborhood is defined by a prenorm $p \in \mathcal{P}(X)$: $V = \{x\colon p(x) < 1\}$. But then, as suggested in remark 1.1, there are a prenorm q, an $r > 0$, and a corresponding compact set $K = S^0(q, r)$ with the properties required: $\mu(B) < \epsilon$ ($B \in \mathcal{U}_X$, $B \cap K = \emptyset$, $\mu \in \mathcal{M}$) if $\delta = \delta(\epsilon)$ is sufficiently small. □

COROLLARY 2.4. *If $\mathcal{M}$ is a set of positive normalized Radon measures in (X', σ) whose set of characteristic functionals $\tilde{\mathcal{M}}$ is equicontinuous at zero in the topology $\tau_S(X, \tau(X, X'))$, then $\mathcal{M}$ is uniformly regular.*

Combined with theorem 2.1, lemma 2.3 leads to

THEOREM 2.6. *Let X be a locally convex space, and $\mathcal{M}$ a set of positive normalized Radon measures in X'_σ. If the set of characteristic functionals $\tilde{\mathcal{M}}$ of these measures is equicontinuous at zero in the topology $\tau_S(X, \tau(X, X'))$, then $\mathcal{M}$ is relatively weakly compact in $\mathfrak{M}(X'_\sigma)$.*

REMARK 2.2. Under the additional assumption of metrizability of compact subspaces of X'_σ, the set $\mathcal{M}$ is relatively sequentially weakly compact.

Now we consider the problem of giving a converse of theorem 2.6. The following assertion is a reformulation of remarks 1.2 and 1.3.

THEOREM 2.7. *Let X be a locally convex space, and $\mathcal{M}$ a bounded, uniformly strongly regular subset in $\mathfrak{M}(X'_\sigma)$. Then the set of characteristic functionals $\tilde{\mathcal{M}}$ of these measures is equicontinuous in the Mackey topology $\tau(X, X')$. If X is barrelled ad $\mathcal{M}$ is only uniformly regular, then $\tilde{\mathcal{M}}$ is equicontinuous in the original topology.*

COROLLARY 2.5. *Let X be a barrelled locally convex space, and suppose that its dual X' has the property R in a certain topology τ' compatible with the duality between X and X'. Then weak sequential precompactness of a set of Radon measures $\mathcal{M} \subset \mathfrak{M}(X', \tau')$ implies that the set of characteristic functionals $\tilde{\mathcal{M}}$ of these measures is equicontinuous at zero.*

In fact, $\mathcal{M}$ is uniformly small on open sets of X', since otherwise there would exist, for some $\epsilon > 0$ and sequence of open sets U_n, a sequence μ_n such that $|\mu_n|(U_n) > \epsilon$. Also, by theorem 2.2, there would be a fundamental subsequence in $\{\mu_n\}$. Since (X', τ') is an R-space, $\mathcal{M}$ is uniformly regular.

If, moreover, compact sets in (X', τ') are metrizable, then by remark 2.1 weak compactness of $\mathcal{M}$ already implies equicontinuity of $\tilde{\mathcal{M}}$.

We use these results to find conditions for weak convergence of a sequence of Radon probability (i.e. nonnegative and normalized) measures.

THEOREM 2.8. *Let X be a locally convex space, and $\{\mu_n\}$ a sequence of Radon probability measures in X'_σ. For weak compactness of this sequence:*

a) *it is sufficient that the sequence* $\{\chi_n(x)\}$ *of characteristic functionals of these measures converges at every* $x \in X$ *and is equicontinuous at zero in the topology* $\tau_S(X, \tau(X, X'))$;

b) *it is necessary that the sequence* $\{\chi_n(x)\}$ *converges at every point* $x \in X$ *and is equicontinuous at zero in the original topology of* X, *provided that* X *is barrelled and* X' *is an* R*-space.*

PROOF. Convergence of $\{\chi_n(x)\}$ follows directly from weak convergence of the sequence of measures. So b) follows from corollary 2.6.

Conversely, suppose the sequence $\{\chi_n(x)\}$ converges to a function $\chi(x)$. Since a measure is uniquely determined by its characteristic functional, the sequence μ_n cannot have more than one limit point. By theorem 2.6, the conditions a) give the relative weak compactness of the sequence under consideration, and since $\mathfrak{M}(X')$ is weakly complete, its convergence. □

COROLLARY 2.6. *(Lévy's theorem.) If* X *is a nuclear barrelled locally convex space and* X' *is an* R*-space, then a necessary and sufficient condition for weak convergence of a sequence of Radon probability measures* $\{\mu_n\}$ *is that the corresponding sequence of characteristic functionals is equicontinuous at zero in* X, *and converges at every point of* X.

COROLLARY 2.7. *Let* (X, β) *be a locally convex space, and* $T_x^{(n)}$ *a sequence of unitary representations of* X *in a Hilbert space* $\mathcal{H}$. *Suppose that the sequence of functions* $(T_x^{(n)}h, h)$ *is equicontinuous at zero in the topology* $\tau_S(X, \beta)$, *for every* $h \in \mathcal{H}$. *By theorem 1.4, the following representation holds:*

$$(T_x^{(n)}h, h) = \int_{X'} e^{i\langle x,y\rangle}(E^{(n)}(dy)h, h) \quad (h \in \mathcal{H}),$$

where $E^{(n)}$ *are orthogonal projector-valued measures supported in* X'.

Suppose that for every $x \in X$ *the following relation holds:*

$$(T_x^{(n)}h, h) \to (T_x h, h) \quad (h \in \mathcal{H}).$$

Then for every bounded continuous function f *on* X' *the corresponding sequence of generators of* $T^{(n)}$ *converges:*

$$\int_{X'} f(y)(E^{(n)}(dy)h, h) \to \int_{X'} f(y)(E(dy)h, h).$$

3. Chapter III: Supplementary remarks and historical comments

The problem of σ-additivity of a cylindrical measure arises naturally in the theory of random functions when constructing a probability space by means of its finite-dimensional distributions. Kolmogorov's general theorem (see chapter I) implies the σ-additivity of such a measure in the space of all functions of the argument values under consideration. However, sometimes it is important to be able to choose a more restricted space of full measure. Already in chapter II have we discussed this problem, for a Gaussian measure. The sufficient condition for continuity of almost all realizations of a stochastic process (and hence for the corresponding probability measure to be supported in the space of continuous functions) was given by A.N. Kolmogorov for the first time (see theorem VII.1.1). This very question also naturally arises for cylindrical measures in an abstract Hilbert space. It was solved independently, and almost simultaneously, in Minlos [1,2] for a countable-Hilbert space and in Sazonov [3] for a Hilbert space.

Minlos started from an equivalent of the notion of Gelfand–Itô random distribution (a generalized stochastic process, see Gelfand–Vilenkin [1]) for a nonnegative cylindrical measure. He showed that a criterion for extendibility of a random distribution in a nuclear countable-Hilbert space X is continuity of this distribution (equivalently, continuity of its characteristic functional, see theorem 1.5) in the topology of X, and also proved that nuclearity is necessary.

On the other hand, Sazonov introduced the topology τ_S in a Hilbert space, and showed that continuity of the characteristic functional of a nonnegative cylindrical measure in the topology τ_S is equivalent to σ-additivity of the measure. Kolmogorov [3] noted that by introducing the topology τ_S in a countable-Hilbert space one can obtain a result generalizing both Minlos' and Sazonov's theorem. He also suggested a simple proof, contained in Prokhorov [1], of a basic inequality of the type (1.1) (these results are presented in detail in Sazonov [4]).

A number of results concerning characteristic functionals on a Hilbert space related to Bochner's theorem and to the extension of measures was obtained later by a novel approach in Gross [3]. Transition to measures in more general spaces was done in Bourbaki [1], Badrikian [1], Schwartz [1]. Our presentation of these results is based on Smolyanov–Fomin [1].

In order to keep the presentation within definite scope we have omitted the generalization of the Minlos–Sazonov theorem to the case of sign-varying measures given in Shavgulidze [1].

The Minlos–Sazonov theorem implies, in particular, the existence of a topology in a Hilbert or countable-Hilbert space with the property that continuity of a positive-definite function in this topology is equivalent to the existence of a measure with this function as characteristic functional. In Vershik–Sudakov [2] it was shown that a topology with such a property does not exist in an arbitrary Banach space.

The spaces having a topology which specifies the characteristic functionals of positive Radon measures were studied in Mushtari [1,2,3].

Another approach to the study of measures in linear spaces, based on properties of the set of measurable linear functionals, was developed in Vershik [2,3]. Various geometrical problems in the theory of measure in infinite-dimensional spaces were investigated in Sudakov [2,3] (see also Dudley [1,2]). For a review of certain results in this direction see Vershik–Sudakov [1].

The above-described results are related to various generalizations of Bochner's theorem on the representability of a positive definite function as a Fourier transform.

The results presented in § 2 are related to another classical theorem in harmonic analysis: Lévy's theorem on convergence of measures. The start for work in this direction was given in Prokhorov [1]. In particular, in this paper a condition was established for weak compactness of a family of measures in a metric space (theorem 2.1). Weak completeness of a space of measures in a complete metric space was proved in Gikhman–Skorokhod [3]. In Sazonov [1], Prokhorov–Sazonov [1], Lévy's theorem was transferred to countable-Hilbert space. Here it was also shown that nuclearity of the space is a necessary condition for the existence of a topology with the property that equicontinuity at zero of the set of characteristic functionals is equivalent to weak compactness of the set of measures in the dual space. In particular, there is no such topology in a Hilbert space.

Our presentation is an improved version of a corresponding section in Smolyanov–Fomin [1] (see also Smolyanov [4]). This improvement is related to the notion of R-space.

CHAPTER IV

Differentiable measures and distributions

In this chapter we introduce the basic concepts of a differential calculus for functions and measures. Further, these concepts are transferred to dual objects: distributions (generalized measures) and generalized functions. The simplest differential operators on functions and distributions are defined. We also consider properties of the Fourier transforms of distributions, as well as related spectral decompositions of generalized kernels which possess the property of invariance or quasi-invariance

1. Differentiable functions, differentiable expressions

1.1. Derivatives of a vector function. Let X and Y be linear topological spaces, and let $\mathcal{L}(X,Y)$ be the linear space of continuous linear maps from X to Y. If it is desirable to stress the topology, we will write it as an index to the corresponding space: X_τ, X_σ, etc. A topology in $\mathcal{L}(X,Y)$ will be given by indicating a system β of bounded sets in X: $\mathcal{L}_\beta(X,Y)$ is obtained by endowing $\mathcal{L}(X,Y)$ with the topology of uniform convergence on the system of sets β. Similarly we introduce the spaces of continuous k-linear maps, $\mathcal{L}^{(k)}(X,Y)$, and the corresponding locally convex spaces $\mathcal{L}_\beta^{(k)}(X,Y)$. The symbols $\mathcal{L}_S^{(k)}$, $\mathcal{L}_{\beta,S}^{(k)}$ will be used for the subspaces of these spaces consisting of the symmetric maps. When $X=Y$ we will abbreviate the notation, to, e.g., $\mathcal{L}^{(k)}(X)=\mathcal{L}^{(k)}(X,X)$.

If X and Y are normed spaces with the strong topology, the most important case is when β is the system of neighborhoods of zero. Thus, $\mathcal{L}_\beta^{(k)}(X,Y)$ is a normed space, with norm

$$\|\phi\| = \sup_{x_1,\dots,x_n\in X} \frac{\|\phi(x_1,\dots,x_n)\|}{\|x_1\|\dots\|x_n\|} \quad (\phi\in\mathcal{L}^{(k)}(X,Y)).$$

If X and Y are Hilbert spaces, then $\mathcal{L}_2^{(k)}(X,Y)$ denotes the subspace of $\mathcal{L}^{(k)}(X,Y)$ consisting of the Hilbert–Schmidt maps, i.e. the maps ϕ with finite norm

$$\sigma_2^2(\phi) = \sum_{j_1,\dots,j_k=1}^{\infty} \|\phi(e_{j_1},\dots,e_{j_k})\|_Y^2,$$

where $\{e_j\}_{j=1}^\infty$ is an (any) orthonormal basis in X.

The space $\mathcal{L}_2^{(k)}(X,Y)$ is itself a Hilbert space, with scalar product

$$(\phi,\psi)_2 = \sum_{j_1,\dots,j_k=1}^{\infty} (\phi(e_{j_1},\dots,e_{j_k}),\psi(e_{j_1},\dots,e_{j_k}))_Y$$
$$(\phi,\psi\in\mathcal{L}_2^{(k)}(X,Y)).$$

Let X be a topological space, $(\mathcal{H},\beta)$ a topological linear subspace of it, with β the system of bounded sets, and let Y be a topological linear space.

DEFINITION 1.1. A map $f\colon X\to Y$ is called *continuous at a point $x\in X$ with respect to the subspace $\mathcal{H}$* (*an $\mathcal{H}$-continuous map*) if the map $h\to f(x+h)$ from $\mathcal{H}$ to Y is continuous at the zero of $\mathcal{H}$. This map is called *$(\mathcal{H},\beta)$-continuous* at x if

$$f(x)=\lim_{\tau\to 0} f(x+\tau h)$$

uniformly with respect to h on every set in β (denoted by $f(x)=\overset{\beta}{\lim}_{\tau\to 0}f(x+\tau h)$).[1]

DEFINITION 1.2. A map $f\colon X\to Y$ is called *$(\mathcal{H},\beta)$-differentiable at a point $x\in X$* if there is a linear map $f'(x)\in\mathcal{L}_\beta(\mathcal{H},Y)$ for which

$$f(x+h)-f(x)=f'(x)h+\delta(x,h)\quad(h\in\mathcal{H}),$$

where

$$\overset{\beta}{\lim}_{\tau\to 0}\frac{\delta(x,\tau h)}{\tau}.$$

The linear map $f'(x)$ is then called the *derivative* (*$(\mathcal{H},\beta)$-derivative*) of f at x.

The systems most often used are:

a) the system β_f of all finite sets; in this case the derivative is called *weak derivative*;
b) the system β_c of all sequentially compact sets; in this case the derivative is called *compact derivative*;
c) the system β_b of all bounded sets; in this case the derivative is called *bounded derivative*;

Clearly, bounded differentiability implies compact differentiability, which, in turn, implies weak differentiability. It is easily seen that for normed spaces the weak derivative and the bounded derivative become the Gâteaux derivative and the Fréchet derivative, respectively.

PROPOSITION 1.1. *$\mathcal{H}$-continuity of a map $f\colon X\to Y$ at a point $x\in X$ implies $(\mathcal{H},\beta)$-continuity at x with respect to any system of bounded sets β.*

PROOF. In fact, $\mathcal{H}$-continuity of f implies that there is for each neighborhood of zero V in Y a neighborhood of zero U in $\mathcal{H}$ for which $f(x+\tau h)-f(x)\in V$ $(h'\in U)$.

Since there is for each set B bounded in $\mathcal{H}$ a $\tau_U>0$ such that $\tau B\subset U$ if $|\tau|<\tau_U$, we have $f(x+\tau h)-f(x)\in V$ $(h\in B,\ |\tau|<\tau_U)$. □

[1] Hereafter the symbol $\overset{\beta}{\lim}$ denotes that the convergence is uniform on each set in the system of sets β.

PROPOSITION 1.2. *$(\mathcal{H}, \beta)$-differentiability of a map $f \colon X \to Y$ at a point $x \in X$ implies $(\mathcal{H}, \beta)$-continuity at x.*

PROOF. Both terms at the righthand side of the equation

$$f(x + \tau h) - f(x) = \tau f'(x)h + \delta(x, \tau h)$$

tend to zero as $\tau \to 0$, uniformly on the sets in β. □

THEOREM 1.1. *If $\mathcal{H}$ is metrizable, then compact $\mathcal{H}$-differentiability of a map f at a point $x \in X$ implies $\mathcal{H}$-continuity at x.*

PROOF. It suffices to show that $(\mathcal{H}, \beta_c)$-continuity of f implies $\mathcal{H}$-continuity (by proposition 1.1 these notions then turn out to be equivalent for $\mathcal{H}$ metrizable).

Let ρ be a metric in $\mathcal{H}$ defining the original topology. It suffices to show that $\rho(h_n, 0) \to 0$ $(h_n \neq 0,\ n \to \infty)$ implies $f(x + h_n) - f(x) \to 0$.

The sequence $\{h_n/\sqrt{\rho(h_n, 0)}\}$ tends to zero, and is hence relatively sequentially compact. Therefore, uniformly in n,

$$\lim_{k \to \infty} \left\{ f\left(\left(x + \sqrt{\rho(0, h_k)} \frac{h_n}{\sqrt{\rho(h_n, 0)}}\right) - f(x)\right\} = 0;$$

hence there is for each neighborhood of zero $V \subset Y$ an N_V such that

$$f(x + h_n) - f(x) \in V \quad (n \geq N_V). \ \square$$

The following formal differentiation rules are obvious, and can be verified in the usual way.

a) The set of maps $f \colon X \to Y$ that are $(\mathcal{H}, \beta)$-differentiable at a given point $x \in X$ form a linear space, and $f \to f'(x)$ is a linear map from this space to $\mathcal{L}_\beta(X, Y)$;

b) Let $\phi \colon \mathcal{H}_1 \times \cdots \times \mathcal{H}_n \to Y$ be a continuous multilinear map, and let $f_j \colon X \to \mathcal{H}_j$ $(j = 1, \ldots, n)$ be maps which are $(\mathcal{H}, \beta)$-differentiable at a given point $x_0 \in X$. Then the map

$$f \colon X \to Y, \quad f(x) = \phi(f_1(x), \ldots, f_n(x))$$

is $(\mathcal{H}, \beta)$-differentiable at x_0, and

$$f'(x_0)h = \sum_{j=1}^{n} \phi(f_1(x_0), \ldots, f_{j-1}(x_0), f_j'(x_0)h, f_{j+1}(x_0), \ldots, f_n(x_0)).$$

c) Let X, Y be linear spaces, $\mathcal{H} \subset X$, $\mathcal{K} \subset Y$, and Z topological linear spaces, $f \colon X \to Y$, $g \colon Y \to Z$, and suppose that $x_1 - x_2 \in \mathcal{H}$ implies $f(x_1) - f(x_2) \in \mathcal{K}$. If f is compact (bounded) $\mathcal{H}$-differentiable at a point x, and g is compact (bounded) $\mathcal{K}$-differentiable at the point $f(x)$, then the composite $g \circ f$ is compact (bounded) $\mathcal{H}$-differentiable at x, and

$$(g \circ f)'(x) = (g' \circ f)(x) f'(x).$$

In the sequel we will need the following version of the mean value theorem.

PROPOSITION 1.3. *Let f be a continuous map from the interval $[a,b]$ on the real axis into a topological linear space Y which is differentiable at every point in this interval. Then*

$$f(b) - f(a) \in \overline{conv}\,\{f'(t) : a \leq t \leq b\} \cdot (b-a)$$

($\overline{conv}\, M$ is the closure of the convex hull of the set M).

The proof follows by the Hahn–Banach theorem and the fact that for every linear functional l on Y we have

$$|l(f(b) - f(a))| \leq \sup_{[a,b]} |l(f'(t))| \cdot (b-a).$$

COROLLARY 1.1. *Let a map f from a linear space into a topological linear space Y be differentiable in a direction h at every point of the line segment connecting points x_0 and $x_0 + h$. Then*

$$f(x_0 + h) - f(x_0) \in \overline{conv}\,\{f'(x_0 + \tau h)h : 0 \leq \tau \leq 1\}, \tag{1.1}$$

and for each point $a \in X$ at which f is differentiable in the direction h we have

$$f(x_0 + h) - f(x_0) - f'(a)h \in \overline{conv}\,\{[f'(x_0 + th) - f'(a)]h : 0 \leq t \leq 1\}. \tag{1.2}$$

The first assertion can be proved by applying the mean value theorem to the function $\phi(t) = f(x_0 + th)$, and for the second assertion we use the function $\psi(t) = f(x_0 + th) - tf'(a)h$.

1.2. Higher order derivatives. There are various, mutually equivalent ways of defining higher order derivatives of a map $f\colon X \to Y$. We use one of these, based on Taylor's formula.

Suppose that there exist symmetric multilinear maps $f^{(k)}(x) \in \mathcal{L}_{\beta,s}(\mathcal{H}, Y)$ ($k = 1, \ldots, n$) and a function $\delta_n(x,h)$ with the properties that the function $\alpha(t) = \delta_n(x, th)$ is n times differentiable at zero, $\alpha^{(k)}(0) = 0$ ($k = 1, \ldots, n$), and

$$f(x+h) - f(x) = \sum_{k=1}^{n} \frac{1}{k!} f^{(k)}(x)(h, \ldots, h) + \delta_n(x,h). \tag{1.3}$$

In this case the map f is called $(\mathcal{H}, \beta)$-*differentiable up to order* n at x, and the maps $f^{(k)}$ are called its *derivatives*.

It can be readily shown that if a map f is differentiable up to order n in the sense of (1.3) in a neighborhood of a point x, then $f'(x)$ is its $(\mathcal{H}, \beta)$-derivative in the sense of definition 1.2. This derivative, regarded as a map $f'\colon X \to \mathcal{L}(\mathcal{H}, Y)$, is differentiable up to order $n-1$, and $(f')^{(k-1)} = f^{(k)}$ ($k = 1, \ldots, n-1$).

Conversely, it is known that n times (successive) differentiability of a map f implies the Taylor formula (1.3).

1.3. Linear differential expressions. We introduce linear differential expressions for maps u from a topological linear space X into a topological linear space Y. Let $\mathcal{H} \subset X$ and let u be $(\mathcal{H}, \beta)$-differentiable at a point $x_0 \in X$ up to order k. The derivative $u^{(k)}(x_0)$ is an element of $\mathcal{L}^{(k)}_{\beta,s}(\mathcal{H}, Y)$. If there exists a continuous linear operator on this space mapping it to a topological linear space Z, then the following map is naturally defined:

$$l^{(k)}_\phi : u \to (\phi(u^{(k)}(x_0)) = l^{(k)}_\phi(u)(x_0);$$

it is called the *linear differential expression of order* k at x_0 determined by the operator ϕ. In general, $\phi = \phi(x_0)$ depends also on the point x_0.

Examples of such linear differential expressions on a Hilbert subspace $\mathcal{H}$ can be easily constructed.

Let $\{e_k\}_{k=1}^\infty$ be an orthonormal basis in $\mathcal{H}$, and $\alpha \in \mathcal{L}^{(k)}(\mathcal{H}, \mathcal{L}(Y, Z))$. Consider the linear differential expression determined by the series

$$\text{(1.4)} \qquad \operatorname{Tr}^{(k)}[\alpha \otimes u^{(k)}(x_0)] \stackrel{\text{def}}{=} \sum_{a_1,\dots,a_k=1}^{\infty} \alpha(e_{j_1}, \dots, e_{j_k}) u^{(k)}(x_0)(e_{j_1}, \dots, e_{j_k}),$$

which is assumed to be absolutely convergent. This expression certainly makes sense if Y and Z are also Hilbert spaces, while α and $u^{(k)}$ are Hilbert–Schmidt maps.

We are mainly interested in linear differential expressions of orders one or two, so we will consider these in more detail. We begin with the scalar case ($Y = Z = \mathbf{C}$).

Since $u'(x_0) \in \mathcal{L}_\beta(\mathcal{H}, \mathbf{C}) = H'$, linear differential expressions of order one are given by the formula

$$l^{(1)}_{\phi(x)}(u)(x) = \langle \phi(x), u'(x) \rangle,$$

where $\phi(x) \in \mathcal{H}$ $(x \in X)$.

Below we will systematically use the symbolic notation $u'(x) = \nabla u(x)$, where $\nabla = \nabla_{\mathcal{H}}$ is the symbol of differentiation along the subspace $\mathcal{H}$, equipped with the algebraic properties of the elements of $\mathcal{H}$. Now we can write $l^{(1)}_{\phi(x)}(u(x)) = \langle \phi(x), u'(x) \rangle = \langle \phi(x), \nabla \rangle u(x)$.

In particular, let

$$\text{(1.5)} \qquad u(x) = \int_{X'} e^{i\langle x,\theta\rangle} \nu(d\theta) \quad (x \in X)$$

be the Fourier transform of a measure ν in X'. If

$$\int_X |\langle h, \theta \rangle| |\nu|(d\theta) < \infty \quad (h \in \mathcal{H}),$$

then

$$l^{(1)}_\phi(u)(x) = i \int_{X'} \langle \phi(x), \theta \rangle e^{i\langle x,\theta\rangle} \nu(d\theta).$$

The function $\alpha(x, \theta) = i\langle \phi(x), \theta \rangle$ on $X \times X'$ has the property

$$\text{(1.6)} \qquad l^{(1)}_\phi e^{i\langle x,\theta\rangle} = \alpha(x, \theta) e^{i\langle x,\theta\rangle},$$

and is called the *symbol* of the linear differential expression $l^{(1)}_{\phi(x)}$. From (1.5), (1.6) the relation

$$l^{(1)}_{\phi(x)} = \langle \phi(x), \nabla \rangle = \alpha\left(x, \frac{1}{i}\nabla\right)$$

follows. Below we will extend this relation to a more general case.

Now we consider linear differential expressions of order two, assuming $\mathcal{H}$ to be a Hilbert space. Hence we may regard X' as being embedded in $\mathcal{H}$: $X' \subset \mathcal{H}' = \mathcal{H}$. The second derivative $u''(x)$ is a symmetric bounded bilinear functional on $\mathcal{H}$, and is canonically identified with a symmetric bounded linear operator on $\mathcal{H}$, denoted by the same symbol $u''(x)$. If $a \in \mathcal{L}^{(2)}(\mathcal{H}, \mathbf{C})$ has the form

$$a(f,g) = (Af, Bg)_{\mathcal{H}} \quad (A, B \in \mathcal{L}(\mathcal{H}, \mathcal{H}) = \mathcal{L}(\mathcal{H})),$$

then it can be readily seen that

$$l^{(2)}_a(u) = \mathrm{Tr}^{(2)}\, a \otimes u'' = \mathrm{Tr}_{\mathcal{H}}\, AU''B^*, \tag{1.7}$$

where the operator $Au''B^*$ is nuclear (the symbol Tr at the righthand side denotes the usual trace operator on the Hilbert space $\mathcal{H}$); in particular, for $A, B \in \mathcal{L}_2(\mathcal{H})$. To compute the symbol of the linear differential expression $l^{(2)}_a$, note that for $u(x) = e^{i\langle x,y\rangle}$ we have $u''(x)(h_1, h_2) = -e^{i\langle x,y\rangle}\langle h_1, y\rangle\langle h_2, y\rangle$, and hence

$$l^{(2)}_a(e^{i\langle x,y\rangle}) = -(Ay, By)e^{i\langle x,y\rangle}.$$

Thus, for this linear differential expression the function $\alpha(x,y)$ satisfying a relation of type (1.6) (the definition of symbol) is of the form

$$\alpha(x,y) = -(A(x)y, B(x)y) = -(B^*(x)A(x)y, y).$$

Note that for $B = A$ this quadratic functional is nonpositive. Such a symbol, as well as the linear differential expression $L^{(2)}_a$ itself, can be naturally called *elliptic*.

Extending the above system of notation to the case under consideration, we put

$$\mathrm{Tr}\, A(x)u''(x)B^*(x) = \alpha\left(x, \frac{1}{i}\nabla\right)u(x) = (B^*(x)A(x)\nabla, \nabla)u(x).$$

For functions of the type (1.5) it is easily checked that

$$\alpha\left(x, \frac{1}{i}\nabla\right)u = \int_{X'} \alpha(x,y)e^{i\langle x,y\rangle}\nu(dy). \tag{1.8}$$

Now we consider vector functions taking values in a real Hilbert space $\mathcal{H}_1 = Y = Z$.

The scalar linear differential expressions introduced above can be obviously extended to such a space by means of the relations

$$\begin{aligned} &(\langle \phi(x), \nabla\rangle u, f) = \langle \phi(x), \nabla\rangle (u, f) \quad (f \in \mathcal{H}_1) \\ &((B^*A\nabla, \nabla)u, f) = (B^*A\nabla, \nabla)(u, f). \end{aligned} \tag{1.9}$$

Further, for a pair $f \in \mathcal{H}_1$, $y \in Y$ we introduce the linear operator $f \otimes g \in \mathcal{L}(\mathcal{H}, Y)$ defined by

$$(f \otimes g)h = (h, f)y.$$

This notation justifies the relation $u'(x) = \nabla \otimes u(x)$. In fact, here we have

$$u'(x)h = (\nabla \otimes u(x))h = \langle h, \nabla \rangle u(x),$$

and for $f \in Y'$ we obtain

$$\langle u(x), f \rangle' h = \langle u'(x)h, f \rangle = \langle h, \nabla \rangle \langle u(x), f \rangle = \langle h, \nabla \langle u(x), f \rangle \rangle.$$

If $\mathcal{H}_1 = \mathcal{H}$ and $u'(x)$ is nuclear, then the trace of $u'(x)$ can be naturally called the *divergence* of the vector function $u(x)$:

$$\operatorname{div} u(x) = \operatorname{Tr} u'(x) = (\nabla, u(x)).$$

We also introduce a notion of divergence for an operator function $A(x) \in \mathcal{L}(\mathcal{K}, \mathcal{H})$ $(x \in X)$, defined as the vector $\operatorname{div} A(x) \in \mathcal{K}$ satisfying the relation

$$(\operatorname{div} A(x), \phi)_{\mathcal{K}} = \operatorname{div}(A(x)\phi) \quad (\phi \in \mathcal{K})$$

in case the operator $(A(x)\phi)' = A'_\phi(x)$ is nuclear on $\mathcal{H}$ for any $\phi \in \mathcal{K}$.

Let now $u\colon X \to \mathbf{R}$. Then $u'\colon X \to \mathcal{H}$, and we can introduce the second-order linear differential expression

$$\operatorname{div} A(x)u'(x),$$

where $A(x) \in \mathcal{L}(\mathcal{H})$ $(x \in X)$. It is not difficult to verify the relation

$$\begin{aligned} \operatorname{div} A(x)u'(x) &= \langle \nabla, A(x)\nabla \rangle u(x) = \\ &(A^*(x)\nabla, \nabla)u(x) + (\operatorname{div} A(x), \nabla)u(x), \end{aligned} \tag{1.10}$$

which is true if all expressions occurring in it make sense.

By similar considerations, for a vector function $u\colon X \to \mathcal{H}$ the linear differential expression

$$(A^*(x)\nabla, u(x)) = \operatorname{Tr} A(x)u'(x) \tag{1.11}$$

is defined, and the formula

$$(A^*(x)\nabla, u(x)) + (\operatorname{div} A(x), u(x)) = \operatorname{div} A(x)u(x) \tag{1.12}$$

holds.

The expression (1.11) makes sense if $A()$ and $u'(x)$ are both Hilbert–Schmidt operators (or if $A(x)$ is nuclear). Formulas (1.10) and (1.12) are true if the operators $A(x)$, $[A(x)\phi]'$ $(\phi \in \mathcal{H})$ are nuclear.

We now consider a bilinear map $a(x)\colon \mathcal{H} \times \mathcal{H}_1 \to \mathcal{H}_2$, and define the first-order linear differential expression

$$a(\nabla, u(x)) \tag{1.13}$$

as follows. Let, for $\phi \in \mathcal{H}_2$, the operator $A_\phi \in \mathcal{L}(\mathcal{H}, \mathcal{H}_1)$, linearly depending on ϕ, be defined by

$$(a(f, g), \phi)_{\mathcal{H}_2} = (A_\phi f, g)_{\mathcal{H}_1} \quad (f \in \mathcal{K}, \quad g \in \mathcal{K}_1).$$

In accordance with (1.11) we put, by definition,

$$(\phi, a(\nabla, u(x))) = (A_\phi \nabla, u(x)) = \operatorname{Tr} A_\phi^*(x)u'(x).$$

This expression makes sense for $u(x) \in \mathcal{L}_2(\mathcal{H}, \mathcal{H}_1)$ if the operator $A_\phi(x) \in \mathcal{L}_2(\mathcal{H}, \mathcal{H}_1)$ depends continuously on ϕ, e.g. for $a(f,g) = c(Bf, g)$ where $B \in \mathcal{L}_2$.

In particular, consider the vector function

$$u(x) = \int_{X'} e^{i\langle x,\theta\rangle} \nu(d\theta) = \int_{X'} e^{i\langle x,\theta\rangle} \rho(\theta)\nu_0(d\theta),$$

which is the Fourier transform of a vector measure ν; for the sake of simplicity, ν is taken to be absolutely continuous with respect to a certain scalar measure ν_0: $\rho(\theta) = \nu(d\theta)/\nu_0(d\theta)$ (unless otherwise stipulated, the integrals are assumed to be absolutely convergent). In this case

$$u'(x)h = i\int_{X'} e^{i\langle x,\theta\rangle} \langle h, \theta\rangle \rho(\theta)\nu_0(d\theta),$$

and hence

$$u'(x) = \nabla \otimes u(x) = i\int_{X'} e^{i\langle x,\theta\rangle} (\theta \otimes \rho(\theta))\nu_0(d\theta) = i\int_{X'} e^{i\langle x,\theta\rangle} (\theta \otimes \nu(d\theta)).$$

The following formulas are now easily obtained:

$$\operatorname{div} u(x) = i\int_{X'} e^{i\langle x,\theta\rangle} (\theta, \nu(d\theta)),$$

and for a scalar function $u(x)$:

$$\operatorname{div} A(x)u'(x) = -\int_{X'} e^{i\langle x,\theta\rangle} (A(x)\theta, \theta)\nu(d\theta).$$

Finally, for the linear differential expression (1.13) we obtain the representation

$$a(x)(\nabla, u(x)) = i\int_{X'} e^{i\langle x,\theta\rangle} a(x)(\theta, \nu(d\theta)).$$

It is natural to regard the operator $\tilde{a}(x,\theta)$ having the property $a(x)(\theta, \rho(\theta)) = \tilde{a}(x,\theta)\rho(\theta)$ (if it exists) as its symbol.

Generalizing these considerations, we can introduce for functions

$$u(x) = \int_{X'} e^{i\langle x,\theta\rangle} \nu(d\theta)$$

the pseudodifferential expression

$$\alpha\left(x, \frac{1}{i}\nabla, u(x)\right) = \int_{X'} e^{i\langle x,\theta\rangle} \alpha(x, \theta, \nu(d\theta))$$

with operator symbol $\tilde{\alpha}(x,\theta)$, $\alpha(x,\theta,h) = \tilde{\alpha}(x,\theta)h$, satisfying the condition

$$\int_{X'} \|\tilde{\alpha}(x,\theta)\| \cdot |\nu|(d\theta) < \infty.$$

If $\tilde{\alpha}(x,\theta)$ is a polynomial in θ, then $\alpha(x,\nabla,u)$ is a differential operator.

1.4. Symmetric and dissipative differential operators. We now turn to a less formal consideration of differential operators related to the linear differential expressions introduced above.

Let $X = \mathcal{H}_- \supset \mathcal{H} \supset \mathcal{H}_+$ be a canonical triple of Hilbert spaces[2], ν a related standard Gaussian measure in X, and $\mathcal{K}$ a Hilbert space.

Consider the Hilbert space $\mathfrak{H} = \mathfrak{H}(X, \mathcal{K}, \nu)$ of square integrable (with respect to ν) maps $u \colon X \to \mathcal{K}$ with norm $|||u|||^2 = \int_X \|u(x)\|^2 \nu(dx)$, and the Hilbert subspace $\mathfrak{H}_r(X, \mathcal{K}, \nu) \subset \mathfrak{H}$ of the maps that are $\mathcal{H}$-differentiable up to order r, with norm

$$|||u|||_r^2 = \int_X \left\{ \|u(x)\|^2 + \sum_{j=1}^{r} \sigma_2^2(u^{(j)}(x)) \right\} \nu(dx) < \infty.$$

Denote by $C_r(X, B)$ the Banach space of maps $u \colon X \to B$ (B a Banach space) that are $\mathcal{H}$-differentiable up to order r and, together with all their derivatives, bounded on X.

Let $A \in \mathcal{L}_2(\mathcal{H})$, $f \in \mathfrak{H}_1(X, \mathcal{L}_2(\mathcal{H}, \mathcal{K}), \nu)$. Then, as shown in theorem II.3.5, there is a measurable function $\beta(A, f)(x) \in \mathfrak{H}(\mathcal{H}_-, \mathcal{K}, \nu)$; for $A \in \mathcal{L}_2(\mathcal{H})$ it is of the form

$$\beta(A, f) = f(x)Ax - \operatorname{div}[f(x)A]^*.$$

Theorem II.3.4 implies that

$$\int_X \beta(A, f)(x)\nu(dx) = 0. \tag{1.14}$$

Let $\alpha \in C_1(X, \mathbf{R})$. Replacing in (1.14) $f(x)$ by $f(x)\alpha(x)$ we obtain the equation

$$\int_X \beta(A, f)(x)\alpha(x)\nu(dx) = \int_X f(x)A\alpha'(x)\nu(dx). \tag{1.15}$$

Let $\{\phi_j\}$ be an orthonormal basis in K, and $v(x) = \sum_j \alpha_j(x)\phi_j$, where $\alpha_j \in C_1(X, \mathbf{R})$. After minor transformation, (1.15) gives a formula for integration by parts ($A = \mathrm{id}$):

$$\int_X (\beta(f)(x), v(x))\nu(dx) - \int_X \operatorname{Tr} f(x)v'^*(x)\nu(dx). \tag{1.16}$$

By limit transition it is extended to $v \in \mathfrak{H}_1(X, \mathcal{K}, \nu)$.

Let $B \in C_1(X, \mathcal{L}(\mathcal{H}))$. Define a differential operator on $\mathfrak{H}_2(X, \mathcal{K}, \nu)$ by

$$\tilde{\mathfrak{U}}_0 u(x) = \beta(u'(x)B(x)).$$

Note that its restriction to the subspace $\tilde{\mathfrak{H}}_2(X, \mathcal{K}, \nu)$ of functions $u(x) \in \mathfrak{H}_2(X, \mathcal{K}, \nu)$ for which $[u^*(x)\phi]'$ ($\phi \in \mathcal{K}$) is a nuclear operator, has the form

$$\begin{aligned}\tilde{\mathfrak{U}}_0 u(x) - \beta(u'(x)B(x)) - u'(x)B(x)x - \operatorname{div}[u'(x)B(x)] = \\ (B(x)x, \nabla)u(x) - (\nabla, B(x)\nabla)u(x).\end{aligned} \tag{1.17}$$

Putting $f(x) = u'(x)B(x)$ in (1.16) we obtain the *Green formula*

$$\int_X (\beta(u'(x)B(x)), v(x))\nu(dx) = \int_X \operatorname{Tr}[u'(x)B(x)v'^*(x)]\nu(dx), \tag{1.18}$$

[2]I.e. a rigged Hilbert space with Hilbert–Schmidt imbeddings.

implying the following result.

THEOREM 1.2. *The $\mathfrak{H}_2(X,\mathcal{K},\nu)$-defined differential operator $\tilde{\mathfrak{U}}_0$, which is the extension of (1.17) on $\tilde{\mathfrak{H}}_2(X,\mathcal{K},\nu)$, is symmetric on $\mathfrak{H}_2(X,\mathcal{K},\nu)$ for $B(x) = B^*(x)$, and is nonpositive for $B(x) \geq 0$:*

$$\int_X (\tilde{\mathfrak{U}}_0 u(x), u(x))\nu(dx) = -\int_X \mathrm{Tr}[u'(x)B(x)u'^*(x)]\nu(dx).$$

We now consider a nondiagonal operator $a(\nabla, u)$. First, let $a(f,g) = (A(x)f,g)$, where $A \in C(X,\mathcal{L}_2(\mathcal{H}))$ takes scalar values. The operator $a(x)(\nabla,u(x)) = (A(x)\nabla,u(x))$ is defined on $\mathfrak{H}_1(X,\mathcal{K},\nu)$, and takes values in $\mathfrak{H}_1(X,\mathbf{R},\nu)$; also

$$\int_X |(A(x)\nabla,u(x))|^2\nu(dx) = \int_X |\,\mathrm{Tr}\, A^*(x)u'(x)|^2\nu(dx) \leq \text{const}\int_X \sigma_2^2(u'(x))\nu(dx).$$

We compute the adjoint operator. It follows from (1.12) that

$$(A(x)\nabla,u(x)) = \mathrm{div}\, A^*(x)u(x) - (\mathrm{div}\, A^*(x),u(x)).$$

Therefore, for $v \in \mathfrak{H}_1(X,\mathbf{R},\nu)$ we have

$$\int_X (A(x)\nabla,u(x))v(x)\nu(dx) =$$
$$\int_X (\nabla, A^*(x)u(x))v(x)\nu(dx) - \int_X (\mathrm{div}\, A^*(x),u(x))v(x)\nu(dx) =$$
$$\int_X (\nabla, v(x)A^*(x)u(x))\nu(dx) - \int_X (\mathrm{div}\, A^*(x),u(x)v(x))\nu(dx) -$$
$$\int_X (A^*(x)u(x),\nabla)v(x)\nu(dx) =$$
$$\int_X (u(x), A(x)xv(x) - \mathrm{div}\, A^*(x)v(x) - A(x)\nabla v(x))\nu(dx),$$

provided $\mathrm{div}\, A^*$ exists. Thus,

$$(A(x)\nabla)^* = A(x)x - \mathrm{div}\, A^*(x) - A(x)\nabla\colon \mathfrak{H}_1(X,\mathbf{R},\nu) \to (X,K,\nu). \tag{1.19}$$

Now we consider the more general case when $a(x) \in L^2(X\times\mathcal{K},\mathcal{K})$ takes vector values. First, let $v(x) = \sum_j a_j(x)\phi_j$. Then

$$\int_X (a(\nabla)(\nabla,u(x)),v(x))\nu(dx) = \sum_j \int_X (A_{\phi_j}(x)\nabla,u(x))a_j(x)\nu(dx) =$$
$$\int_X \left(u(x), \sum_j [A_{\phi_j}(x)x - \mathrm{div}\, A^*_{\phi_j}(x) - A_{\phi_j}(x)\nabla]a_j(x)\right)\nu(dx).$$

Introducing the notation $(a(f,g),h) = (\tilde{a}(f,h),g)$, we obtain

$$\left(\sum_j a_jA_{\phi_j}f,g\right) = \sum_j a_j(a(f,g),\phi_j) = (\tilde{a}(f,v),g),$$

and hence

$$\sum_j (u, A_{\phi_j}(x-\nabla)a_j) = (u, \tilde{a}(x-\nabla)v).$$

Moreover, if $\{e_k\}$ is an orthonormal basis in X, then

$$(f, \operatorname{div} A_\phi^*(x)) = \sum_k (f, [A_\phi^{*\prime} e_k]e_k) = \sum_k ((A_\phi' e_k)f, e_k) =$$
$$\sum_k ((\tilde{a}(f,\phi), e_k)', e_k) = \operatorname{Tr} \tilde{a}'(x)(f,\phi),$$

and consequently,

$$\sum_j (u, \operatorname{div} A_{\varphi_j}^*)a_j = \sum_j \operatorname{Tr} \tilde{a}'(u, \phi_j a_j) = \operatorname{Tr} \tilde{a}'(u,v).$$

Thus, the adjoint of $l_a(u) = a(\nabla, u)$ in $\mathcal{L}(X, \mathcal{K}, \nu)$ is

$$l_a^*(v) = \tilde{a}(x-\nabla, v) - \operatorname{Tr} \tilde{a}'(\cdot, v) \tag{1.20}$$

(the last term means that if $b(u,v)$ is a bilinear functional, then $b(\cdot, v)$ is a vector satisfying the condition $(u, b(\cdot, v)) = b(u,v)$).

Using this expression, sufficient conditions for an operator $l_a(u) = a(\nabla, u)$ to be dissipative on $\mathfrak{H}(X, \mathcal{H}, \nu)$ can be readily found. The formula

$$l_a(u) + l_a^*(u) = a(\nabla, u) + \tilde{a}(x, u) - \tilde{a}(\nabla, u) - \operatorname{Tr} \tilde{a}'(\cdot, u)$$

leads to the following assertion.

THEOREM 1.3. *Let the conditions*

$$(a(f,g), h) = (a(f,h), g) \tag{1.21}$$

and

$$(f, a(x,f)) - \operatorname{Tr} \tilde{a}'(x)(f,f) \leq 0 \tag{1.22}$$

hold. Then

$$(l_a(u), u) \leq 0.$$

The important property of a dissipative operator is its essential maximality (in particular, the essential selfadjointness of a symmetric operator).

In the following chapters we will obtain conditions for maximality of infinite-dimensional differential operators. These conditions will be based on a general assertion stated at the end of the present section.

THEOREM 1.4. *Let $\mathcal{H}$ be a Hilbert space, $\mathcal{D}$ a Banach space with a stronger topology which is densely imbedded in $\mathcal{H}$. Let $u(t)$ be a strongly continuous semigroup of operators on $\mathcal{D}$ with generating operator A which leaves invariant a linear set $\mathcal{D}_0$ in the domain of definition $\mathcal{D}_A$ of A; moreover, let $\mathcal{D}_0$ be dense in $\mathcal{H}$. If the restriction $A_{\mathcal{D}_0}$ is dissipative (symmetric) on $\mathcal{H}$, then its closure is a maximal dissipative (selfadjoint) operator on $\mathcal{H}$.*

PROOF. It suffices to show that the condition

$$((A-I)\phi, f) = 0 \quad (\phi \in \mathcal{D}_0)$$

implies $f = 0$.

Consider the function $\alpha(t) = e^{-t}(u(t)\phi, f)$. It has the properties $\lim_{t\to 0}\alpha(t) = 0$ and $\alpha'(t) = e^{-t}((A-I)u(t)\phi, f) = 0$. This implies $(f,\phi) = \lim_{t\to 0}\alpha(t) = 0$ $(\phi \in \mathcal{D}_0)$. □

2. Differentiable measures

2.1. Derivative of a measure. Let $(X,\mathcal{U})$ be a linear space with σ-algebra of subsets which is invariant under shift by an element of a topological linear subspace $\mathcal{K} \subset X$, and let Y be a topological linear space. Consider a linear space $\mathcal{M}$ of measures on $(X,\mathcal{U})$ with values in Y, equipped with a linear topology δ. This topology is called *compatible* with $\mathcal{H}$ if the shift map $\mu \to \mu_h$ $(\mu_h(A) = \mu(A+h))$ is a linear homomorphism from $(\mathcal{M},\delta)$ onto itself, for every $h \in \mathcal{H}$.

A measure $\mu \in \mathcal{M}$ determines a map from $\mathcal{H}$ into $(\mathcal{M},\delta)$,

$$h \to \mu_h. \tag{2.1}$$

DEFINITION 2.1. A measure μ is called $(\mathcal{M},\delta)$-*differentiable along* $(\mathcal{H},\beta)$, where β is the system of bounded sets in $\mathcal{H}$, if the map (2.1) is $(\mathcal{H},\beta)$-differentiable at $h = 0$.

In other words, a measure μ is $(\mathcal{M},\delta)$-differentiable along $(\mathcal{H},\beta)$ if there are a continuous linear map

$$\mu' : \mathcal{H} \to \mathcal{M}$$

and a function $r(A,h) \in (\mathcal{M},\delta)$ $(h \in \mathcal{H})$ with the property

$$\overset{\beta}{\lim}_{t\to 0}\frac{1}{t}r(A,th) = 0$$

(uniformly over each set in β) such that

$$\mu(A+h) - \mu(A) = \mu'(A)h + r(A,h) \quad (A \in \mathcal{U}, \quad h \in \mathcal{H}). \tag{2.2}$$

The map μ' is called the $\mathcal{H}$-*derivative* of μ, and the measure $\mu'(A)h$ is the derivative of the measure μ in the direction h.

Clearly, if $\mathcal{H}$ is one-dimensional, the $\mathcal{H}$-derivative becomes the derivative in its direction.

We are mainly interested in the following three topologies:

a) the topology τ_f of weak convergence of measures:

$$\mu_n \to \mu \quad \text{if} \int_X \phi(x)\mu_n(dx) \to \int_X \phi(x)\mu(dx)$$

for each bounded continuous realvalued function;

b) the topology τ_S of convergence on each set $A \in \mathcal{U}$;

c) the topology τ_v of convergence in variation (if the latter is defined).

Note that weak convergence in the space of all realvalued measures with the topology τ_v as strong topology is equivalent to τ_S-convergence of bounded sequences.

We now clarify some connections between the types of differentiability related to the choices of topologies.

First of all, the well-known Nikodým theorem readily implies the following assertion.

PROPOSITION 2.1. *A necessary and sufficient condition for a measure μ in $\mathcal{U}$ with values in a locally convex space Y to be τ_S-differentiable in a direction h is that the map*

$$t \to \mu(A + th) \quad (\mathbf{R} \to Y)$$

is differentiable at $t = 0$ for every $A \in \mathcal{U}$.

In the sequel we will need some consequences of the mean value theorem (proposition 1.3).

PROPOSITION 2.2. *Let δ be a locally convex topology in $\mathcal{M}$, and let μ be a measure which is $(\mathcal{M}, \delta)$-differentiable in a direction h. Then*

$$\mu_h - \mu \in \overline{conv}\{\mu' h|_{\theta h} : 0 < \theta < 1\}.$$

PROOF. It suffices to apply proposition 1.3 to the map $\psi \colon t \to \mu_{th}$:

$$\mu_h - \mu = \psi(1) - \psi(0) \in \overline{\text{conv}}\{\psi'(\theta) : 0 < \theta < 1\} =$$
$$\overline{\text{conv}}\{(\mu_{\theta h})' h : 0 < \theta < 1\} = \overline{\text{conv}}\{[\mu' h]_{\theta h} : 0 < \theta < 1\}. \quad \square$$

COROLLARY 2.1. *Let $\mathcal{M}$ be the space of all realvalued measures on $(X, \mathcal{U})$, and let δ be a linear topology in $\mathcal{M}$ compatible with $\mathcal{H}$ and with the property that the unit ball of the Banach space $(\mathcal{M}, \tau_v)$ is closed in $(\mathcal{M}, \delta)$.*

Then, under the conditions of proposition 2.2, the following estimates hold:

$$\|\mu_h - \mu\| < \|\mu' h\|, \tag{2.3}$$

$$\|\mu_{th} - \mu - t\mu' h\| \leq \|t\| \sup_{0 \leq s \leq t} \{\|(\mu' h)_{sh} - \mu' h\|\}. \tag{2.4}$$

These results are obtained by applying the estimates (1.1) and (1.2) to the map ψ, taking into account that the set $\{(\mu' h)_{\theta h}\}$ belongs to the ball of radius $\|\mu' h\|$.

THEOREM 2.1. *If a realvalued measure $\mu \in \mathcal{M}$ is τ_S-differentiable in a direction h, then it is also τ_v-differentiable in this direction.*

PROOF. It suffices to show that

$$\left\| \mu' h - \frac{\mu_{th} - \mu}{t} \right\| \to 0 \quad (t \to 0).$$

By requirement, for each $A \in \mathcal{U}$,

$$\frac{\mu_{t_0 h + th}(A) - \mu_{t_0 h}(A)}{t} \to \mu' h(A + t_0 h) = (\mu' h)_{t_0 h}(A).$$

Therefore the function $\phi_\mu: t \to \mu_{th}$ is weakly differentiable, and hence its derivative $\psi_\mu: t \to (\mu'h)_{th}$ is weakly measurable in $\mathcal{M}_v(X,\mathcal{U})$. The estimate $\|\mu_{th} - \mu\| \leq \|t\| \cdot \|\mu'h\|$ implies that $\phi_\mu(t)$ is continuous, and hence the minimal linear space containing its values is separable. Since the latter contains the values of $\phi(t)$, this function is strongly measurable. By invariance under shifts, the following relation holds:

$$\|\psi_\mu(t_0 + \epsilon) - \psi_\mu(t_0)\| = \int_0^1 \|\psi_\mu(t_0 + \epsilon) - \psi_\mu(t_0)\| \, d\epsilon =$$
$$\int_0^1 \|\psi_\mu(t_0 + \tau + \epsilon) - \psi_\mu(t_0 + \tau)\| \, d\tau.$$

Since the righthand side tends to zero as $\epsilon \to 0$, ψ_μ is continuous. Therefore, by (2.4),

$$\left\| \frac{\mu_{th} - \mu}{t} - \mu'h \right\| \leq \sup_{|\tau|<t} \|(\mu'h)_{\tau h} - \mu'h\| \to 0 \quad (t \to 0). \quad \square$$

2.2. The logarithmic derivative.

THEOREM 2.2. *Let μ be a nonnegative measure on $(X,\mathcal{U})$ which is τ_S-differentiable in a direction h. Then its derivative $\mu'h$ in this direction is a measure which is absolutely continuous with respect to μ; consequently, there is a μ-integrable function $f_h(x)$ (the logarithmic derivative of μ in the direction h) such that*

$$(\mu'h)(A) = \int_A f_h(x)\mu(dx). \tag{2.5}$$

PROOF. The function $\phi(t) = \mu(A + th)$ is nonnegative and everywhere differentiable. If $\mu(A) = 0$, then $\phi(\theta) = 0$, and $t = 0$ is a minimum for $\phi(t)$. So,

$$(\mu'h)(A) = \phi'(0) = 0. \quad \square$$

This assertion can also be extended to more general measures, first of all due to the following result.

THEOREM 2.3. *Let μ be a realvalued measure on $(X,\mathcal{U})$, and let $\mu = \mu^+ - \mu^-$ be its Jordan–Hahn decomposition. If μ is τ_S-differentiable in a direction h, then the nonnegative measures μ^+, μ^- also have this property.*

PROOF. There is an $S \in \mathcal{U}$ such that

$$\mu^+(A) = \mu(A \cap S), \quad \mu^-(A) = -\mu(A \setminus S).$$

Clearly, the function $\phi(t) = \mu(S + th) = \mu(S_{th}) = \mu(S \cap S_{th}) + \mu(S_{th} \setminus S)$ has a maximum at $t = 0$ and is everywhere differentiable. So,

$$\frac{\phi(t) - \phi(0)}{t} = \frac{\mu(S_{th} \setminus S)}{t} + \left(-\frac{\mu(S - S_{th})}{t}\right) \to 0 \quad (t \to 0).$$

Since both terms are nonnegative, each tends to zero as $t \to 0$. The relation $S = (S_{th} \cup (S - S_{th})) \setminus (S_{th} \setminus S)$ implies that

$$\mu^+(A_{th}) - \mu^+(A) = \frac{\mu(A_{th} \cap S) - \mu(A \cap S)}{t} =$$

$$\frac{\mu(A_{th} \cap S) - \mu(A \cap S)}{t} + \frac{\mu(A \cap (S \setminus S_{th})) - \mu(A \cap (S_{th} \setminus S))}{t} \to \mu'_h(A \cap S) \quad (t \to 0),$$

since the second term tends to zero. This implies

$$(\mu^+)'_h(A) = (\mu' h)(A \cap S). \tag{2.6}$$

Also,

$$(\mu^-)'_h(A) = -(\mu' h)(A \setminus S). \quad \square \tag{2.7}$$

COROLLARY 2.2. *A realvalued differentiable measure μ on $(X, \mathcal{U})$ has a logarithmic derivative, given by*

$$(\mu' h)(A) = \int_A f_h(x)\mu(dx), \tag{2.8}$$

where $f_h(x) = f_h^{\pm}(x)$, for $x \in S$ and $x \notin S$, respectively, $f_h^{\pm}$ being the logarithmic derivatives of $\mu^{\pm}$, respectively.

THEOREM 2.4. *Let X be a locally convex space, X' its dual, and let the σ-algebra of subsets of X generated by the X'-cylinder sets coincide with the Borel σ-algebra. If a realvalued measure μ is τ_f-differentiable in a direction h and the derivative $\mu' h$ is a measure which is absolutely continuous with respect to μ, then μ is τ_v-differentiable in the direction h.*

PROOF. Every ball in $(\mathcal{M}, \tau_v)$ is closed in $(\mathcal{M}, \tau_f)$. Therefore, by (2.4), the following estimate holds

$$\left\| \frac{\mu_{th} - \mu}{t} - \mu' h \right\| \leq \sup_{0<\theta<1} \{ \| (\mu' h)_{\theta th} - \mu' h \| \}.$$

Hence it suffices to show that $\lim_{\theta \to 0} (\mu' h)_{\theta h} - \mu' h = 0$. From (2.3) it follows that $\lim_{t \to 0} \|\mu_{th} - \mu\| = 0$, and thus the proof reduces to that of the following lemma.

LEMMA 2.1. *Let μ, ν be realvalued measures on the σ-algebra $\mathcal{U} = \sigma(\mathcal{U}_{X'})$ of a locally convex space X generated by the X'-cylinder sets, and let $\nu < \mu$. Then $\lim_{t \to 0} \|\mu_{th} - \mu\| = 0$ implies that $\lim \|\nu_{th} - \nu\| = 0$.*

PROOF. Let $\nu(A) = \int_A f(x)\mu(dx)$, and let f^ϵ be a simple $\mathcal{U}_{X'}$-measurable function for which $\int_X |f(x) - f^\epsilon(x)| |\mu|(dx) < \epsilon$. The relations

$$\|\nu_{th} - \nu\| = \|f_{th}\mu_{th} - f\mu\| \leq$$
$$\|f_{th}\mu_{th} - f^\epsilon \mu_{th}\| + \|f^\epsilon_{th}\mu_{th} - f^\epsilon \mu\| + \|f^\epsilon \mu - f\mu\| \leq$$
$$2\epsilon + \|f^\epsilon_{th}\mu_{th} - f^\epsilon \mu\|$$

imply that, without loss of generality, f can be taken a $\mathcal{U}_{X'}$-measurable finitely valued function; moreover, since such a function is a finite linear combination of indicators of cylinder sets, f can be taken such an indicator. In this case

$$\begin{gathered}\|f_{th}\mu_{th} - f\mu\| \leq \\ \|f_{th}\mu_{th} - f_{th}\mu\| + \|f_{th}\mu - f\mu\| \leq \\ \|\mu_{th} - \mu\| + \|(f_{th} - f)\mu\|,\end{gathered}$$

and the proof is finished by noting that $f_{th} \xrightarrow{\mu} f$. □ □

2.3. The derivative of a measure as an element of the dual space. Up till now we have considered derivatives in a fixed direction h. If a measure μ is differentiable in every direction $h \in \mathcal{H}$ and, moreover, the map $h \to \mu'(A)h$ is linear in h for each $A \in \mathcal{U}$, then μ is τ_f-differentiable along $\mathcal{K}$. However, in general this does not yet imply that the linear map $\mu'(A)\colon \mathcal{K} \to \mathbf{R}$ is continuous. Nevertheless, if $\mu'(A) \in \mathcal{K}'$, then the following result can be obtained.

THEOREM 2.5. *If a measure μ is τ_S-differentiable along $\mathcal{K}$, and $\mathcal{K}$ is a semireflexive space (i.e. is algebraically isomorphic to its second dual), then the $\mathcal{K}$-derivative μ' is a σ-additive function on $\mathcal{U}$ with values in $\mathcal{K}'$.*

PROOF. By the well-known Pettis theorem it suffices to prove weak additivity of μ'. The semireflexivity of $\mathcal{K}$ implies that for any $\phi \in \mathcal{K}''$ there is an $h \in \mathcal{H}$ such that $\phi(u) = u(h)$. Then $\phi(\mu'(A)) = \mu'(A)h$ is a σ-additive measure. □

REMARK 2.1. $(\mathcal{K}, \tau_S)$-differentiability of a measure μ implies its continuity, and hence boundedness of the map $\mu'\colon \mathcal{K} \to \mathcal{M}_w(X,\mathcal{U})$, where w is the weak topology of the Banach space $\mathcal{M}_v$, and consequently boundedness of $\mathcal{K} \to \mathcal{M}_v(X,\mathcal{U})$, since the families of bounded sets in these spaces coincide. If $\mathcal{K}$ is bornological, then the last map is also continuous. The estimate $\|\mu_h - \mu\| \leq |\mu' h|$ implies for a weakly differentiable measure continuity in variation of the shift operator.

2.4. Higher order derivatives. The higher order derivatives of a measure $\mu \in \mathcal{M}$ are determined in the same way as the first order derivative, i.e. as corresponding derivatives of the map $\mu_h\colon \mathcal{K} \to \mathcal{M}$. Thus, the $(\mathcal{M}, \delta)$-derivative of order n of μ along $(\mathcal{K}, \beta)$ is a multilinear map $\mu^{(n)}\colon \mathcal{K} \times \ldots \mathcal{K} \to \mathcal{M}$ of class $\mathcal{L}^{(n)}(\mathcal{K}, \mathcal{M})$. A measure μ which is differentiable up to order n along a subspace $(\mathcal{K}, \beta)$ satisfies the relation

$$\mu(A+h) - \mu(A) = \sum_{k=1}^{n} \frac{1}{k!}\mu^{(k)}(h,\ldots,h)(A) + \xi(A,t,h) \quad (A \in \mathcal{U}). \tag{2.9}$$

Here the map $\phi(t) = \xi(A,t,h)\colon (-\epsilon,\epsilon) \to (\mathcal{M},\delta)$ is n times differentiable, and $\phi^{(k)}(0) = 0$ $(k = 1,\ldots,n)$ uniformly with respect to the system of sets β.

Relation (2.9) implies that the measure $\mu' h$ $(h \in \mathcal{H})$ is differentiable up to order $n-1$ in the direction h, its derivatives being $(\mu' h)^{(k-1)}(h,\ldots,h) = \mu^{(k)}(h,\ldots,h)$.

Moreover,

$$\mu^{(k)}(A)(h_1,\ldots,h_k) = \frac{\partial^k}{\partial \tau_1 \ldots \partial \tau_k}\mu(A + \tau_1 h_1 + \cdots + \tau_k h_k)|_{\tau_1=\cdots=\tau_k=0}.$$

This expression is called the *derivative of order k in the direction* $(h_1, \ldots, h_k)$. Clearly,

$$\mu^{(k)}(A)(h_1, \ldots, h_k) = [\mu^{(k-1)}(A)(h_1, \ldots, h_{k-1})]'h_k \quad (h_1, \ldots, h_k \in \mathcal{H}),$$

so this formula can be taken as a basis for determining higher derivatives.

Let X be a separable topological linear space, $\mathfrak{M}_X$ the space of all Radon measures in X, $C(X)$ the space of all bounded continuous functions on X, τ_f the weak topology in $\mathfrak{M}_X$ (which is the dual of $C(X)$), and $\mathcal{U}$ the σ-algebra of Borel subsets of X. The following proposition follows directly from the definitions.

PROPOSITION 2.3. *A necessary condition for a measure $\mu \in \mathfrak{M}_X$ to be τ_f-differentiable in a direction h is that for every function $\phi \in C(X)$ the limit*

$$\lim_{t \to 0} \frac{1}{t} \left\{ \int_X \phi(x) \mu_{th}(dx) - \int_X \phi(x) \mu(dx) \right\} \tag{2.10}$$

exists. (This condition is also sufficient if $(\mathfrak{M}_X, \tau_f)$ is a sequentially complete space, cf. theorem III.3.2.)

τ_f-differentiability is weaker than τ_v-differentiability. However, there is the following result.

THEOREM 2.6. *If $\mu \in \mathfrak{M}_X$ is twice τ_f-differentiable in a direction (h, h), then it is τ_v-differentiable in the direction h.*

PROOF. The unit ball of $(\mathfrak{M}_X, \tau_v)$ is closed in τ_f. Therefore, by (2.3),

$$\|(\mu'h)_{th} - \mu'h\| \le t\|(\mu'h)'h\|,$$

and hence the map $t \to (\mu'h)_{th} \colon \mathbf{R} \to (\mathfrak{M}_X, \tau_v)$ is continuous (here τ_f-derivatives are meant). Further, by (2.4) the estimate

$$\left\| \frac{\mu_{th} - \mu}{t} - \mu'h \right\| \le \sup_{0<\theta<1} \|(\mu'h)_{\theta h} - \mu'h\|$$

holds, implying, combined with the continuity of the map $t \to (\mu'h)_{th}$ the result required. □

COROLLARY 2.3. *For any n and set of elements $h_1, \ldots, h_n$ of a subspace $\mathcal{K} \subset X$, the class of measures in $\mathfrak{M}_X$ that are τ_f-differentiable in the direction $(h_1, \ldots, h_n)$ up to order n coincides with the class of measures that are τ_v-differentiable in each of these directions.*

REMARK 2.2. It can be proved that if $\mathcal{K}$ is a Fréchet space, then τ_S-differentiability up to order n in any direction among $h_1, \ldots, h_n \in \mathcal{K}$ of a measure μ implies that the map

$$(h_1, \ldots, h_n) \to \mu^{(n)}(h_1, \ldots, h_n) \colon \mathcal{K} \times \cdots \times \mathcal{K} \to (\mathfrak{M}_X, \tau_v)$$

is multilinear and continuous (for more details see Smolyanov [4]).

This assertion implies that a sufficient condition for a measure μ to be infinitely τ_v-differentiable along a given subspace $\mathcal{K}$ is that for any continuous function $f \in C(X)$ and set $h_1, \ldots, h_n \in \mathcal{K}$, the function

$$\psi \colon (t_1, \ldots, t_n) \to \int_X f(x + t_1 h_1 + \cdots + t_n h_n) \mu(dx)$$

is n times differentiable at $t = 0$ $(n = 1, 2, \dots)$.

REMARK 2.3. Let X be a locally convex space with topology given by a collection of Hilbert prenorms, and let $C^\infty(X)$ be the linear space of infinitely differentiable functions on X that, together with their derivatives, are bounded on X. Let δ_c be the weak topology in $\mathfrak{M}_X$ determined by the duality associated with the bilinear form

$$\langle f, \mu \rangle = \int_X f(x)\mu(dx) \quad (f \in C^\infty(X)).$$

It can be shown (see Smolyanov [4]) that theorem 2.6 and corollary 2.3 can be generalized to this case. This follows from the fact that the unit ball in $(\mathfrak{M}_X, \tau_v)$ is closed in δ_c.

3. Distributions and generalized functions

3.1. Test functions and measures. Let X, Y be linear spaces with duality relation $\langle x, y \rangle$ $(x \in X, y \in Y)$. Consider a topological linear space Φ of complexvalued functions on Y which are infinitely-differentiable on Y in any finite set of directions from a linear subset $Y_0 \subset Y$ (possibly coinciding with Y) and, together with all derivatives, bounded: for all $k = 1, 2, \dots,$

$$|\phi^{(k)}(y)(h_1, \dots, h_k)| \leq C_{k;h_1,\dots,h_k} \quad (y \in Y, \quad h_1, \dots, h_k \in Y_0).$$

This space Φ is called a *test space* (or *space of test functions*) if the following conditions hold:

T1) for any σ-additive complexvalued measure μ in Y, convergence $\phi_n \to \phi$ in Φ implies

$$\int_Y \phi(y)\mu(dy) = \lim_{n\to\infty} \int_Y \phi_n(y)\mu(dy); \tag{3.1}$$

T2) for any $h \in Y_0$ the shift operation

$$\phi \to S_h\phi, \quad (S_h\phi)(y) = \phi(y - h),$$

is defined and continuous in Φ;

T3) for any collection $h_1, \dots, h_k \in Y_0$ the differentiation operation $D^{(k)}_{h_1,\dots,h_k}\phi(y) = \phi^{(k)}(y)(h_1, \dots, h_k)$ is continuous in Φ.

A function $\psi(y)$ $(y \in Y)$ is called a *multiplier* on Φ if multiplication by it defines a continuous linear operator on Φ:

$$(T_\psi\phi)(y) = \psi(y)\phi(y).$$

The set of multipliers on Φ, which is clearly an algebra, will be denoted by $\check{\Phi}$. Note that if Φ is an algebra with continuous product, then $\Phi \subset \check{\Phi}$, i.e. the elements of Φ are themselves multipliers.

Below we might need the following additional conditions:

T4) the function $\phi_x(y) = e^{i\langle x, y \rangle}$ $(x \in X)$ is a multiplier on Φ which continuously depends on $x \in X$;

T4′) in addition to T4, $1 \in \Phi$. (Then $\check{\Phi} \subset \Phi$; in particular, $e^{i\langle x, y \rangle} \in \Phi$ depends continuously on $x \in X$ as an element of Φ.)

The following case is important: $X = \mathcal{H}_+ = \overleftarrow{\lim}\mathcal{H}_\alpha$ is the projective limit of a family of Hilbert spaces $\mathcal{H}_\alpha$ and $Y = X' = \mathcal{H}_- = \overrightarrow{\lim}H_{-\alpha}$ is the inductive limit of the dual spaces $\mathcal{H}_{-\alpha} = \mathcal{H}'_\alpha$. Some space in this family can always be identified with its dual, say $\mathcal{H}_0 = \mathcal{H}'_0$. Then also

$$X \subset \mathcal{H}_0 = \mathcal{H}'_0 \subset X' \text{ and } \mathcal{H}_\alpha \subset \mathcal{H}_0 = \mathcal{H}'_0 \subset \mathcal{H}_{-\alpha}. \tag{3.2}$$

In this case Y_0 can be taken one of $\mathcal{H}_\alpha$, e.g. $\mathcal{H}_0$.

This case will be frequently considered below. In it, one additional condition is used:

T5) there is a complete set $\mathcal{P}$ of finite-dimensional orthoprojectors on $\mathcal{H}_0$, with values in X, for which $\phi \in \Phi$ implies $\phi_P \in \Phi$, where $\phi_P(y) = \phi(Py)$ $(P \in \mathcal{P})$.

In a sense, the following condition is a weak version of this:

T5′) for any $P \in \mathcal{P}$ and $\phi, \psi \in \Phi$ we have $\phi_P \psi_{I-P} \in \Phi$.

We consider some examples of test spaces.

EXAMPLE 3.1. The space Φ_1 (Φ_1^α) of all complexvalued functions that are infinitely-differentiable in directions from Y ($\mathcal{H}_\alpha$) and, together with their derivatives, are bounded. Its topology is given by uniform convergence on compact subsets of Y of bounded sequences $\psi_n^{(k)}(y)(h_1, \dots, h_k)$, for any $k, h_1, \dots, h_k \in Y$ $(\mathcal{H}_\alpha)$.

This space satisfies the conditions T1), T2), T3), T4′), and T5).

EXAMPLE 3.2. The space $\Phi_0^{(\mathcal{P})} \subset \Phi$ of all $\mathcal{P}$-cylindrical functions $\phi(y) = \phi(Py)$ $(P \in \mathcal{P})$ in Φ, with a topology which is stronger than that induced from Φ_1: convergence of a sequence of functions is defined by including that they also have a common support.

The conditions T1), T2), T3), T4′), and T5) hold.

EXAMPLE 3.3. For a Hilbert space $X = X'$, the space of all functions in Φ_1 having compact support, with the additional requirement on a convergent sequence that there is a common support.

This space does not satisfy T5), but T5′) holds.

Consider the topological linear space $\mathfrak{M} = \mathfrak{M}(Y)$ of all complexvalued measures on $(Y, \sigma(\mathcal{U}_X))$. The space $\mathfrak{M}$ is called a *space of test measures* if the following conditions hold:

M1) the topology in $\mathfrak{M}$ is not weaker than the topology τ_f of weak convergence, i.e. convergence $\mu_n \longrightarrow \mu$ in $\mathfrak{M}$ implies

$$\int_{X'} f(y)\mu(dy) = \lim_{n\to\infty} \int_{X'} f(y)\mu_n(dy) \quad (f \in C(X'));$$

M2) for each $x \in X$ it follows from $\mu \in \mathfrak{M}$ that $\langle x, y\rangle\mu \in \mathfrak{M}$, this product operation being continuous with respect to x;

M3) for each $x \in X$ it follows from $\mu \in \mathfrak{M}$ that $e^{i\langle x,y\rangle}\mu \in \mathfrak{M}$, this product operation being continuous with respect to x.

In some situations we assume one additional condition to hold:

M4) for any $h \in Y_0$ (where Y_0 is a subspace of Y) the shift operation

$$\mu \to \mu_h, \quad \mu_h(A) = \mu(A+h),$$

is defined and continuous on $\mathfrak{M}$.

EXAMPLE 3.4. The space M_1 of all complexvalued measures satisfying the condition

$$\|\langle x_1, y\rangle \dots \langle x_n, y\rangle \mu\| < \infty$$
$$(\forall n, x_1, \dots, x_n \in X), \quad n = 1, 2, \dots,$$

with topology determined by the system of prenorms

$$\| \int_{X'} \langle x_1, y\rangle \dots \langle x_n, y\rangle f(y)\mu(dy)\|$$
$$(x_1, \dots, x_n \in X, \quad f \in C(X'), \quad n = 1, 2, \dots).$$

A variant is obtained by replacing the topology by the stronger topology determined by the system of prenorms

$$\|\langle x_1, y\rangle \dots \langle x_n, y\rangle \mu\| < \infty.$$

For both spaces the conditions M1), M2), M3), and M4) hold.

EXAMPLE 3.5. The subspace $M_{1,\infty}$ of M_1 consisting of all measures in M_1 that are infinitely differentiable in directions from Y_0, together with their derivatives. The topology is determined by prenorms of the same type as in example 3.4, but taking into account derivatives.

PROPOSITION 3.1. *If $M(X)$ is a space of test measures in X, then the set of Fourier transforms,*

$$\Phi_M = \left\{\phi : \phi(y) = \int_X e^{i\langle x,y\rangle}\mu(dx), \quad \mu \in M(X)\right\},$$

of elements of $M(X)$ is a space of test functions on Y, with topology generated by the map $\mu \to \int_X e^{i\langle x,y\rangle}\mu(dx)$.

PROOF. Condition T1) follows from M1); T2) from M3); T3) from M2). Note that T4) follows from M4). □

3.2. Distributions. Operations on distributions. Let Φ be a space of test functions on Y. Its dual space Φ', equipped with the topology of weak convergence, is called the corresponding *space of distributions* (*generalized measures*) on Y. The value of a distribution $\xi \in \Phi'$ on an element $\phi \in \Phi$ is denoted by $\langle \phi, \xi\rangle$. Condition T1) implies that each complexvalued measure μ in X' gives rise to a distribution (which will be identified with this measure and denoted by the same symbol), by means of the relation

$$\langle \phi, \mu\rangle = \int_Y \phi(y)\mu(dy).$$

Sometimes it is convenient to use the integral expression at the righthand side of this formula as a notation for $\langle \phi, \xi\rangle$ (replacing μ by ξ, of course, so $\langle \phi, \xi\rangle \stackrel{\text{def}}{=} \int_Y \phi(y)\xi(dy)$).

We will now consider some basic operations on distributions. These operations are determined, as usual, as dual to the corresponding operations on Φ. Then their restriction to the set of measures must coincide with the operations which are naturally defined for measures. A number of formulas below can be established by means of a standard (hence not given) argument.

A) The shift operator is defined by

$$\langle \phi, \xi_h \rangle = \langle \phi_h, \xi \rangle, \quad \phi_h(y) = \phi(y - h) \quad (h \in \mathcal{H}).$$

Clearly, if μ is a measure, then so is μ_h, and $\mu_h(A) = \mu(A + h)$.

B) Let $\phi \in \Phi$, and let μ be a measure which is differentiable in a direction $h \in \mathcal{H}$. Differentiating with respect to t at $t = 0$ the relation

$$\int_Y \phi(y)\mu_{th}(dy) = \int_Y \phi(y - th)\mu(dy),$$

we obtain a formula for integration by parts:

$$\int_Y \phi(y)(\mu' h)(dy) = -\int_Y \phi'(y)h\mu(dy). \tag{3.3}$$

This formula gives a clue for defining the derivative $\xi' h$ of a distribution ξ in a direction h:

$$\langle \phi, \xi' h \rangle = -\langle \phi' h, \xi \rangle \quad (\phi \in \Phi). \tag{3.4}$$

In a similar way we define the derivative of order m in a direction $(h_1, \ldots, h_m)$ as the distribution $\xi^{(m)}(h_1, \ldots, h_m)$ defined by the relation

$$\langle \phi, \xi^{(m)}(h_1, \ldots, h_m) \rangle = (-1)^m \langle \phi^{(m)}(h_1, \ldots, h_m), \xi \rangle. \tag{3.5}$$

Multilinearity of $\phi^{(m)}(y)$ implies multilinearity of the distribution $\xi^{(m)}(h_1, \ldots, h_m)$ in $h_1, \ldots, h_m$. The symbol $\xi^{(m)}$ will denote the corresponding multilinear map

$$\xi^{(m)} \colon \mathcal{H} \times \cdots \times \mathcal{H} \to \Phi'.$$

C) For a pair $\psi, \xi \in \Phi$ we define the distribution $\psi\xi$ by

$$\langle \phi, \psi\xi \rangle = \langle \phi\psi, \xi \rangle \quad (\phi \in \Phi).$$

It is obvious that the bilinear expression $\psi\xi$ is continuous in each of its arguments. The relation

$$(\psi_1\psi_2)\xi = \psi_1(\psi_2\xi) \quad (\psi_1, \psi_2 \in \Phi)$$

holds, and if $\psi' h \in \Phi$, then

$$(\psi\xi)' h = (\psi' h)\xi + \psi(\xi' h). \tag{3.6}$$

D) Let $\Phi^{(1)}$, $\Phi^{(2)}$ be spaces of test functions on Y_1, Y_2, respectively, and let $\Phi = \Phi^{(1)} \otimes \Phi^{(2)}$ be the tensor product; it is a test space on $Y_1 \times Y_2$. The elements of Φ' will be called *bidistributions*.

Suppose that the following condition holds:

For any $\phi \in \Phi$ and $y_2 \in Y_2$ we have $\phi_{y_2} \in \Phi^{(1)}$, where

$$\phi_{y_2}(y_1) = \phi(y_1, y_2) \quad (y_1 \in Y_1, \quad y_2 \in Y_2). \tag{3.7}$$

If for some $\xi \in \Phi^{(1)\prime}$ we have $\langle \phi_{y_2}, \xi\rangle \in \Phi^{(2)}$ $(\forall \phi \in \Phi)$, then for any $\eta \in \Phi^{(2)\prime}$ the product $\xi \times \eta \in \Phi^{(1)}$ is defined:

$$\langle \phi, \xi \times \eta\rangle = \langle\langle \phi_{y_2}, \xi\rangle, \eta\rangle. \tag{3.8}$$

For $\phi(y_1, y_2) = \phi_1(y_1)\phi_2(y_2)$ formula (3.8) takes the simple form

$$\langle \phi_1 \otimes \phi_2, \xi \times \eta\rangle = \langle \phi_1, \xi\rangle\langle \phi_2, \eta\rangle. \tag{3.9}$$

It follows from (3.9) that if $\xi = \mu$, $\eta = \nu$ are measures in Y_1, Y_2, respectively, then $\mu \times \nu$ coincides with their product in $T_1 \times Y_2$.

From (3.8) we can derive a formula for the derivative of a product:

$$(\xi \times \eta)'(h_1, h_2) = (\xi' h_1) \times \eta + \xi \times (\eta' h_2). \tag{3.10}$$

Note that the first term is defined, since under the conditions imposed the following relation is true:

$$\langle \phi_{y_2}, \xi' h\rangle = \langle (\phi' h_1)_{y_2}, \xi\rangle \in \Phi^{(2)}.$$

E) Let A be a linear map from Y_1 to Y_2 which is compatible with a pair of test spaces $\Phi^{(1)}$, $\Phi^{(2)}$ in the sense that

$$\phi \in \Phi^{(2)} \to \phi_A(x) = \phi(Ax) \in \Phi^{(1)}. \tag{3.11}$$

Then to every $\xi \in \Phi^{(1)}$ there corresponds a distribution $\xi_A \in \Phi^{(2)\prime}$:

$$\langle \phi, \xi_A\rangle = \langle \phi_A, \xi\rangle \quad (\phi \in \Phi^{(2)}).$$

The corresponding formula for derivatives is

$$(\xi' h)_A = \xi_A'(Ah). \tag{3.12}$$

In particular, let $X = \mathcal{H}_+ \subset \mathcal{H}$, and P a finite-dimensional projection operator on $Y = \mathcal{H}_- \supset \mathcal{H}$ with range in $\mathcal{H}_+$. It is compatible with the pair (Φ, Φ) if T5) holds. The distribution ξ_P is called a *finite-dimensional projection* of the distribution ξ. The derivative of a finite-dimensional projection can be expressed by the formula $(\xi' h)_P = (\xi_P')(Ph)$.

F) We introduce the operation of convolution of distributions. Let Φ be a test space on Y, and $\Phi^{(2)} = \Phi \times \Phi$ its tensor square. $\Phi^{(2)}$ is supposed to be defined as a test space on $Y \times Y$. Further, let Ψ be a test space on Y with the property that the map

$$(y_1, y_2) \to (y_1 + y_2) \colon Y \times Y \to Y$$

is compatible with the pair $(\Phi^{(2)}, \Psi)$.

Then there is defined for every bidistribution $\beta \in \Phi^{(2)}$ a distribution $\tilde{\beta} \in \Psi$:

$$\langle \phi, \tilde{\beta}\rangle_\Psi = \langle \phi(y_1 + y_2), \beta\rangle_{\Phi^{(2)}}. \tag{3.13}$$

If $\beta = \xi \times \eta$, then $\tilde{\beta}$ is called the *convolution* of the distributions ξ and η, denoted by $\xi * \eta$. For symmetry reasons it follows that taking convolution is a commutative operation. Equality (3.13) for measures μ, ν leads to the usual relation

$$(\mu * \nu)(B) = \int_Y \mu(B - y)\nu(dy). \tag{3.14}$$

It is easily verified that the differentiation formulas

$$(\xi * \nu)'h = \xi'h * \nu = \xi * \nu'h \tag{3.15}$$

and

$$(\xi * \nu)'(h_1 + h_2) = (\xi'h_1) * \nu + \xi * (\nu'h_2) \tag{3.16}$$

hold.

G) We consider the transform of a distribution under a nonlinear transformation of space. Let $Y = \mathcal{H}_- \supset \mathcal{H} \supset \mathcal{H}_+ = X$, let $z = f(y)$ be an invertible map from Y onto Y, and let $y = F(z)$ be its inverse. Assume that they are both differentiable along $\mathcal{H}_+$, and that the derivatives of $f(y)$, $F(z)$ leave invariant the space $\mathcal{H}_+$. We say that f is compatible with Φ if $\phi^f(y) = \phi(f(y)) \in \Phi$ when $\phi \in \Phi$. We then define the map ξ^f by

$$\langle \phi, \xi^f \rangle = \langle \phi^f, \xi \rangle \quad (\phi \in \Phi).$$

We compute the derivative of a distribution in a direction $h \in \mathcal{H}_+$. By definition,

$$\langle \phi, \xi^{f\prime} h \rangle = -\langle \phi' h, \xi^f \rangle = -\langle (\phi' h)^f, \xi \rangle,$$

where

$$(\phi' h)^f = (\phi', h)^f = (\phi'^f, h) = (\phi f\prime, (f')^{-1} h) = \sum_k (\phi^{f\prime}, e_k)(e_k, (f')^{-1} h),$$

where $\{e_k\}$ is an orthonormal basis in $\mathcal{H}$, $e_k \in \mathcal{H}_+$ $(k = 1, 2, \dots)$.

Each term in this sum can be written as

$$\{\phi(f(y))(e_k, [f'(y)]^{-1} h)\}' e_k - \phi(f(y))(e_k, [f'(y)]^{-1} h)' e_k.$$

The map $y \to f'(y)^{-1} h$ acts from $\mathcal{H}_-$ to $\mathcal{H}_+$: $[f'(y)]^{-1} h \colon \mathcal{H}_- \to \mathcal{H}_+$. Suppose that this operator is nuclear (this is true if $\mathcal{H}_+$ is a nuclear space, or, more generally, if for some α, $\mathcal{H}_\alpha \to \mathcal{H}_0$ is a Hilbert–Schmidt imbedding).

Then

$$\operatorname{div}[f'(y)]^{-1} h = \operatorname{Tr}([f'(y)]^{-1} h)' = \sum_k (e_k, ([f'(y)]^{-1} h)' e_k)$$

exists, and after summation we arrive at

$$\langle (\phi', h)^f, \xi \rangle = -\left\langle \phi^f, \sum_k (e_k, f'^{-1}(y) h)(\xi', e_k) \right\rangle - \langle \phi^f \operatorname{div}[f'(y)]^{-1} h, \xi \rangle.$$

Regarding ξ' as a vector distribution with the property that

$$\langle \phi, (\xi' h) \rangle = \langle \phi h, \xi' \rangle,$$

we can write

$$\sum_k \langle \phi^f (e_k, [f'(y)]^{-1} h), (\xi' e_k) \rangle =$$

$$\left\langle \phi^f \sum_k (e_k, [f'(y)]^{-1} h) e_k, \xi' \right\rangle =$$

$$\langle \phi^f [F'(z) h]^f, \xi' \rangle = \langle \phi, (\xi'^f F'(z) h) \rangle.$$

So finally we obtain the following formula:

$$\langle \phi, \xi^{f\prime} h \rangle = \langle \phi, (\xi'^{f} F'h) \rangle + \langle \phi, \operatorname{div}[f'^{-1}h]^F \xi^f \rangle, \tag{3.17}$$

which can be written as

$$(\xi^f)'h = \xi'^f (F'h) + [\operatorname{div}(F'(f))h]^F \xi^f. \tag{3.17'}$$

Note the important case when ξ is determined by a measure μ on Y which is differentiable along $\mathcal{H}_0$. In this case formula (3.17') gives the weak derivative of the transformed measure μ^f. By theorem 2.4, however, μ^f is strongly differentiable (in the sense of τ_v-convergence) as well, since the righthand side of (3.17') is absolutely continuous with respect to μ^f.

3.3. Generalized functions and kernels. If M is a space of test measures in Y, then its dual M', equipped with the topology of weak convergence, is called a *space of generalized functions* on Y.

Condition M1) implies that the set of bounded continuous functions lies in this space.

Operations on generalized functions are defined in the same way as those on distributions. So we restrict ourselves to a brief indication.

a) The shift operator. Suppose M4) is fulfilled. Then the formula

$$\langle \mu, \alpha_h \rangle = \langle \mu_h, \alpha \rangle \quad (\mu \in M, \quad h \in Y)$$

defines, for every $\alpha \in M'$, a generalized function $\alpha_h \in M'$.

b) Differentiation. If the operation of differentiation in a direction $h \in Y$ is defined and continuous on M, then it is also defined for generalized functions, by

$$\langle \mu, \alpha'_h \rangle = -\langle \mu'_h, \alpha \rangle.$$

c) Multiplication by a function. If a function ϕ is a multiplier on M, i.e. the operation $\mu \to \phi\mu$ is continuous in M, then the operation $\alpha \to \phi\alpha$ is defined on M':

$$\langle \mu, \phi\alpha \rangle = \langle \phi\mu, \alpha \rangle.$$

d) Convolution of a generalized function and a measure. Suppose that the operation of convolution of measures leaves invariant M and is continuous with respect to each factor. Put $\nu^*(A) = \nu(-A)$. If $\alpha \in M'$, $\mu \in M$, then the formula

$$\langle \mu * \nu^*, \alpha \rangle = \langle \mu, \beta \rangle$$

defines a generalized function $\beta = \alpha * \nu$, called the *convolution* of α and ν.

It is readily verified that if α is a bounded continuous function, then

$$(\alpha * \nu)(y) = \int_Y \alpha(y - z)\nu(dz).$$

e) Change of variables. Let $y = f(x)$ be a map on Y which is compatible with M in the sense that the map

$$\mu \to \mu^f : \mu^f(A) = \mu(f^{-1}(A))$$

leaves invariant M and is continuous on M. Then for every $\alpha \in M'$ the generalized function α^f is defined by

$$\langle \mu, \alpha^f \rangle = \langle \mu^f, \alpha \rangle.$$

f) Tensor product. Generalized kernels. Let M be a space of test measures in Y, and $M^{(2)}$ a space of test measures in $Y \times Y$ having the property

for every $A \in \mathcal{B}(Y)$, $\mu \in M^{(2)}$ the measure $\mu_A(B) = \mu(A \times B) \in M$ and the measure $\mu^{(\alpha)}(A) = \langle \mu_A, \alpha \rangle \in M$ for $\alpha \in M'$ continuously depend on μ.

So, for $\beta \in M'$ the generalized function $\alpha \otimes \beta \in M^{(2)}$ is defined by

$$\langle \mu, \alpha \otimes \beta \rangle = \langle \mu^{(\alpha)}, \beta \rangle \quad (\mu \in M^{(2)}).$$

It is called the tensor product of the generalized functions α and β.

Suppose $M^{(2)}$ contains the product $\mu_1 \times \mu_2$ of all pairs $\mu_1, \mu_2 \in M$, and suppose that this product operation depends continuously on each factor. Clearly, for $\mu = \mu_1 \times \mu_2$ $(\mu_1, \mu_2 \in M)$:

$$\langle \mu_1 \times \mu_2, \alpha \otimes \beta \rangle = \langle \mu_1, \alpha \rangle \langle \mu_2, \beta \rangle.$$

Elements of $M^{(2)}$ are called *generalized kernels*.

Under the above conditions, convolution of a generalized kernel and an element of M is defined. Suppose $\alpha \in M^{(2)\prime}$, $\mu \in M$. Then $\mu * \alpha$ is regarded as an element of M', defined by

$$\langle \mu, \alpha * \nu \rangle = \langle \mu \times \nu, \alpha \rangle.$$

Under the conditions of subsection 3.2, a generalized function $\alpha \in M'$ gives rise to a generalized kernel $\tilde{\alpha}$, which is defined (at least) on products of measures:

$$\langle \mu \times \nu, \tilde{\alpha} \rangle = \langle \mu \times \nu^*, \alpha \rangle.$$

Such a kernel is called a *difference kernel.*

It is easily verified to have the invariance property:

$$\tilde{\alpha}_h = \tilde{\alpha},$$

where shift of a kernel is defined by

$$\langle \mu \times \nu, \tilde{\alpha}_h \rangle = \langle \mu_h \times \nu_h, \tilde{\alpha} \rangle.$$

Clearly, the convolution $\tilde{\alpha} * \mu$ of a difference kernel $\tilde{\alpha}$ induced by a generalized function α and a measure μ coincides with the convolution $\alpha * \mu$ of this generalized function α and the measure μ.

g) Let α be a Y-invariant generalized function, $\alpha = \alpha_h$ $(h \in Y)$, that is,

$$\langle \mu, \alpha \rangle = \langle \mu_h, \alpha \rangle \quad (h \in Y).$$

In certain cases we assume that the following condition holds:

M5) every invariant test function coincides with a constant, i.e.

$$\langle \mu, \alpha \rangle = C\mu(Y).$$

We consider some situations in which M5) holds. Suppose the convolution of measures

$$\mu \to \mu * \nu^* = S_\nu(\mu)$$

is a continuous operation on M, while the integral in the representation

$$(\mu * \nu^*)(A) = \int_Y \mu_{-y}(A)\nu(dy)$$

converges on M. Then

$$\langle \mu * \nu^*, \alpha \rangle = \int_Y \langle \mu_{-y}, \alpha, \rangle \nu(dy) = \int_Y \langle \mu, \alpha_{-y} \rangle \nu(dy) = \langle \mu, \alpha \rangle \nu(Y),$$

provided the generalized function α is invariant. Replacing ν^* by ν, we find

$$\langle \mu * \nu, \alpha \rangle = \langle \mu, \alpha \rangle \nu(Y) = \langle \nu, \alpha \rangle \mu(Y),$$

for $\nu^*(Y) = \nu(Y)$ and for symmetry reasons.

Applying this relation for a probability measure ν_0, we obtain $\langle \mu, \alpha \rangle = \langle \nu_0, \alpha \rangle \mu(Y)$, and so the constant C in the formula $\langle \mu, \alpha \rangle = C\mu(Y)$ equals $\langle \nu_0, \alpha \rangle$, the value of the invariant generalized function considered on a test probability measure.

3.4. Fourier transformation of distributions. Suppose, first, that the test space Φ satisfies T4), i.e. $e^{i\langle x,y \rangle} \in \Phi$ and the map $x \to e^{i\langle x,y \rangle}$ $(X \to \Phi)$ is continuous.

Then the function

$$\chi_\xi(x) = \langle e^{i\langle x,y \rangle}, \xi \rangle \quad (\xi \in \Phi') \tag{3.18}$$

is continuous on X. If μ is a measure, $\chi_\mu(x)$ is its characteristic functional.

The map $\xi \to \chi_\xi$ is called *Fourier transformation* of the map ξ; the phrase *Fourier transform* is reserved for the function $\chi_\xi(x)$ itself. We consider some properties of Fourier transformation (and presume a mild form of T4) to hold in each case).

a) Let $A\colon X_1 \to X_2$ be a linear map, and $A^*\colon X_1' = Y_1 \to X_2' = Y_2$ its adjoint, which is assumed to be compatible with the pair Φ_1, Φ_2 of test spaces. To each $\xi \in \Phi_1'$ we associate $\xi_{A^*} \in \Phi_2'$, and

$$\chi_{\xi_{A^*}}(x) = \langle e^{i\langle x,y \rangle}, \xi_{A^*} \rangle = \langle e^{i\langle x,A^*y \rangle}, \xi \rangle = \langle e^{i\langle Ax,y \rangle}, \xi \rangle = \chi_\xi(Ax). \tag{3.19}$$

In particular, let $X' = \mathcal{H}_- \supset \mathcal{H} \supset \mathcal{H}_+ = X$, and let $P\colon X' \to X_P'$ be a finite-dimensional orthoprojector on $\mathcal{H}_-$ mapping it to the finite-dimensional subspace $X_P' \subset \mathcal{H}$. It is obvious that $P = (P|_X)^*$ is the adjoint of the restriction of P to X. The map $\xi \to \xi_P$ associates to a distribution ξ on X' a distribution ξ_P on X_P', the corresponding space of test functions being $\Phi_P = \{\phi_p\colon \phi_P(y) = \phi(Py)\}$. The Fourier transform of the finite-dimensional projection ξ_P of the distribution ξ is the function $\chi_{\xi_P}(Px) = \chi_\xi(Px)$, i.e. the restriction of χ_ξ to X_P'.

b) Differentiation. Under differentiation of a distribution in a direction h, its Fourier transform is multiplied by a linear function:

$$\chi_{\xi'h}(x) = \langle e^{i\langle x,y \rangle}, \xi'h \rangle = -\langle (e^{i\langle x,y \rangle})'h, \xi \rangle =$$
$$-i\langle x,h \rangle \langle e^{i\langle x,y \rangle}, \xi \rangle = -i\langle x,h \rangle \chi_x i(x).$$

By repeated differentiation we obtain the formula

$$\chi_{\xi^{(m)}(h_1,\dots,h_m)} = (-i)^m \langle x, h_1\rangle \dots \langle x, h_m\rangle \chi_\xi(x).$$

c) Product and convolution. Formula (3.9) immediately implies that

$$\chi_{\xi*\eta}(x_1,x_2) = \langle e^{i\langle x_1,y_1\rangle + i\langle x_2,y_2\rangle}, \xi\times\eta\rangle =$$
$$\langle e^{i\langle x_1,y_1\rangle}, \xi\rangle \langle e^{i\langle x_2,y_2\rangle}, \eta\rangle = \chi_\xi(x_1)\chi_\eta(x_2).$$

Using (3.19) with a map $A^*\colon X'\times X' \to X'$, $A^*(y,z) = y+z$, adjoint to the map $A\colon X \to X\times X$, $Ax = (x,x)$, we obtain a formula for the Fourier transform of a convolution:

$$\chi_{\xi*\eta}(x) = \chi_\xi(x)\chi_\eta(x).$$

Now we consider the case when the space Φ of test functions is constructed, as in proposition 3.1, from a space of measures M on X (i.e. it consists of the Fourier transforms of these measures).

Let $\phi(y) = \int_X e^{i\langle x,y\rangle}\nu(dx)$ and $\xi' \in \Phi$. We define the Fourier transform of this distribution as the generalized function $\hat{\xi}'$ defined by

$$(\nu, \hat{\xi}) = \langle \phi, \xi\rangle. \tag{3.20}$$

If the distribution ξ is defined on bounded functions, in particular when $\xi = \mu$ is a measure, then its Fourier transform $\chi_\xi(x)$, which is a continuous function, is defined. Also,

$$\langle \phi, \xi\rangle = \int_X \langle e^{i\langle x,y\rangle}, \xi\rangle \nu(dx) = \int_X \chi_\xi(x)\nu(dx),$$

and thus the generalized function $\hat{\xi}$ can be identified with $\chi_\xi(\omega)$.

Some of the properties described above hold also, under certain conditions, in this version of Fourier transformation (in particular, a) and c)). However, we omit the simple argumentation. We only note a formula for the value of the differential kernel $\tilde{\hat{\xi}}$ determined by a generalized function $\hat{\xi}$ on a product of test measures:

$$\langle \mu * \nu, \tilde{\hat{\xi}}\rangle = \langle \mu * \nu^*, \hat{\xi}\rangle = \langle \chi_\mu \chi_\nu, \xi\rangle. \tag{3.21}$$

3.5. Differential expressions for distributions. Differential expressions for distributions can be defined as expressions dual to differential operations on test functions. In particular, certain operations on measures give as result distributions. However, if the result turns out to be a measure, then the corresponding differential expression can, as a rule, be defined directly by means of differential operations on measures. If a differential expression is defined on a space of test measures, it can be transferred to generalized functions by means of passing again to the dual operation.

We consider some examples.

Let $X \subset \mathcal{H} \subset X'$ be a rigged Hilbert space, and $\Phi(X')$ a space of test functions. Consider the differential expression $l_\phi(u(y)) = (\phi(y), u'(y))$, where $\phi\colon X' \to \mathcal{H}$ and u' is the $\mathcal{H}$-derivative of the function $u\colon X' \to \mathcal{H}$. Assume that the operation l_ϕ leaves invariant Φ. Then the operator $_\phi^*$ is defined on Φ':

$$\langle l_\phi u, \xi\rangle = \langle u, l_\phi^* \xi\rangle \quad (u \in \Phi). \tag{3.22}$$

Using expansion in an orthogonal basis $\{e_k\}$ in $\mathcal{H}$, it is readily seen that if $\phi_k(y) = (\phi(y), e_k) \in \Phi$ $(k = 1, 2, \dots)$ and $\operatorname{div} \phi \in \Phi$, then

$$l_\phi^*(\xi) = -\xi_\phi' - \operatorname{div} \phi\xi.$$

Formula (3.22) can now be written as

$$\langle (\phi, \nabla) u, \xi \rangle = -\langle u, (\phi, \nabla)\xi \rangle - \langle u, \operatorname{div} \phi \cdot \xi \rangle. \tag{3.23}$$

Consider a second-order differential operator,

$$l_a(u) = \sum_{j,k=1}^{\infty} a(e_j, e_k) u''(y)(e_j, e_k) = \operatorname{Tr} A u''(y),$$

where $a(f, g) = (Af, g)$ is a bilinear form with nuclear operator A. In this case,

$$l_a^*(\xi) = \sum_{j,k=1}^{\infty} a(e_j, e_k) \xi''(e_j, e_k), \tag{3.24}$$

the series being weakly convergent.

If $a = a(y)$, the formula becomes more cumbersome, due to the addition of a first-order differential operator.

The condition that A be nuclear can be weakened to $A \in \mathcal{L}_2(\mathcal{H})$ if the test space consists of functions $u(x)$ for which $u''(x) \in \mathcal{L}_2(\mathcal{H})$.

Higher-order expressions of the type (3.24) can be similarly considered.

4. Positive definiteness. Quasi-invariant distributions and bidistributions

4.1. Positive distributions. Consider a test space Φ satisfying T5), and hence containing sufficiently smooth cylindrical functions. For a distribution ξ on Φ its Fourier transform $\chi_\xi(x) = \langle e^{i\langle x, y\rangle}, \xi \rangle$ is defined.

A distribution ξ is called *positive* if $\phi \geq 0$ implies $\langle \phi, \xi \rangle \geq 0$.

The Fourier transform of a positive distribution is positive definite:

$$\sum_{j,k=1}^{\infty} \chi_\xi(x_k - x_j) c_k \bar{c}_j = \sum_{j,k=1}^{n} \langle e^{i\langle x_k - x_j, y\rangle}, \xi \rangle c_k \bar{c}_j = \left\langle \left| \sum_{j,k=1}^{n} c_k e^{-i\langle x_k, y\rangle} \right|^2, \xi \right\rangle \geq 0.$$

The Fourier transform of a finite-dimensional projection ξ_P $(P \in \mathcal{P})$ is the function $\chi_\xi(Px)$.

PROPOSITION 4.1. *A necessary and sufficient condition for the Fourier transform $\chi_\xi(x)$ of a positive distribution ξ to have continuous restrictions to all finite-dimensional subspaces $X_P = PX$ $(P \in \mathcal{P})$ is that this distribution is a nonnegative $\mathcal{P}$-cylindrical measure.*

PROOF. The function $\chi_\xi(Px)$ is positive definite and continuous on X_P, hence it is the Fourier transform of a finite positive measure, which is naturally identified with the finite-dimensional projection ξ_P (see proposition 3.3). The converse is obvious. □

REMARK 4.1. If T4) holds, then $\chi_\xi(x)$ is continuous on X, while if X is also nuclear, then by theorem III.1.3 the distribution ξ can be identified with a measure supported in X'.

REMARK 4.2. For distributions that are not positive an exact description of the finite-dimensional projections is not known. However, if also

$$\int_{X_P} |\chi_\xi(Px)|^2 \, d_P x < \infty \quad (P \in \mathcal{P}), \tag{4.1}$$

where $d_P x$ is Lebesgue measure in X_P, then in X_P the inverse Fourier transformation makes sense:

$$\frac{1}{(2\pi)^{n/2}} \int_{X_P} \chi_{\xi,P}(x) e^{-i\langle x,y\rangle} \, d_P x = \rho_P(y).$$

It is locally integrable, and gives rise to a measure μ_P: $\mu_P(dy) = \rho_P(y)\, d_P y$. It is not difficult to state sufficient conditions for $\int |\rho_P(y)|\, d_P y$ to exist (e.g. the conditions (4.1) for derivatives $\chi^{(n)}(x)$ of sufficiently high order). So, finite-dimensional distributions ξ_P can be identified with (in general, nonpositive) measures in X'.

We now weaken T5) to T5′), and assume that T4) holds.

We first consider the finite-dimensional case ($\dim X < \infty$). The function

$$\langle \phi(y) e^{i\langle x,y\rangle}, \xi\rangle = \chi_{\xi,\phi}(x)$$

is continuous, for every $\phi \in \Phi$. If $\xi \geq 0$ and $\phi \geq 0$, then also $\xi_\phi = \phi\xi \geq 0$, and since its Fourier transform is continuous, ξ_ϕ is a nonnegative measure. Thus there is defined a map

$$\phi \to \xi_\phi,$$

a associating to a nonnegative test function ϕ a nonnegative measure $\xi_\phi \in X'$. This map is linear, and has the property

$$\xi_{\phi\psi} = \phi\xi_\psi. \tag{4.2}$$

Using this property it can be easily shown that there is a nonnegative, in general σ-finite, measure ξ in X' such that $\xi_\phi = \phi\xi$. This measure is defined by the formula

$$\xi(A) = \int_A \frac{1}{\phi(y)} \xi_\phi(dy) \tag{4.3}$$

for sets A on which $\phi(y) > 0$. It follows from (4.2) that the defining equation (4.3) does not depend on the choice of ϕ. In fact, for any ψ,

$$\int_A \psi(y)\xi(dy) = \int_A \frac{\psi(y)}{\phi(y)} \xi_\phi(dy) = \int_A \frac{\xi_\psi \phi(dy)}{\phi(y)} = \int_A \chi_\psi(dy) = \chi_\psi(A),$$

does not depend on ϕ.

We now return to the infinite-dimensional case. The considerations above remain rue if X is a nuclear space, since in this case continuity of $\chi_{\xi,\phi}(x)$ implies σ-additivity of the measure $\xi_\phi = \phi \circ \xi$ in X'. If X is not nuclear, we need continuity of $\chi_{\xi,\phi}$ in the, stronger, τ-topology.

Let $\phi \in \Phi$ and $P \in \mathcal{P}$. The formula

$$\langle \phi_P, \xi_{P,\psi}\rangle = \langle \phi_P \psi_{I-P}, \xi\rangle$$

associates to a $\xi \in \Phi'$ a distribution $\xi_{P,\psi}$ in the finite-dimensional space X_P. For $\psi \geq 0$, as shown above, $\xi_{P,\psi}$ is a σ-finite measure in X'_P. The family $\{\xi_{P,\psi}\}$ satisfies the consistency conditions

$$\xi_{P,\psi}(A) = \int_{X'_Q - X'_P} \psi_{Q-P}(y)\xi_{Q,\psi}(A \times dy),$$

(4.4)

$$P \subset Q, \quad \psi(y) = \psi(Qy)\psi((I-Q)y).$$

In a certain sense this family is similar to the system of finite-dimensional distributions of a finite σ-additive measure. It is of interest to study the question whether this system is determined by a σ-finite measure.

4.2. Integral representations of invariant generalized kernels. A generalized kernel β is called *positive definite* if

$$\langle \mu \times \hat{\mu}, \beta \rangle \geq 0 \quad (\mu \in M).$$

A generalized function α is called *positive definite* if the difference kernel determined by it is such, i.e. $\langle \mu \times \bar{\mu}^*, \tilde{\alpha} \rangle \geq 0$.

The Fourier transform of a positive distribution is positive definite. In fact, by (3.21),

$$\langle \mu \times \bar{\mu}^*, \tilde{\hat{\xi}} \rangle = \langle |\chi_\mu|^2, \xi \rangle \geq 0.$$

Consider a symmetric (i.e. $\langle \mu_1 \times \bar{\mu}_2, \beta \rangle = \overline{\langle \mu_2 \times \bar{\mu}_1, \beta \rangle}$) positive definite generalized kernel β which is invariant under shifts by elements of Y:

$$\langle \mu_{1y} \times \mu_{2y}, \beta \rangle = \langle \mu_1 \times \mu_2, \beta \rangle.$$

Introduce the scalar product

$$(\mu_1, \mu_2)_\beta = \langle \mu_1 \times \mu_2, \beta \rangle$$

in the space M of test measures, and construct the Hilbert space M_β by the usual procedure of completion and factorization.

In this space the operator group

$$U_y \mu = \mu_y \quad (y \in Y)$$

acts, and invariance of the kernel readily implies that the operators U_y are unitary: $U_y^* = U_{-y}$. Thus we obtain a representation of Y by unitary operators on M_β.

If M4) holds for $Y = Y_0$, then the function $\langle \mu_y \times \bar{\mu}, \beta \rangle = (U_y \mu, \mu)_\beta$ is continuous on Y.

Let Y be a nuclear space and $X = Y'$. By theorem III.1.4, the following representation holds:

$$(U_y \mu_1, \mu_2)_\beta = \int_X e^{i\langle x, y \rangle} (E(dx)\mu_1, \mu_2)_\beta,$$

where $E(\Delta)$ is the spectral measure supported in $X = Y'$ with as values orthoprojectors on M_β.

Suppose there is a Hilbert space M_+ for which $M \subset M_+ \subset M_\beta$, with imbedding $J \in \mathcal{L}_2(M_+, M_\beta)$. Since we can make the identification $M_\beta = M'_\beta$, we have

$$M \subset M_+ \subset M_\beta = M'_\beta \subset M'_+ \subset M'.$$

As pointed out in remark III.1.4, also in this case there is a measure ν on Y with respect to which $J^*E(\Delta)J$ can be differentiated, and such that the following representation holds:

$$(U_y\mu_1, \mu_2)_\beta = \int_X e^{i\langle x,y\rangle} K_x(\mu_1, \mu_2)\nu(dx) \quad (\mu_1, \mu_2 \in M), \tag{4.5}$$

where K_x is a ν-almost everywhere defined bilinear form on $M_+ \subset M$.

For $y = 0$ we obtain

$$(\mu_1, \mu_2)_\beta = \int_X K_x(\mu_1, \mu_2)\nu(dx). \tag{4.6}$$

Formulas (4.5) and (4.6) imply (see (III.1.14) and (III.1.15)):

$$K_x(U_y\mu_1, \mu_2) = e^{i\langle x,y\rangle} K_x(\mu_1, \mu_2); \quad K_x(U_y\mu_1, U_y\mu_2) = K_x(\mu_1, \mu_2); \tag{4.7}$$
$$(y \in X'; \quad x \in X \bmod \nu).$$

Consider the generalized function $\alpha_{x,\mu}$: $\langle \mu, \alpha_{x,\mu}\rangle = \overline{K_x(\mu, \mu_1)}$, and let $\gamma_{x,\mu} = e^{i\langle x,y\rangle}\alpha_{x,\mu}$. Now (4.6) implies invariance of $\gamma_{x,mu}$:

$$\begin{aligned}\langle \mu_1, \gamma_{x,\mu,h}\rangle &= \langle \mu_{1h}, e^{i\langle x,y\rangle}\alpha_{x,\mu}\rangle = \langle e^{i\langle x,y\rangle}\mu_{1h}, \alpha_{x,\mu}\rangle = \\ &\overline{K_x(\mu, e^{i\langle x,y\rangle}\mu_{1h})} = e^{-i\langle x,h\rangle}\overline{K_x(\mu, [e^{i\langle x,y\rangle}\mu_1]_h)} = \\ &\overline{K_x(\mu, e^{i\langle x,y\rangle}\mu_1)} = \langle \mu_1, \gamma_{x,\mu}\rangle.\end{aligned}$$

Suppose that M5) holds, i.e. only constants can be invariant. Then $e^{i\langle x,y\rangle}\alpha_{x,\mu} = \overline{C_x(\mu)}$, C_x depending linearly and continuously on μ, i.e. $C_x(\mu) = \langle \mu, \lambda_x\rangle$ ($\lambda_x \in M'$).

So we obtain the formula

$$K_x(\mu_1, \mu) = \langle \bar{\mu}_1, C_x(\mu)e^{i\langle x,y\rangle}\rangle = \langle \mu, \lambda_x\rangle\overline{\chi_{\mu_1}(-x)},$$

and for symmetry reasons, $K_x(\mu, \mu_1) = \overline{\langle \mu_1, \lambda_x\rangle}\chi_\mu(-x)$. Here, for some $\mu_0 \in M$:

$$\langle \mu, \lambda_x\rangle = \frac{\overline{\langle \mu_0, \lambda_x\rangle}\chi_\mu(-x)}{\overline{\chi_{\mu_0}(-x)}} = k(x)\chi_mu(-x), \quad k(x) = \frac{\overline{\langle \mu_0, \lambda_x\rangle}}{\overline{\chi_{\mu_0}(-x)}};$$

and finally,

$$K_x(\mu, \mu_1) = k(x)\chi_\mu(-x)\overline{\chi_{\mu_1}(-x)}, \tag{4.8}$$

where $k(x)$ is a measurable function on X. (It can be easily found that $k(x) = K_x(\mu_0, \mu_0)/|\chi_{\mu_0}(x)|^2$, where μ_0 is a probability measure.)

Putting $k(x)\nu(dx) = \Lambda(-dx)$ we obtain the representation

$$(U_y\mu_1, \mu_2)_\beta = \int_X e^{-i\langle x,y\rangle}\chi_{\mu_1}(x)\overline{\chi_{\mu_2}(x)}\Lambda(dx).$$

which for $y = 0$ takes the form

$$\langle \mu_1 \times \bar{\mu}_2, \beta \rangle = int_X \left[\int_Y e^{i\langle x,y \rangle} \mu_1(dy) \right] \overline{\left[\int_Y e^{i\langle x,y \rangle} \mu_2(dy) \right]} \Lambda(dx). \tag{4.9}$$

This formula shows that β is a generalized difference kernel, being the Fourier transform of the (in general, σ-finite) measure Λ: $\beta(y_1, y_2) = \int_X e^{i\langle x, y_1 - y_2 \rangle} \Lambda(dx)$.

Note that the assumption that there exists a Hilbert space M_+ with appropriate properties is fulfilled if β is continuous in the nuclear topology τ_S of M; in particular, if M is nuclear.

This implies

THEOREM 4.1. *Let M be a space of test measures on a nuclear space Y, suppose M4) holds for all $h \in Y$, and suppose that also M5) holds. Let β be an invariant positive definite generalized kernel for which the bilinear form $\langle \mu_1 \times \bar{\mu}_2, \beta \rangle$ is continuous in the nuclear topology of M.*

Then the representation

$$\beta(y_1, y_2) = \int_X e^{i\langle x, y_1 - y_2 \rangle} \Lambda(dx)$$

holds, where Λ is a σ-additive, in general σ-finite, measure in $X = Y'$.

In particular, β is a difference kernel.

REMARK 4.3. The nuclearity of Y can be replaced by continuity of the function $y \to \langle \mu_y \times \mu, \beta \rangle$ in the nuclear topology $\tau_S(Y)$.

4.3. Quasi-invariant distributions. A distribution ξ is called *$\mathcal{H}$-quasi-invariant* ($\mathcal{H} \subset Y$) if there is a function $\rho_h \in \Phi$ ($h \in \mathcal{H}$), $\rho_h(y) = \rho(h, y)$ ($y \in Y$) such that

$$\langle \phi_h, \xi \rangle = \langle \rho_h \phi, \xi \rangle \quad (h \in \mathcal{H}, \quad \phi \in \Phi).$$

In other words, ξ is $\mathcal{H}$-quasi-invariant if

$$\xi_h = \rho_h \xi. \tag{4.10}$$

It follows from (4.10) that for any $k \in \mathcal{H}$ the distribution ξ_k is also $\mathcal{H}$-quasi-invariant: $(\xi_k)h = [\rho_h]_k \xi_k$. Here is is necessary that

$$[(\rho_h)_k \rho_k - \rho_{h+k}]\xi = 0.$$

This is certainly true if

$$\rho(k + h, y) = \rho(k, y)\rho(h, y + k) \quad (y \in Y \quad h, k \in \mathcal{H}). \tag{4.11}$$

Everywhere below we assume that $\rho(h, y)$ (the *quasi-invariance factor*) has this property.

If ξ is a nonnegative measure, then $\rho(h, y) \geq 0$, and (4.10) means that ξ_h is absolutely continuous with respect to ξ.

THEOREM 4.2. *Let Y be a barrelled space. There does not exist a nonzero Y-quasi-invariant Radon measure in Y.*

PROOF. Let μ be a Radon measure in Y. There is a compact set $K \subset Y$ such that $|\mu|(K) > 0$. Let $\Gamma(K)$ be the closed balanced convex hull of K. The set $\Gamma(K)$ cannot be absorbing in Y, since otherwise it would be a barrel in Y and hence a neighborhood of zero in Y. However, together with K, $\Gamma(K)$ is compact, which is impossible because Y is infinite dimensional.

This implies that the linear set $L = \cup_{n=1}^{\infty} n\Gamma(K)$ does not coincide with Y. Let $y \in Y \setminus L$. Assume μ to be Y-quasi-invariant. Then for every $t \in \mathbf{R}$, $|\mu|(L + ty) > 0$ and $(L + t_1 y) \cap (L + t_2 y) = \emptyset$ $(t_1 \neq t_2)$.

However, the finite measure $|\mu|$ cannot be positive on the family $L_t = L + ty$ $(t \in \mathbf{R})$ of disjoint sets, since the cardinality of this family is that of the continuum. □

REMARK 4.4. It can be readily shown that for a distribution not determined by a σ-additive measure this result need not hold. In fact, let Y be a Hilbert space, Φ a space of smooth cylindrical functions, and ξ a Gaussian cylindrical measure with the identity as correlation operator. Then for $\phi = \phi_P$,

$$\langle \phi, \xi \rangle = \frac{1}{(2\pi)^{n/2}} \int_{Y_P} \phi(y) e^{-\|y\|^2/2}\, dy \quad (Y_P = PY, \quad \dim Y_P = n),$$

and it is readily verified that the quasi-invariance condition holds for all $h \in Y$, while also

$$\rho(h, y) = e^{(h,y) - \|h\|^2/2}.$$

We will clarify the relations between quasi-invariance and differentiability of a measure.

Let μ be a $\mathcal{H}$-quasi-invariant measure in Y, $A \in \mathcal{U}$. The relation

$$\mu(A + h) = \mu_h(A) = \int_A \rho(h, y)\mu(dy)$$

implies that

$$\frac{\mu(A + th) - \mu(A)}{t} = \int_A \frac{\rho(th, y) - \rho(0, y)}{t} \mu(dy).$$

If $\rho(h, y)$ is differentiable at zero with respect to h, then the derivative

$$\mu'(A)h = \int_A [\rho'(0, y)h]\mu(dy) \tag{4.12}$$

exists. Thus, in this case μ is differentiable on every set $A \in \mathcal{U}$. Its derivative is then absolutely continuous with respect to itself. By theorem 2.4, μ is differentiable in variation as well.

EXAMPLE 4.1. Consider the Gaussian measure with zero mean and correlation operator B on the Hilbert space $\mathcal{H}$.

It follows from (II.4.2) that this measure is $B^{1/2}\mathcal{H}$-quasi-invariant, and also

$$\rho(h, y) = \exp\left\{(B^{-1}h, y) - \frac{1}{2}\|B^{-1/2}h\|^2\right\}. \tag{4.13}$$

The function (4.13) is differentiable for $h \in B^{1/2}\mathcal{H}$, and

$$\rho_h'(0, y)h = (B^{-1}h, y). \tag{4.14}$$

Therefore μ is differentiable in variation along $B^{1/2}\mathcal{H}$. By similar considerations we can show that it is infinitely differentiable, and compute higher derivatives. E.g.,

$$\mu''(A)(h_1, h_2) = \int_A \{(B^{-1}h_1, x)(B^{-1}h_2, x) - (B^{-1}h_1, h_2)\}\mu(dx).$$

Let ξ be an $\mathcal{H}$-quasi-invariant distribution with quasi-invariance factor $\rho(h,x)$, and let $\phi \in \Phi$ be a function on Y ($\phi(y) = 0$, $y \in Y$). Then the distribution $\xi_\phi = \phi\xi$ is also $\mathcal{H}$-quasi-invariant, with factor

$$\rho_\phi(h, y) = \rho(h, y)\frac{\phi(y+h)}{\phi(y)}. \tag{4.15}$$

In fact,

$$\langle \psi_h, \xi_\phi \rangle = \langle \psi_h \phi, \xi \rangle = \langle (\psi\phi_{-h})_h, \xi \rangle = \langle \psi\phi_{-h}\rho_h, \xi \rangle = \left\langle \psi\rho_h \frac{\phi_{-h}}{\phi}, \xi \right\rangle. \tag{4.16}$$

It can be seen from (4.15) that for a function ϕ which vanishes at certain points, ξ_ϕ is not, in general, $\mathcal{H}$-quasi-invariant, although it is a differentiable measure if ξ is differentiable and ϕ is smooth. Thus there exist differentiable measures which are not $\mathcal{H}$-quasi-invariant. In this connection, the following result is of interest.

THEOREM 4.3. *Let μ be a measure in Y which is τ_S-differentiable along a separable locally convex bornological topological linear space $\mathcal{H}$. Then there is a $\mathcal{H}$-quasi-invariant measure λ, also τ_S-differentiable along $\mathcal{H}$, with respect to which μ is absolutely continuous: $\mu < \lambda$.*

PROOF. Let h_n be a dense sequence in $\mathcal{H}$ ($h_0 = 0$), and put

$$\nu = \sum_{n=0}^{\infty} \frac{1}{2^n}\mu_{h_n}, \qquad \lambda = \frac{1}{\nu(\mathcal{H})}\nu.$$

It is readily verified that ν is differentiable along $\mathcal{H}$, as is μ. Since $\mathcal{H}$ is bornological, $\nu'(A) \in \mathcal{H}'$ (see remark 2.1). It follows from $\nu(A) = 0$ that $\mu_0(A) = \mu(A) = 0$, i.e. $\mu < \nu$. It remains to show the $\mathcal{H}$-quasi-invariance of ν. If $\nu(A) = 0$, then $\mu_{h_n}(A) = 0$, and hence $\mu(A + h_n) = 0$ ($\forall n$). Because $\mathcal{H}$ is bornological, the map $h \to \mu(A+h)$ is continuous. Therefore $\mu(A+h) = 0$ ($\forall h \in \mathcal{H}$) and $\mu_{h_n}(A+h) = \mu(A+h+h_n) = 0$ ($\forall h \in \mathcal{H}$). Finally $\nu(A+h) = 0$. $\square$

We now clarify some conditions under which the quasi-invariance factor uniquely determines a distribution (up to a constant factor).

Suppose T4) holds, implying $\check{\Phi} \subset \Phi$.

PROPOSITION 4.2. *Suppose two nonzero distributions ξ, η are $\mathcal{H}$-quasi-invariant with factor $\rho(h,y)$. If the linear hull of the set of functions $\{\rho_h(y)\colon h \in \mathcal{H}\}$ is dense in Φ, then there is a constant C such that $\eta = C\xi$.*

PROOF. Suppose $\langle 1, \xi\rangle \neq 0$, and put $C = \langle \eta, 1\rangle / \langle \xi, 1\rangle$. Then

$$\langle \rho_h, \eta - C\xi\rangle = \langle \rho_h \cdot 1, \eta\rangle - C\langle \rho_h \cdot 1, \xi\rangle = 0,$$

and thus, under the conditions stated, $\eta - C\xi = 0$. Similar considerations show that $\langle \xi, 1\rangle = 0$ implies $\xi = 0$. □

EXAMPLE 4.2. Let $\mathcal{H}_+ \subset \mathcal{H}_0 \subset \mathcal{H}_-$ be a standard triple of Hilbert spaces, and let μ be a canonical Gaussian measure in $\mathcal{H}_-$ with zero mean and correlation operator which is the identity on $\mathcal{H}_0$. It follows from (II.4.2) that for $h \in \mathcal{H}_0$ we have

$$\rho_h(y) = e^{(h,y) - \|h\|^2/2} \quad (h \in \mathcal{H}_0, \quad y \in \mathcal{H}_-).$$

Consider the test space of example 3.2. The linear hull of the set of functions $\rho_n(y)$ contains polynomials in the linear functionals (h, y), and hence all $\mathcal{P}$-cylindrical functions. This means that in the corresponding space of distributions the Gaussian measure μ determines a unique distribution with corresponding quasi-invariance factor.

PROPOSITION 4.3. *Suppose that a distribution $\xi \in \Phi'$ is $\mathcal{H}$-quasi-invariant with factor $\rho(h, y)$, and suppose that a function $\phi \in \Phi$ has the property*

$$\langle \psi\phi, \xi_h\rangle = \langle \psi_h \phi, \xi\rangle \quad (\psi \in \Phi), \tag{4.17}$$

i.e. $(\phi\xi)_h = \phi\xi_h$. Then the distribution $\xi_\phi = \phi\xi$ is $\mathcal{H}$-quasi-invariant with the same factor ρ. Conversely, from the last assertion (4.17) follows.

PROOF. Since $\langle \psi\phi, \xi_h\rangle = \langle \psi\phi\rho_h, \xi\rangle = \langle \psi\rho_h, \xi_h\rangle$, (4.17) is equivalent to

$$\langle \psi_h, \xi_\phi\rangle = \langle \psi\rho_h, \xi_\phi\rangle. \quad \square$$

COROLLARY 4.1. *Suppose that (4.17) implies $\phi \equiv$ const. Then the distribution $\eta = \xi_\phi$ is $\mathcal{H}$-quasi-invariant with the same quasi-invariance factor as ξ precisely if $\phi \equiv$ const.*

For instance, in a finite-dimensional space two equivalent Lebesgue measures which are quasi-invariant with the same factor, can differ by a constant factor only.

4.4. Integral representations for quasi-invariant bidistributions. A bidistribution $\beta \in \Phi \otimes \Phi$ is called *positive* if

$$(\phi \otimes \bar{\phi}, \beta \geq 0 \quad (\phi \in \Phi).$$

A bidistribution β is called $\mathcal{H}$-quasi-invariant with quasi-invariance factor $\rho(h, y)$, where $\rho_h(y) = \rho(h, y) \in \Phi$, if

$$\langle \phi_h \otimes \psi_h, \beta\rangle = \langle \phi\rho_h \otimes \psi\rho_h, \beta\rangle \quad (\phi, \pi \in \Phi). \tag{4.18}$$

Everywhere below we assume that $\rho(h, y) > 0$ and that (4.11) holds.

We introduce a scalar product in Φ by

$$(\phi, \psi)_\beta = \langle \phi \otimes \bar{\psi}, \beta\rangle, \tag{4.19}$$

and consider the Hilbert space $\mathcal{H}_\beta$ obtained from Φ by standard completion.

If β is continuous in the nuclear topology $\tau_S(\Phi)$, in particular if Φ is nuclear, then there is a Hilbert space $\mathcal{H}_+ \subset \Phi$ which is Hilbert–Schmidt imbedded in $\mathcal{H}_\beta$.

The family of operators

$$(U_h\phi)(y) = \rho(h, y)\phi(y + h) = \rho_h\phi_{-h} \quad (h \in \mathcal{H})$$

acts in Φ. These operators are unitary in the sense of the scalar product $(\cdot, \cdot)_\beta$:

$$(U_h\phi, U_h\psi)_\beta = \langle \rho_h\phi_{-h} \otimes \rho_h\bar{\psi}_{-h}, \beta\rangle = \langle \phi \otimes \bar{\psi}, \beta\rangle = (\phi, \psi)_\beta,$$

and hence can be extended to $\mathcal{H}_\beta$ as unitary operators. It follows from (4.10) that the set $\{U_h\colon h \in \mathcal{H}\}$ is a group:

$$(U_{h+\tau}\phi)(y) = \rho(h + \tau, y)\phi_{-(h+\tau)}(y) = \rho(h, y[\rho(\tau, y)\phi_{-\tau}(y)]_{-h} = U_hU_\tau\phi(y).$$

Thus, we obtain a representation of the additive group of $\mathcal{H}$ by unitary operators on $\mathcal{H}_\beta$.

Let now $X = \mathcal{H})+ \subset \mathcal{H} \subset \mathcal{H}_- = Y$, where $\mathcal{H}_+$ is a nuclear space and $\rho(h, y)$ is continuous in $\mathcal{H}_+$, as an element of $\check{\Phi}$. By theorem III.1.4 and the remark contained in it, we have the representation:

$$(U_xf, g)_\beta = \int_Y e^{i\langle x,y\rangle}\langle f \otimes \bar{g}, \beta_y\rangle\nu(dy) \quad (f, g \in \Phi, \quad x \in X), \tag{4.20}$$

where β_y is the positive distribution determined by $\langle f \otimes \bar{g}, \beta_y\rangle = (E(dy)f, g)_\beta$ and having the properties (ν-almost everywhere with respect to y)

$$\begin{gathered}\langle U_xf \otimes \overline{U_xg}, \beta_y\rangle = \langle f \otimes g, \beta_y\rangle, \\ \langle U_xf \otimes \bar{g}, \beta_y\rangle = e^{i\langle x,y\rangle}\langle f \otimes \bar{g}, \beta_y\rangle,\end{gathered} \tag{4.21}$$

We assume that T4) holds. Then for $f = g = 1$,

$$(U_x1, 1)_\beta = \langle \rho_x 1 \otimes 1, \beta_y\rangle = \int_Y e^{i\langle x,y\rangle}\langle f \otimes \bar{g}, \beta_y\rangle\nu(dy). \tag{4.22}$$

PROPOSITION 4.4. *Suppose that the function $(\rho_x, 1)$ is continuous in the topology $\tau_S(\mathcal{H})$. Then $\tilde{\nu}(dy) = \langle 1 \otimes 1, \beta_y\rangle$ is a finite measure in $\mathcal{H}$, and the following representation holds:*

$$(U_hf, g)_\beta = \int_{\mathcal{H}} e^{i(h,y)}\frac{\langle f \otimes \bar{g}, \beta_y\rangle}{\langle 1 \otimes 1, \beta_y\rangle}\tilde{\nu}(dy) \quad (h \in \mathcal{H}, \quad f, g \in \Phi). \tag{4.23}$$

In particular,

$$(f, g)_\beta = \int_{\mathcal{H}} \frac{\langle f \otimes \bar{g}, \beta_y\rangle}{\langle 1 \otimes 1, \beta_y\rangle}\tilde{\nu}(dy) \quad (f, g \in \Phi). \tag{4.24}$$

PROOF. Theorem III.1.3 implies that $(\rho_x, 1)_\beta$ is the Fourier transform of a Radon measure in $\mathcal{H}$. Then (4.22) implies that $\tilde{\nu}$, and hence ν, are supported in $\mathcal{H}$. So (4.20) implies (4.23) for $x \in X$. The righthand side of this equality is also defined for $x \in \mathcal{H}$, and since both sides are continuous, they coincide for $x = h \in \mathcal{H}$. □

Introducing the notation $\beta * g \in \Phi'$ for the distribution $\langle f, \beta * g\rangle = \langle f * \bar{g}, \beta\rangle$, (4.22) can be rewritten:

$$[\rho_x(\beta_\theta * g)]_{-x} = e^{i\langle x,\theta\rangle}(\beta_\theta * g) \quad (\theta, x \in \mathcal{H}),$$

or, equivalently

$$(\beta_\theta * g)_x = \rho_x e^{i\langle x,\theta\rangle}(\beta_\theta * g). \tag{4.25}$$

This shows that the distribution $\beta_\theta * g$ is $\mathcal{H}$-quasi-invariant with factor $\rho(x,\theta)e^{-i\langle x,\theta\rangle}$.

Suppose that the following condition holds:

K) there is a probability measure μ which is $\mathcal{H}$-quasi-invariant with quasi-invariance factor $\rho(x, y)$.
Moreover, suppose that the linear hull of the set $\{\rho_x\}_{x\in\mathcal{H}}$ is dense in Φ (so that μ is the only $\mathcal{H}$-quasi-invariant distribution in Φ with factor ρ_x). Further, extend, for each $h \in \mathcal{H}$, the scalar product (h, y) as a measurable linear functional on Y which is square integrable with respect to μ.

Under the condition K) the function

$$\rho_x^{(\theta)}(y) = \frac{e^{-i(y+x,\theta)}}{e^{-i(y,\theta)}}\rho(x,y) = e^{-i(x,\theta)}\rho(x,y)$$

is the quasi-invariance factor of the complexvalued measure $\mu^{(\theta)} = e^{-i(\theta,y)}\mu$, and $\mu^{(\theta)}$ is uniquely determined by it in the class Φ'.

In this case, (4.24) implies

$$\beta_\theta * g = c(\theta, g)\mu^{(\theta)} = c(\theta, g)e^{-i(\theta,y)}\mu,$$

and by the symmetry of β_θ:

$$\langle f * \bar{g}, \beta_\theta\rangle = \langle f, \beta_\theta * g\rangle = c(\theta, g)\langle f, \mu^{(\theta)}\rangle = c(\theta, f)\bar{\langle g, \mu^{(\theta)}\rangle}.$$

In particular, for $f = 1$ we obtain

$$c(\theta, g) = \frac{\overline{c(\theta,1)\langle g, \mu^{(\theta)}\rangle}}{\langle 1, \mu^{(\theta)}\rangle}.$$

Therefore,

$$\langle f * \bar{g}, \beta_\theta\rangle = \frac{\overline{c(\theta,1)}}{\chi_\mu(-\theta)}\langle f, \mu^{(\theta)}\rangle\overline{\langle g, \mu^{(\theta)}\rangle} \quad (\theta \in \mathcal{H}, \quad f, g \in \Phi),$$

and then

$$\frac{\langle f * \bar{g}, \beta_\theta\rangle}{\langle 1 \times 1, \beta_\theta\rangle} - \langle f, \mu^{(\theta)}\rangle\overline{\langle g, \mu^{(\theta)}\rangle}\frac{1}{|\chi_\mu(\theta)|^2}.$$

This and proposition 4.4 imply

THEOREM 4.4. *Let $X = \mathcal{H}_+ \subset \mathcal{H} \subset \mathcal{H}_- = Y$ be a rigging of the Hilbert space $\mathcal{H}$ with a nuclear space $\mathcal{H}_+$ and its dual $\mathcal{H}_-$. Let Φ be a test space satisfying T4′. Let β be a positive definite bidistribution which is $\mathcal{H}$-quasi-invariant with factor $\rho(x,y)$, where $\rho(x,y)$ is an element of Φ which is continuous with respect to x. If the function*

$\langle \rho_x \otimes 1, \beta \rangle$ is continuous on $\mathcal{H}$ in the topology $\tau_S(\mathcal{H})$, then there is a measure ν, supported in $\mathcal{H}$, such that

$$\beta = \int_{\mathcal{H}} \beta_\theta \nu(d\theta),$$

where β_θ ($\theta \in \mathcal{H}$) are positive definite bidistributions satisfying (4.25) (and defined ν-almost everywhere with respect to θ).

If, moreover, condition K) holds, then the following representation is true:

$$\langle f * \bar{g}, \beta_\theta \rangle = \frac{\langle 1 \times 1, \beta_\theta \rangle}{|\chi_\mu(\theta)|^2} \langle f, \mu^{(\theta)} \rangle \overline{\langle g, \mu^{(\theta)} \rangle},$$

where $\mu^{(\theta)} = e^{i(\theta, y)}\mu$, μ is the unique $\mathcal{H}$-quasi-invariant measure in Y with factor $\rho(x, y)$, and (θ, y) ($\theta \in \mathcal{H}$, $y \in Y$) is a measurable linear functional on Y (which exists by condition K)).

Thus, in this case we have the following representation:

$$\langle f * \bar{g}, \beta \rangle = \int_{\mathcal{H}} \langle f, \mu^{(\theta)} \rangle \overline{\langle g, \mu^{(\theta)} \rangle} \hat{\nu}(d\theta),$$

where

$$\hat{\nu}(d\theta) = \frac{\hat{\nu}(d\theta)}{|\chi_\mu(\theta)|^2}$$

is, in general, a σ-finite measure in $\mathcal{H}$.

5. Chapter IV: Supplementary remarks and historical comments

There is a number of versions of differential calculus in infinite-dimensional spaces. A systematic and sufficiently detailed review of these theories in contained in Averbukh-Smolyanov [1,2], which we have followed in § 1. Second-order differential operators of the type (1.7) for functions in abstract spaces were introduced in Gross [4] and Dalecky [6,7,8]. In concrete sequence or function spaces such operators become operators involving infinitely many partial derivatives, or operators involving functional derivatives. In the computation of adjoint operators on the space of functions that are square integrable with respect to a Gaussian measure we have used the formula for integration by parts (chapter II) from Dalecky–Paramonova [1,2]. The investigation of selfadjointness of an operator of the type (1.17) was first undertaken in Dalecky [8], using a technique based on theorem 1.4. The corresponding operator semigroup was then constructed by methods presented in chapter VII below.

Subsequently conditions for selfadjointness of differential operators on functions of an infinite-dimensional argument were studied, using another technique and under various assumptions regarding the coefficients, in Reed [1], Berezansky [1], Berezansky–Mikhailyuk [1], Berezansky–Us [1], Berezansky–Sazonov [1], Frolov [5], Marchenko [1], Simon–Höegh-Krohn [1], and Dalecky [14].

The last paper studied dissipative operators on vector functions, probably for the first time ever. For other works in the field of differential equations with infinite-dimensional argument see also the comments to chapters V and VII.

Unfortunately, we have not succeeded in reporting the interesting direction in this field initiated in Lévy [1] and continued by E.M. Polishchuk, M.N. Feller, G.E. Shilov,

I.Ya. Dorfman, A.S. Nemirovsky, Yu.V. Bogdansky, et al., connected with the so-called Laplace–Lévy and similar operators (which are essentially infinite-dimensional in the sense of Bogdansky [1]).

The notion of a differentiable measure was introduced in Fomin [1], and studied by Averbukh–Smolyanov–Fomin [1] (see also Skorokhod [4]). Further progress in this theory has been reported in M.P. Kac [3], Uglanov [1,2,3,4], and Smolyanov–Fomin [1]. Our presentation in § 2 follows in essence Averbukh–Smolyanov–Fomin [1]. The theorems expressing relations between the various types of differentiability were first obtained by O.G. Smolyanov and M.P. Kac.

In Fomin [1,2] it was, for the first time ever, pointed out that in the distribution theory on infinite-dimensional spaces there arises a ramification of notions, since the latter spaces do not have a standard invariant measure. Linear functionals on the space of smooth functions can be generated by measures, but not by functions. The latter determine, in turn, functionals on smooth measures.

Thus there emerge two kinds of distributions: generalized measures (which are still called 'distributions') and generalized functions.

Various linear operations on distributions can be introduced as dual operations to those on test elements. For the sake of brevity, here we have presented operations on smooth measures as a special case of operations on generalized functions. Various spaces of test and generalized functions and measures have been considered in Dudin [1], Uglanov [3,4], Dalecky–Fomin [1].

The methods for the spectral analysis of positive definite generalized kernels are based on ideas of M.G. Krein. They were developed in Berezansky [1], and applied to functions of an infinite-dimensional argument in Berezansky–Gali [1], Berezansky–Gali–Zhuk [1], Kostyuchenko–Mityagin [1] Theorems 4.1 and 4.4, which are in some sense generalizations of theorems of Bochner and Schwartz (see, e.g., Gelfand–Vilenkin [1] and Minlos–Sazonov (see chapter III), are due to Dalecky–Samoilenko [1].

Quasi-invariance of Gaussian measures can already be found in papers by Cameron and Martin (in the special case of the Wiener measure). In Girsanov–Mityagin [1] and Sudakov [1] it is shown that in an infinite-dimensional space not all shifts of a σ-additive measure lead to absolutely continuous transforms (theorem 4.2). Following a proposal by I.M. Gelfand, quasi-invariant measures are taken to be those σ-additive measures that have a dense set of shifts, admissible in the sense mentioned. These measures were studied from different points of view in Gelfand–Vilenkin [1], Vershik [2], Vershik–Sudakov [1], Skorokhod [4], M.P. Kac [3]. Theorem 4.3 was proved in M.P. Kac [3]. The notion of a quasi-invariant kernel was introduced in Dalecky–-Samoilenko [1].

CHAPTER V

Evolution differential equations

In this chapter we consider various aspects related to the Cauchy problem for differential equations involving differential operators acting on functions of an infinite-dimensional argument, introduced in chapter IV.

In section 1 we present simple formal notions concerning evolution equations in spaces of test and generalized elements. The main part of this chapter (section 2) is devoted to the study of second order parabolic equations and systems. They are studied by projecting the argument into finite-dimensional spaces. We will obtain a fundamental solution of our problem by limit transition from the thus constructed, inconsistent, system of measures in these finite-dimensional spaces.

Using a special transformation, an equation involving a vector function is reduced to a scalar equation involving an argument in a space of higher dimension.

In chapter VII we will consider a number of questions from another point of view, based on techniques from the theory of stochastic processes.

1. Weak solutions of evolution equations

1.1. Fundamental distributions. We use the notations and concepts from the previous chapter.

Let (Ω, X) be a dual pair of linear topological spaces, $\Phi = \Phi(X)$ a space of test functions on X, and Φ' the corresponding space of distributions.

We denote by $\hat{\Phi}_k = \hat{\Phi}_k(T)$ the class of functions $u(x,t)$ of two variables ($x \in X$, $t \in T_\tau = [a,\tau] \subset [a,b] = T \subset \mathbf{R}^1$) such that for each $t \in T$ the function $u_t(x) = u(x,t) \in \Phi$ is continuous as an element of Φ (for $k = 0$), or (for $k > 0$) is k times continuously differentiable with respect to the parameter t.

Similarly, $\hat{\Phi}'_k$ is the class of distributions $\mu_t \in \Phi'$ ($t \in T$), depending on a parameter t, that are continuous (for $k = 0$) or (for $k > 0$) k times continuously differentiable with respect to t.

Consider a linear operator $\mathcal{A}(t)$ on Φ, depending on a parameter $t \in T = [a,b]$ but with dense, fixed (i.e. independent of t) domain of definition $\mathcal{D}$, and the related Cauchy problem

$$\frac{\partial u(x,t)}{\partial t} + \mathcal{A}(t)u(x,t) = v(x,t), \quad u(x,\tau) = u_0(x) \quad (t \leq \tau), \tag{1.1}$$

where $u_0 \in \Phi$, $v \in \hat{\Phi}_0$. A *solution* of this problem is a function $u(x,t)$ of class $\hat{\Phi}_1(T_\tau)$ with the property $u_t \in \mathcal{D}$ ($t \in T_\tau$) and identically satisfying equation (1.1).

As a rule, throughout this chapter and the next chapters, we consider the Cauchy problem in inverse form, by specifying an initial condition at the right end of the interval and by writing the generating operator at the lefthand side of the equation. On the one hand, this does not restrict the generality (the usual notation can be obtained by replacing t by $-t$). On the other hand, this notation has a certain advantage: the direct form is reserved for the dual equation. It is of utmost convenience that the direct forms express characteristic equations (when they make sense), in particular Itô stochastic equations, giving an analog to the characteristic equations for diffusion equations (see chapter VII).

The problem (1.1) is said to be *well-posed* (or *correct*) if the corresponding homogeneous problem

$$\frac{\partial u}{\partial t} + \mathcal{A}(t)u = 0, \quad u(x,\tau) = u_0(x) \quad (t \in T_\tau, \quad \tau \in T) \tag{1.1'}$$

has a unique solution, for each $u_0 \in \mathcal{D}$, and this solution depends continuously on t (in the space Φ), i.e.

$$u(x,t) = (U(t,\tau)u_0)(x),$$

where $U(t,\tau)$ $(t \in T_\tau)$ is a continuous linear operator on Φ.

This definition implies that the family of operators $U(t,\tau)$ leaves invariant the domain $\mathcal{D}$ and has the *evolution property* $U(t,s)U(s,\tau) = U(t,\tau)$, $U(t,t) = \text{id}$ $(t \leq s \leq \tau)$.

Let $\mathcal{A}^*(t)$ be the operator on Φ' adjoint to $\mathcal{A}(t)$:

$$\langle \mathcal{A}(t)\phi, \psi\rangle = \langle \phi, \mathcal{A}^*(t)\psi\rangle \quad (\phi \in \mathcal{D}, \quad \psi \in \mathcal{D}_{\mathcal{A}^*(t)}).$$

Consider the Cauchy problem dual to (1.1):

$$\frac{\partial \mu_\tau}{\partial \tau} = \mathcal{A}^*(\tau)\mu_\tau + \xi_\tau, \quad \mu_\tau = \nu. \tag{1.2}$$

A *solution* of this problem is a family of distributions $\mu_t \in \hat{\Phi}_1'([t,b])$ satisfying

$$\frac{\partial}{\partial \tau}\langle \phi, \mu_\tau\rangle \equiv \langle \mathcal{A}(t)\phi, \mu_\tau\rangle + \langle \phi, \xi_\tau\rangle, \quad \mu_t = \nu, \quad t \leq \tau, \tag{1.3}$$

for every $\phi \in \mathcal{D}$, or the equivalent integral equation

$$\langle \phi, \mu_s\rangle = \langle \phi, \nu\rangle + \int_t^s \langle \mathcal{A}(\tau)\phi, \mu_\tau\rangle\, d\tau + \int_t^s \langle \phi, \xi_\tau\rangle\, d\tau. \tag{1.3'}$$

A family $\mu_\tau \in \hat{\Phi}_0'$ is called a *weak solution* of (1.2) if it satisfies the integral relation

$$\langle \alpha_s, \mu_s\rangle - \langle \alpha_t, \nu\rangle = \int_t^s \left\langle \frac{\partial \alpha_\tau}{\partial \tau} + \mathcal{A}(\tau)\alpha_\tau, \mu_\tau \right\rangle d\tau + \int_t^s \langle \alpha_\tau, \xi_\tau\rangle\, d\tau \tag{1.4}$$

for every function $\alpha(x,t) \in \hat{\Phi}_1 \cap \mathcal{D}$.

PROPOSITION 1.1. *A necessary and sufficient condition for a family of distributions to be a solution of the problem (1.2) is that it is a weak solution of this problem.*

PROOF. If μ is a solution of (1.2), then (1.3) implies

$$\frac{\partial}{\partial \tau}\langle \alpha_\tau, \mu_\tau\rangle = \left\langle \alpha_\tau, \frac{\partial \mu_\tau}{\partial \tau}\right\rangle + \left\langle \frac{\partial \alpha_\tau}{\partial \tau}, \mu_\tau\right\rangle = \left\langle \frac{\partial \alpha_\tau}{\partial \tau} + \mathcal{A}(\tau)\alpha_\tau, \mu_\tau\right\rangle + \langle \alpha_\tau, \xi_\tau\rangle$$

and (1.4) follows by termwise integration.

Conversely, let (1.4) be fulfilled. Put $\alpha(x,\tau) = \phi(x)b(\tau)$, with $b(\tau)$ a continuously differentiable function satisfying the condition $b(s) = 0$, and $\phi \in \mathcal{D}$.

Integration by parts gives (1.4) the form

$$\int_t^s b'(\tau)\left\{\langle\phi,\mu_\tau\rangle - \langle\phi,\nu\rangle - \int_t^\tau [\langle\mathcal{A}(\theta)\phi,\mu_\theta\rangle + \langle\phi,\xi_\theta\rangle]\,d\theta\right\} = 0.$$

Since $b'(\tau)$ is arbitrary, the integrand vanishes, i.e. (1.3) holds. □

A family of distributions $\mu_x \in \Phi'$ $(x \in X)$ is called a Φ-*family* if it determines a continuous linear operator on Φ:

$$f(x) \mapsto \mu(f)(x) = \langle f, \mu_x\rangle.$$

Let, for $t \le \tau$, $\mu_{x,t,\tau}$ be a Φ-family for which the corresponding operators are strongly continuous with respect to t and τ and leave invariant a linear set $\mathcal{D}$.

If the relation

$$\mu_{t,\tau}(f) = \mu_{t,s}(\mu_{s,\tau}(f)), \quad \mu_{t,t}(f) = f \quad (t \le s \le \tau) \tag{1.5}$$

holds, then the family $\mu_{x,t,\tau}$ is called a ($\mathcal{D}$-invariant) *evolution family of distributions*.

A *fundamental distribution* of a dual pair of problems (1.1), (1.2) is a family of distributions $\mu_{x,t,\tau}$ $(x \in X,\ t \le \tau)$ satisfying

$$\frac{\partial \mu_{x,t,\tau}}{\partial\tau} = \mathcal{A}^*(\tau)\mu_{x,t,\tau}, \quad \mu_{x,t,t} = \nu_x, \tag{1.6}$$

where ν_x is a normalized measure concentrated at $x \in X$.

THEOREM 1.1. *A necessary and sufficient condition for the problem (1.1) to be well-posed is that it has a $\mathcal{D}$-invariant fundamental distribution $\mu_{x,t,\tau}$ having the evolution property in the space Φ.*

In this case $U(t,\tau)f(x) = \mu_{t,\tau}(f)(x) = \langle f, \mu_{x,t,\tau}\rangle$, and the unique solutions of (1.1) and (1.2) are given by

$$\begin{aligned} u(x,t) &= \langle u_0, \mu_{x,t,\tau}\rangle - \int_t^\tau \langle v_s, \mu_{x,t,\tau}\rangle\,ds = \\ &U(t,\tau)u_0(x) - \int_t^\tau U(t,s)v(x,s)\,ds, \end{aligned} \tag{1.7}$$

$$\langle\phi,\mu_\tau\rangle = \langle U(t,\tau)\phi,\nu\rangle + \int_t^\tau \langle U(s,\tau)\phi,\xi_s\rangle\,ds. \tag{1.8}$$

PROOF. Suppose (1.1) is well-posed in Φ. Consider a solution $u(x,t) = U(t,\tau)u_0(x)$ of the homogeneous problem (1.1'). The formula $\langle u_0, \mu_{x,t,\tau}\rangle = U(t,\tau)u_0(x)$ defines the distribution $\mu_{x,t,\tau}$ for any $x \in X$, $\tau \ge t$. Also, $\mu_{t,\tau}(f) = U(t,\tau)f$, and the evolution property for the family $\mu_{x,t,\tau}$ follows from the analogous property of the family of operators $U(t,\tau)$. It remains to examine the fundamentality of the family $\mu_{x,t,\tau}$.

Differentiating the relation

$$\mu_{t,\tau}(f) = \mu_{t,s}(\mu_{s,\tau}(f))$$

termwise with respect to s, we obtain

$$\frac{\partial}{\partial s}\mu_{t,s}(\mu_{s,\tau}\phi) + \mu_{t,s}\left(\frac{\partial}{\partial s}\mu_{s,\tau}\phi\right) = 0, \tag{1.9}$$

or

$$\left\langle \langle \phi, \mu_{\cdot,s,\tau}\rangle, \frac{\partial}{\partial s}\mu_{x,t,s}\right\rangle + \left\langle \left\langle \phi, \frac{\partial}{\partial s}\mu_{\cdot,s,\tau}\right\rangle, \mu_{x,t,s}\right\rangle = 0.$$

Since $\mu_{x,s,\tau}\phi = \langle \phi, \mu_{x,s,\tau}\rangle$ is a solution of the homogeneous problem (1.1') coinciding with ϕ for $s = \tau$, it follows from (1.9) that for $s = \tau$,

$$\left\langle \phi, \frac{\partial}{\partial \tau}\mu_{x,t,\tau}\right\rangle = \langle \mathcal{A}(\tau)(\phi), \mu_{x,t,\tau}\rangle,$$

i.e., since $\phi \in \mathcal{D}(\mathcal{A})$ is arbitrary,

$$\frac{\partial}{\partial \tau}\mu_{x,t,\tau} = \mathcal{A}^*(\tau)\mu_{x,t,\tau}. \tag{1.10}$$

We now turn to the proof of sufficiency. Suppose $\mu_{x,t,\tau}$ is a fundamental distribution having the evolution property. Set $u_\tau(x,t) = \langle u_0, \mu_{x,t,\tau}\rangle$. By requirement, this function belongs to Φ and depends continuously on u_0 for any $t \leq \tau$. Similar to the above this leads to (1.9):

$$0 = \left\langle \frac{\partial u_\tau(x,s)}{\partial s}, \mu_{x,t,s}\right\rangle + \left\langle u_\tau(\cdot,s), \frac{\partial \mu_{x,t,s}}{\partial s}\right\rangle =$$
$$\left\langle \frac{\partial u_\tau(\cdot,s)}{\partial s}, \mu_{x,t,s}\right\rangle + \langle u_\tau(\cdot,s), \mathcal{A}^*(s)\mu_{x,t,s}\rangle,$$

which for $t = s$ implies

$$\frac{\partial u_\tau(x,s)}{\partial s} + \mathcal{A}(s)u_\tau(x,s) = 0.$$

Since clearly $u_\tau(x,\tau) = u_0$, to prove well-posedness we need only show that the solution is unique. Let $v(x,t)$ be some solution of (1.1') satisfying the condition $v(x,\tau) = u_0$. Then, for any $t < s < \tau$,

$$\frac{\partial}{\partial s}\langle v_s, \mu_{x,t,s}\rangle = \left\langle \frac{\partial v_s}{\partial s}, \mu_{x,t,s}\right\rangle + \left\langle v_s, \frac{\partial \mu_{x,t,s}}{\partial s}\right\rangle =$$
$$-\langle \mathcal{A}(s)v_s, \mu_{x,t,s} + \langle v_s, \mathcal{A}^*(s)\mu_{x,t,s}\rangle = 0,$$

and consequently $\langle v_s, \mu_{x,t,s}\rangle = \text{const}$.

So,

$$v(x,t) = \langle v_t, \mu_{x,t,t}\rangle = \langle v_\tau, \mu_{x,t,\tau}\rangle = \langle u_0, \mu_{x,t,\tau}\rangle = u(x,t).$$

Formula (1.7) for the solution of the homogeneous problem has already been established. For the inhomogeneous problem it follows from general semigroup considerations.

Let now $\mu(t)$ be a solution of (1.2), and put

$$\alpha(x,s) = \langle \phi, \mu_{x,t,s}\rangle.$$

Then $\frac{\partial \alpha_s}{\partial s} + \mathcal{A}(s)\alpha_s = 0$, and (1.8) follows from (1.4). □

REMARK 1.1. Suppose (1.9) has a unique solution $\mu_{x,t,\tau}$ which is a Φ-family for every pair $t < \tau$. Moreover, suppose $\mu_{x,t,\tau}(f) \in \mathcal{D}$ for $f \in \mathcal{D}$. Then the evolution equation (1.5) holds. Consequently, problem (1.1) is well-posed, and (1.7)–(1.8) hold.

In fact, consider a family of distributions $\mu(t)$ coinciding with $\mu_{x,t,\tau}$ for $s \le t \le \tau$ and defined for $t \le s$ by the relation

$$\langle \phi, \mu(t)\rangle = \langle\langle \phi, \mu_{\cdot,s,t}\rangle, \mu_{x,t,s}\rangle.$$

On each interval $t \le s$ and $t \ge s$ it is a solution of (1.6), and these two solutions coincide at $t = s$. By uniqueness it coincides with $\mu_{x,t,\tau}$ also for $t < s$.

1.2. Systems of equations. These considerations can be easily transferred to systems of equations of the form

$$\frac{\partial u_j(x,t)}{\partial t} + \sum_{k=1}^{n} \mathcal{A}_{jk}(t) u_k(x,t) = v_j(x,t) \quad (t \le \tau), \tag{1.11}$$
$$u_j(x,\tau) = u_{0j}(x) \quad (j = 1, \dots, n),$$

where the $\mathcal{A}_{jk}(t)$ are operators on Φ.

A *fundamental distribution* is a matrix of distributions $\mu^{jk}_{x,t,\tau}$ that is a solution of the dual problem

$$\frac{\partial}{\partial \tau}\mu^{jk}_{x,t,\tau} = \sum_{r=1}^{n} \mathcal{A}_{rk}(\tau)\mu^{jr}_{x,t,\tau}; \quad \mu^{jk}_{x,t,t} = \delta_{jk}\nu_x \quad (t \le \tau). \tag{1.12}$$

If the system (1.1) is well-posed, this matrix gives rise to a family of evolution operators,

$$(U(t,\tau)f)_j(x) = \sum_k \langle f_k, \mu^{jk}_{t,\tau}\rangle,$$

on the vector function space $u(x,t) = (u_1(x,t), \dots, u_n(x,t))$.

Conversely, the existence of a fundamental matrix with the evolution property implies that (1.11) is a well-posed problem.

A *weak solution* of the problem

$$\frac{\partial \mu^*_\tau}{\partial \tau} = \sum_r \mathcal{A}^*_{rk}(\tau)\mu^r_\tau + \xi^k_\tau; \quad \mu_\tau|_{\tau-t} = \mu_t \quad (t \le \tau), \tag{1.13}$$

which is dual to (1.11), is defined by the relation

$$\sum_k \{\langle \alpha^k_s, \mu^k_s\rangle - \langle \alpha^k_t, \mu^k_t\rangle\} =$$
$$\int_t^s \sum_k \left\langle \frac{\partial \alpha^k_\tau}{\partial \tau} + \sum_j \mathcal{A}_{kj}(\tau)^{\alpha^j_\tau}, \mu^k_\tau \right\rangle d\tau + \int_t^s \sum_k \langle \alpha^k_\tau, \xi^k_\tau\rangle\, d\tau, \tag{1.14}$$

where $\alpha^k_s(x) = \alpha^k(x,s) \in \hat{\Phi}_1 \cap \mathcal{D}$.

Formulas (1.7) and (1.8) are also readily verified. Proofs of them do not differ from the proofs given earlier for a single equation, so we omit them.

There is a trick allowing one to reduce a system of equations to a single equation. Consider the system

$$\frac{\partial u_j}{\partial t} + \mathcal{A}(t)u_j + \sum \mathcal{A}_{jk}(t)u_k = v_j(x,t) \quad (t \leq \tau), \tag{1.15}$$

obtained from (1.11) by inserting

$$\mathcal{A}_{jk} \to \mathcal{A}\delta_{jk} + \mathcal{A}_{jk},$$

where $\mathcal{A}(t)$ is an operator on Φ.

Introduce new independent variables $y = (y_1, \ldots, y_n)$ and the functions

$$z(x,y,t) = \sum_{j=1}^{n} y_j u_j(x,t). \tag{1.16}$$

Multiplying the equations (1.15) termwise by y_j, adding them, and taking into account that $u_j = \partial z/\partial y_j$, we obtain an equation in z:

$$\frac{\partial z}{\partial t} + \mathcal{A}(t)z + \sum_{j,k} y_j \mathcal{A}_{jk}(t)\frac{\partial z}{\partial y_k} = \sum_j y_j v_j(x,t). \tag{1.17}$$

This equation is of type (1.1), with operator

$$\mathcal{U}(t) = \mathcal{A}(t) + \sum y_j \mathcal{A}_{jk}(t)\frac{\partial}{\partial y_k} \tag{1.18}$$

on the space $\Phi(X \times \mathbf{R}^n)$, but is distinguished by the property that its elements can linearly increase with respect to the variables $y_1, \ldots, y_n$.

Equation (1.17) is a consequence of (1.15).

Conversely, if (1.17) has a solution (1.16), linear with respect to y_k, then the coefficient at y_k of this solution satisfies (1.15).

Note that the operator (1.18) leaves invariant the space of functions that are linear with respect to y. The more complicated operator

$$\mathcal{U}_1(t) = \mathcal{A}(t) + \sum y_j \mathcal{A}_{jk}(t)\frac{\partial}{\partial y_k} + \sum \gamma_{jk}^{lr} y_r y_l \frac{\partial^2}{\partial y_j \partial y_k} \tag{1.19}$$

also has this property. Here, in the class of functions that are linear in y,

$$\mathcal{U}_1(t)z = \mathcal{U}(t)z,$$

and so, in this class of functions the equation

$$\frac{\partial z}{\partial t} + \mathcal{U}_1(t)z = \sum y_j v_j(x,t) \tag{1.20}$$

is equivalent to (1.15), as before. Note, however, that the last equation can also be considered in the class of functions that are quadratic in y, which results in systems more complicated then (1.15). Such notions will not be developed here.

Let us examine a relation between the fundamental distributions for equations (1.20) and (1.11).

Suppose $\mu_{x,y,t,\tau}$ is a system of distributions which is fundamental for (1.20) in $\Phi(X \times \mathbf{R}^n)$. It is described by the integral relation

$$\langle \alpha_s, \mu_{x,y,t,s}\rangle - \alpha_t(x,y) = \int_t^s \left\langle \frac{\partial \alpha_\tau}{\partial \tau} + \mathcal{U}_1(\tau)\, d_\tau, \mu_{x,y,t,\tau} \right\rangle d\tau. \tag{1.21}$$

Insert in it a function $\alpha_t(x,y) = \sum \phi_j(t,x) y_j$, where $\phi_j \in \hat{\Phi}$, and introduce distributions in $\Phi(X)$ by means of the formula

$$\langle \psi, \mu^{rj}_{x,t,\tau}\rangle = \langle \psi y_j, \mu_{x,e_r,t,\tau}\rangle, \tag{1.22}$$

where $e_r = (0, \dots, 1_r, \dots, 0)$ is a basis vector in $\mathbf{R}^n$.

For distributions $\mu^{rj}_{x,t,\tau}$, (1.21) implies

$$\sum_k \langle \phi_{ks}, \mu^{rk}_{x,t,s}\rangle - \phi_r(t,x) =$$

$$\int_t^s \sum_k \left\{ \left\langle \frac{\partial \phi_{k\tau}}{\partial \tau}, \mu^{rk}_{x,t,\tau} \right\rangle + \langle \mathcal{A}\phi_{k\tau}, \mu^{rk}_{x,t,\tau}\rangle + \sum_j \langle \mathcal{A}_{kj}\phi_j, \mu^{rk}_{x,t,\tau}\rangle \right\} d\tau.$$

By (1.14), the system of distributions $\mu^{rj}_{x,t,\tau}$ is a weak solution of the system

$$\frac{\partial \mu^{rk}_{x,t,\tau}}{\partial \tau} = \mathcal{A}^*(t)\mu^{rk}_{x,t,\tau} + \sum_\rho \mathcal{A}^*_{\rho k}\mu^{r\rho}_{x,t,\tau};$$

$$\mu^{rk}_{x,t,\tau} = \delta_{rk}\nu_x,$$

i.e., by (1.12) it is a fundamental matrix of the system (1.15).

Thus, the following result holds.

THEOREM 1.2. *A solution of the system (1.15) determines, by (1.16), a solution of the equation (1.17). Conversely, a fundamental distribution of (1.17) determines, by (1.22), a matrix of fundamental distributions of (1.15).*

REMARK 1.2. For the sake of clearness only we have considered a system with a finite-dimensional vector function. In the sequel we will use similar considerations also for the infinite-dimensional situation.

1.3. Equations with constant operator. In case the operator $\mathcal{A}(t) = \mathcal{A}$ does not depend on t, the evolution operators $U_{t,t_0} = U_{t-t_0}$ depend (for a well-posed problem) on the difference of the arguments, and form a semigroup: $U_{t+\tau} = U_t U_\tau$.

In this case a fundamental family also depends on the difference:

$$\mu_{x,t,t_o} = \mu_{x,t-t_0},$$

and satisfies the equation

$$\frac{\partial}{\partial t}\mu_{x,t} = \mathcal{A}^*\mu_{x,t}; \quad \mu_{x,0} = \nu_x. \tag{1.23}$$

Conversely, a family of solutions of this equation having the evolution (semigroup) property

$$\mu_{x,t+\tau}(f) = \mu_{x,t}(\mu_{\cdot,\tau}(f))$$

gives rise to a fundamental family

$$\mu_{x,t,\tau} = \mu_{x,t-\tau}$$

for the problem (1.1) with constant operator $\mathcal{A}$.

Suppose that for some $\lambda \geq 0$ and each $f \in \Phi$ the following integral converges:

$$\int_0^\infty e^{-\lambda t} \langle f, \mu_{x,t} \rangle \, dt = S_f(x, \lambda).$$

If $S_f(x, \lambda)$ is continuous with respect to $f \in \Phi$, then it determines a distribution:

$$\int_0^\infty e^{-\lambda t} \mu_{x,t} \, dt = \theta_{\lambda,x} : \quad S_f(x, \lambda) = \langle f, \theta_{\lambda,x} \rangle.$$

Integrating termwise the relation

$$\frac{\partial e^{-\lambda t} \mu_{x,t}}{\partial t} = (\mathcal{A}^* - \lambda I) e^{-\lambda t} \mu_{x,t},$$

we obtain the equation (in $\theta_{\lambda,x}$)

$$(\mathcal{A}^* - \lambda I) \theta_{\lambda,x} = -\nu_x.$$

On the other hand, since $\mathcal{A}\langle f, \mu_{x,t} \rangle = \langle \mathcal{A} f, \mu_{x,t} \rangle$ for $f \in \mathcal{D}_{\mathcal{A}}$, the similar relation for $\theta_{\lambda,x}$ holds, and thus $S_f(x, \lambda)$ satisfies the equation

$$(\mathcal{A} - \lambda I) S_f(x, \lambda) = -f(x). \tag{1.24}$$

Similar relations are easily proved for systems of equations of the type considered in the previous section. A fundamental matrix μ^j makes it possible to derive a potential matrix

$$\theta_{\lambda,x}^{jk} = \int_0^\infty e^{-\lambda t} \mu_{x,t}^{jk} \, dt$$

and a vector function

$$\sum_j \langle f^j, \theta_{\lambda,x}^{jk} \rangle = S_f^{(k)}(x, \lambda),$$

satisfying the system

$$\sum_k (\mathcal{A}^{jk} - \lambda I) S_f^{(k)}(x, \lambda) = -f^j(x).$$

1.4. Operators with space-homogeneous symbol. Consider, in particular, the rigged Hilbert space

$$\Omega = \mathcal{H}_+ \subset \mathcal{H} \subset \mathcal{H}_- = X,$$

and the following differential operator on $\Phi(X)$ with coefficients independent of x:

$$\mathcal{A}(t)u = \sum_{r=0}^{n} \operatorname{Tr}[a_r(t) \otimes u^{(r)}(x)] =$$

$$\sum_{r=0}^{n} \sum_{k_1,\ldots,k_r} a_r(t)(e_{k_1}, \ldots, e_{k_r}) u^{(r)}(x)(e_{k_1}, \ldots, e_{k_r}), \tag{1.25}$$

where $a_r(t)$ is an r-linear functional on $\mathcal{H}$, $u^{(r)}$ is the rth order derivative of $u(x)$ along $\mathcal{H}$, and $\{e_k\}_{k=1}^{\infty}$ is an orthonormal basis in $\mathcal{H}$. The expression (1.25) is defined if, e.g., a_r and $u^{(r)}$ are multilinear Hilbert–Schmidt functionals.

Suppose the space Φ contains the functions $e^{i(\omega,x)}$ $(\omega \in \Omega)$. Then the characteristic functional $\chi_\mu(\omega) = \langle e^{i(\omega,\cdot)}, \mu\rangle$, which represents a complexvalued function on Ω, is defined for distributions $\mu \in \Phi'$.

In particular, the characteristic functional

$$\chi(x,t,\tau,\omega) = \langle e^{i(\omega,\cdot)}, \mu_{x,t,\tau}\rangle$$

of the fundamental distribution $\mu_{x,t,\tau}$ of the problem (1.1) satisfies the differential equation

$$\frac{\partial\chi(x,t,\tau,\omega)}{d\tau} = \langle \mathcal{A}(\tau)e^{i(\omega,\cdot)}, \mu_{x,t,\tau}\rangle, \qquad (1.26)$$
$$\chi(x,t,\tau,\omega)|_{\tau=t} = e^{i(\omega,x)}.$$

Suppose that $\mathcal{A}(\tau)$ has a homogeneous symbol, i.e. the function

$$\phi(\tau,\omega) = e^{-i(\omega,x)}\mathcal{A}(\tau)e^{i(\omega,x)} \qquad (1.27)$$

exists. Then equation (1.26) becomes

$$\frac{\partial\chi(x,t,\tau,\omega)}{d\tau} = \phi(\tau,\omega)\chi(x,t,\tau,\omega), \qquad (1.28)$$

and its solution, given an appropriate initial condition, is the function

$$\chi(x,t,\tau,\omega) = \exp\left\{i(\omega,x) + \int_t^\tau \phi(s,\omega)\,ds\right\}. \qquad (1.29)$$

For the operator (1.25) we can easily evaluate the symbol:

$$\phi(\tau,\omega) = \sum_{r=0}^{n} i^r a_r(\tau)(\omega,\ldots,\omega). \qquad (1.30)$$

If the function $\chi(x,t,\tau,\omega)$ is bounded and integrable with respect to ω on every finite dimensional space, then, as was shown in chapter I § 3, it determines a quasi-measure $\mu_{x,t,\tau}$, which is precisely the fundamental distribution under consideration. For any cylindrical function $u_0(x)$, a solution of the Cauchy problem (1.1) can then be written as

$$u(x,t) = \int_X u_0(y)\mu_{x,t,\tau}(dy).$$

However, the class of admissible initial functions can be enlarged.

Consider functions of the form

$$f(x) = \int_\Omega e^{i(\omega,x)}\nu(d\omega), \qquad (1.31)$$

where ν is a measure on Ω such that

$$\int_\Omega |\phi(\tau,\omega)||\nu|(d\omega) < \infty. \qquad (1.32)$$

For such functions,

$$\mathcal{A}(\tau)\phi(x) = \int_\Omega \phi(\tau,\omega)e^{i(\omega,x)}\nu(d\omega).$$

Consider a function

$$u(x,t,\tau) = \int_X f(y)\mu_{x,t,\tau}(dy) = \int_\Omega \chi(x,t,\tau,\omega)\nu(d\omega). \tag{1.33}$$

Termwise integration with respect to ν of the equality

$$\chi(x,t,\tau,\omega) - e^{i(\omega,x)} = \int_t^\tau \phi(s,\omega)\chi(x,s,\tau,\omega)\,ds$$

gives

$$u(x,t,\tau) - f(x) = \int_t^\tau \int_\Omega \phi(s,\omega)\chi(x,s,\tau,\omega)\nu(d\omega)\,ds = \int_t^\tau \mathcal{A}(s)u(x,s,\tau)\,d\omega,$$

since by (1.29) and (1.27),

$$\mathcal{A}(s)u(x,s,\tau) = \int \mathcal{A}(s)\chi(x,s,\tau,\omega)\nu(d\omega) = \int \phi(s,\omega)\chi(x,s,\tau,\omega)\nu(d\omega).$$

Thus, the function (1.33) is a solution of the problem

$$\frac{\partial u(x,t,\tau)}{\partial t} + \mathcal{A}(t)u(x,t,\tau) = 0 \quad (t \le \tau), \qquad u(x,\tau,\tau) = f(x). \tag{1.34}$$

We state the result.

THEOREM 1.3. *Let $\mathcal{A}(t)$ be an operator with symbol $\phi(t,\omega)$ such that the function (1.29) is bounded and integrable with respect to ω along every finite-dimensional subspace in Ω. Then the fundamental family for the problem (1.1) is the quasimeasure with characteristic functional (1.29).*

For functions (1.31) which can be represented as Fourier transforms of measures satisfying (1.32), the solution of the homogeneous problem (1.34) is given by (1.33).

1.5. Second order equations related to Gaussian measures. Consider, in particular, the second order differential operator

$$\mathcal{A}(t)u = \frac{1}{2}\operatorname{Tr} A(t)u'' + (\alpha(t), u') \tag{1.35}$$

obtained from (1.25) for $n = 2$ when

$$a_2(t)(f,g) = \frac{1}{2}(A(t)f,g), \quad a_1(t)(f) = (a(t),f), \quad a_0(t) = 0,$$

where $a(t) \in X$, and $A(t)$ is an operator acting from Ω into X which is positive definite in the sense that $(A(t)f,f) > 0$ ($f \in \Omega$, $f \neq 0$). The symbol of the operator (1.35) is

$$\phi(s,\omega) = -\frac{1}{2}(A(s)\omega,\omega) + i(a(s),\omega) \quad (\omega \in \Omega),$$

and the characteristic functional of the fundamental distribution is

$$\chi(x,t,\tau,\omega) = \exp\left\{ i\left(\omega, x + \int_t^\tau a(s)\,ds\right) - \frac{1}{2}\int_t^\tau (A(s)\omega,\omega)\,ds\right\}.$$

This function is the characteristic functional of the Gaussian cylindrical measure $\mu_{x,t,\tau}$ with correlation operator $B_{t,\tau} = \int_t^\tau A(s)\,ds$ and mean $x + \int_t^\tau a(s)\,ds$.

Thus,

$$\mu_{x,t,\tau}(\mathcal{M}) = \mu_{t,\tau}\left(\mathcal{M} - x - \int_t^\tau a(s)\,ds\right), \tag{1.36}$$

where $\mu_{t,\tau}$ is the centered Gaussian measure with correlation operator $B_{t,\tau}$.

It follows from theorem 1.3 that the formula

$$u(x,t) = \int_X u_0(y)\mu_{x,t,\tau}(dy) = \int_X u_0\left(x + y + \int_t^\tau a(s)\,ds\right)\mu_{t,\tau}(dy) \tag{1.37}$$

is a solution of the Cauchy problem

$$\frac{\partial u}{\partial t} + \mathcal{A}(t)u = 0, \quad u(x,\tau) = u_0(x) \quad (t \le \tau), \tag{1.38}$$

at least for cylindrical function $u_0(x)$ and the wide class of initial conditions given by (1.31). In the case under consideration, however, this class can be enlarged, given certain additional assumptions.

Suppose we have a representation $A(s) = V^*(s)V(s)$, where $V(s)\colon \mathcal{H}_+ \to \mathcal{H}_0$, $V^*(s)\colon \mathcal{H}_0 \to \mathcal{H}_-$ for each s. Assume that $V(s)$ can be extended to a Hilbert space. So, from the very beginning it is assumed that $\mathcal{H}_+$ is a Hilbert space which is densely imbedded in $\mathcal{H}_0$, with norm $\|x\|_+ = \|Sx\|_0$, where S is an operator on $\mathcal{H}_0$. If the condition

$$\sigma_2^2(V(t)S^{-1}) \le c < \infty$$

holds, then the operator $S^{-1}B_{t,\tau}S^{-1}$ is nuclear on $\mathcal{H}_0$, and hence $\mu_{t,\tau}$ is supported in $\mathcal{H}_-$. If also

$$a(s) \in \mathcal{H}_-,$$

then the measure $\mu_{0,t,\tau}$ is supported in the same Hilbert space.

A change of variables, which will not be given in detail here, brings (1.38) to the Cauchy problem

$$\frac{\partial u(s,t)}{\partial t} + \frac{1}{2}\operatorname{Tr} V(s)u''(s,x)V^*(s) + (a(s), u'(s,x)) = 0 \tag{1.39}$$

with respect to the function $u(s,x)$ on $T \times X$.

The operator $\mathcal{A}(t)u = \frac{1}{2}\operatorname{Tr} V(t)u''(x)V^*(t) + (a(t), u'(t))$ is defined (at least) on the domain $\mathcal{D} = C_2(X)$ of functions that are bounded and continuous together with their derivatives $u'(x) \in X$, $u''(x) \in L(X)$ $(x \in X)$, provided the coefficients satisfy the conditions $a(t) \in X$, $V(t) \in L_2(X)$ and are continuous with respect to the parameter t.

The Cauchy problem is well-posed in this class and its solution is given by (1.37).

Using the properties of integral with respect to a Gaussian measure, the class of initial conditions for which the problem is well-posed can be enlarged (see Gross [4], Kuo [1]).

We consider one simple equation with variable coefficients whose solution can also be obtained by means of Gaussian measures. We have in mind an equation of the form

$$\frac{\partial u}{\partial t} + \frac{1}{2}\operatorname{Tr} V^* u'' V + (u', Bx) = 0, \quad u(x,0) = u_0(x), \tag{1.40}$$

where $A = VV^*$ and B are constant bounded operators. We seek its solution as

$$u(t,x) = v(t, e^{-Bt}x).$$

In this case

$$\frac{\partial u(t,x)}{\partial t} = \frac{\partial v(t, e^{-Bt}x)}{\partial t} - (v'_x, Be^{-Bt}x),$$
$$u'_x = e^{-B^*t}v'_y, \quad u''_{xx} = e^{-B^*t}v''_{yy}e^{-Bt},$$

since $v(t,y)$ satisfies the equation

$$\frac{\partial v}{\partial t} + \frac{1}{2}\operatorname{Tr} Ve^{-B^*t}v''e^{-Bt}V^* = 0, \quad v(x,0) = u_0(x), \tag{1.41}$$

which is of the type considered above.

The solution of (1.41) can be presented by the formula

$$v(t,y) = \int_X f(x+y)\mu_{t,\tau}(dy),$$

where $\mu_{t,\tau}$ is the centered Gaussian measure with correlation operator

$$B_{t,\tau} = \int_\tau^t e^{-Bs}V^*Ve^{-B^*s}\, ds. \tag{1.42}$$

We distinguish the important particular case when $B = A/2 = VV^*/2$. Here the integral (1.42) can be computed explicitly:

$$B_{0,t} = \int_0^t Ae^{-As}\, ds = (I - e^{-At}).$$

THEOREM 1.4. *Let $\phi(x)$ be a function of class $C_2(X)$, V a Hilbert–Schmidt operator on $\mathcal{H}$, and $A = V^*V$. The Cauchy problem*

$$\frac{\partial u}{\partial t} + \frac{1}{2}\operatorname{Tr} Au'' + \frac{1}{2}(Ax, u') = 0, \quad u(x,0) = \phi(x) \tag{1.43}$$

has the unique solution

$$u(x,t) = v(t, e^{-At}x),$$

where

$$v(t,y) = \int_X f(x+y)\mu_{t,\tau}(dy)$$

and $\mu_{t,\tau}$ is the centered Gaussian measure with correlation operator $I - e^{-At}$.

REMARK 1.3. The constraints on the initial function can be substantially weakened, since the operator

$$\mathcal{A}u = \frac{1}{2}[\operatorname{Tr} Au'' - (Ax, u')]$$

is also defined on functions u for which Au'' is only a Hilbert–Schmidt operator (see chapter II § 3).

2. Second order equations with variable coefficient

2.1. Statement of the problem. Let $\mathcal{H}_+ \subset \mathcal{H}_0 \subset \mathcal{H}_- = X$ be a rigged Hilbert space, and $\mathcal{K}$ a Hilbert space. Here we consider the Cauchy problem

$$\frac{\partial u(t,x)}{\partial t} + \mathcal{A}u(t,x) = 0, \quad u(\tau, x) = u_0(x) \quad (t < \tau), \tag{2.1}$$

involving a vector function $u(t,x)$ ($t \geq 0$, $x \in X$) taking values in $\mathcal{K}$, where the second order differential operator is

$$\mathcal{A}u(x) = \frac{1}{2}(A(x)\nabla, \nabla)u + (a(x), \nabla)u + b(x)(\nabla, u) + c(x)u. \tag{2.2}$$

The coefficients of this operator are functions on X whose values are described as follows ($x \in X$):

a) $A(x)$ is a linear operator from $\mathcal{H}_+$ into $\mathcal{H}_-$ which is nonnegative in the sense that

$$(A(x)\phi, \phi)_{\mathcal{H}_0} \geq 0 \quad (\phi \in \mathcal{H}_+).$$

As a rule we assume that

$$A(x) = V^*(x)V(x), \tag{2.3}$$

where $V(x)\colon \mathcal{H}_+ \to \mathcal{H}_0$, $V^*(x)\colon \mathcal{H}_0 \to \mathcal{H}_-$;

b) $a(x) \in \mathcal{H}_-$, $b(x) \in \mathcal{L}_2(\mathcal{H}_0 \times \mathcal{K}, \mathcal{K})$, $c(x) \in \mathcal{L}(\mathcal{K})$.

Differential operators of this type are described in chapter IV.

According to the method presented in subsection 1.2, we should compare the problem (2.1) with the associated Cauchy problem

$$\frac{\partial \tilde{u}}{\partial t} + \tilde{\mathcal{A}}\tilde{u} - 0, \quad \tilde{u}(0, \tilde{x}) = (y, u_0(x))_{\mathcal{K}}, \tag{2.4}$$

involving a scalar function with argument $\tilde{x} = (x, y) \in \mathcal{H}_+ \oplus \mathcal{K}$,

$$\tilde{u}(t, \tilde{x}) = (u(t,x), y)_{\mathcal{K}}, \tag{2.5}$$

the operator being

$$\begin{aligned}\tilde{\mathcal{A}}\tilde{u}(\tilde{x}) = \Big[\frac{1}{2}(A(x)\nabla_x, \nabla_x) + (y, b(x)(\nabla_x, \nabla_y))_{\mathcal{K}} + \frac{1}{2}(\gamma(\tilde{x})\nabla_y, \nabla_y)_{\mathcal{K}}\Big]\, \tilde{u}(\tilde{x}) +\\ [(a(x), \nabla_x)_{\mathcal{H}_0} + (c^*(x)y, \nabla_y)_{\mathcal{K}}]\tilde{u}(\tilde{x}),\end{aligned} \tag{2.6}$$

where $\gamma(\tilde{x}) \in \mathcal{L}(\mathcal{K})$ ($\tilde{x} \in \mathcal{H}_+ \oplus \mathcal{K}$).

A straightforward calculation shows that

$$(y, \mathcal{A}u(x))_{\mathcal{K}} = \tilde{\mathcal{A}}(y, u(x))_{\mathcal{K}} \tag{2.7}$$

and this leads to the following result.

PROPOSITION 2.1. *In a certain class of functions S the Cauchy problem (2.1) is equivalent to the associated problem (2.4) for the class of functions (2.5) with $u \in S$.*

Clearly, for $b = 0$, $c = 0$ the operator $\tilde{\mathcal{A}}$ has the form (2.2),

$$\tilde{\mathcal{A}}u = \left[\frac{1}{2}(\tilde{A}(\tilde{x})\nabla, \nabla) + (\tilde{a}(x), \nabla)\right] u, \tag{2.8}$$

the coefficients being determined by the relations

$$\begin{gathered}(\tilde{A}(\tilde{x})\tilde{\theta}, \tilde{\theta})_{\mathcal{H}_0 \oplus \mathcal{K}} = (A(x)\theta, \theta)_{\mathcal{H}_0} + 2(y, b(\theta, \eta))_{\mathcal{K}} + (\gamma(\tilde{x})\eta, \eta)_{\mathcal{K}}, \\ \tilde{a}(x) = (a(x), c^*(x)y), \quad \tilde{\theta} = (\theta, \eta).\end{gathered} \tag{2.9}$$

PROPOSITION 2.2. *Let A satisfy the condition (2.3), and let*

$$b(\theta, \eta) = \alpha(V\theta, \eta), \tag{2.10}$$

where $\alpha(\theta, \eta) \in \mathcal{L}_2(\mathcal{H}_0 \times \mathcal{K}, \mathcal{K})$. Then we can choose

$$\gamma = \alpha_y(x)\alpha_y^*(x) + \rho\rho^*,$$

where $\rho \in \mathcal{L}(\mathcal{K})$ and $(\alpha_y h, k)_{\mathcal{K}} = (y, \alpha(h, k))_{\mathcal{K}}$, and where $\tilde{A}$ also satisfies the conditions (2.3),

$$\tilde{A} = \tilde{V}^* V,$$

where

$$\tilde{V}(h, k) = (Vh + \alpha_y^* k, \rho^* k). \tag{2.11}$$

PROOF. We compute the adjoint of (2.11). We have

$$\begin{gathered}(\tilde{V}(h, k), (\theta, \eta)) = (Vh + \alpha_y^* k, \theta)_{\mathcal{H}_0} + (\rho^* k, h) = \\ (h, V^*\theta) + (k, \alpha_y\theta) + (k, \rho\eta).\end{gathered}$$

This implies

$$\tilde{V}^*(\theta, \eta) = (V^*\theta, \alpha_y\theta + \rho\eta).$$

Also,

$$\tilde{V}^*[\tilde{V}(h, k)] = \tilde{V}^*(Vh + \alpha_y^* k, \rho^* k) = (V^*Vh + V^*\alpha_y^* k, \alpha_y Vh + \alpha_y\alpha_y^* k + \rho\rho^* k),$$

and finally

$$(\tilde{V}^*\tilde{V}\tilde{h}, \tilde{h}) = (V^*Vh, h) + (V^*\alpha_y^* k, h) + (\alpha_y Vh, k) + (\alpha_y\alpha_y^* k, k) + (\rho\rho^* k, k). \tag{2.12}$$

This expression coincides with (2.9), since

$$(\alpha_y Vh, k) = (y, \alpha(Vh, k)) = (y, b(h, k)). \quad \square$$

REMARK 2.1. The choice of the operator γ does not affect the quantity (2.7), since the second order derivative of the linear function (2.5) equals zero. In particular, (2.12) implies that

$$\|\tilde{V}\tilde{h}\|^2 = \|Vh + \alpha_y^* k\|^2 + \|\rho^* k\|^2,$$

and for ρ^* nondegenerate (i.e. $\rho^* k = 0 \Rightarrow k = 0$) the operator $\tilde{V}$ is nondegenerate.

2.2. Dimension independent a priori estimates. For the time being we will concentrate especially on the case when all spaces $\mathcal{H}_+ = \mathcal{H}_0 = \mathcal{H}_-$ and $\mathcal{K}$ are finite dimensional. Then the operator (2.2) becomes

$$z_s(x) = (\mathcal{A}u)_s(x) =$$

$$\frac{1}{2}\sum_{j,k=1}^{n} a_{jk}(x)\frac{\partial^2 u_s}{\partial x_j \partial x_k} + \sum_{j=1}^{n} a_j(x)\frac{\partial u_s}{\partial x_j} + \sum_{\sigma=1}^{m}\sum_{j=1}^{n} b_{k\sigma}^s(x)\frac{\partial u_\sigma}{\partial x_k} + \sum_{\sigma=1}^{m} c_{s\sigma}(x)u_\sigma \tag{2.13}$$

$$(s = 1, \ldots, m),$$

and the relations (2.3) and (2.10) become

$$a_{jk}(x) = \sum_{s=1}^{n} v_{sj}(x)v_{sk}(x), \quad b_{k\sigma}^s(x) = \sum v_{rk}\alpha_{r\sigma}^s. \tag{2.14}$$

Our purpose is to obtain a priori estimates for solutions of Cauchy problems, regardless the dimensions of the spaces $\mathcal{H}_0$ and $\mathcal{K}$. By proposition 2.1, for a general system with operator (2.13) this problem can be reduced to the related problem for the simpler equation

$$\frac{\partial u}{\partial t} + \frac{1}{2}\sum_{j,k=1}^{n} a_{jk}(x)\frac{\partial^2 u}{\partial x_j \partial x_k} + \sum_{j=1}^{n} a_j(x)\frac{\partial u}{\partial x_j} = 0, \tag{2.15}$$

or, in vector form,

$$\frac{\partial u}{\partial t} + \frac{1}{2}(A(x)\nabla, \nabla)u + (a(x), \nabla)u = 0,$$

where $A(x) = \|a_{jk}(x)\|$ is a nonnegative symmetric matrix, $a(x) = (a_1(x), \ldots, a_n(x))$.

DEFINITION 2.1. Equation (2.15) is called *well-posed* (or *correct*) if, at least for any twice continuously differentiable initial function $u_0(x)$ satisfying the estimate

$$|u_0^{(k)}(x)| \le c(1 + \|x\|^2) \quad (k = 0, 1, 2),$$

the corresponding Cauchy problem has a unique twice continuously differentiable solution which is, moreover, representable as

$$u(x,t) = \int_{\mathbf{R}^n} u_0(y)\mu_{x,T-t}(dy), \tag{2.16}$$

where, for all $x \in \mathbf{R}^n$, $t > 0$, $\mu_{x,t}$ is a normalized nonnegative measure.

Conditions under which the well-posedness condition holds are well-known. They include, e.g., each of the following:

A) The coefficients $A(x)$ and $a(x)$ are bounded, satisfy a Hölder condition, and, moreover, the uniform parabolicity condition

$$(A(x)\theta,\theta) \geq \lambda_0\|\theta\|^2 \tag{2.17}$$

(see Friedman [1]).

B) The coefficients $V(x)$ and $a(x)$ have bounded continuous first and second order derivatives, and satisfy the growth condition

$$\|V(x)\| \leq \alpha\|x\| + \beta; \quad \|\alpha(x)\| \leq \alpha\|x\| + \beta$$

(see chapter VII).

However, in the sequel we will, as a rule, use well-posedness of the equation under consideration, and not conditions A) or B).

In a similar manner we define well-posedness for systems with an operator (2.13) at the righthand side.

Proposition 2.1 implies that well-posedness of a system is implied by well-posedness of the associated equation in the class of functions linearly depending on y. Conversely, well-posedness of a system implies well-posedness of the associated equation.

THEOREM 2.1. *Let (2.15) be a well-posed equation. If the inequality*

$$\operatorname{Tr} A(x) + 2(x, a(x)) \leq C\|x\|^2 + K \tag{2.18}$$

holds, then we have the estimate

$$\int_{\mathbf{R}^n} \|y\|^2 \mu_{x,t}(dy) \leq (\|x\|^2 + Kt)e^{Ct}. \tag{2.19}$$

PROOF. Consider the function $\phi(x) = \|x\|^2$, and let $\psi(x,t) = \int_{\mathbf{R}^n} |y\|^2 \mu_{x,t}(dy)$. Put $\psi(x,t) = z(x,t) + \phi(x)$. We obtain the following Cauchy problem for the function $z(x,t)$:

$$\frac{\partial z}{\partial t} + \frac{1}{2}(A\nabla,\nabla)z + (a,\nabla)z + g(x) = 0, \quad z(x,0) = 0, \tag{2.20}$$

where

$$g(x) = \frac{1}{2}(A\nabla,\nabla)\phi + (a,\nabla)\phi = \operatorname{Tr} A + 2(x, a(x)).$$

The solution of the inhomogeneous problem (2.20) can be written as

$$z(t) = \int_0^t \int_{\mathbf{R}^n} g(y)\mu_{x,t-s}(dy)\, ds = \int_0^t \int_{\mathbf{R}^n} [\operatorname{Tr} A(y) + 2(y, a(y))]\mu_{x,t-s}(dy)\, ds.$$

By (2.18) we arrive at the estimate

$$z(t) \leq \int_0^t \left\{K + C\int_{\mathbf{R}^n} \|y\|^2 \mu_{x,t-s}(dy)\right\} ds,$$

implying that

$$\psi(x,t) \leq Kt + \|x\|^2 + C\int_0^t \psi(x,t-s)\, ds = Kt + \|x\|^2 + C\int_0^t \psi(x,s)\, ds.$$

By the well-known Gronwall lemma, this implies (2.19). □

REMARK 2.2. Let B be an operator on $\mathbf{R}^n$. If the norm $\|x\|_B = \|Bx\|$ is introduced in this space, then the theorem can be stated as follows: The inequality

$$\operatorname{Tr} BA(x)B^* + 2(Bx, Ba(x)) \le C\|Bx\|^2 + K \tag{2.18'}$$

implies the estimate

$$\int_{\mathbf{R}^n} \|By\|^2 \mu_{x,t}(dy) \le (\|Bx\|^2 + Kt)e^{Ct}. \tag{2.19'}$$

In particular, let P be an orthogonal projector, and let $\mu_{x,t}^{(P)}$ be the image of the measure $\mu_{x,t}$ under projection of $\mathbf{R}^n$ onto $\mathcal{L}_P = P\mathbf{R}^n$. Then

$$\operatorname{Tr} PA(x)P + 2(Px, a(x)) \le C\|Px\|^2 + K \tag{2.18''}$$

implies the estimate

$$\int_{calL_P} \|y\|^2 \mu_{x,t}^{(P)}(dy) \le (\|Px\|^2 + Kt)e^{Ct}. \tag{2.19''}$$

We apply these results to the system (2.1) with operator (2.13). We assume that the associated problem is well-posed. This is true if, e.g., the following assumptions about the coefficients hold:

B_1) The operator function $A(x)$ can be represented as (2.14). The operator function $V(x)$ and vector function $a(x)$ satisfy the estimate

$$\|a(x)\| + \|V(x)\| \le C_1 + C_2\|x\|$$

and have bounded continuous first and second order derivatives in $\mathbf{R}^n$.

B_2) The operator function $b(x)$ can be represented as (2.14). The operator functions $\alpha(x)$ and $c(x)$ are bounded and continuous, as are their first and second order derivatives.

In fact, under these conditions the functions $\tilde{V}(x)$ and $\tilde{\alpha}(x)$ determined from (2.9) and (2.11) satisfy the above mentioned condition B).

Well-posedness of the associated problem implies existence of smooth solutions for it, corresponding to initial conditions of the kind $(u_0(x), y)$.

Since the original problem can be reduced to the associated one, its solution $u(x,t)$ is unique for a fixed bounded initial condition $u_0(x)$, and is representable as

$$(u(x,t), y) = \int (u_0(x'), y')\mu_{x,y,t}(dx' \times dy'), \tag{2.20'}$$

where $\mu_{x,y,t}$ is the fundamental distribution for the dual problem. On the other hand, the fundamental distribution of the associated problem determines (see § 1) the matrix of fundamental distributions of the initial problem, which, in the present case, has the required smoothness properties and so defines a smooth solution of the initial problem.

Using formula (2.20') and theorem 2.1, we can obtain certain estimates for these solutions.

Let R be an invertible operator on $\mathbf{R}^n$. Then, from (2.20'),

$$\|Ru(x,t)\|^2 = \sup_{\|R^{-1}y\|\leq 1} (Ru, R^{-1}y)^2 =$$

$$\sup_{\|R^{-1}y\|\leq 1} \left| \int_{\mathbf{R}^{2n}} (Ru_0(x'), R^{-1}y')\mu_{x,y,t}(dx' \times dy') \right|^2 \leq$$

$$\sup_{x'\in\mathbf{R}^n} \|Ru_0(x')\|^2 \sup_{\|R^{-1}y\|\leq 1} \int_{\mathbf{R}^{2n}} \|R^{-1}y'\|^2 \mu_{x,y,t}(dx' \times dy'). \tag{2.21}$$

We now use the estimates (2.18')–(2.19'). Let the operator be defined by

$$B(x,y) = (0, R^{-1}y).$$

The lefthand side of (2.18') for coefficients $\tilde{V}$ (at $\rho = 0$) and $\tilde{\alpha}$ as in (2.11) and (2.9) is then written as

$$\operatorname{Tr} B\tilde{A}B + 2(B\tilde{x}, B\tilde{a}) = \sigma_2^2(\tilde{V}B) + 2(B\tilde{x}, B\tilde{a}) =$$
$$\sigma_2^2(\alpha_y^* R^{-1}) + (R^{-1}y, R^{-1}c^*y),$$

and so the condition itself becomes

$$\sigma_2^2(\alpha_y^*(x)R^{-1}) + (R^{-1}y, R^{-1}c^*(x)y) \leq C\|R^{-1}y\|^2, \tag{2.22}$$

since $K = 0$ because of the bilinearity of the lefthand and righthand sides. This implies that the estimate

$$\int_{\mathbf{R}^{2n}} \|R^{-1}y'\|^2 \mu_{x,y,t}(dx' \times dy') \leq \|R^{-1}y\|^2 e^{Ct}, \tag{2.23}$$

which follows from (2.19'), holds. This and (2.21) imply the following assertion.

THEOREM 2.2. *Let the system*

$$\frac{\partial u}{\partial t} + \frac{1}{2}(V^*(x)V(x)\nabla, \nabla)u + (a(x), \nabla)u + \alpha(x)(V(x)\nabla, u) + c(x)u = 0 \tag{2.24}$$

be well-posed. If (2.22) holds, then the solution of the Cauchy problem (2.24) under the condition $u(x,\tau) = u_0(x)$ satisfies the estimate

$$\sup_x \|Ru(x,t)\|^2 \leq e^{C(\tau - t)} \sup_x \|Ru_0(x)\|^2. \tag{2.25}$$

In particular, under the condition

$$\sigma_2^2(\alpha_y^*(x)) + (c^*(x)y, y) \leq C\|y\|^2 \tag{2.22'}$$

the following estimate holds:

$$\sup_x \|u(x,t)\|^2 \leq e^{C(\tau - t)} \sup_x \|u_0(x)\|^2. \tag{2.25'}$$

2.3. A priori estimates for derivatives of solutions of the Cauchy problem, independent of the number of arguments. Termwise differentiation of (2.15) and use of (2.14) leads to a system of equations for the quantities $\dot{u}_s = \partial u/\partial x_s$ (at this point it is sometimes convenient to denote differentiation by a dot):

(2.26)

$$\frac{\partial \dot{u}_r}{\partial t} + \frac{1}{2}\sum_{j,k} a_{jk}\frac{\partial^2 \dot{u}_r}{\partial x_j \partial x_k} + \sum_j a_j(x)\frac{\partial \dot{u}_r}{\partial x_j} + \sum_{s,k,j} v_{sj}\frac{\partial \dot{u}_k}{\partial x_j}\dot{v}_{skr} + \sum_j \dot{a}_{jr}(x)\dot{u}_j = 0,$$

where $\dot{v}_{skr} = \partial v_{sk}/\partial x_r$, $\dot{a}_{jr} = \partial a_j/\partial x_r$. This system takes the form (2.24) for $c(x) = \dot{a}^*(x)$ and $a^r_{sk}(x) = \dot{v}_{skr}(x)$, i.e. $\alpha^*_y = \dot{V}y$.

For an operator function $\phi(x)$ defined on $\mathbf{R}^n$ we introduce the following notations:

$$m(\phi) = \sup_{\substack{\|y\|=1 \\ x\in\mathbf{R}^n}} (\phi(x)y, y), \quad \tilde{\sigma}_2^2(\dot{\phi}) = \sup_{\substack{\|y\|=1 \\ x\in\mathbf{R}^n}} \sigma_2^2(\dot{\phi}(x)y).$$

The following result is a consequence of theorem 2.2.

PROPOSITION 2.3. *The following estimate holds:*

(2.27)

$$\sup_x \|R\dot{u}(x,t)\|^2 \leq \sup_x \|R\dot{u}_0(x)\|^2 \exp\{\tilde{\sigma}_2^2((VR^{-1})^{\cdot}R) + m(R^{-1}\dot{a}R))\}(\tau - t).$$

In the same manner we can estimate higher order derivatives. We will only study second order derivatives $\ddot{u}_{jk} = \partial^2 u/\partial x_j \partial x_k$ in detail. Termwise differentiation of (2.26) gives the system

$$\frac{\partial \ddot{u}_{r_1r_2}}{\partial t} + \frac{1}{2}\sum_{j,k} a_{jk}\frac{\partial^2 \ddot{u}_{r_1r_2}}{\partial x_j\partial x_k} + \sum_j a_j\frac{\partial \ddot{u}_{r_1r_2}}{\partial x_j} + \sum_{s,k,j} v_{sj}\frac{\partial \ddot{u}_{r_1k}}{\partial x_j}\dot{v}_{skr_2} + \sum v_{sj}\frac{\partial \ddot{u}_{kr_2}}{\partial x_j}\dot{v}_{skr_1} +$$
$$\sum \dot{a}_{jr_2}\ddot{u}_{r_1j} + \sum \dot{v}_{sjr_2}\dot{v}_{skr_1}\ddot{u}_{kj} + \sum v_{sj}\ddot{v}_{skr_1r_2}\ddot{u}_{kj} + \sum \dot{a}_{jr_1}\ddot{u}_{jr_2} + \sum \ddot{a}_{jr_1r_2}\dot{u}_j = 0.$$

This can also be written as

$$\frac{\partial \ddot{u}_{r_1r_2}}{\partial t} + \frac{1}{2}\sum_{j,k} a_{jk}\frac{\partial^2 \ddot{u}_{r_1r_2}}{\partial x_j\partial x_k} + \sum_j a_j\frac{\partial \ddot{u}_{r_1r_2}}{\partial x_j} + \sum_{s,k,j} v_{sj}\frac{\partial \ddot{u}_{k_1k_2}}{\partial x_j}(\delta_{k_1r_1}\dot{v}_{sk_2r_2} + \delta_{k_2r_2}\dot{v}_{sk_1r_1} +$$

(2.28)
$$\sum_s(\dot{v}_{sk_2r_2}\dot{v}_{sk_1r_1} + \dot{a}_{k_1r_1}\delta_{k_2r_2})\ddot{u}_{k_1k_2} + \sum \ddot{a}_{jr_1r_2}\dot{u}_j = 0.$$

Let a basis in $\mathbf{R}^n$ be chosen such that $\|Rf\|^2 = \sum \lambda_k^2 f_k^2$. Put $p_j = \lambda_j\dot{\mu}_j$ and $q_{jk} = \lambda_j\lambda_k\ddot{u}_{jk}$. We can reduce (2.26) and (2.28) to the equation (2.24) with respect to $w = p \oplus q \in \mathbf{R}^{n+n^2}$:

(2.29)
$$\frac{\partial w}{\partial t} + \frac{1}{2}(A\nabla, \nabla)w + (a, \nabla)w + \alpha(V\nabla, w) + c(x)w = 0,$$

where the bilinear map α and the linear map c are determined by matrices:

$$\alpha = \begin{pmatrix} \alpha^r_{sk} & 0 \\ \alpha^{(r_1,r_2)}_{sk} & \alpha^{(r_1,r_2)}_{s,(k_1,k_2)} \end{pmatrix}, \quad c = \begin{pmatrix} c_{sk} & 0 \\ c_{(r_1,r_2),k} & c_{(r_1,r_2),(k_1,k_2)} \end{pmatrix}.$$

Here,

$$\alpha'_{sk} = \dot{v}_{skr}\frac{\lambda_r}{\lambda_k}; \quad \alpha_{sk}^{(r_1,r_2)} = \ddot{v}_{sk(r_1,r_2)}\frac{\lambda_{r_1}\lambda_{r_2}}{\lambda_k};$$

$$\alpha_{s,(k_1,k_2)}^{(r_1,r_2)} = \delta_{k_1r_1}\dot{v}_{sk_2r_2}\frac{\lambda_{r_2}}{\lambda_{k_2}} + \delta_{k_2r_2}\dot{v}_{sk_1r_1}\frac{\lambda_{r_1}}{\lambda_{k_1}};$$

$$c_{rk} = a_{rk}\frac{\lambda_r}{\lambda_k}; \quad c_{(r_1,r_2)k} = \ddot{a}_{kr_1r_2}\frac{\lambda_{r_1}\lambda_{r_2}}{\lambda_k};$$

$$c_{(r_1,r_2)(k_1,k_2)} = \dot{a}_{k_1r_1}\delta_{k_2r_2}\frac{\lambda_{r_1}}{\lambda_{k_1}} + \dot{a}_{k_2r_2}\delta_{k_1r_1}\frac{\lambda_{r_2}}{\lambda_{k_2}} + \sum \dot{v}_{sk_1r_1}\frac{\lambda_{r_1}}{\lambda_{k_1}}\dot{v}_{sk_2r_2}\frac{\lambda_{r_2}}{\lambda_{k_2}}.$$

After some simple (but awkward, hence omitted) calculations we arrive at the following estimates for the quantities in (2.22'):

$$\sup_{\|y\|=1} \sigma_2^2(\alpha_y^*(x)) \le 5\sigma_2^2((VR^{-1})'R) + \sigma_2^2((VR^{-1})''(R\otimes R)),$$

$$\sup_{\|y\|=1} (c^*(x)y,y) \le 3m(R^{-1}\dot{a}R) + \sigma_2^2((VR^{-1})'R) + \sigma_2^2(R^{-1}\ddot{a}(R\otimes R)). \tag{2.30}$$

For a tensor $K = \{k_{j_1\dots j_p}\}$ we put

$$K(R\otimes\cdots\otimes R) = \{k_{j_1\dots j_p}\lambda_{j_1}\dots\lambda_{j_p}\}.$$

As a result, the following estimate of the second order derivative follows from (2.25'):

$$\sup_x\{\sigma_2^2[\ddot{u}(x,t)(R\otimes R)] + \sigma_2^2[(\dot{u}(x,t))(R)]\} \le$$

$$e^{C(\tau-t)}\left\{\sup_x \sigma_2^2[\ddot{u}(x,\tau)(R\otimes R)] + \sup_x \sigma_2^2[(\dot{u}(x,t))(R)]\right\}, \tag{2.31}$$

where C depends on the quantities at the righthand side of (2.29) and related to the derivatives of the coefficients a and V, but does not depend on n.

We state a general proposition, neglecting details.

PROPOSITION 2.4. *Suppose the coefficients of equation (2.15) have continuous derivatives up to order N (inclusive), and*

$$\sup_{\substack{x\\k=1,\dots,N}} \{\sigma_2^2[(VR^{-1})^{(k)}(R\otimes\cdots\otimes R)], \sigma_2^2[(R^{-1}a)^{(k)}(R\otimes\cdots\otimes R)] \le C. \tag{2.32}$$

Then

(2.33)

$$\sum_{k=1}^N \sup_x \sigma_2^2[u^{(k)}(x,t)(R\otimes\cdots\otimes R)] \le \sum_{k=1}^N e^{C\rho_k t}\sup \sigma_2^2[u_0^{(k)}(x)(R\otimes\cdots\otimes R)],$$

where ρ_k is a positive integer depending only on k.

In a similar manner we can carry out computations for the system

$$\frac{\partial u}{\partial t} + \frac{1}{2}(A\nabla,\nabla)u + (a,\nabla)u + \alpha(V\nabla,u) + cu = 0. \tag{2.34}$$

Termwise differentiation of this system and association of the system obtained with the initial one leads to a system of type (2.34) for the element $w = u \oplus \dot{u}R$, with elements

$$\alpha_{sk}^{(r_1,r_2)} = \dot{\alpha}_{skr_1}^{r}\lambda_{r_1}; \quad \alpha_{s,(k_1,k_2)}^{(r_1,r_2)} = \dot{v}_{sk_1r_1}\frac{\lambda_{r_1}}{\lambda_{k_1}}\delta_{r_2k_2} + \alpha'_{sk} + \dot{v}_{sk_1r_1}\frac{\lambda_{r_1}}{\lambda_{k_1}}\alpha_{sj}^{r};$$

$$c_{(r,r_1,k)} = \dot{c}_{rkr_1}\lambda_{r_1}; \quad c_{(r_2,r_1)(k_2,k_1)} = c_{jr}\delta_{k_1r_1} + \dot{a}_{k_1r_1}\frac{\lambda_{r_1}}{\lambda_{k_1}}\delta_{rj}. \tag{2.35}$$

Thus, the form of the system of differential equations (2.34) turns out to be invariant under termwise differentiation. Using (2.35) we can write down matrices α and c for subsequent systems describing higher order derivatives of the solution (2.34). Using these expressions we can obtain, as above, the following result.

PROPOSITION 2.5. *Suppose the coefficients of the system (2.13) have continuous derivatives up to order N (inclusive), and the conditions (2.32) and*

$$\sup_{\substack{x \\ k=1,\dots,N}} \{\sigma_2^2[\alpha^{(k)}(R\otimes\cdots\otimes R)], \sigma_2^2[c^{(k)}(R\otimes\cdots\otimes R)] \le C$$

hold. Then the derivatives of a solution of the system (2.34) satisfy the estimate (2.33).

2.4. Equations with cylindrical coefficients. We turn to the consideration of equations in an infinite-dimensional space. First we will consider the Cauchy problem for a single scalar equation:

$$\frac{\partial u}{\partial t} + \mathcal{A}u = \frac{\partial u}{\partial t} + \frac{1}{2}(A(x)\nabla,\nabla)u + (a(x),\nabla)u = 0, \quad u(0,x) = u_0(x) \quad (x \in X). \tag{2.36}$$

Let $\mathcal{P}$ be a complete set of finite-dimensional orthoprojectors on $\mathcal{H}_0$ with ranges in $\mathcal{H}_+$:

$$\mathcal{L}_P = P\mathcal{H}_0 = PX \subset \mathcal{H}_+, \quad \mathcal{L} = \cup_{P\in\mathcal{P}}\mathcal{L}_P, \quad \overline{\mathcal{L}} = \mathcal{H}_0.$$

The coefficients of equation (2.36) determine operator and vector functions in each of the $\mathcal{L}_P$:

$$A_P(x) = PA(Px)P, \quad a_P(x) = Pa(Px), \tag{2.37}$$

as well as a corresponding Cauchy problem:

$$\frac{\partial u}{\partial t} + \mathcal{A}_P u = \frac{\partial u}{\partial t} + \frac{1}{2}(A_P(x)\nabla,\nabla)u + (a_P(x),\nabla)u = 0, \quad u(x,\tau) = u_{0,P}(x) = u_0(Px). \tag{2.38}$$

Below, all these finite-dimensional problems are assumed to be well-posed.

Let $u_P(x,t)$ $(x \in \mathcal{L}_P)$ be a solution of (2.38). Consider a function

$$u(x,t) = u_P(Px,t) \quad (x \in \mathcal{D}'). \tag{2.39}$$

Since clearly $u'(x,t) = Pu'_P(Px,t)$, $u''(x,t) = Pu''_P(Px,t)P$, this function satisfies

$$\frac{\partial u(x,t)}{\partial t} + \frac{1}{2}\operatorname{Tr} A_P(Px)u''(x,t) + (a_P(Px), u'(x,t)) =$$
$$\frac{\partial u_P(Px,t)}{\partial t} + \frac{1}{2}\operatorname{Tr} A_P(Px)u''_P(x,t) + (a_P(Px), u'_P(Px,t)) = 0.$$

This equation coincides with (2.36) if the cylindricity conditions

$$PA(Px)P = PA(x)P, \quad Pa(Px) = Pa(x) \tag{2.40}$$

hold.

Thus, provided (2.40) holds, the solution of (2.38) determines, by (2.39), the solution of (2.36) with cylindrical initial condition $u_0(Px)$. Let $Q \supset P$. Regarding X as a finite-dimensional space $\mathcal{L}_Q$ and the problem (2.36) as (2.38) with projection operator Q instead of P, we can apply the considerations above. Then (2.39) yields the consistency relation

$$u_Q(x,t) = u_P(Px,t) \quad (x \in \mathcal{L}_Q), \tag{2.41}$$

provided $u_Q(x,\tau) = u_Q(Px,\tau) = u_0(Px)$.

Formula (2.41) implies that the relation

$$u(x,t) = u_P(Px,t)$$

well-posedly determines a function $u(x,t)$ on X with cylindrical initial value $u(x,\tau) = u_0(Px)$.

Let $\mu_{x,t}^{(P)}$ be a fundamental distribution for the problem (2.38); i.e.

$$u_P(t,x) = \int_{\mathcal{L}_Q} u_P(\tau,y)\mu_{x,\tau-t}^{(P)}(dy).$$

Equation (2.41) now becomes

$$\int_{U_P} u_0(y)\mu_{(Px),\tau-t}^{(P)}(dy) = \int_{\mathcal{L}_Q} u_0(Py)\mu_{x,\tau-t}^{(Q)}(dy) = \int_{\mathcal{L}_Q} u_0(y)\mu_{x,\tau-t}^{(Q,P)}(dy),$$

where $\mu_{x,\tau-t}^{(Q,P)}$ is the image of the measure $\mu_{x,\tau-t}^{(Q)}$ under P.

Since $u_0(y)$ is arbitrary, this implies

$$\mu_{x,\tau-t}^{(Q,P)} = \mu_{(Px),\tau-t}^{(P)} \quad (x \in \mathcal{L}_Q).$$

In particular, the following consistency condition holds for the measure:

$$\mu_{Px,t}^{(P)} = \mu_{Px,t}^{(Q,P)} \quad (x \in X, \quad t \geq 0, \quad P, Q \in \mathcal{P}). \tag{2.42}$$

Let $x \in \mathcal{L}$, i.e. $x = P_0x$ for some $P_0 \in \mathcal{P}$. For $x \in \mathcal{L}$ and $t > 0$ fixed, the consistent set of measures $\mu_{x,t}^{(P)}$ $(P \supset P_0)$ determines a cylindrical measure in X with, in general, a σ-additive extension to some extension $\tilde{X} \supset X$. If, however, $\mathcal{L} = \mathcal{H}_+$, then Kolmogorov's theorem implies $\tilde{X} = X$, and thus $\mu_{x,t}$ is a measure in X.

Note that this situation occurs if $X = \mathbf{R}^\infty$, the space of all realvalued sequences.

For every $P \in \mathcal{P}$, $x \in \mathcal{L}_P$, $t \geq 0$ the measure $\mu_{x,t}^{(P)}$ is a weak solution of the corresponding finite-dimensional problem:

$$\langle \alpha_\tau, \mu_{x,\tau}^{(P)} \rangle - \alpha(x,t) = \int_t^\tau \left\langle \frac{\partial \alpha_s}{\partial s} + \mathcal{A}^{(P)} \alpha_s, \mu_{x,s}^{(P)} \right\rangle ds, \tag{2.43}$$

where

$$\langle \alpha_s, \mu_{x,s}^{(P)} \rangle = \int_{\mathcal{L}_P} \alpha(s,y) \mu_{x,s}^{(P)}(dy) = \int_X \alpha(s, Py) \mu_{x,s}(dy).$$

Now (2.40) implies that $\mathcal{A}^{(P)} \alpha_t(Px) = \mathcal{A}\alpha_t(Px)$, and thus (2.43) can be written as

$$\langle \alpha_\tau, \mu_{x,\tau} \rangle - \alpha(x,t) = \int_t^\tau \left\langle \frac{\partial \alpha_s}{\partial s} + \mathcal{A}\alpha_s, \mu_{x,s} \right\rangle ds,$$

for any cylindrical function $\alpha(x,t) = \alpha(Px,t)$.

It can be readily seen that limit transition is allowed in this equality, extending it to functions of class Φ_1 (i.e. to bounded functions with bounded derivatives). For the case under consideration, this shows that the measure $\mu_{x,t}$, regarded as a distribution in Φ_1', is a weak solution to the problem under consideration.

Note that (2.40) holds for equations with constant coefficients. This case was studied in the previous section.

2.5. Measures determined by sets of finite-dimensional equations. More involved is the case when the cylindricity condition (2.40) is violated, since then the fundamental distributions $\mu_{x,t}^{(Q)}$ $(Q \in \mathcal{P})$ for problems (2.38) are not, in general, projections of a single measure in X. It will be shown below that the fundamental distribution of a problem (2.36) can nevertheless be constructed, given some assumptions regarding the coefficients, and then, turning to associated equations, we can study the Cauchy problem for systems (2.2).

Consider a more general situation.

Suppose we are given, for each $Q \in \mathcal{P}$, a well-posed equation in $\mathcal{L}_Q$:

$$\frac{\partial u}{\partial t} + \mathcal{A}_Q u = \frac{\partial u}{\partial t} + \frac{1}{2}(A_Q(x)\nabla, \nabla)u'' + (a_Q(x), \nabla)u' = 0, \tag{2.44}$$

where $A_Q(x) = V_Q^*(x)V_Q(x)$, and $V_Q \in L(\mathcal{L}_Q)$, $a_Q(x) \in \mathcal{L}_Q$ are continuous functions, not necessarily satisfying (2.37). If (2.37) holds, the set of equations (2.44) will be called *consistent*.

Let $\mu_{x,t}^{(Q)}$ be a fundamental distribution of equation (2.44) in $\mathcal{L}_Q$, and let $\mu_{x,t}^{(P,Q)}$ be its projection to $\mathcal{L}_P$:

$$\mu_{x,t}^{(P,Q)}(\mathfrak{M}) = \mu_{x,t}^{(Q)}(P^{-1}(\mathfrak{M}) \cap \mathcal{L}_Q) \quad (P \subset Q).$$

PROPOSITION 2.6. *Suppose the condition*

$$\operatorname{Tr} PA_Q(x)P + 2(Px, a_Q(x)) \leq C_P \|Px\|^2 + K_P \quad (P \subset Q) \tag{2.45}$$

is fulfilled, where the constants C_P and K_P are independent of Q. Then the set

$$\mathfrak{M}(P, R, \tau) = \{\mu_{x,t}^{(P,Q)} : x \in \mathcal{L}_Q,\ \|Px\| \leq R,\ 0 \leq t \leq \tau,\ Q \supset P\}$$

is compact in the sense of weak convergence of measures in $\mathcal{L}_P$, *for every* $P \in \mathcal{P}$, $R > 0$, $\tau > 0$.

PROOF. A sufficient condition for weak compactness of a set of measures in a finite-dimensional space $\mathcal{L}_P$ is uniform boundedness of their second moments:

$$\int_{\mathcal{L}_P} \|x'\|^2 \mu_{x,t}^{(P,Q)}(dx') \leq \text{const}.$$

This is a consequence of the estimate (2.19"), given the condition (2.45). □

REMARK 2.3. This assertion remains true if (2.45) is replaced by:

$$\text{Tr}\, A_Q(x) + 2(x, a_Q(x)) \leq c\|Qx\|^2 + K \quad (x \in \mathcal{L}_Q) \tag{2.45'}$$

The result then follows from (2.19), since

$$\int_{\mathcal{L}_P} \|x'\|^2 \mu_{x,t}^{(P,Q)}(dx') \leq \int_{\mathcal{L}_Q} \|x'\|^2 \mu_{x,t}^{(Q)}(dx').$$

For fixed $t > 0$, $x \in \mathcal{L} = \cup \mathcal{L}_Q$, and $P \in \mathcal{P}$, let $S_{x,t}^{(P)}$ be the set of limit points of the set of measures $\{\mu_{x,t}^{(P,Q)}\}$ as $Q \to I$. It is easily seen that the elements of $S_{x,t}^{(P)}$ have the properties:

a) $S_{x,t}^{(P)}$ is compact in the sense of weak convergence of measures in $\mathcal{L}_P$;
b) if $\mu \in S_{x,t}^{(P)}$ and $P' \subset P$, then the P'-projection $\mu_{P'}$ has the property $\mu_{P'} \in S_{x,t}^{(P')}$;
c) $\mu(\mathcal{L}_P) = 1$ for $\mu \in S_{x,t}^{(P)}$.

THEOREM 2.3. *Under the conditions (2.45) or (2.45') there is a normalized positive cylindrical measure* $\tilde{\mu}_{x,t}$ *in* X *whose projection has the property* $\tilde{\mu}_{x,t}^{(P)} \in S_{x,t}^{(P)}$. *If* $\tilde{\mathcal{L}} \cup P_k X = \mathcal{H}_+$, *then* $\mu_{x,t}$ *is* σ*-additive in* X.

PROOF. Consider a monotone increasing sequence of projection operators $P_k \to I$ ($P_k \in \mathcal{P}$). Using the diagonal technique we can extract a subsequence k_s such that the sequence of measures $\mu_{x,t}^{(P_{k_s}, P_{k_j})}$ ($j > s$) converges weakly for each s as $j \to \infty$. Put

$$\tilde{\mu}_{x,t}^{P_{k_s}} = \lim_{j \to \infty} \mu_{x,t}^{(P_{k_s}, P_{k_j})}.$$

This gives a set of measures in the finite-dimensional sets $\mathcal{L}_{P_{k_s}}$ whose compatibility follows from: $\mu_{x,t}^{(P_{k_s}, P_{k_j})} = (\mu_{x,t}^{(P_{k_s}, P_{k_j})})_{P_{k_s}}$ ($k_s < k_j$). This set determines the cylindrical measure $\tilde{\mu}_{x,t}$. By Kolmogorov's theorem this cylindrical measure has, in general, a σ-additive extension to an extension $\tilde{X} \supset X$; this last extension coincides with X if $\tilde{\mathcal{L}} = \mathcal{H}_+$. □

REMARK 2.4. For a finite set of values of x and t the union $\cup_{x,t} S_{x,t}^{(P)}$ is weakly compact. Therefore we can determine, by the diagonal technique, cylindrical measures $\tilde{\mu}_{x,t}$ at once for countable sets $Z = \{x_1, x_2, \dots\}$ and $\theta = \{t_1, t_2, \dots\}$ using the same sequence of projections:

$$\tilde{\mu}_{x,t}^{P_{k_s}} = \lim_{j \to \infty} \mu_{x,t}^{(P_{k_s}, P_{k_j})} \quad (x \in Z, \quad t \in \theta).$$

Here we can take $\bar{Z} = \mathcal{H}_0$, $\bar{\Theta} = \mathbf{R}_+$.

REMARK 2.5. Let $\phi(x)$ be a bounded function having the property that $\phi(Px)$ is continuous in $\mathcal{L}_P$ $(P \in \mathcal{P})$ and such that its integral with respect to the cylindrical measure $\tilde{\mu}_{x,t}$ exists:

$$\int_X \phi(y)\tilde{\mu}_{x,t}(dy) = \lim_{P\to I} \int_{\mathcal{L}_P} \phi(Py)\tilde{\mu}^{(P)}_{x,t}(dy).$$

Since according to weak convergence of the measures,

$$\int_{\mathcal{L}_P} \phi(Py)\tilde{\mu}^{(P)}_{x,t}(dy) = \lim_{Q_k\to I} \int_{\mathcal{L}_P} \phi(Py)\mu^{(P,Q_k)}_{x,t}(dy) = \lim_{Q_k\to I} \int_{\mathcal{L}_{Q_k}} \phi(Py)\mu^{(Q_k)}_{x,t}(dy),$$

where Q_k is the corresponding sequence of projection operators, the following formula hold:

$$\int_X \phi(y)\tilde{\mu}_{x,t}(dy) = \lim_{P\to I}\lim_{Q_k\to I} \int_{\mathcal{L}_{Q_k}} \phi(Py)\mu^{(Q_k)}_{x,t}(dy). \tag{2.46}$$

THEOREM 2.4. *Suppose that under the conditions of theorem 2.3 there is an operator* $B\colon \mathcal{H}_+ \to \mathcal{H}_0$ *such that*

$$\operatorname{Tr} B_Q A_Q(x) B_Q^* + 2(B_Q x, B_Q a_Q(x)) \le C\|B_Q x\|^2 + K, \tag{2.47}$$

where $B_Q = BQ$.

Then the cylindrical measure $\tilde{\mu}_{x,t}$ $(x \in \tilde{\mathcal{L}},\ t > 0)$ *has a* σ*-additive extension to the Hilbert space* $\mathcal{H}_B$ *which is the completion of* $\mathcal{H}_+$ *with respect to the norm* $\|x\|_B = \|Bx\|$. *Also, the second moment*

$$\int_{\mathcal{H}_B} \|y\|^2_b \tilde{\mu}_{x,t}(dy) < \infty$$

is finite.

PROOF. It suffices (see chapter III § 1) to prove the estimate

$$\int_{\mathcal{L}_P} \|y\|^2_B \tilde{\mu}^{(P)}_{x,t}(dy) \le C_{x,t} < \infty,$$

uniformly with respect to P.

From (2.47) and (2.19') we find

$$\int_{\mathcal{L}_P} \|BPy\|^2 \mu^{(P,Q)}_{x,t}(dy) \le (\|Px\|^2 + Kt)e^{Ct},$$

which leads after the limit transition $Q \to I$ to the desired result. To justify this limit transition, we have to choose a monotone increasing sequence of bounded continuous functions $\phi_n(y) \to \|BPy\|^2$. Then

$$\int_{\mathcal{L}_P} \phi_n(y)\tilde{\mu}^{(P)}_{x,t}(dy) = \lim_{Q\to I} \int_{\mathcal{L}_P} \phi_n(y)\tilde{\mu}^{(P,Q)}_{x,t}(dy) \le C,$$

and it remains to pass to the limit as $n \to \infty$. □

THEOREM 2.5. *Under the conditions of theorem 2.4 the measure $\tilde{\mu}_{x,t}$ can be defined for all $t > 0$ so that it depends weakly continuously on t, that is, for every $\phi \in C_2(\mathcal{H}_B)$ and $x \in \tilde{L}$ the function $\int \phi(y)\tilde{\mu}_{x,t}(dy)$ is continuous in t.*

PROOF. There is a sequence $P_k \to I$ $(k \to \infty)$ such that $\tilde{\mu}_{x,t}^{(P)} = \lim_{k\to\infty} \mu_{x,t}^{(P,P_k)}$ for each $t \in \theta$, where $\overline{\theta} = \mathbf{R}_+$.

Let $\phi \in C_2(\mathcal{H}_B)$ and $\phi_P(x) = \phi(Px)$. Then

$$u^{(P,P_k)}(x,t) = \int_{\mathcal{L}_P} \phi_P(y)\mu_{x,t}^{(P,P_k)}(dy)$$

is a solution of the Cauchy problem (in $\mathcal{L}_P$)

$$\frac{\partial u}{\partial t} + (A_{P_k}\nabla, \nabla)u + (a_{P_k}, \nabla)u = \frac{\partial u}{\partial t} + \mathcal{A}_k u = 0,$$
$$u(x,\tau) = \phi(x).$$

Its time derivative $v^{(P,P_k)} = \partial u^{(P,P_k)}/\partial t$ satisfies the same equation, with initial condition

$$v^{(P,P_k)}(x,\tau) = \frac{\partial u^{(P,P_k)}(x,\tau)}{\partial \tau} = -\mathcal{A}_k\phi(x).$$

Thus, for $t \le \tau$,

$$|v^{(P,P_k)}(x,t)| \le \sup_{x,k} |\mathcal{A}_k\phi(x)| \le C_\phi.$$

where C_ϕ depends on ϕ only.

This implies that for every x the sequence $u^{(P,P_k)}(x,t)$ is compact in the sense of uniform convergence on every finite interval of values of t. It converges on the dense set θ,

$$\lim_{k\to\infty} u^{(P,P_k)}(x,t) = \int_{\mathcal{H}_B} \phi(Py)\mu_{x,t}^{(P)}(dt). \tag{2.48}$$

Hence it converges for all t to a continuous (even Lipschitz) function with respect to t.

Furthermore, the sequence of measures $\mu_{x,t}^{(P,P_k)}$ is weakly compact for any t, and so weakly converges to a measure $\tilde{\mu}_{x,t}^{(P)}$ for which (2.48) is true for all t. Thus the theorem has been proved for cylindrical ϕ.

It can be proved for the general case by limit transition as $P \to I$, since the collection of measures under consideration is 'almost concentrated' on a compact set if $\phi(Py)$ converges uniformly to $\phi(x)$. □

2.6. Solution of the Cauchy problem for a single equation in an infinite-dimensional space. We turn to the consideration of the Cauchy problem for the infinite-dimensional equation (2.36),

$$\frac{\partial u}{\partial t} + \mathcal{A}u = \frac{\partial u}{\partial t} + \frac{1}{2}(A(x)\nabla, \nabla)u + (a(x), \nabla)u = 0, \tag{2.36}$$

$$u(x,0) = u_0(x) \quad (x \in X).$$

This equation determines a consistent (in the sense of (2.37)) collection of finite-dimensional equations (2.44) under projection into finite-dimensional subspaces. Without especially saying so, we will henceforth assume that all these equations are well-posed. Besides, we assume there is an operator B for which (2.47) is satisfied. Then the measures $\tilde{\mu}_{x,t}$ ($x \in \tilde{\mathcal{L}}$, $t \geq 0$) described in theorems 2.3 and 2.4 are supported in the Hilbert space $\mathcal{H}_B$, which contains $\mathcal{H}_+$ and is its completion with respect to the norm $\|x\|_B = \|Bx\|$ ($x \in \mathcal{H}_+$). Hereafter, these measures are related to equation (2.36).

THEOREM 2.6. *Suppose the conditions of theorem 2.4 hold for the consistent system of equations (2.44) determined by equation (2.36). Moreover, suppose that the coefficients of the equation satisfy the following uniform continuity conditions:*

$$\lim_{P,Q\to I} \sup_{\|x\|\leq R} |a(Px) - a(Qx)| = 0, \tag{2.48$'$}$$

$$\lim \sup |V(Px) - V(Qx)| = 0.$$

Then for every $x \in \tilde{\mathcal{L}}$ the family of measures $\tilde{\mu}_{x,t}$ ($t \geq 0$), which is weakly continuous in t in the sense of theorem 2.5, is a weak solution of the Cauchy problem (2.36).

PROOF. For $Q \in \mathcal{P}$ the measure $\mu^Q_{x,t}$ is a fundamental distribution for the corresponding finite-dimensional problem (2.44). So, for every function α of class $\hat{\Phi}_1$ and of bounded support in $\mathcal{H}_B$ the relation

$$\langle \alpha^P_t, \mu^Q_{x,t}\rangle|^T_s = \int_s^T \left\langle \frac{\partial \alpha^P_t}{\partial t} + A^Q \alpha^P_t, \mu^Q_{x,t}\right\rangle dt =$$

$$\int_s^T \left\langle \frac{\partial \alpha^P_t}{\partial t} + A^P \alpha^P_t, \mu^{P,Q}_{x,t}\right\rangle dt + \int_s^T \langle (A^Q - A^P)\alpha^P, \mu^Q_{x,t}\rangle\, dt$$

holds, where $P \subset Q$, $\alpha^P_t(x) = \alpha(Px, t)$.

We can pass to the limit in the lefthand side and in the first term at the righthand side of this equation, first as $Q \to I$, and then as $P \to I$. If the second term at the righthand side turns to zero, then by remark 2.5 the relation

$$\langle \alpha_t, \tilde{\mu}_{x,t}\rangle|^T_s = \int_s^T \left\langle \frac{\partial \alpha}{\partial t} + A\alpha, \tilde{\mu}_{x,t}\right\rangle dt$$

holds. It is equivalent to the required assertion, since the additional assumption that α be of bounded support can be easily dropped by another limit transition.

The integrand in the second term is

$$(A^{(Q)} - A^{(P)})\alpha^{(P)} =$$

$$\mathrm{Tr}[V(Qx)P\alpha''(Px)PVQ(x) - V(Px)P\alpha''(Px)PV(Px)] + (a(Qx) - a(Px), P\alpha'(Px)).$$

Since α is of bounded support, (2.48') is a necessary condition in order that this expression tends to zero. □

REMARK 2.6. Condition (2.48) can be verified as follows, e.g. Suppose a vector function $h(x)$ with values in a Banach space $\mathcal{H}$ satisfies the estimate

$$\|h(x) - h(y)\|_{\mathcal{H}} \leq C\|SB(x-y)\|_{\mathcal{H}_0} \quad (x, y \in \mathcal{H}_+), \tag{2.49}$$

with S a completely continuous operator on $\mathcal{H}_0$.

Here

$$\|h(Px) - h(Qx)\|_{\mathcal{H}} \leq C\|SB(P-Q)x\|_{\mathcal{H}_0},$$

and since S is compact, as $P, Q \to I$ the righthand side tends to zero uniformly on balls in $\mathcal{H}_B$.

The estimate (2.49) holds if $h(x)$ is differentiable along $\mathcal{H}_+$ and the operator $h'(x)B^{-1}S^{-1}$ is bounded.

Thus, for x fixed we have constructed measures $\tilde{\mu}_{x,t}$ which are weak solutions of the Cauchy problem under consideration.

However, the construction depends on x, and can up till now only be consistently carried out for a countable set $\mathcal{D}$ which is dense in $\mathcal{H}_+$.

Note that the conditions under which the proofs of the last theorems were carried out led to measures supported in a Hilbert subspace $\mathcal{H}_B$ of $\mathcal{H}_-$.

We now impose further restrictions on the coefficients of the equation, leading in fact to the consideration of a Cauchy problem in this space.

In this connection, the version when $\mathcal{H}_+ = \mathcal{H}_0 = \mathcal{H}_-$ are the same Hilbert space is very useful.

We give all considerations for this version first, and then indicate the modifications necessary to treat certain cases in which the Hilbert space $\mathcal{H}_B \subset X$ is not fixed beforehand.

We denote by $C_\infty(\mathcal{H}, \mathcal{K})$ the collection of functions $f(x)$ in $\mathcal{H}$, with values in $\mathcal{K}$, that are infinitely differentiable along $\mathcal{H}$ and satisfy estimates of the type

$$\|f(x)\| \leq C_1\|x\| + C_2,$$

$$\sup_x \sigma_2^2(f^{(r)}(x)) = \sup_x \sum_{k_1,\dots,k_r} \|f^{(r)}(x)(e_{k_1}, \dots, e_{k_r})\|^2 < \infty$$

($1 \leq r < \infty$, $\{e_k\}$ is an orthonormal basis in $\mathcal{H}$).

Let $C_\infty^0(\mathcal{H}, \mathcal{K}) \subset C_\infty(\mathcal{H}, \mathcal{K})$ be the subspace of bounded functions: $\|f(x)\| \leq C_2$.

THEOREM 2.7. *In equation (2.36), let $a \in C_\infty(\mathcal{H}, \mathcal{K})$, $V \in C_\infty(\mathcal{H}, L_2(\mathcal{H}))$, and suppose that the following estimate holds:*

$$\sigma_2^2(V(x)) + 2(x, a(x)) \leq C_1\|x\|^2 + C_2. \tag{2.50}$$

Then the Cauchy problem (2.36) is well-posed in the class $D = C_\infty^0(\mathcal{H}, \mathbf{R}^1)$. Its solution can be represented as

$$u(x,t) = \int_{\mathcal{H}} u_0(y)\mu_{x,\tau-t}(dy), \tag{2.51}$$

where $\mu_{x,t}$ ($x \in \mathcal{H}$, $t \geq 0$) are normalized nonnegative measures in $\mathcal{H}$. This solution can be obtained as the limit of the solutions of the finite-dimensional problems (2.44) resulting when (2.36) is projected into finite-dimensional subspaces.

PROOF. Apply theorem 2.4 with $B = I$. Condition (2.50) implies the estimate (2.47) for $Q \in \mathcal{P}$, with the same constants C, C_1, independent of Q. Thus, the measures $\tilde{\mu}_{x,t}$ ($t \geq 0$, $x \in Z$) are defined on the Borel σ-algebra of the space $\mathcal{H}$ for a dense set in Z (by means of the limit procedure described in theorem 2.3 with the same sequence of projection operators $P_k \to I$).

Let $f \in \mathfrak{H}_r(\mathcal{H}, \mathbf{R}^1)$. Consider the sequence of functions

$$u_{jk}(x,t) = u^{(P_j,P_k)}(x,t) = \int f(P_j y)\mu_{P_k x,t}^{(P,P_k)}(dy) \quad (x \in \mathcal{H}, \quad t \geq 0).$$

By the above,

$$\lim_{k\to\infty} u^{(P_j,P_k)}(x,t) = u_{P_j}(x,t) = \int f(P_j y)\tilde{\mu}_{x,t}(dy)$$
$$(x \in Z \cap \cup \mathcal{L}_k, \quad \mathcal{L}_k = P_k\mathcal{H}).$$

From proposition 2.4 (with $R = I$) it follows that $u_t^{(P_j,P_k)} \in \mathfrak{H}_r(\mathcal{H}, \mathbf{R}^1)$, with uniform estimates with respect to j and k.

In particular, uniform boundedness of the first order derivatives implies a uniform Lipschitz condition for $u_{jk}(x,t)$, and thus a uniform Lipschitz condition for the limit functions $u_j(x,t)$ on the dense set Z. Hence these functions can be extended to all of $\mathcal{H}$, preserving the Lipschitz property, and thus,

$$u_j(x,t) = \lim_{k\to\infty} u_{jk}(x,t) \quad (x \in \mathcal{H}, \quad t \geq 0). \tag{2.52}$$

On every ball in a finite-dimensional subspace $\mathcal{L}_m$, compactness in the sense of uniform convergence of the family of derivatives $u_{jk}^{(s)}(x,t)$ ($s = 1, \ldots, r-1$) along any fixed direction in $\mathcal{L}_m$ follows from uniform estimates for derivatives. Choosing a subsequence we may assume that uniform convergence on every ball in every $\mathcal{L}_m$ for derivatives up to order $r-1$, along any direction in $\mathcal{L}$, holds in (2.52). The estimates being uniform for all derivatives $u_{jk}^{(s)}$, they must be uniform for the functions $u_j^{(s)}$, and subsequently for the functions $u(x,t) = \lim_{j\to\infty} u_j(x,t)$.

Thus, the operators

$$V_t f(x) = \int f(y)\tilde{\mu}_{x,t}(dy)$$

leave invariant the linear set $\mathcal{D} = \cap_{r=1}^{\infty}\mathfrak{H}_r(\mathcal{H}, \mathbf{R}^1)$.

By theorem 2.6 (which holds since the required continuity condition follows from complete continuity of the derivatives, see remark 2.6), $\tilde{\mu}_{x,t}$ is a weak solution of the dual problem. Therefore, as follows from theorem 1.1 and remark 1.1, to prove well-posedness of the Cauchy problem (2.36) we need only prove the semigroup relation (1.5), or the equivalent uniqueness property of the weak solution of the dual problem. Both these properties can be obtained by limit transition as $P \to I$, by considerations close to those in the proof of theorem 2.6.

E.g., we show uniqueness of the weak solution. It suffices to show (see (1.4)) that if the equality

$$\langle u_T, \mu_T \rangle = \int_0^T \left\langle \frac{\partial u_t}{\partial t} + \mathcal{A}(u_t), \mu_t \right\rangle dt \tag{2.53}$$

is valid for each $u_t \in \Phi(t, C_1)$, then $\mu_t = 0$.

Let $u(x,t) = u(Px,t)$. Rewrite (2.53) as

$$(2.54)\quad \langle u_T, \mu_T \rangle = \int_0^T \left\langle \frac{\partial u_t}{\partial t} + \mathcal{A}^{(P)}(u_t), \mu_t \right\rangle dt + \int_0^T \langle \mathcal{A}(u_t) - \mathcal{A}^{(P)}(u_t), \mu_t \rangle \, dt,$$

where

$$\mathcal{A}(u) - \mathcal{A}^{(P)}(u) = \operatorname{Tr}[A(x) - A_P(x)]u''.$$

Let u_t be a solution of the finite-dimensional Cauchy problem

$$\frac{\partial u_t}{\partial t} + \mathcal{A}^{(P)}(u_t) = 0, \quad u(x,\tau) = \phi(x).$$

Then (2.54) implies

$$\langle \phi^{(P)}, \mu \rangle = \int_0^T \langle \mathcal{A}(u_t) - \mathcal{A}^{(P)}(u_t), \mu_t \rangle \, dt;$$

the righthand side of this equation tends to zero as $P \to I$, since the set $(u^{(P)})''$ is uniformly bounded, proving what was required. □

REMARK 2.7. Using boundedness of the support of the second moments of the measure $\mu_{x,t}$ we can show that (2.51) is a solution also in case the initial function is of class $C_\infty(\mathcal{H}, \mathbf{R}^1)$.

REMARK 2.8. In case the equation under consideration is given in a space $X = \mathcal{H}_- \supset \mathcal{H} \supset \mathcal{H}_+$ that is not a Hilbert space, the considerations are along the same lines.

Instead of condition (2.50) we have to require the existence of an operator B satisfying

$$(2.55)\quad \sigma_2^2(BV^*(x)) + 2(Bx, Ba(x)) \le C\|Bx\|^2 + C_1, \quad Ba, BV^* \in \hat{\mathfrak{H}},$$

where the class $\mathfrak{H}$ is replaced by the class $\hat{\mathfrak{H}}$ of functions f for which the quantities

$$(2.56)\quad \sup_x \sigma_2^2(f^{(j)}(x)(B^{-1}, \dots, B^{-1}))$$

are bounded.

If there also exists a sequence of orthoprojectors P_k commuting with B, theorem 2.4 is still applicable, and uniform estimates for derivatives follow.

EXAMPLE 2.1. Consider the space $X = \mathcal{H}_- = \mathbf{R}^\infty$ of all realvalued sequences $x = (x_1, x_2, \dots)$, let $\mathcal{H} = \ell_2$, and let $\mathcal{H}_+ = \mathbf{R}_0^\infty$ be the space of sequences such that $a_n = 0$ from a certain n onwards.

For B we choose a multiplication operator: $Bx = (\lambda_1 x_1, \lambda_2 x_2, \dots)$, where $\{\lambda_k\}_{k=1}^\infty$ is a sequence of positive numbers. Let $P_n x = x^{(n)} = (x_1, \dots, x_n, 0, 0, \dots)$, so that $BP_n = P_n B$. The space $\mathcal{H}_B$ consists of all sequences satisfying the condition $\|x\|_B^2 = \sum \lambda_k^2 |x_k|^2 < \infty$. The Cauchy problem (2.36) becomes

$$(2.57)\quad \frac{\partial u(x,t)}{\partial t} + \frac{1}{2} \sum_{j,k=1}^\infty a_{jk}(x) \frac{\partial^2 u}{\partial x_j \partial x_k} + \sum_{k=1}^\infty a_k(x) \frac{\partial u}{\partial x_k} = 0, \quad u(x,\tau) = u_0(x),$$

where $a_{jk}(x) = \sum_{s=1}^{\infty} v_{sj}v_{sk}$.

The restrictions imposed on the coefficients are:

$$\sum_{k,j_1,\dots,j_s} \left| \frac{\lambda_k}{\lambda_{j_1}\dots\lambda_{j_r}} \cdot \frac{\partial^r a_k}{\partial x_{j_1}\dots\partial x_{j_r}} \right|^2 < \infty,$$

$$\sum \left| \frac{\lambda_s}{\lambda_{j_1}\dots\lambda_{j_r}} \cdot \frac{\partial^r v_{sk}}{\partial x_{j_1}\dots\partial x_{j_r}} \right|^2 < \infty, \quad \sum_{s,k} |\lambda_s v_{sk}|^2 + 2\sum_s \lambda_s^2 x_s a_s < C\sum \lambda_s^2 x_s^2 + C_1.$$

Under these conditions the Cauchy problem is well-posed in the class $C_\infty(\mathcal{H})$, and its solution is given by the formula

$$u(x,t) = \int_{\mathcal{H}_B} u_0(y)\mu_{x,\tau-t}(dy) \quad (x \in \mathcal{H}_B).$$

2.7. Systems of equations in an infinite-dimensional space. We consider the system (2.1)–(2.2) with respect to a vector function $u(x)$ taking values in a Hilbert space,

(2.58)

$$\frac{\partial u}{\partial t} + \frac{1}{2}(V^*(x)V(x)\nabla, \nabla)u + (a(x), \nabla)u + \alpha(x)(V(x)\nabla, u) + C(x)u = 0 \quad (t \le \tau),$$
$$u(x,\tau) = u_0(x).$$

To prove well-posedness of the Cauchy problem (2.58) it suffices to show well-posedness of the corresponding associated problem, i.e. of the Cauchy problem for the single equation

$$\frac{\partial \tilde{u}}{\partial t} + \tilde{A}(\tilde{x})\tilde{u}(\tilde{x}) = 0, \quad \tilde{u}(\tilde{x},\tau) = (u_0(x), y) \tag{2.59}$$

with respect to the function $\tilde{u}(\tilde{x}) = (u(x), y)_{\mathcal{K}}$ ($\tilde{x} = (x,y)$), and with operator (2.6).

Here some additional remarks are appropriate, since well-posedness of the problem (2.59) does not follow formally from the previous section. The mater is that theorem 2.7 assumes boundedness of the derivatives of the coefficients of the equation, while functions that are linear in y have linear growth with respect to y after differentiation with respect to x.

However, by an additional reasoning we can establish uniform boundedness of the derivatives of the solution for the equations obtained by projecting the initial system into finite-dimensional spaces, and this is the basis for all subsequent developments.

The reasoning referred to is as follows. After differentiation of the expression $(u(x), y)$ with respect to y we obtain a function $u(x)$ whose boundedness follows from boundedness on the ball $\|y\| \le 1$ of the solution $(u(x), y)$ of the associated equation. After differentiation with respect to x we obtain a function of the type $(u'(x)h, y)$, which can be obtained by differentiation of the initial system and subsequent consideration of the new associated equation.

In this way we can obtain, as before, uniform boundedness conditions for derivatives of solutions of equations approximating the equation (2.59) associated with (2.58), and so prove well-posedness of its Cauchy problem.

Now the solutions of the approximating equations are linear in y, and hence the solution of (2.59) also has this property.

By proposition 2.1 this leads to well-posedness of the problem (2.58). We state the result without additional details.

THEOREM 2.8. *If, in addition to the conditions of theorem 2.7, $\alpha \in C^0_\infty(\mathcal{H}, L_2(\mathcal{H} \times \mathcal{K}, \mathcal{K}))$, $c \in C^0_\infty(\mathcal{H}, L_2(\mathcal{H}, \mathcal{K}))$, then the problem (2.58) is well-posed in $\mathcal{D} = C^0_\infty(\mathcal{H}, \mathcal{K})$, and its solution $u(x)$ is given by the formula*

$$(y, u(x,t)) = \int_{\mathcal{H}\times\mathcal{K}} (y', u(x'))\mu_{x,y,\tau-t}(dx' \times dy'),$$

where $\mu_{x,y,t}$ is a fundamental distribution of the problem (2.59), representing a family of normalized nonnegative measures in $\mathcal{H} \times \mathcal{K}$.

COROLLARY 2.1. *Theorems 2.7 and 2.8 and theorem IV.1.4 imply conditions for essential selfadjointness or maximal dissipativity of differential operators having the corresponding formal properties (see IV § 1) in the space L_2 of functions which are square integrable with respect to a Gaussian measure.*

More general results related to this can be found in chapter VII.

3. Chapter V: Supplementary remarks and historical comments

Differential equations involving functions of an infinite-dimensional argument are widespread in various branches of theoretical physics, but only recently became the subject of a systematic mathematical treatment.

Diverse, though not rigorous, material in this area was gathered in the well-known book by Lévy [1]. This initial stage of investigation of equations in functional derivatives, analogous to ordinary differential equations, was marked by the papers Sherstnev [1], Novikov [1], and M.N. Feller [6]. The papers by Dalecky–Kukharchuk [1] and Averbukh [1] deal with an infinite-dimensional analog of of the theory of first order quasilinear equations. In the comments to chapter IV we have pointed to a number of papers on the Laplace–Lévy operator. Second order parabolic partial differential equations in functional derivatives were treated as Kolmogorov equations for infinite-dimensional stochastic processes in Baklan [2,3], and Chantladze [1], who generalized some results of Gikhman–Skorokhod [1] to the infinite-dimensional case.

Second order parabolic equations in an abstract space and of type (1.35), (2.1)–(2.2) were first considered, independently, in Gross [4] and Dalecky [6,7,8]. Gross has made a rather comprehensive study of the equation with constant operator coefficient (for a review of his results with subsequent developments see Kuo [1]). Subsequently, in Piech [1] certain equations with variable coefficients were studied using a generalization of the 'parametrix' technique. In Dalecky [6,7,8] techniques from the theory of stochastic processes were used (see chapter VII). The presentation of this chapter is based on Dalecky–Fomin [2,4] (see also Dalecky [11]). The technique of projecting into finite-dimensional subspaces was inspired by Foias [1], which is devoted to other questions (the study of Hopf equations). These results were transferred to systems of equations in Bokhonov–Dalecky [1], in which the method of reducing a system to a single equation involving a higher-dimensional argument was applied. This method has a probabilistic meaning, and is connected with the notion of operator multiplicative functional (see chapter VII).

Other methods for studying differential equations involving an infinite-dimensional argument were used in Frolov [1,2,3,4,5]. M.I. Vishik et al. (see Vishik [1], Blecker–Vishik [1], Vishik–Marchenko [1]) have transferred a number of results in the theory of pseudodifferential operators to the infinite-dimensional case. In the comments to chapter IV we did describe certain results related to the study of selfadjointness of differential operators involving an infinite-dimensional argument and related to those considered here.

As has been suggested in chapter IV, in the infinite-dimensional case, regardless of an analysis of functions, an analysis of measures must be constructed. Here we have not studied differential equations involving measures in particular, and have considered these only in § 1 in the more general framework of differential equations involving distributions, dual to differential equations involving functions. A study of various kinds of differential equations involving measures was made in Uglanov [1,2,3,4].

CHAPTER VI

Integration in path space

In this chapter we consider so-called functional integrals, i.e. integrals with respect to a quasimeasure in path space determined by a transition measure. In case this transition measure is positive, the functional integral becomes the mean over a Markov stochastic process. This aspect will be considered in more detail in the next chapter. In the general case the role of transition measure is taken by a fundamental solution of a certain evolution equation. In this connection, functional integrals are closely connected with representations of solutions of evolution equations. Therefore, in this chapter we consider certain special multiplicative representations of solutions of linear and nonlinear evolution equations. These representations are used to construct path integrals in case the corresponding quasimeasure does not, in general, have bounded variation. For nonlinear equations this approach leads to integrals in the space of branching paths.

1. Markov quasimeasures

1.1. Transition measures. A boundedness condition for a Markov quasimeasure. Let $(\tilde{X}_T, \tilde{\mathcal{U}}) = \prod_{t\in T}(X_t, \mathcal{U}_t)$ $(T = [\tau, \tau_1] \subset \mathbf{R})$ be a product of measurable spaces.

A complexvalued function $P(t_1, x_1, t_2, A)$ $(\tau \leq t_1 \leq t_2 \leq \tau_1,\ x_1 \in X_{t_1},\ A \in \mathcal{U}_{t_2})$ is called a *transition measure* if the following conditions hold:

a) the set function $\nu_{x_1,t_1,t_2}(A) = P(t_1, x_1, t_2, A)$ is a measure on $(X_{t_2}, \mathcal{U}_{t_2})$;

b) the function $\alpha_{t_1,t_2,A}(x_1) = P(t_1, x_1, t_2, A)$ is $\mathcal{U}_{t_1}$-measurable;

c) the relation

$$P(t_1, x_1, t_2, A) = \int_{X_s} P(t_1, x_1, s, dy) P(s, y, t_2, A) \quad (t_1 \leq s \leq t_2) \tag{1.1}$$

holds, the integral being absolutely convergent.

A transition measure is called *normalized* if

$$P(t_1, x_1, t_2, X_{t_2}) = 1 \quad (t_1, t_2 \in [\tau, \tau_1]). \tag{1.2}$$

If we are given a transition measure, then we can construct a quasimeasure μ on the product as follows, using finite-dimensional projections.

For each ordered set $q = (t_1, \ldots, t_n)$, where $\tau = t_0 < t_1 < \cdots < t_n < \tau_1 = t_{n+1}$, and each point $x_0 \in X_\tau$ we define a measure in $X_q \times X_\tau$ $(X_q = X_{t_1} \times \cdots \times X_{t_n})$ by the formula

(1.3)

$$\mu^{(q)}_{x_0,\tau}(B) = \int_B \prod_{k=1}^{n+1} P(t_{k-1}, x_{k-1}, t_k, dx_k), \quad B \in \mathcal{U}_q \times \mathcal{U}_{\tau_1}, \quad \mathcal{U}_q = \mathcal{U}_{t_1} \times \cdots \times \mathcal{U}_{t_n}.$$

(If necessary we will indicate the final point of the interval, as in $\mu^{(q)}_{x_0,\tau,\tau_1}$.)

These measures form a consistent set. In fact, let $B = B_1 \times X_{t_j} \times B_2$, where $B_1 \in \mathcal{U}_{q_1}$, $B_2 \in \mathcal{U}_{q_2} \times \mathcal{U}_{\tau_1}$; $q_1 = (t_1, \dots, t_{j-1})$, $q_2 = (t_{j+1}, \dots, t_n)$. Then, by (1.1),

$$\mu^{(q)}_{x_0,\tau}(B) = \int_{B_1 \times B_2} \prod_{k=1}^{j-1} P(t_{k-1}, x_{k-1}, t_k, dx_k) \times$$

$$\left[\int_{X_{t_j}} P(t_{j-1}, x_{j-1}, t_j, dx_j) \prod_{k=j+1}^{n+1} P(t_{k-1}, x_{k-1}, t_k, dx_k)\right] =$$

$$\int_{B_1 \times B_2} \prod_{k=1}^{j-1} P(t_{k-1}, x_{k-1}, t_k, dx_k) P(t_{j-1}, x_{j-1}, t_{j+1}, dx_{j+1}) \times$$

$$\prod_{k=j+2}^{n+1} P(t_{k-1}, x_{k-1}, t_k, dx_k) = \mu^{(q_1 \cup q_2)}_{x_0,\tau}(B_1 \times B_2). \tag{1.4}$$

It readily follows from (1.4) that the measure $\mu^{(q)}_{x_0,\tau}$ becomes $\mu^{(q')}_{x_0,\tau}$ under the natural projection $\pi_{q'q} \colon X_q \to X_{q'}$, where $q' < q$:

$$[\mu^{(q)}_{x_0,\tau}]^{\pi_{q'q}} = \mu^{(q')}_{x_0,\tau}.$$

Note that if the transition measure $P(t, x_1, t_2, dx_2)$ is normalized and $\tilde{B} = B \times X_{\tau_1}$ ($B \in X_q$), then the measure

$$\tilde{\mu}^{(q)}_{x_0,\tau}(\tilde{B}) = \mu^{(q)}_{x_0,\tau}(B \times X_{\tau_1}) = \int_B \prod_{k=1}^{n} P(t_{k-1}, x_{k-1}, t_k, dx_k) \tag{1.5}$$

does not depend on the final point of the interval $[\tau, \tau_1]$.

The quasimeasures (1.3), (1.5) constructed from transition measures will be called *Markov quasimeasures*.

Obviously, for a normalized nonnegative transition measure P the quasimeasure $\tilde{\mu}^{(q)}_{x_0,\tau}$ is bounded. But this is not true in the general case, so we will consider conditions under which a Markov quasimeasure is bounded.

PROPOSITION 1.1. *If a normalized transition measure P satisfies the condition*

$$C = \sup_{x,t_1,t_2} \frac{1}{t_2 - t_1} \{\|\nu_{x,t_1,t_2}\| - 1\} < \infty, \tag{1.6}$$

then the Markov quasimeasure $\tilde{\mu}_{x_0,\tau}$ is bounded.

PROOF. If $B \in \tilde{\mathcal{U}}$, then $B \in \mathcal{U}_q$ for some $q = (t_1, \dots, t_n)$. Therefore,

$$|\tilde{\mu}_{x_0,\tau_1}(B)| = |\mu^{(q)}_{x_0,\tau_1}(B)| \le \int_B \prod_{k=1}^{n} |\nu_{x_{k-1},t_{k-1},t_k}|(dx_k) \le$$

$$\prod_{k=1}^{n} \sup_x \|\nu_{x,t_{k-1},t_k}\| \le \prod_{k=1}^{n} \{1 + c(t_k - t_{k-1})\} \le e^{C(\tau_1 - \tau)}. \quad \square$$

Under certain additional restrictions, condition (1.6) turns out to be close to necessary.

PROPOSITION 1.2. *Let $(X_t, \mathcal{U}_t)$ $(t \in T)$ be topological Radon spaces, and let $P(t_1, x_1, t_2, A)$ be a realvalued normalized transition measure which is continuous with respect to x_1 for fixed t_1, t_2, A. If the condition*

$$\lim_{t_2 \to t_1} \frac{1}{t_2 - t_1} [\nu^+_{x,t_1,t_2}(X_{t_2}) - 1] = \infty \quad (x \in X_{t_1}) \tag{1.7}$$

holds, then the Markov quasimeasure (1.5) has unbounded variation on every set $B \in \mathcal{U}_q$; $q = (t_1, \ldots, t_n)$, $\tau < t_1 < \cdots < t_n < \tau_1$.

PROOF. It follows from (1.7) that for each triple (t_1, t_2, x) there is a compact set $\delta(t_1, t_2, x) \in \mathcal{U}_{t_2}$ on which the measure ν_{x,t_1,t_2} has positive restriction and $P(t_1, x_1, t_2, \delta(t_1, t_2, x)) \geq 1 + c(t_2 - t_1)$., the constant c being sufficiently large for $t_2 - t_1$ sufficiently small. It follows from the continuity condition imposed that under small deviations of x the set $\delta(t_1, t_2, x)$ can be regarded as being fixed. For certain s, s_1 and $x \in X_s$ we consider the set $\delta(s, s_1, x)$, and represent it as a union of nonintersecting sets, $\delta(s, s_1, x) = \cup_{j_1=1}^{r} \gamma_{j_1}$ such that for each γ_{j_1} and $s_2 > s_1$ there is a set δ_{j_1} satisfying

$$P(s_1, x_1, s_2, \delta_{j_1}) \geq 1 + c(s_2 - s_1) \quad (x_1 \in \gamma_{j_1}).$$

We similarly represent $\delta_{j_1} = \cup_{j_2=1}^{r_2} \gamma_{j_1 j_2}$, so that for $s_3 > s_2$ there is a set $\delta_{j_1 j_2}$ for which

$$P(s_2, x_2, s_3, \delta_{j_1 j_2}) \geq 1 + c(s_3 - s_2),$$

etc. Now we consider cylinder sets

$$\Gamma^{s,x_0}_{j_1,\ldots,j_n} = \mathcal{U}^{s,x_0}_{s_1,\ldots,s_{n+1}}(\gamma_{j_1}, \ldots, \gamma_{j_1 \ldots j_n}, \delta_{j_1,\ldots,j_n}) =$$
$$\{x : x(s) = x_0, \; x(s_1) \in \gamma_{j_1}, \ldots, x(s_n) \in \delta_{j_1,\ldots,j_n}, \; x(s_{n+1}) \in \gamma_{j_1,\ldots,j_n}\},$$

and let

$$\Gamma^{s,x_0} = \cup_{j_1,\ldots,j_n} \Gamma^{s,x_0}_{j_1,\ldots,j_n}.$$

Subsequently, using the properties of the sets δ, γ we estimate from below the quasimeasure:

$$\tilde{\mu}_{x_0,s}(\Gamma^{s,x_0}) = \sum_{j_1,\ldots,j_n} \tilde{\mu}_{x_0,s}(\Gamma^{s,x_0}_{j_1,\ldots,j_n}) =$$
$$\sum_{j_1,\ldots,j_n} \int_{\delta_{j_1 \ldots j_n}} \int_{\gamma_{j_1 \ldots j_n}} \cdots \int_{\gamma_{j_1}} \prod_{k=1}^{n+1} P(s_{k-1}, x_{k-1}, s_k, dx_k) \geq$$
$$(1 + c\Delta s_n) \sum_{\gamma_{j_1 \ldots j_{n-1}}} \int_{\delta_{j_1 \ldots j_{n-1}}} \cdots \int_{\gamma_{j_1}} \prod_{k=1}^{n} P(s_{k-1}, x_{k-1}, s_k, dx_k) \geq$$
$$\prod_{k=1}^{n} (1 + c\Delta s_k) \int_{\delta(s,s_1,x)} P(s, x_0, s_1, dx_1) \geq \prod_{k=0}^{n} (1 + c\Delta s_k)$$

Choosing $\Delta_k = (\tau_1 - s)/n$ we arrive at

$$\tilde{\mu}_{x_0,s}(\Gamma^{s,x_0}) \geq \left(1 + c\frac{\tau_1 - s}{n}\right)^{n+1} \approx e^{c(\tau_1 - s)} \tag{1.8}$$

for n large while the constant c can be made sufficiently large. Let now $B \in \mathcal{U}_q$, $q = (t_1, \ldots, t_m = s < \tau_2)$. There always exists a set of type $B_1 \times B_2 \subset B$ with $B_1 \in \mathcal{U}_{q_1}$, $q_1 = (t_1, \ldots, t_{m-1})$, $B_2 \in \mathcal{U}_s$ such that the measure $m(A) = \mu_{x_1,\tau_1}(B_1 \times A)$ is sign-definite on $\mathcal{U}_s$ for $A \subset B_2$. It follows from (1.8) that for sufficiently large n the measure of the cylinder set $B_1 \times \cup_{x_0 \in B_2} \Gamma^{s,x_0}$ is sufficiently large. □

In particular, consider the case when $X_t = X$ $(t \in T)$ is a linear space and the transition measure is defined by

$$P(t_1, x_1, t_2, A) = P(t_2 - t_1, A - x_1), \tag{1.9}$$

where $P(t, A)$ is a measure on $(X, \mathcal{U})$ satisfying the conditions

$$P(t, X) = 1, \quad P(t_1 + t_2, A) = \int_X P(t_1, dy) P(t_2, A - y). \tag{1.10}$$

Such a transition measure will be called *homogeneous.*

For a homogeneous transition measure the expressions (1.6) and (1.7) do not depend on x_1, and as can be seen from the proof of proposition 1.2, the limit in (1.7) can be replaced by limes superior. This leads to the following result.

THEOREM 1.1. *Let $(X, \mathcal{U})$ be a topological linear Radon space. A Markov quasimeasure constructed from a realvalued homogeneous transition measure P which is continuous on every $A \in \mathcal{U}$ with respect to shifts, is bounded if and only if*

$$\overline{\lim_{t \to 0}} \frac{1}{t} [P^+(t, x) - 1] < \infty. \tag{1.11}$$

REMARK 1.1. It follows from the proof of proposition 1.2 that under condition (1.7) (respectively, if (1.11) is not true), the quasimeasure under consideration is discontinuous on the algebra of cylinder sets, and hence cannot be σ-additive.

REMARK 1.2. Using a more sophisticated reasoning proposition 1.2 can be transferred to complexvalued quasimeasures as well (see Feldman [3]). In a particular case we can modify the reasoning above. In fact, let the sets δ described in the proof have the property

$$\Re P(t_1, x_1, t_2, \delta) \geq 1 + c(t_2 - t_1), \quad |\Im P(t_1, x_1, t_2, \delta)| < \epsilon,$$

where ϵ is sufficiently small. Then we can estimate from below, as above, the realvalued quasimeasure $\Re \mu$.

REMARK 1.3. It follows from the relation $\nu^+ = (1 + |\nu|)/2$ that condition (1.7) is equivalent to

$$\lim_{t_2 \to t_1} \frac{1}{t_2 - t_1} \{|\nu_{x,t_1,t_2}|(X_{t_2}) - 1\} = \infty.$$

Thus, (1.11) can be rewritten as

$$\overline{\lim_{t\to 0}}\frac{1}{t}[\|P(t,\cdot)\|-1]<\infty. \tag{1.11'}$$

In particular, for $X=\mathbf{R}^m$ and a homogeneous transition measure with density with respect to Lebesgue measure,

$$P(t,A)=\int_A \rho(t,x)\,dx.$$

We have

$$\|P(t,\cdot)\|=\int_{\mathbf{R}^m}|\rho(t,x)|\,dx,$$

and (1.11) does not hold if $\|P(t,\cdot)\|\geq 1+\alpha$ ($\alpha>0$).

E.g., (1.11) holds precisely when $P\geq 0$ and $\|P(t,\cdot)\|=\text{const}\geq 1$, since then $\|P(t,\cdot)\|=1$.

EXAMPLE 1.1. Consider the fundamental solution $\rho(t,x)$ of the Cauchy problem

$$\frac{\partial u}{\partial t}=(-1)^{k+1}\sigma^{2k}\frac{\partial^{2k}u}{\partial x^{2k}}.$$

It determines a homogeneous transition measure $P(t,A)=\int_A\rho(t,x)\,dx$ satisfying the conditions (1.10). Applying Fourier transformation we can easily compute that

$$\rho(t,x)=\frac{1}{2\pi}\int_{-\infty}^{\infty}\exp\{i\omega x-(\sigma\omega)^{2k}t\}\,d\omega.$$

After the change of variables $x\to x\sigma\sqrt[2k]{t}$, $\omega\to\omega/(\sigma\sqrt[2k]{t})$ we find

$$\|P(t,\cdot)\|=\frac{1}{2\pi}\int_{-\infty}^{\infty}\left|\int_{-\infty}^{\infty}\exp\{i\omega x-(\sigma\omega)^{2k}t\}\,d\omega\right|\,dx=$$
$$\frac{1}{2\pi}\int_{-\infty}^{\infty}\left|\int_{-\infty}^{\infty}\exp\{i\omega x-\omega^{2k}t\}\,d\omega\right|\,dx=\text{const}\geq 1.$$

Thus, the corresponding Markov quasimeasure has bounded variation if and only if it is positive. This is known to hold only if $k=1$.

EXAMPLE 1.2. Consider the differential equation

$$\frac{\partial u}{\partial t}=\sigma^2\frac{\partial^2 u}{\partial x^2}$$

with complex parameter $\sigma^2=\sigma_0^2(i+\eta)$ ($\sigma_0^2>0$). Its fundamental solution is

$$\rho(t,x)=\frac{1}{\sqrt{4\pi\sigma^2 t}}\exp\left\{-\frac{x^2}{4\sigma^2 t}\right\}.$$

For $\eta=0$,

$$\Re\rho(t,x)=\frac{1}{\sqrt{4\pi\sigma_0^2 t}}\sin\left\{\frac{x^2}{4\sigma_0^2 t}-\frac{\pi}{4}\right\},\quad \Im\rho(t,x)=\frac{1}{\sqrt{4\pi\sigma_0^2 t}}\cos\left\{\frac{x^2}{4\sigma_0^2 t}-\frac{\pi}{4}\right\},$$

and it is easily checked that there is a set δ such that

$$\Re \int_\delta \rho(t,x)\,dx \geq 1 + \text{const} > 1, \qquad \left|\Im \int_\delta \rho(t,x)\,dx\right| \leq \epsilon.$$

For reasons of continuity, such a set exists also for $\eta > 0$ small. By remark 1.2, the corresponding Markov quasimeasure has unbounded variation.

1.2. Integration with respect to a Markov quasimeasure. For computing integrals with respect to a Markov quasimeasure we can use the technique of projecting (see (I.3.5)) as follows.

Suppose all spaces $X_t = X$ $(t \in T)$ coincide. Consider the subset $\tilde{X}_{\tau,x_0} \subset \tilde{X}$ of functions $x(t) \in X$ $(t \in T)$ satisfying the relation $x(\tau) = x_0$. We define a projection operation on $\tilde{X}_{\tau,x_0}$ by

$$\pi_q\colon x \to x_q; \quad x_q(t) = x(t_{k-1}) \quad (t_{k-1} \leq t < t_k),$$
$$q = (\tau = t_0 < t - 1 < \cdots < t_{n+1} = \tau_1).$$

Here, to every complexvalued function $F(x)$ on $\tilde{X}$ we associate the function $(S_qF)(x) = F(x_q) = F_q(x_0,\ldots,x_n)$ $(x_k = x(t_k))$ on the space $X_q = \prod_{k=1}^n X_{t_k}$. Note that the projection operator S_q is defined for those F which are given (at least) on the set $\tilde{X}^0_{\tau,x_0} = \cup_q \pi_q \tilde{X}_{\tau,x_0}$ of all possible step functions.

We denote by $\mathcal{F}$ the linear space of bounded functions $F(x)$ for which all operators S_q are defined and for which the functions S_qF are $\mathcal{U}_q$-measurable. For $F \in \mathcal{F}$, the following expression makes sense

$$J_q(F) = J_q^{(\tau,x_0)}(F) = \int_{X_q} (S_qF)(x_0,\ldots,x_n) \prod_{k=1}^{n+1} P(t_{k-1},x_{k-1},t_k,dx_k). \tag{1.12}$$

DEFINITION 1.1. A function F is said to be *integrable with respect to the Markov quasimeasure* $\mu_{x_0,\tau}$ if the limit

$$J(F) = J^{(\tau,x_0)}(F) = \lim_q J_q^{(\tau,x_0)}(F) \tag{1.13}$$

along the (generalized) net given by the partitions q exists. This limit will be called a (*functional*) *integral* with respect to the quasimeasure:

$$J^{(\tau,x_0)}(F) = \int_{\tilde{X}_{\tau,x_0}} F(x)\mu_{x_0,\tau}(dx)$$

We now consider the case when X is a locally convex space. Let $P(t_1,x,t_2,A) = P(t_1,t_2,A-x)$ be a spatially homogeneous transition measure. Then (1.1) becomes

$$P(t_1,t_2,A) = \int_X P(t_1,s,dy)P(s,t_2,A-y). \tag{1.14}$$

If $\phi(t_1,t_2,y) = \int_X e^{i\langle x,y\rangle} P(t_1,t_2,dx)$ $(y \in X')$ is the characteristic functional of a transition measure, then (1.14) implies that

$$\phi(t_1,t_2,y) = \phi(t_1,s,y)\phi(s,t_2,y).$$

In particular, suppose the following representation holds:

$$\phi(t_1, t_2, y) = \exp\left\{\int_{t_1}^{t_2} \alpha(s, y)\, ds\right\}, \tag{1.15}$$

where $\alpha(s, y)$ ($s \in T$, $y \in X'$) is a complexvalued function. Under these assumptions we can compute the characteristic functional of a Markov quasimeasure

$$\chi_{\mu_\tau, x_0}(y(\cdot)) = \int_{\tilde{X}_{\tau, x_0}} e^{i \int_\tau^{\tau_1} \langle x(t), y(t)\rangle\, dt} \mu_{\tau, x_0}(dx). \tag{1.16}$$

If $y(t)$ is a weakly integrable function with values in X', the integrand

$$F(x, y) = \exp\left\{i \int_\tau^{\tau_1} \langle x(t), y(t)\rangle\, dt\right\}$$

is determined by

$$(S_q F)(x) = F(x_q, y) = \exp\left\{i \sum_{k=1}^{n+1} \langle x_{k-1}, y_{k-1}\rangle\right\}, \quad y_{k-1} = \int_{t_{k-1}}^{k} y(t)\, dt.$$

Using (1.15) we obtain

$$\int_X e^{i\langle x_n, y_n\rangle} P(t_{n-1}, x_{n-1}, t_n, dx_n) = e^{i\langle x_{n-1}, y_n\rangle} \phi(t_{n-1}, t_n, y_n).$$

Furthermore,

$$\int_X e^{i\langle x_{n-1}, y_{n-1}\rangle} e^{i\langle x_{n-1}, y_n\rangle} \phi(t_{n-1}, t_n, y_n) P(t_{n-2}, x_{n-2}, t_{n-1} dx_{n-1}) =$$
$$e^{i\langle x_{n-2}, y_{n-1} + y_n\rangle} \phi(t_{n-1}, t_n, y_n) \phi(t_{n-2}, t_{n-1}, y_{n-1} + y_n),$$

and finally, under the condition (1.2),

$$J_q(F) = \exp\left\{i \left\langle x_0, \sum_{k=1}^{n} y_k\right\rangle\right\} \prod_{k=1}^{n} \phi\left(t_{k-1}, t_k, \sum_{j=k}^{n} y_j\right) =$$
$$\exp\left\{i \left\langle x_0, \int_{t_1}^{\tau_1} y(t)\, dt\right\rangle + \sum_{k=1}^{n} \int_{t_{k-1}}^{t_k} \alpha\left(s, \int_{t_k}^{\tau_1} y(t)\, dt\right) ds\right\}. \tag{1.17}$$

If $\alpha(s, y)$ is a bounded function, integrable with respect to s for every $y \in X'$ and continuous with respect to y for every $s \in T$, then limit transition with respect to partitions q is possible in the latter formula. This leads to the formula

$$\chi_{\mu_{x_0, \tau}}(y) = \exp\left\{i \left\langle x_0, \int_\tau^{\tau_1} y(t)\, dt\right\rangle + \int_\tau^{\tau_1} \alpha\left(s, \int_s^{\tau_1} f(t)\, dt\right) ds\right\}. \tag{1.18}$$

Now we can use the arguments applied in proposition 1.3.4 to compute integrals of functions

$$G(x) = \int \exp\left\{i \int_{\tau_1}^{\tau_2} \langle x(t), y(t)\rangle\, dt\right\} \nu(dy), \tag{1.19}$$

where ν is a complexvalued measure in $\tilde{X}' = \prod_{t\in T} X'_t$. Integrating the expression (1.17) termwise and passing to the limit, we obtain

$$J(G) = \lim_q \int_{\tilde{X}'} J_q(F)\nu(dy) = \int_{\tilde{X}'} \chi_{\mu_{x_0,\tau}}(y)\nu(dy) =$$

$$\int_{\tilde{X}'} \exp\left\{i\left\langle x_0, \int_\tau^{\tau_1} y(t)\,dt\right\rangle + \int_\tau^{\tau_1} \alpha\left(s, \int_s^{\tau_1} \phi(t)\,dt\right) ds\right\} \nu(dy). \tag{1.20}$$

2. Evolution families of operators

2.1. Construction of the evolution family. Let X be a complete metric space with metric ρ. An operator on X is taken to mean a map $y = A(x)$ from X into itself and satisfying a Lipschitz condition,

$$\rho(A(x), A(y)) \leq C(A, S)\rho(x, y) \quad (x, y \in S),$$

on each ball $S = S(x_0, r) = \{x\colon \rho(x, x_0) \leq r\}$.

For two operators A and B their superposition, which is clearly again an operator, will be denoted (as usual for linear operators) by their product: $(AB)(x) = A(B(x))$.

A family of operators A_λ ($\lambda \in \Lambda$) is said to be *uniformly bounded* if by all operators of the family each ball is mapped into some ball, and *uniformly Lipschitz* if the set of Lipschitz constants of the operators in the family is bounded on each ball: $C(A_\lambda, S) \leq C(S)$ ($\lambda \in \Lambda$). If a family satisfies both these requirements, it is called a *uniform family*. The uniform Lipschitz condition clearly implies the uniform boundedness condition if there is at least one x_0 at which $\rho(x_0, A_\lambda(x_0)) \leq \text{const}$ ($\lambda \in \Lambda$); e.g., if there is a fixed point $x_0 = A_\lambda(x_0)$.

Below we will consider two-parameter families of operators ($U(t, s)$ ($t, s \in T = [\tau, \tau_1]$). We assume that

$$U(t, t) = I \quad (t \in T). \tag{2.1}$$

Such a family of operators $U(t, s)$ ($t, s \in T$) is called an *evolution family* if it has the *evolution property*

$$U(t, s) = U(t, \theta)U(\theta, s) \quad (t, s, \theta \in T, \quad t \leq \theta \leq s). \tag{2.2}$$

We present a simple device for constructing, under certain conditions, evolution families.

Let $q = \{\tau = t_0 < t_1 < \cdots < t_n = \tau_1\}$ be a partition of T, and let $\{q\}$ be the totality of all partitions. We consider a family of operators $U(t, s)$ on T, and construct for each q a new family by setting for $t_{j-1} < t \leq t_j < t_m \leq s < t_{m+1}$:

$$U_q(t, s) = U(t, t_j) \prod_{k=j}^{m-1} U(t_k, t_{k+1}) U(t_m, s). \tag{2.3}$$

The family $U_q(t, s)$ has the *partial evolution property*

$$U_q(t, s) = U_q(t, t_r)U_q(t_r, s) \quad (t_r \in q \cap [t, s]). \tag{2.4}$$

Below, the relation

$$\tilde{U}(t,s) = \lim_q U_q(t,s) \tag{2.5}$$

is to be understood as the limit on each element $x \in X$ along the (generalized) net formed by the partitions q of T: for each pair $x \in X$, $\epsilon > 0$ there is a q such that any refinement $q' \supset q$ of it satisfies the estimate $\rho(\tilde{U}(t,s)(x), U_{q'}(t,s)(x)) < \epsilon$.

PROPOSITION 2.1. *If $U_q(t,s)$ ($q \in \{q\}$, $t,s \in T$) is a uniform family and the limit (2.5) exists, then the operators $U(t,s)$ form a uniform evolution family.*

PROOF. Let $t < \theta < s$ and q be chosen so that $\theta \in q$. By (2.4),

$$\rho(\tilde{U}(t,s)(x), \tilde{U}(t,\theta)\tilde{U}(\theta,s)(x)) \leq$$
$$\rho(\tilde{U}(t,s)(x), U_q(t,s)(x)) + \rho(U_q(t,\theta)U_q(\theta,s)(x), U_q(t,\theta)\tilde{U}(\theta,s)(x)) +$$
$$\rho(U_q(t,\theta)\tilde{U}(\theta,s)(x), \tilde{U}(t,\theta)\tilde{U}(\theta,s)(x))$$

and the righthand side tends to zero since U_q is a uniform family. So, $\tilde{U}(t,s) = \tilde{U}(t,\theta)\tilde{U}(\theta,s)$. Uniformity of the limiting family is obvious. □

Now we state conditions for the existence of the limit (2.5).

LEMMA 2.1. *Let $V(t,s)$, $W(t,s)$ ($s,t \in T$) be two families of operators such that $V_q(t,s)$, $W_q(t,s)$ are uniform families and $C(V_q(t,s),S) \leq C(S)$ for every ball S.*

Moreover, suppose there is a set D which is invariant under the family $W(t,s)$ and such that

$$\rho(V(t,s)(x), W(t,s)(x)) \leq (s-t)\epsilon(x, s-t) \quad (x \in D). \tag{2.6}$$

Then for $x \in D$ the following estimate holds:

$$\rho(V_q(t,s)(x), W_q(t,s)(x)) \leq C(S)(s-t)\sup_{W_x} \epsilon(y, d_q), \tag{2.7}$$

where $d_q = \max_q |t_{k+1} - t_k|$,

$$W_x = \{y : y = W_q(t,s)(s),\ q \in \{q\},\ t,s \in T\},$$
$$S \supset \{V(t,s)W_x\}.$$

PROOF. It follows from the triangle inequality that

$$\rho(V_q(t,s)(x), W_q(t,s)(x)) \leq$$
$$\sum_{k=j}^{m-1} \rho(V_q(t,t_k)W(t_k,t_{k+1})W_q(t_{k+1},s)(x), V_q(t.t_k)V(t_k,t_{k+1})W_q(t_{k+1},s)(x)) \leq$$
$$C(S)\sum_{k=j}^{m-1} \rho(W(t_k,t_{k+1})W_q(t_{k+1},s)(x), V(t_k,t_{k+1})W_q(t_{k+1},s)(x)) \leq$$
$$C(S)\sup_{q,k} \epsilon(W_q(t_k,s)(x), \Delta t_k)\sum_{k=j}^{m-1} \Delta t_k. \quad \square$$

We apply lemma 2.1 to the families $V(t,s) = U(t,s)$, $W(t,s) = U_q(t,s)$.

THEOREM 2.1. *Let $U_q(t,s)$ ($q \in \{q\}$, $t,s \in T$) be a uniform family of operators and suppose there is a set D, dense in X and invariant under the family $U(t,s)$, on which the condition*

$$\rho(U(t,s)(x), U(t,\theta)U(\theta,s)(x)) \leq K(x)(\theta - t)(s-\theta)^\alpha \quad (0 < \alpha \leq 1) \tag{2.8}$$

holds, where K is a function satisfying

$$\sup_{q,t,s} K(U_q(t,s)(x)) = k(x) < \infty \quad (x \in D). \tag{2.9}$$

Then the uniform evolution family

$$\tilde{U}(t,s) = \lim_q U_q(t,s)$$

exists.

PROOF. Since U_q is a uniform family and condition (2.9) holds, we have

$$\rho(U(t,s)(x), U_q(t,s)(x)) \leq$$
$$\rho(U(t,s)(x), U(t,t_j)U(t_j,s)) +$$
$$\sum_{k=j}^{m-1} \rho(U_q(t,t_k)U(t_k,s)(x), U_q(t,t)_k)U(t_k,t_{k+1})U(t_{k+1},s)(x)) \leq$$
$$K(x)(t_j - t)(s-t_j)^\alpha + \sum_{k=j}^{m-1} CK(x)(t_{k+1} - t_k)(s - t_{k+1})^\alpha \leq$$
$$CK(x)(s-t)^{1+\alpha}.$$

Now apply lemma 2.1 with $V(t,s) = U(t,s)$ and $W(t,s) = U_{q'}(t,s)$. It follows from (2.7) that

$$\rho(U_q(t,s)(x), U_{q\cap q'}(t,s)(x)) \leq \mathrm{const}(s-t) \sup_{y \in U_x} K(y)[d(q)]^\alpha,$$

and so, since X is complete, for $x \in D$ the limit $\tilde{U}(t,s)(x) = \lim_q U_q(t,s)(x)$ exists. Uniformity of U_q implies uniformity of the limit family. The uniform Lipschitz condition implies that limit transition is possible on all of X.

Finally, by proposition 2.1 the family $\tilde{U}$ is an evolution family. □

REMARK 2.1. A similar result can be obtained by replacing (2.8) with

$$\rho(U(t,s)(x), U(t,\theta)U(\theta,s)(x)) \leq K(x)(\theta - t)^\alpha(s-\theta).$$

Then we can use the estimate

$$\rho(U(t,s)(x), U_q(t,s)(x)) \leq \sum_{k=j}^{m+1} \rho(U(t,t_k)U_q(t_k,s)(x), U(t,t_{k-1})U_q(t_{k-1},s)(x).$$

REMARK 2.2. It follows from the uniform Lipschitz condition that under the conditions of the theorem the convergence is uniform on compact sets in X.

Now we state sufficient conditions for the family $W_q(t,s)$ to converge to an evolution family $V(t,s)$, for some $W(t,s)$.

THEOREM 2.2. *Let $V(t,s)$, $W_q(t,s)$ $(q \in \{q\},\ t,s \in T)$ be uniform families of operators on X and suppose that for some set D which is dense in X and invariant under the evolution family V the estimate*

$$\rho(V(t,t+\Delta t)(x), W(t,t+\Delta t)(x)) \leq \Delta t \cdot \epsilon(x,\Delta t)$$

holds, where

$$\lim_{\Delta t\to 0} \sup_{t,s} \epsilon(V(t,s)(x),\Delta t) = 0 \quad (x \in D).$$

Then

$$V(t,s)(x) = \lim_q W_q(t,s)(x) \quad (x \in X). \tag{2.10}$$

If also the metric space (X,ρ) is densely imbedded in a metric space (X_0,ρ_0) such that $\rho \geq \rho_0|_X$ and the estimates

$$\begin{gathered}\rho(V(t,s)(x), V(t,s)(y)) \leq C(t,s)\rho_0(x,y);\\ \rho(W_q(t,s)(x), W_q(t,s)(y)) \leq C(t,s)\rho_0(x,y);\\ (t < s, \quad q \in \{q\}, \quad x,y \in X)\end{gathered} \tag{2.11}$$

hold, then for $t < s$ the operators $W_q(t,s)$ can be continuously extended to X_0, map X_0 to X, and the relation (2.10) is true in the metric ρ for every $x \in X_0$. The convergence is uniform on compact sets in X_0.

PROOF. Lemma 2.1 implies that

$$\rho(V(t,s)(x), W_q(t,s)(x)) \leq C(s-t) \sup_\theta \epsilon(V(\theta,s)(x), d(q)),$$

which proves (2.10), first for $x \in D$ and then (since all families are uniform Lipschitz) for $x \in X$.

Suppose (2.11) is true. Then $V(t,s)$ and $W_q(t,s)$ can be extended to X_0 with preservation of these estimates, and for $x \in X$, $x_0 \in X_0$,

$$\begin{gathered}\rho(V(t,s)(x_0), W_q(t,s)(x_0)) \leq\\ \rho(V(t,s)(x_0), V(t,s)(x)) + \rho(W_q(t,s)(x_0), W_q(t,s)(x)) + \rho(V(t,s)(x), W_q(t,s)(x)) \leq\\ 2C(t,s)\rho_0(x_0,x) + C(s-t)\sup_\theta \epsilon(V(\theta,s)(x), d(q)).\end{gathered}$$

Choosing first x sufficiently close to x_0, we can make the first term sufficiently small; subsequently, x being fixed, we make the second term small for appropriate $d(q)$. Hence, for $x_0 \in X_0$, (2.10) follows. The uniform convergence on compact sets follows from the uniform Lipschitz property, by Arzelà's theorem. □

REMARK 2.3. The requirement that the Lipschitz constants satisfy an estimate of type

$$C(V(t,s),S) \leq e^{\alpha(s-t)} \tag{2.12}$$

(α being independent of the ball S) obviously leads to the same estimate for the family V_q: $C(V_q(t,s)) \le e^{\alpha(s-t)}$, since the corresponding constants are multiplied under superposition of Lipschitz maps.

REMARK 2.4. Condition (2.11) can be verified also by the following considerations.

Suppose X is equipped with a scale of metrics ρ_λ ($\lambda \in [0,1]$) such that $\rho_1 = \rho$, $\rho_{\lambda_1} \ge \rho_{\lambda_2}$ ($\lambda_1 \ge \lambda_2$), and suppose that for some monotone function $\beta(\theta)$ mapping $[t,s]$ to $[0,1]$, $\beta(t) = 1$, $\beta(s) = 0$, and some monotone function $k(\theta)$ the following estimate holds:

(2.13)
$$\rho_{\beta(\theta_2)}(U(\theta_2,\theta_1)(x), U(\theta_2,\theta_1)(y)) \le e^{k(\theta_2)-k(\theta_1)}\rho_{\beta(\theta_1)}(x,y) \quad (x,y \in X, \quad \theta_2 < \theta_1).$$

Then

$$\begin{aligned}\rho_1(U_q(t,s)(x), U_q(t,s)(y)) &\le \\ e^{k(t)-k(t_1)}\rho_{\beta(t_1)}(U_q(t_1,s)(x), U_q(t_1,s)(y)) &\le \\ e^{k(t)-k(t_2)}\rho_{\beta(t_2)}(U_q(t_2,s)(x), U_q(t_2,s)(y)) &\le \cdots \le \\ e^{k(t)-k(s)}\rho_0(x,y)&,\end{aligned}$$

independently of the partition q.

Evolution families on linear spaces arise when solving the Cauchy problem for differential equations. E.g., let Y be a Banach space, D a dense set in it, X a neighborhood of zero in Y, $D_X = X \cap D$, and $A\colon T \times D_X \to Y$ a map such that the Cauchy problem

$$\frac{dx}{dt} + A(t,x) = 0, \quad x(s) = x_0, \quad t \le s \quad (t,s \in T), \quad x(t) \in D_X, \tag{2.14}$$

is well-posed on the interval $T = [\tau, \tau_1]$. We denote the solution of this problem by

$$x(t) = U(t,s)s_0.$$

The map $U(t,s)\colon D_X \to D_X$ has the evolution property, i.e.

$$U(t,s) = U(t,\theta)U(\theta,s) \quad (t \le \theta \le s),$$

since the solution of (2.14) is unique. Sometimes these maps turn out to be Lipschitz, and can be extended to all of X. Then they are operators (in our sense) on the metric space X. The family $U(t,s)$ is called the *evolution family for the problem* (2.14), and $A(t,x)$ is called the *generating map* of this family. If (2.14) can be integrated termwise, it is equivalent to the integral equation

$$U(t,s)(x) = x + \int_t^s A(\theta, U(\theta,s)(x))\,d\theta \quad (t \le s, \quad t,s \in T). \tag{2.15}$$

Conversely, suppose the evolution family $U(t,s)(x)$, acting in a domain X in a Banach space Y, be differentiable for $x \in D_X$, the set D_X being invariant under this family. Then

$$\frac{U(t-\Delta t,s)(x) - U(t,s)(x)}{-\Delta t} = \frac{U(t-\Delta t,t)(x) - I}{-\Delta t}U(t,s)(x) \to U'_\theta(\theta,t)|_{\theta=t}U(t,s)(x),$$

and thus the generating operator of $U(t,s)$ is $A(t) = -U'_\theta(\theta,t)|_{\theta=t}$.

PROPOSITION 2.2. *Suppose that under the conditions of theorem 2.1 the domain X is a set in a Banach space Y, and suppose that for $x \in D_X$ the function $U(t,s)(x)$ is continuously differentiable with respect to t: $dU(t,s)(x)/dt = U'_1(t,s,x)$. Let D_X be invariant under U, and put $\tilde U - \lim_q U_q$.*

Then the generating map of the family $\tilde U$ is $A(t) = -U'_1(t,t)$.

PROOF. For $x \in D$, the relation

$$\|U_q(t-\Delta t,t)(x) - x - \Delta t A(t,x)\| \leq$$

$$\int_{t-\Delta t}^{t} \|U'_1(\theta,t)(x) - U'_1(t,t)(x)\|\,d\theta + \|U_q(t-\Delta t,t)(x) - U(t-\Delta t,t)(x)\| \leq$$

$$\Delta t\left\{\sup_{\theta\in[t-\Delta t,t]} \|U'_1(\theta,t)(x) - U'_1(t,t)(x)\| + CK(x)\Delta t^\alpha\right\}$$

implies an estimate independent of q, leading after limit transition with respect to q to a similar estimate for $\tilde U(t-\Delta t,t)$. This implies

$$\frac{d}{d\theta}\tilde U(\theta,t)(x)|_{\theta-t} = U'_1(t,t)(x).$$

Since $\tilde U(t,s)(x) \in D_X$ for $x \in D_X$, we have

$$\frac{d\tilde U(s,t)(x)}{dt} = \lim_{\Delta t\to 0}\frac{\tilde U(t-\Delta t,s)(x) - \tilde U(t,s)(x)}{-\Delta t} =$$

$$\lim_{\Delta t\to 0}\frac{\tilde U(t-\Delta t,s) - I}{-\Delta t}\tilde U(t,s)(x) = -A(t,\tilde U(t,s,x)). \quad \square$$

2.2. The chronological product of evolution families. Let $V(t,s)$, $Z(t,s)$ be two evolution families on a space X. The family of operators $U(t,s) = V(t,s)Z(t,s)$ does not have the evolution property, in general. If, however, there the evolution family $\tilde U(t,s) = \lim_q U_q(t,s)$ exists, we call it the *chronological product* of the families V and Z, and denote it by

$$\tilde U(t,s) = (V \otimes Z)(t,s). \tag{2.16}$$

Using theorem 2.1 we can prove that the chronological product exists. Thus the question arises of the validity of the conditions of this theorem. We can derive the uniformity condition from (2.12), as suggested in remark 2.3. In verifying the condition we have to estimate the quantity

$$\rho(U(t,s)(x), U(t,\theta)U(\theta,s)(x)) =$$
$$\rho(V(t,s)Z(t,s)(x), V(t,\theta)Z(t,\theta)V(\theta,s)Z(\theta,s)(x)) \leq$$
$$C(V)\rho(V(\theta,s)Z(t,\theta)Z(\theta,s)(x), Z(t,\theta)V(\theta,s)Z(\theta,s)(x)).$$

Hence the following condition would suffice:

$$\rho(V(\theta,s)Z(t,\theta)(y), Z(t,\theta)V(\theta,s)(y)) \leq K(y)(\theta-t)(s-\theta)^\alpha \quad (0<\alpha\leq 1),$$

where $\sup_{t,s,q} K(U_q(t,s)(x)) < \infty$.

In the sequel we will consider evolution families on a Banach space X determined by differential equations, or, strictly speaking, integral equations like (2.15). Let

$$V(t,s)(x) = x + \int_t^s A(\theta, V(\theta,s)(x))\, d\theta, \tag{2.17}$$

$$Z(t,s)(x) = x + \int_t^s B(\theta, Z(\theta,s)(x))\, d\theta. \tag{2.18}$$

Consider the expression

$$R(x) = Z(t,\theta)V(\theta,s)(x) - V(\theta,s)Z(t,\theta)(x) =$$

$$V(\theta,s)(x) + \int_t^\theta B(\sigma, Z(\sigma,\theta)V(\theta,s)(x))\, d\sigma - V(\theta,s)\left[x + \int_t^\theta B(\sigma, Z(\sigma,\theta)(x))\, d\sigma\right] +$$

$$V(\theta,s)\int_t^\theta B(\sigma, Z(\sigma,\theta)(x))\, d\sigma - V(\theta,s)\int_t^\theta B(\sigma, Z(\sigma,\theta)(x))\, d\sigma.$$

Using formulas (2.17) and (2.18) we can represent $R(x)$ as the sum of two terms, $R = R_1 + R_2$, where

$$R_1(x) =$$

$$\int_t^\theta B(\sigma, Z(\sigma,\theta)V(\theta,s)(x))\, d\sigma - V(\theta,s)\int_t^\theta B(\sigma, Z(\sigma,\theta)(x))\, d\sigma -$$

$$\int_\theta^s A\left(r, V(r,s)\left(\int_t^\theta B(\sigma, Z(\sigma,\theta)(x))\, d\sigma\right)\right) dr =$$

$$-\int_t^\theta [B(\sigma, Z(\sigma,\theta)(x)) - B(\sigma, Z(\sigma,\theta)V(\theta,s)(x))]\, d\sigma,$$

$$R_2(x) =$$

$$V(\theta,s)(x) - V(\theta,s)\left(x + \int_t^\theta B(\sigma, Z(\sigma,\theta)(x))\, d\sigma\right) +$$

$$V(\theta,s)\left(\int_t^\theta B(\sigma, Z(\sigma,\theta)(x))\, d\sigma\right) -$$

$$\int_\theta^s A\left(r, V(r,s)\left(\int_t^\theta B(\sigma, Z(\sigma,\theta)(x))\, d\sigma\right)\right) dr =$$

$$[V(\theta,s)(x) - x] - V(\theta,s)\left(x + \int_t^\theta B(\sigma, Z(\sigma,\theta)(x))\, d\sigma\right) +$$

$$\left[x + \int_t^\theta B(\sigma, Z(\sigma,\theta)(x))\, d\sigma\right] =$$

$$\int_\theta^s \left[A\left(r, V(r,s)(x) - A(r, V(r,s)\left(x + \int_t^\theta B(\sigma, Z(\sigma,\theta)(x))\, d\sigma\right)\right)\right] dr.$$

Under certain restrictions we can estimate these expressions, and obtain the following result.

THEOREM 2.3. *Let X_0 be a Banach space, D a dense set in it, and suppose there is a collection of densely imbedded Banach spaces X_λ ($\lambda \in [0,1]$): $X_{\lambda_1} \subset X_{\lambda_2}$ ($\lambda_1 > \lambda_2$), $X_1 \supset D$, with norms $\|x\|_{\lambda_1} \geq \|x\|_{\lambda_2}$ ($\lambda_1 > \lambda_2$), which are invariant under the families V and Z. Suppose the following conditions hold:*

a) *the maps A and B are homogeneous, in the sense that $A(t,0) = 0$, $B(t,0) = 0$;*
b) *the following estimates hold:*

$$\|A(s_2, V(s_2,s_1)(x_2)) - A(s_2, V(s_2,s_1)(x_1))\|_{\lambda_2} \leq \frac{C\|x_2 - x_1\|_{\lambda_1}}{(s_2 - s_1)^{1+\lambda_2-\lambda_1}} \quad (\lambda_2 \leq \lambda_1); \tag{2.19}$$

$$\|B(s_2, Z(s_2,s_1)(x_2)) - B(s_2, Z(s_2,s_1)(x_1))\|_\lambda \leq C\|x_2 - x_1\|_{\lambda+\alpha} \quad (\alpha, \lambda \geq 0, \quad \lambda + \alpha \leq 1); \tag{2.20}$$

c) *the Lipschitz constants for the families V and Z are bounded and for each of the spaces X_0, X_1 satisfy the estimate*

$$C(V(t,s)) \leq e^{\gamma(s-t)}, \quad C(V(t,s)) \leq e^{\gamma(s-t)}. \tag{2.21}$$

Then there is an evolution family $\tilde{U}(t,s)$ that is the chronological product of V and Z: $\tilde{U} = V \otimes Z$.

PROOF. Using the estimates (2.19) and (2.20), we obtain for $x \in D$,

$$\|R_1(x)\|_\lambda \leq C(\theta - t)\|x - V(\theta,s)(x)\|_{\lambda+\alpha} =$$
$$C(\theta - t)\left\|\int_\theta^s A(\sigma, V(\sigma,s,x))\,d\sigma\right\|_{\lambda+\alpha} \leq$$
$$C^2\|x\|_{\lambda_1}(\theta - t)\int_\theta^s \frac{d\sigma}{(s-\sigma)^{1+\lambda+\alpha-\lambda_1}} = \frac{C^2\|x\|_{\lambda_1}(\theta - t)(s-\theta)^{\lambda_1-(\lambda+\alpha)}}{\lambda_1 - (\lambda+\alpha)},$$

and

$$\|R_2(x)\|_\lambda \leq C\left\|\int_t^\theta B(\sigma Z(\sigma,\theta)(x))\,d\sigma\right\|_{\lambda_1-\alpha}\int_\theta^s \frac{dr}{(s-r)^{1+\lambda-(\lambda_1-\alpha)}} \leq$$
$$C^2(\theta - t)\|x\|_{\lambda_1}\frac{(s-\theta)^{\lambda_1-(\lambda+\alpha)}}{\lambda_1 - (\lambda+\alpha)} \quad (1 \geq \lambda_1 > \lambda + \alpha \geq 0).$$

Note that we have used the relations

$$\|A(\sigma, V(\sigma,s)(x))\|_{\lambda+\alpha} \leq \frac{C\|x\|_{\lambda_1}}{(s-\sigma)^{1+\lambda-\alpha-\lambda_1}},$$
$$\|B(\sigma, Z(\sigma,s)(x))\|_{\lambda_1-\alpha} \leq C\|x\|_{\lambda_1},$$

which follow from a). Thus, by (2.16) the condition (2.8) in theorem 2.1 holds:

$$\|U(t,s)(x) - U(t,\theta)U(\theta,s)(x)\|_\lambda \leq e^{\gamma(s-t)}(\|R_1(x)\|_\lambda + \|R_2(x)\|_\lambda) \leq$$
$$2C^2e^{\gamma(s-t)}\frac{1}{\lambda_1 - (\lambda+\alpha)}(\theta - t)(s-\theta)^{\lambda_1-(\lambda+\alpha)},$$

where $K(x) = \text{const}\|x\|_1$ and $k(x) < \infty$ $(x \in D)$, by b). Uniformity of the family $U_q(s,t)$ follows from a) and b) (see remark 2.1). $\square$

REMARK 2.5. By means of proposition 2.2 we can find the generating operator of the family $\tilde{U}(t,\tau)$:

$$C(t,x) = -U'_\theta(\theta,t)(x)|_{\theta=t} =$$
$$-V'_\theta(\theta,t,Z(\theta,t)(x))|_{\theta=t} - V'_x(\theta,t,Z(\theta,t)(x))|_{\theta=t} \cdot Z'_\theta(\theta,t)(x)|_{\theta=t} =$$
$$A(t,x) + B(t,x).$$

Thus, taking the chronological product $V \otimes Z$ leads to addition of the generating operators.

REMARK 2.6. Condition (2.20) is fulfilled if the family Z is Lipschitz on X_λ and B satisfies the estimate

$$\|B(s,x_2) - B(s,x_1)\|_\lambda \le C\|x_2 - x_1\|_{\lambda+\alpha}. \tag{2.22}$$

In particular, for $\alpha = 0$ this means uniformity of $B(y)$ on X_λ. It suffices to consider the value $\lambda = 1$ only. Here (2.19) becomes

$$\|A(s_2, V(s_2,s_1)(x_2)) - A(s_2, V(s_2,s_1)(x_1))\|_0 \le C\|x_2 - x_1\|_1.$$

Since V is assumed to be Lipschitz in X_1, it suffices to require that

$$\|A(s_2,x_2) - A(s_1,x_1)\|_0 \le C\|x_2 - x_1\|_1.$$

2.3. Linear evolution families with constant generating operator. The above described convergence conditions for multiplicative expressions are simplified when applied to linear operators on a Banach space X. In this case the Lipschitz constant coincides with the operator norm, and the uniformity condition for the family of operators $V_q(s,t)$ follows (see remark 2.3) from the estimate

$$\|V(t,s)\| \le e^{\alpha(s-t)}.$$

We now consider some particular cases which will be of use in the next section. We do not aim at complete generality (see S. Krein [1], Dalecky [5]).

Recall that an operator function $V\colon \mathbf{R}_+ \to \mathcal{L}(X)$ (where $\mathcal{L}(X)$ is the space of bounded linear operators on X) is called a *strongly continuous semigroup* if

$$V(s+t) = V(s)V(t) \quad (s,t \ge 0)$$

and

$$\lim_{t\to 0} V(t)x = x \quad (x \in X).$$

A strongly continuous semigroup has a closed generating operator

$$A = \lim_{t\to 0} \frac{V(t) - I}{t},$$

defined on a linear set D which is dense in X and invariant under $V(t)$. Below we will use the well-known notation $V(t) = e^{At}$. In the simplest case, when A is a bounded operator, e^{At} is defined for all $t \in \mathbf{R}$ by $e^{At} = \sum_{k=0}^\infty A^k t^k/k!$, which converges in the operator norm.

Also, $\|e^{At}\| \leq e^{t\|A\|}$. The operators $U(t,s) = e^{A(s-t)}$ form an evolution family.

Necessary and sufficient conditions for a closed operator to be the generating operator of a strongly continuous semigroup are well known. Here we give some particular cases, of use in the sequel.

The condition

$$\|(A-\lambda I)^{-1}\| \leq \frac{1}{\lambda-\omega} \quad (\lambda \in \mathbf{R}, \quad \lambda > \omega) \tag{2.23}$$

is sufficient for the existence of a strongly continuous semigroup satisfying the estimate

$$\|e^{At}\| \leq e^{\omega t}. \tag{2.24}$$

Condition (2.23) (hence the estimate (2.24)) holds if, in particular, A is a normal operator on a Hilbert space X whose spectrum σ_A lies in the halfplane

$$\Re\lambda \leq \omega \quad (\lambda \in \sigma_A).$$

In this case the following spectral representation holds:

$$A = \int_{\sigma_A} \lambda E(dx): \quad e^{At} = \int_{\sigma_A} e^{\lambda t} E(dx), \tag{2.25}$$

where $E(\Delta)$ is a measure in the plane (on the real axis for $A = A^*$) whose vales are orthoprojectors which map disjoint Borel sets to disjoint projections.

Without loss of generality we can regard σ_A to lie inside the left halfplane ($\omega < 0$), and hence $A^{-1} \in \mathcal{L}(X)$.

For operators (2.25) we can define the powers

$$A^\alpha = \int_{\sigma_A} \lambda^\alpha E(d\lambda) \quad (\alpha \in \mathbf{R}), \tag{2.25'}$$

which are bounded operators for $\alpha \leq 0$.

From the formulas (2.25) and (2.25') we derive

$$\|A^\alpha e^{At}\| = \sup_{\lambda \in \sigma_A} |\lambda^\alpha e^{\lambda t}| = C(A,\alpha,t). \tag{2.26}$$

If $t > 0$ this quantity can be finite also for $\alpha > 0$, provided the spectrum of A is properly located, e.g., if the spectrum belongs to the domain $\pi \geq |\operatorname{Arg}\lambda| \geq \epsilon + \pi/2$ ($\epsilon > 0$).

In particular, we will consider operators of the type

$$A = zA_0$$

with A_0 a nonnegative operator on a Hilbert space X, $z = \rho e^{i\phi}$, $\pi \geq |\phi| \geq \epsilon + \pi/2$. It can be readily computed that

$$C(A,\alpha,t) = \frac{\alpha}{(\alpha t|\cos\phi|e)^\alpha}. \tag{2.27}$$

In this case a new norm can be introduced in the domain of definition D_α:

$$\|x\|_\alpha = \|A^\alpha x\|, \tag{2.28}$$

and so a family of densely imbedded Hilbert spaces is obtained. Each such space is invariant under e^{At}, and for $t > 0$,

$$e^{At}: X \to D_\alpha,$$

because of (2.28). Property (2.23) (with $\omega + \|B\|$ instead of ω) is readily verified to be preserved when passing from the generating operator A to an operator $A+B$ with B a bounded operator.

If $A = A_0$ is nonnegative on the Hilberts pace X, then we can turn to the more general case of an operator B subordinate to a fractional power of A, i.e. for some $\gamma \in (0,1)$, $D_B \supset D_{A_0^\gamma}$ and $\|BA_0^{-\gamma}\| < \infty$. Suppose that the operator B^* also has this property. It is then readily verified that the operators A_0+B and $(A_0+B)^* = A_0+B^*$ are dissipative on D_{A_0}, i.e.

$$\Re\left((A_0 + B)\psi, \psi\right) \le \text{const}\|\psi\|^2,$$
$$\Re\left((A_0 + B^*)\psi, \psi\right) \le \text{const}\|\psi\|^2 \quad (\psi \in D_{A_0}).$$

Hence the semigroup $e^{(A_0+B)t}$ exists and satisfies the condition (2.24). This semigroup is governed by the differential equation

$$\frac{d}{dt}e^{(A_0+B)t}\psi = (A_0 + B)e^{(A_0+B)t}\psi = A_0 e^{(A_0+B)t}\psi + Be^{(A_0+B)t}\psi$$

and hence by the integral equation

$$e^{(A_0+B)t} = e^{A_0 t} + \int_0^t e^{A_0(t-s)}Be^{(A_0+B)s}\,ds,$$

which is obtained from it by variation of constants. Using the latter equation we can obtain the following estimates for the semigroup $e^{(A_0+B)t}$:

$$\|A_0^r e^{(A+B)t}A_0^{-s}\| \le \frac{\text{const}}{t^{r-s}},$$

(2.29)

$$\|A_0^{-s} e^{(A+B)t}A_0^{r}\| \le \frac{\text{const}}{t^{r-s}}.$$

Suppose that B is also the generating operator of a semigroup. Under rather wide assumptions a representation in the for of a chronological product is true:

$$e^{(A+B)t} = e^{At} \otimes e^{Bt}, \tag{2.30}$$

i.e.

$$e^{(A+B)t} = \lim_{\substack{\max|\Delta t_k|\to 0\\ \sum \Delta t_k = t}} \prod_{k=1}^n e^{A\Delta t_k}e^{B\Delta t_k}.$$

Here we state just a few sufficient conditions under which this representation follows directly from theorem 2.3.

THEOREM 2.4. *Formula (2.30) is true under each of the following conditions:*

1) *A is the generating operator of a strongly continuous semigroup on X and B is a bounded operator on X which leaves invariant the domain D_A.*

2) *In the case of a Hilbert space* X, $A = zA_0$, *where* A_0 *is a positive definite operator,* $\pi \geq |Arg\, z| \geq \epsilon + \pi/2$, B *is the generating operator of a semigroup* e^{Bt}, *is subordinate to a fractional power of* A_0*:*

$$\|A_0^{\lambda} B A_0^{-\gamma-\lambda}\| < \infty,$$

and also satisfies the estimate

$$\|A_0 e^{Bt} A_0^{-1}\| \leq C < \infty.$$

PROOF. To prove the first part we put $\alpha = 0$, $\lambda_2 = 0$, $\lambda_1 = 1$, and $\|x\|_1 = \|Ax\|_0$. Note that under these conditions we can do without the intermediate spaces X_λ described in theorem 2.3 (see remark 2.6). Then the estimates (2.19) and (2.20) are evident, as is the estimate (2.21) for $V(t,s) = e^{A(s-t)}$. For $Z(t,s) = e^{B(s-t)}$, (2.21) clearly holds in X_0, and by

$$\|Ae^{Bt}A^{-1}\| = \|e^{ABA^{-1}t}\| \leq e^{t\|ABA^{-1}\|}, \tag{2.31}$$

also in X_1.

Turning to the second part, (2.19) follows from (2.26) and (2.27).

To obtain (2.20) we note that

$$\|Be^{Bt}x\|_\lambda = \|A_0^{\lambda} B A_0^{-\lambda-\gamma} A_0^{\lambda+\gamma} e^{Bt} A_0^{-(\lambda+\gamma)} A_0^{\lambda+\gamma} x\| \leq$$
$$\|A_0^{\lambda} B A_0^{-\lambda-\gamma}\| \|A_0^{\lambda+\gamma} e^{Bt} A_0^{-(\lambda+\gamma)}\| \|x\|_{\lambda+\gamma}.$$

From the well-known inequality of E. Heinz (see Dalecky [5]) the estimate

$$\|A_0^{\lambda+\gamma} e^{Bt} A_0^{-(\lambda+\gamma)}\| \leq \max\{\|e^{Bt}\|, \|A_0 e^{Bt} A_0^{-1}\|\} \quad (\gamma + \lambda < 1) \tag{2.32}$$

follows.

The estimates (2.21) are obtained in the same way as above, since $\|A_0 e^{Bt} A_0^{-1}\|$ is bounded. □

REMARK 2.7. Let $A = iA_0$. For a bounded operator B the representation (2.30) follows from the first part of the theorem.

The situation $B = iA_1$, with A_1 essentially selfadjoint on D_A and subordinate to a fractional power of A_0, is also of interest. The operator $A = A_0 + A_1$ is selfadjoint and the unitary operator $e^{i(A_0+A_1)t}$ makes sense. The second part of the theorem implies the decomposition

$$e^{(i-\epsilon)A_0 t + iA_1 t} = e^{(i-\epsilon)A_0 t} \otimes e^{+iA_1 t}.$$

Using a similar reasoning we can show that the lefthand side is strongly continuous with respect to ϵ, also for $\epsilon = 0$. So,

$$e^{iA_0 t + iA_1 t} = \lim_{\epsilon \to 0} e^{(i-\epsilon)A_0 t} \otimes e^{+iA_1 t}.$$

2.4. Linear evolution families with variable generating operator. Using theorem 2.1 we can construct evolution families that are not time-homogeneous.

Let $A(t)$ be the generating operator of a strongly continuous semigroup $e^{A(t)s}$ in X satisfying, for every $t \in T = [\tau, \tau_1]$ the conditions

C1) the domain of definition D of $A(t)$ does not depend on $t \in T$. The operator function $A(t)A^{-1}(s)$ is norm-continuous with respect to s and t, and hence
$$\|A(t)A^{-1}(s)\| \leq C_1 \quad (s, t \in T).$$
Moreover we assume that
$$\|A(t_2)A^{-1}(s) - A(t_1)A^{-1}(s)\| \leq C_2(t_2 - t_1).$$

C2) The following uniform estimate holds:
$$\|e^{A(t)s}\| \leq e^{\omega s}.$$

Without loss of generality we may regard that the constant ω is negative (replacing, if necessary, $A(t)$ by $A(t) - (\omega + \epsilon)I$). Then $\|e^{A(t)s}\| \leq 1$, and there is a bounded inverse operator $A^{-1}(t)$.

PROPOSITION 2.3. *Under the conditions C1) and C2) the family of operators* $U(s,t) = e^{A(s)(s-t)}$ *satisfies the conditions of theorem 2.1. Therefore the evolution family*
$$\tilde{U}(t,s) = \lim_q U_q(t,s) = e^{A(t_j - t)A(t_j)} \prod_{k=j}^{m-1} e^{(t_{k+1} - t_k)A(t_{k+1})} e^{(s - t_m)A(s)}$$
exists.

This evolution family is called a *multiplicative integral*, and is denoted by
$$\tilde{U}(t,s) = \widehat{\exp}\left\{-\int_s^t A(\theta)\, d\theta\right\} = \widehat{\exp} \int_t^s A(\theta)\, d\theta.$$

PROOF. It suffices to establish the estimates (2.8) and (2.9). The relation
$$e^{\theta A(s_2)} - e^{\theta A(s_1)} = \int_0^\theta e^{tA(s_2)}[A(s_2) - A(s_1)]e^{(\theta - t)A(s_1)}\, dt =$$
$$\int_0^\theta e^{tA(s_2)}[A(s_2)A^{-1}(s_1) - I]e^{(\theta - t)A(s_1)}A(s_1)\, dt$$
implies that
$$\|[e^{\theta A(s_2)} - e^{\theta A(s_1)}]A^{-1}(s_1)\| \leq C_2(s_2 - s_1)\theta.$$
So, (2.8) holds for $x \in D$:
$$\|[U(t,s) - U(t,\theta)U(\theta,s)]x\| \leq$$
$$\|[e^{(\theta - t)A(s)} - e^{(\theta - t)A(\theta)}]A^{-1}(s)\| \cdot \|e^{(s-\theta)A(s)}A(s)x\| \leq C_2(\theta - t)(s - \theta)K(x),$$
where $K(x) = \|A(s)x\|$.

Now notice that
$$\|A(\theta)A^{-1}(t)\| \leq 1 + \|[A(t) - A(\theta)]A^{-1}(t)\| \leq 1 + C_2(\theta - t) \leq e^{C_2(\theta - t)}.$$

Therefore (2.9) also holds, because

$$K(U_q(t,s)x) = \|A(s)U_q(t,s)(x)\| \leq$$
$$C_1\|A(t_j)U_q(t,s)A^{-1}(s)\|K(x) \leq C_1 e^{C_2(s-t)}K(x),$$

since

$$\|A(t)U_q(t,s)A^{-1}(s)\| =$$
$$\left\| U(t,t_j)\prod_{k=j}^{m-1}[A(t_k)A^{-1}(t_{k+1})]U(t_k,t_{k+1})U(t_m,s)\right\| \leq e^{C_2(s-t)}. \quad \square$$

We apply proposition 2.3 to the operator $A(t) = zA_0(t)$ with $A_0(t)$ a positive definite operator ($z \in \mathbf{C}$, $|\arg z| \in (\pi/2+\epsilon,\pi)$) on the Hilbert space X.

In this case we can estimate the operator

$$V_{\alpha,\beta}(t,s) = A_0^{\alpha}(t)\widehat{\exp}\left\{z\int_t^s A(\theta)\,d\theta\right\}A^{-\beta}(s). \tag{2.33}$$

In fact, consider the operator

$$V^q_{\alpha,\beta}(t,s) = A_0^{\alpha}(t)e^{(t_j-t)A(t_j)}\prod_{k=j}^{m-1}e^{(t_{k+1}-t_k)A(t_{k+1})}e^{(s-t_m)A(t_s)}A_0^{-\beta}(t) =$$
$$A_0^{\alpha}(t)A_0^{-\alpha}(t_j)e^{(t_j-t)A(t_j)}A_0^{\alpha-\alpha_j}(t_j)\times$$
$$\prod_{k=j}^{m-1}A_0^{\alpha_k}(t_k)A_0^{-\alpha_k}(t_{k+1})e^{(t_{k+1}-t_k)A(t_k)}A_0^{\alpha_k-\alpha_{k+1}}(t_k)A_0^{\alpha_m}(t_m)A_0^{-\alpha_m}(t_s)e^{(s-t_m)A(s)}A_0^{(\alpha_m-\beta)}(s).$$

It is well known (see Krein [1]) that

$$\|A_0^{\alpha}(s)A_0^{-\alpha}(s_1)\| \leq \|A_0(s)A_0^{-1}(s_1)\|^{\alpha}.$$

Moreover, for $\alpha_k = \beta + (\alpha-\beta)(t_k-t)/(\varepsilon-t)$ it follows from (2.27) that

$$\|e^{(t_{k+1}-t_k)A(t_k)}A_0^{\alpha_k-\alpha_{k+1}}(t_k)\| =$$
$$\|e^{z(t_{k+1}-t_k)A_0(t_k)}A_0(t_k)^{(\alpha-\beta)(t_{k+1}-t_k)/(s-t)}\| =$$
$$\left[\frac{\alpha-\beta}{e(s-t)|\Re z|}\right]^{(\alpha-\beta)(t_{k+1}-t_k)/(s-t)}$$

Therefore

$$\|V^q_{\alpha,\beta}(t,s)\| \leq \left[\frac{\alpha-\beta}{e(s-t)|\Re z|}\right]^{(\alpha-\beta)}e^{C_2\alpha(s-t)}.$$

This uniform estimate and the convergence of the multiplicative integral allow us to infer boundedness of the operator (2.23) and to obtain the estimate

$$\left\|A_0^{\alpha}\widehat{\exp}\left\{\int_t^s A(\theta)\,d\theta\right\}A_0^{-\beta}(s)\right\| \leq \frac{e}{|s-t|^{\alpha-\beta}}. \tag{2.34}$$

This estimate implies that the family (2.23) under consideration satisfies (2.19) in the norm (2.28).

By a reasoning similar as in the proof of theorem 2.4 these estimates imply an analogous result in the non-uniform case.

THEOREM 2.4'. *The formula*

$$\widehat{\exp}\int_{\tau}^{\tau_1}[A(s)+B(s)]\,ds=\widehat{\exp}\int_{\tau}^{\tau_1}A(s)\,ds\otimes\widehat{\exp}\int_{\tau}^{\tau_1}B(s)\,ds$$

holds under any of the following conditions:

1) *the operator function $A(t)$ satisfies the conditions of proposition 2.3, $B(t)$ ($t\in[\tau,\tau_1]$) is a bounded operator on X, and*

$$\|A(t)B(t)A^{-1}(t)\|\le const;$$

2) *in the case of a Hilbert space X, $A(t)=zA_0(t)$ with $A_0(t)$ a positive definite operator satisfying condition C1), $|\arg z|\in[\epsilon+\pi/2,\pi]$.*
The operator $B(t)$ satisfies the conditions of proposition 2.3 and the estimates

$$\|A_0^{\lambda}(t)B(t)A_0^{-\gamma-\lambda}\|\le const,$$

$$\|A_0(t)\widehat{\exp}\int_s^t B(\theta)\,d\theta A_0^{-1}(s)\|\le const.$$

If then $B(t)=iA_1(t)$ and $A_0(t)+A_1(t)$ is selfadjoint, we have

$$\widehat{\exp}i\int_s^t[A(\theta)+B(\theta)]\,d\theta=\lim_{\epsilon\to0}\widehat{\exp}(i-\epsilon)\int_s^t A(\theta)\,d\theta\otimes\widehat{\exp}\int_s^t B(\theta)\,d\theta. \tag{2.35}$$

These results facilitate a number of examples; we briefly describe some typical situations.

EXAMPLE 2.1. Let $X=L_2(\mathbf{R}^n)$, and $A=l(t,x,\nabla_x)$ a linear elliptic differential operator of order $2p$ for which the Cauchy problem

$$\frac{\partial u}{\partial t}+l(t,x\nabla_x)u=0$$

is well-posed. The conditions of theorems 2.4 and 2.4' are fulfilled if B is the operator of multiplication by a function $v(x,t)$ which is sufficiently smooth with respect to x and continuous with respect to t. Then the operators $A(t)S(t)A^{-1}(t)$ are bounded, according to classical coerciveness inequalities (see, e.g, Ladyzhenskaya–Solonnikov–-Uraltseva [1]). Under additional assumptions regarding the relation between the behavior of this function on the boundary and the boundary conditions, we can also consider boundary value problems with an elliptic operator A on $L_2(G)$¡ $G\subset\mathbf{R}^n$.

EXAMPLE 2.2. In the same situation as in example 2.1 a more general case is obtained if we take for B a first order differential operator,

$$Bu=\sum_{j=1}^{r}b_j(x,t)\frac{\partial u}{\partial x_j}.$$

The equation

$$\frac{\partial u}{\partial t} + Bu = 0$$

can be solved by the method of characteristics, and the estimates used in theorem 2.4 and 2.4' can be verified to hold, since B is of fractional order with respect to A (see Krein [1] and the references in it).

EXAMPLE 2.3. Under the conditions of example 2.1, if an operator l is selfadjoint on X, the Schroedinger equation

$$\frac{1}{j}\frac{\partial u}{\partial t} = l(t, x, \nabla_x)u + v(x, t)u$$

can be examined in a similar manner, and its solutions can be represented as chronological products.

In the more general case of example 2.2, if the operator

$$l(t, x, \nabla_x) + \sum_j b_j(x, t)\nabla_j$$

is selfadjoint, we have to use remark 2.7. Then a representation of type (2.35) is obtained.

EXAMPLE 2.4. Consider the infinite-dimensional case. Let $X = \mathcal{L}_2(\mathcal{H}, \mu)$ be the Hilbert space of functions on the Hilbert space $\mathcal{H}$ that are square integrable with respect to a Gaussian measure μ.

Consider the second order differential operator

$$A = (a(x)\nabla, \nabla) + (b(x), \nabla)$$

under conditions guaranteeing that the Cauchy problem (see chapter V)

$$\frac{\partial u}{\partial t} + Au = 0$$

is well-posed. Coerciveness inequalities for such operators were established by Frolov [3]. This makes it possible to apply theorem 2.4 to the equation

$$\frac{\partial u}{\partial t} + (a(x)\nabla, \nabla)u + (b(x), \nabla)u + v(x, t)u = 0$$

with a smooth bounded function $v(x, t)$. If A is also selfadjoint (see (IV.1.4), (V.2.8)), then we can consider the analogous Schroedinger equation

$$\frac{1}{i}\frac{\partial u}{\partial t} + (a(x)\nabla, \nabla)u + (b(x), \nabla)u + v(x, t)u = 0.$$

2.5. Evolution families related to quasilinear equations. Consider in a Banach space the equation

$$\frac{\partial u}{\partial t} + A(t)u + B(t,u) = 0, \tag{2.36}$$

where $A(t)$ is a linear operator satisfying the conditions of the previous section and the $B(t,\cdot)$ are nonlinear operators forming a uniform family.

Necessary conditions for applying theorem 2.3 were studied above.

We determine an evolution family Z from the requirement that the nonlinear equation

$$\frac{\partial \psi}{\partial t} + B(t,\psi) = 0, \quad \psi(t) = Z(t,s)\psi(s), \tag{2.37}$$

be well-posed, at least locally.

It suffices to establish exponential bounds for the Lipschitz constants of the operators Z and $Z_1 = A_0 Z A_0^{-1}$.

For the first, Z, such a bound follows from

$$\|\psi_2(t) - \psi_1(t)\| \le$$
$$\|\psi_2(s) - \psi_1(s)\| + \int_t^s \|B(\theta,\psi_2(\theta)) - B(\theta,\psi_1(\theta))\|\, d\theta \le$$
$$|\psi_2(s) - \psi_1(s)\| + C\int_t^s \|\psi_2(\theta) - \psi_1(\theta)\|\, d\theta,$$

after applying the Gronwall inequality.

The operator Z_1 is determined by the equation

$$\frac{\partial \psi}{\partial t} + A_0 B(t, A_0^{-1}\psi) = 0,$$

and can be studied in the same way, if the family of operators $A_0 B(t, A_0^{-1})$ is uniform.

Thus we have the following result.

THEOREM 2.5. *Suppose the linear operator $A(t)$ satisfies the conditions of proposition 2.3, and suppose the families of linear operators $B(t)$ and $A_0 B(t) A_0^{-1}$ are uniform. Then the evolution family $\tilde{V}$ determined by equation (2.36) can be represented as a chronological product:*

$$\tilde{V}(t,\tau) = e^{A(\tau - t)} \otimes Z(t,\tau),$$

where $Z(t)$ is the evolution family determined by (2.37).

EXAMPLE 2.5. Let $A(t) = l(t,x,\nabla)$ be an elliptic differential operator generating an evolution family $U(t,\tau)$ in some space $X = L_p(Q)$ or $C(Q)$, where $Q \subseteqq \mathbf{R}^n$. Then conditions of the type of the theorem are fulfilled for equations of the form

$$\frac{\partial u}{\partial t} + l(t,x,\nabla)u + f(t,u) = 0,$$

where $f(t,u)$ is a function which is bounded and continuous together with its derivatives of sufficiently high order with respect to x and u. The latter condition may hold in a certain domain of X only, provided the solutions to equation (2.38) under consideration are known not to leave this domain.

For example, in $C(Q)$, the family U can be contractive for a second order elliptic operator: $\|U(t,\tau)\| \leq 1$. If the equation

$$\frac{\partial u}{\partial t} + f(t,x,u) = 0$$

has the dissipativeness property, so that its solutions do not leave certain balls, it suffices that f satisfies the boundedness and smoothness condition on these balls.

EXAMPLE 2.6. Under similar conditions and by similar means we can consider integro-differential equations

$$\frac{\partial u}{\partial t} + l(x,t,\nabla)u + \int_Q K(t,x,y,u(y))\,dy = 0.$$

3. Linear evolution families and functional integrals

3.1. Transition measures and evolution families. Integration of multiplicative functionals. Let $(X,\mathcal{U})$ be a measurable space, and $P(t,x,s,dy)$ $(t,s \in T;$ $t \leq s)$ a transition measure on this space.

On the space of bounded measurable functions $\mathcal{B}(X)$ there is defined a family of bounded linear operators

$$U(t,s)f(x) = \int_X P(t,x,s,dy)f(y). \tag{3.1}$$

It follows from (1.1) that this family has the evolution property:

$$U(t,s)f(x) = \int_X \left\{\int_X P(t,x,\theta,dz)P(\theta,z,s,dy)\right\} f(y) =$$
$$\int_X P(t,x,\theta,dz)\left\{P(\theta,z,s,dy)f(y)\right\} = U(t,\theta)U(\theta,s)f(x).$$

On the other hand, suppose we are given a linear evolution family $U(t,s)$, consisting of continuous operators on a space Φ of test functions on X. On the corresponding space of distributions Φ' the adjoint family $U^*(t,s)$ is defined by

$$\langle f, U^*(t,s)\mu\rangle = \langle U(t,s)f,\mu\rangle \quad (f \in \Phi, \quad \mu \in \Phi').$$

Suppose that for each $x \in X$ the normalized measure ν_x concentrated at x belongs to Φ'. The distribution

$$P(t,x,s) = U^*(t,s)\nu_x \tag{3.2}$$

is called a *generalized transition measure*. By definition we have

$$U(t,s,\cdot)f(x) = \langle U(t,s)f,\nu_x\rangle = \langle f, P(t,x,s)\rangle.$$

The family of distributions $P(t,x,s)$ $(x \in X,\ t \leq s)$ is an evolution family in the sense described in chapter V (see (V.1.5)).

In the sequel we will sometimes use the notation

$$\langle f, P(t,x,s)\rangle = \int_X f(y)P(t,x,s,dy), \tag{3.3}$$

although for a generalized transition measure the righthand integral does not immediately make sense. The evolution property (V.1.5) coincides in form with (1.1):

$$\int_X f(y)P(t,x,s,dy) = \int_X P(t,x,\theta,dz)\int_X P(\theta,z,s,dy)f(y). \tag{3.4}$$

If the operators $U(t,s)$ can be extended to the space $\mathcal{B}(X)$ and are continuous for limit transition with respect to monotone sequences of functions, then the distributions $P(t,x,s)$ are determined by measures on $(X,\mathcal{U})$. However, this does not yet imply absolute convergence in (1.1).

It will be seen below that for a certain class of functions we can give a construction similar to that of integral with respect to a Markov quasimeasure (see section 1) also if the transition measure is replaced by a generalized transition measure determined by a family of evolution operators.

Again, let $P(t,x,s,dy)$ be a transition measure on $(X,\mathcal{U})$, and μ the Markov quasimeasure constructed from it. We consider some formal properties of integrals with respect to this quasimeasure, assuming the integrals to exist.

If $x(t)$ is a function on T and $\Delta\subset T$, then $x_\Delta(t)$ denotes the restriction $x(t)|_\Delta$. If a functional f on $\mathcal{B}(X)$ depends only on the restriction $x_\Delta(t)$ of $x(t)$, it will be denoted by f_Δ:

$$f_\Delta(x) = f_\Delta(x_\Delta). \tag{3.5}$$

The following formulas follow immediately from the definitions:

$$\int_{\tilde{X}_{\tau,x_0}} f_{[t,s]}(x)\phi(x(\tau_1))\mu_{x_0,t}(dx) =$$
$$\int_X P(\tau,x_0,t,dx)\int_{\tilde{X}_{t,x}} f_{[t,s]}(z)\left[\int_X P(s,z(s),\tau_1,dy)\phi(y)\right]\mu_{x,t}(dz), \tag{3.6}$$

$$\int_{\tilde{X}_{[\tau,\tau_1]}} f_{[\tau,\theta]}(x)f_{[\theta,\tau_1]}(x)\phi(x(\tau_1))\mu_{x_0,\tau}(dx) =$$
$$\int_{\tilde{X}_{[\tau,\theta]}} f_{[\tau,\theta]}(x)\left\{\int_{\tilde{X}_{[\theta,\tau_1]}} f_{[\theta,\tau_1]}(x)\phi(x(\tau_1))\mu_{x(\theta),\theta}(dx)\right\}\mu_{x_0,\tau}(dx). \tag{3.7}$$

DEFINITION 3.1. Let Z be a space of functions $x(t)$, $t\in T$. A set of complexvalued functions $f_\Delta(x)$ ($\Delta=(t,s]\subset T$) (see (3.5)) on Z satisfying the condition

$$f_{\Delta_1\cup\Delta_2}(x) = f_{\Delta_1}(x)f_{\Delta_2}(x) \quad (\Delta=\Delta_1\cup\Delta_2;\quad \Delta_1\cap\Delta_2=\emptyset) \tag{3.8}$$

will be called a *multiplicative functional* on Z.

PROPOSITION 3.1. *Let $\{f_\Delta(x)\}$ be a multiplicative functional on a space $\tilde{X}^0(T)$ of step functions. The linear operators on $\mathcal{B}(X)$ defined by*

$$U^f(t,s)\phi(x_0) = \int_{X_{[t,s]}} f_{[t,s]}(x(\cdot))\phi(x(s))\mu_{x_0,t}(dx) \tag{3.9}$$

(provided that the integrals exist) form an evolution family.

PROOF. This follows from (3.7). □

Let us clarify the structure of a multiplicative functional on step functions. Let $\Delta = (t, s]$, and $x(\theta)$ a function with only step on Δ at the point s: $x(\theta) = x$ $(\theta < s)$, $x(s) = y$. Put $g(t, s, x, y) = f_\Delta(x)$. Then (3.8) implies

$$g(t, s, x, y) = g(t, \theta, x, x)g(\theta, s, x, y) \quad (t < \theta < s).$$

Setting $g(t, s, x, y) = e^{\alpha(t,s,x,y)}$ we obtain

$$\alpha(t, s, x, y) = \alpha(t, \theta, x, x) + \alpha(\theta, s, x, y). \tag{3.10}$$

In particular, for $y = x$ and $t = \tau$,

$$\alpha(t, \theta, x, x) = \gamma(s, x) - \gamma(\theta, x), \tag{3.11}$$

where $\gamma(\theta, x) = \alpha(\tau, \theta, x, x)$. From (3.10) and (3.11) we infer

$$\alpha(t, s, x, y) + \gamma(t, s) = \alpha(\theta, s, x, y) + \gamma(\theta, x) = \beta(s, x, y),$$

suggesting that the function β is independent of t. Since clearly $\beta(t, x, x) = \gamma(t, x)$, we have

$$\alpha(t, s, x, y) = \beta(s, x, y) - \gamma(t, s) = \beta(s, x, y) - \beta(t, x, x). \tag{3.12}$$

Since (3.10) follows from (3.12) for any function $\beta(t, x, y)$, we find

$$f_\Delta(x) = e^{\beta(s,x,y)-\beta(t,x,x)} \tag{3.13}$$

for the above described function $x(t)$.

Let now the step function $x(t)$ be determined by a partition $\tau = t_0 < t_1 < \cdots < t_n < \tau_1 = t_{n+1}$ and take the values

$$x(\theta) = x_n \quad (t_n \le \theta \le \tau_1), \qquad x(\theta) = x_k \quad (t_k \le \theta < t_{k+1}).$$

Then from (3.13) and (3.8), for $t \in [t_{j-1}, t_j]$, $s \in [t_r, t_{r+1}]$ we find

$$f_{(t,s]}(x) = f_{(t,t_j)}(x) \prod_{k=j}^{r-1} f_{(t_k,t_{k+1})}(x) f_{(t_r,s)}(x) =$$

$$\exp\{[\beta(t_j, x_{j-1}, x_j) - \beta(t_j, x_{j-1}, x_{j-1})] + \sum_{k=j}^{r-1} [\beta(t_{k+1}, x_k, x_{k+1}) - \beta(t_k, x_k, x_k)] +$$
$$[\beta(s, x_r, x_r) - \beta(t_r, x_r, x_r)]\}. \tag{3.14}$$

Setting $\beta(t, z, x(t)) = \beta(z, t)$ we can write this expression as an integral:

$$f_{(t,s]}(x) = \exp\left\{\int_t^s d_\theta \tilde{\beta}(x(\theta - 0, \theta)\right\}, \tag{3.15}$$

(3.14) being its definition.

A straightforward computation and application of proposition 2.2 leads to

THEOREM 3.1. *Let*

$$V(t,s)\phi(x) = e^{-\beta(t,x,x)} \int e^{\beta(s,x,y)} \phi(y) P(t,x,s,dy) \tag{3.16}$$

be a family of operators on $\mathcal{B}(X)$. *Then*

$$V_q(t,s)\phi(x) = \int_{X^q} f_{[t,s]}(\tilde{x})\phi(\tilde{x}(s))\mu_{x,t}^{(q)}(d\tilde{x}), \tag{3.17}$$

where f *is the multiplicative functional (3.15) and* q *is the partition defining the step function* $\tilde{x}$.

So the existence of the evolution family

$$\tilde{V}(t,s) = \lim_q V_q(t,s) \tag{3.18}$$

is equivalent to the existence of the functional integral (3.9), and both families coincide. The generating operator of the evolution family $\tilde{V}(t,s)$ is

$$\tilde{A}(t)\phi(x) = -\frac{d}{dt}V(t,s)\phi(x)|_{s=t} =$$

$$e^{-\beta(t,x,x)}A(t)[e^{\beta(t,x,y)}\phi(y)]_{y=x} - \frac{\partial\beta(t,x,x)}{\partial t}\phi(x), \tag{3.19}$$

where $A(t)$ is the generating operator of the evolution family (3.1).

REMARK 3.1. Formulas (3.16) and (3.17) will also make sense if the transition measure P is replaced by a generalized transition measure determined by an evolution family $V_0(t,\tau)$ of operators on Φ.

Formula (3.16) becomes

$$V(t,s)\phi = e^{-\beta(t,x,x)}V_0(t,s)[e^{\beta(\tau,x,\cdot)}\phi], \tag{3.20}$$

and a sufficient condition for its truth is that the functions $e^{-\beta(t,x,x)}$ and $e^{\beta(\tau,x,y)}$ ($t \in T$, $y \in X$) are multipliers on Φ. The limit evolution family $\tilde{V}(t,s) = \lim_q V_q(t,s)$ (if it exists) will be called a *functional integral with respect to the generalized transition measure* $P = P_{V_0}$ and will be denoted in the same way as in (3.9). Also, the assertion of the theorem remains true.

3.2. Some particular cases. We consider some important cases when the solution of a differential equation can be represented as a functional integral.

Let $\beta(t,x,y) = a(t,x)$. Then (3.16) becomes

$$V(t,s)\phi(x) = e^{a(s,x)-a(t,x)} \int_X \phi(y)P(t,x,s,dy) = Z(t,s)U(t,s)\phi(x),$$

where $U(t,s)$ is the evolution family corresponding to the (generalized) transition measure P, $Z(t,s)\phi(x) = e^{a(s,x)-a(t,x)}$ is the evolution family with $-Z'(t,t)\phi(x) = a'_t(t,x)\phi(x)$ as generating operator, where $a(t,x)$ is a function which is differentiable with respect to t.

Thus, the evolution family (3.18), if existing, is a chronological product,

$$\tilde{V}(t,s) = Z(t,s) \otimes U(t,s), \tag{3.21}$$

and existence of the functional integral (3.9) is equivalent to that of the chronological product. Then the multiplicative functional is

$$f_{[t,s]}(x) = \exp \int_s^t d_\theta(a(\theta, x(\theta))),$$

or

$$f_{[t,s]}(x) = \exp \int_s^t a_\theta(\theta, x(\theta))\, d\theta,$$

for a differentiable function a. The generating operator of the family (3.21) is

$$A(t) = A_0(t) - a_t'(t, x).$$

This reasoning can now be applied to any of the examples in section 2 for which convergence of the chronological product is established.

EXAMPLE 3.1. Let $P(t, x, s, dy) = f(t, x, s, y)\, dy$ be the transition measure on $\mathbf{R}^n$ whose density f with respect to Lebesgue measure is a fundamental solution of the Cauchy problem for the parabolic equation

$$\frac{\partial u}{\partial t} + l(t, x, \nabla_x)u = 0.$$

Under the conditions of example 2.1, the functional integral

$$\tilde{u}(t, s, x_0) = \int_{X_{[t,s],x_0}} e^{\int_t^s v(\theta, x(\theta))\, d\theta} \phi(x(s)) \mu_{x_0,t}^{(q)}(dx) =$$

$$\lim \int_X \cdots \int_X e^{\sum_k \int_t^s v(\theta, x_k)\, d\theta} \prod_k P(t_k, x_k, t_{k+1}, dx_{k+1}), \tag{3.22}$$

representing the solution of the Cauchy problem

$$\frac{\partial \tilde{u}}{\partial t} + l\tilde{u} + v(t, x)\tilde{u} = 0, \quad \tilde{u}(s, s, x) = \phi(x), \tag{3.23}$$

converges. In general, the quasimeasure $\mu^{(l)}$ is not σ-additive. However, for a second order elliptic operator it is positive, and hence σ-additive, by Kolmogorov's theorem.

In this case the existence of the integral (3.22) can be obtained by independent considerations, under much wider assumptions concerning $v(t, x)$. Then theorem 3.1 will imply convergence of the Cauchy problem (3.21) corresponding to (3.23). We will return to this question in chapter VII, where a probabilistic approach will be used.

EXAMPLE 3.2. Consider the Schroedinger type equation

$$\frac{1}{i}\frac{\partial u}{\partial t} + l(t, x, \nabla_x)u + v(t, x)u = 0 \tag{3.24}$$

with a selfadjoint operator l.

The fundamental solutions of the nonperturbed equation

$$\frac{1}{i}\frac{\partial u}{\partial t} + l(t, x, \nabla_x)u = 0 \tag{3.25}$$

are not absolutely integrable in $\mathbf{R}^n$.

Therefore, the straightforwardly written righthand side of (3.22) does not make sense. It can make sense in two ways. The first is to replace the factor i by $i+\epsilon$ in (3.24), with subsequent limit transition as $\epsilon \to 0$. Then we obtain a representation for the solution of (3.24):

$$\tilde{u}(t,s,x_0) = \lim_{\epsilon\to 0} \int_X e^{\int_t^s v(\theta,x(\theta))\,d\theta} \phi(x(s)) \mu^{(l,\epsilon)}(dx),$$

where $\mu^{(l,\epsilon)}$ is the quasimeasure constructed from the fundamental solutions of the equation

$$\frac{\partial u}{\partial t} + (i+\epsilon) l(t,x,\nabla_x) u = 0.$$

If the functions occurring in the integral have appropriate analyticity properties, then, along the lines suggested in chapter II, the integral can be reduced by a complex change of variables, to an integral with respect to a positive measure.

Another way is to consider the integral (3.22) in the sense pointed out in remark 3.1, P being taken a generalized transition measure.

EXAMPLE 3.3. Now consider the second order infinite-dimensional equation

$$\frac{\partial u}{\partial t} + (a(x)\nabla,\nabla)u + (b(x),\nabla)u + v(x)u = 0$$

described in example 2.4. The transition measure $P(t,s,x,\cdot)$ corresponding to the principal part ($v=0$) of this equation is nonnegative and determines a σ-additive measure in the path space with values in $\mathcal{H}$. The solution of the Cauchy problem can be represented as an integral with respect to this measure,

$$u(t,s,x_0) = \int_X e^{\int_t^s v(\theta,x(\theta))\,d\theta} \phi(x(s)) \mu_{x_0,t}(dx),$$

which can be understood as the limit of finite-dimensional integrals related to corresponding Cauchy problems.

For the Schroedinger equation

$$\frac{1}{i}\frac{\partial u}{\partial t} + (a(x)\nabla,\nabla)u + (b(x),\nabla)u + v(x)u = 0$$

with a selfadjoint operator l, a representation as Feynman integrals with infinite-dimensional phase space is then obtained, as in example 3.2.

Another instance of interest is when X is a Hilbert space and the multiplicative functional (3.14) is determined by the function

$$\beta(t,x,y) = (a(x),y)_X$$

and, for step functions, takes the form

$$f_{[t,s)}(x) = \exp\left\{\sum_{j=1}^{n} (a(x_{j-1}), x_j - x_{j-1})\right\} = \exp\left\{\int_t^s (a(x(\theta)), dx(\theta))\right\}. \tag{3.26}$$

Suppose the evolution family (3.1) has generating operator

$$A_x\phi(x) = Q(x,\nabla)\phi(x) = \sum_{k=0}^{n} \alpha_k(x)(\nabla,\dots,\nabla)\phi(x), \tag{3.27}$$

where $\alpha_k(x)$ ($x \in X$) is a k-linear symmetric functional on X.

Suppose the functional integral (3.9) determined by this evolution family and multiplicative functional exists. It determines, by theorem 3.1, an evolution family with generating operator

$$\begin{gathered}\tilde{A}\phi(x) = e^{-(a(x),x)} Ay[e^{(a(x),y)}\phi(y)]_{y=x} = \\ e^{-(a(x),x)}Q(x,\nabla_y)[e^{(a(x),y)}\phi(y)]_{y=x} = Q(x,a(x)+\nabla_y)\phi(y)|_{y=x}.\end{gathered} \tag{3.28}$$

In particular, let $n = 2$ and

$$A\phi = \frac{1}{2}\alpha(x)(\nabla,\nabla)\phi(x). \tag{3.29}$$

Then

$$\begin{gathered}\tilde{A}\phi(x) = \frac{1}{2}\alpha(x)(\nabla_y + a(x), \nabla_y + a(x))\phi(y)|_{y=x} = \\ A\phi(x) + \alpha(x)(a(x),\nabla)\phi(x) + \frac{1}{2}\alpha(x)(a(x),a(x))\phi(x) = \\ (A + A_1)\phi(x).\end{gathered}$$

It is readily verified that the more complicated multiplicative functional

$$f_{[t,s)}(x) = \exp\left\{\int_t^s (a(x(\theta)), dx(\theta)) + \int_t^s v(\theta, x(\theta))\,d\theta\right\} \tag{3.30}$$

leads to the generating operator

$$\tilde{A}\phi = (A + A_2)\phi(x), \quad A_2 = A_1 + v(\theta, x).$$

In particular, take $v(x) = -\alpha(x)(a(x),a(x))/2$, $\tilde{A}\phi = A\phi + A_2\phi = \frac{1}{2}\alpha(x)(\nabla,\nabla)\phi + \alpha(a(x),\nabla)\phi$.

For this case we examine the convergence behavior of the chronological product

$$\tilde{V} = e^{At} \otimes e^{A_2 t}.$$

To compute the operator $e^{A_2 t}$ we have to solve the Cauchy problem for the first order differential equation

$$\frac{\partial u}{\partial t} + \alpha(x)(a(x),\nabla)u = 0.$$

Its solution can be found by the method of characteristics. It is readily computed that

$$(e^{A_2(\tau - t)}\phi)(x) = \phi(\xi(t,x,\tau)),$$

where $\xi(t,x,\tau)$ is the solution of

$$\frac{d}{d\tau}(\xi,h) = \alpha(\xi)(a(\xi),h) \quad (h \in X), \quad \xi(t,x,t) = x. \tag{3.31}$$

Let $P(t, x, \tau, \cdot)$ be the transition measure determined by the semigroup e^{At}, i.e. by the differential equation

$$\frac{\partial u}{\partial t} + \frac{1}{2}\alpha(x)(\nabla, \nabla)u = 0.$$

Then

$$V(t,\tau)\phi = e^{A_2(\tau-t)A(\tau-t)}\phi(x) = \int \phi(y)P(t, \xi(t, x, \tau), \tau, dy) =$$
$$\int \phi(y)g(t, \tau, x, y)P(t, x, \tau, dy),$$

if the density

$$g(t, \tau, x, y) = \frac{P(t, \xi(t, x, \tau), \tau, dy)}{P(t, x, \tau, dy)}$$

exists. Then also

$$V_q(t,\tau)\phi(x) = \int \cdots \int \prod_j g(t_j, t_{j+1}, x_j, x_{j+1})\phi(y)\mu^{(q)}(dx) \quad (y = x_n),$$

and the integrand can be written as a multiplicative functional.

The computations below are carried out for one special case only. A more complete result will be obtained by other means in chapter VII.

EXAMPLE 3.4. Let

$$A = \frac{1}{2}(B\nabla, \nabla),$$

where B is a nonnegative nuclear operator on X. In this case,

$$P(t, \tau, x, dy) = P(\tau - t, dy - x),$$

where $P(t, \cdot)$ is a Gaussian measure with correlation operator Bt.

Formulas for the density of a shifted Gaussian measure with respect to the initial measure imply that $\xi(t, x, \tau) - x \in D_{B^{-1}}$, and the following formula is true:

$$g(t, \tau, x, y) =$$
$$\exp\left\{-\frac{1}{\tau - t}(B^{-1}(\xi(t, x, \tau) - x), y - x) - \right.$$
$$\left.\frac{1}{2(\tau - t)}(B^{-1}(\xi(t, x, \tau) - x), \xi(t, x, \tau) - x)\right\}.$$

The equation (3.31) takes the form $d\xi/dt = Ba(\xi)$, and thus,

$$\xi(t, x, \tau) - x = B\int_t^\tau a(\xi(t, x, s))\, ds = x + Ba(x)(\tau - t) + o(\tau - t).$$

Below we neglect terms of higher order of smallness, and consider an approximate expression for g:

$$g(t,\tau,x,y) =$$
$$\exp\left\{-\frac{1}{2(\tau-t)}\left(B\int_t^\tau a(\xi(t,x,s))\,ds, \int_t^\tau a(\xi(t,x,s))\,ds\right) - \right.$$
$$\left.\frac{1}{\tau-t}\left(\int_t^\tau a(\xi(t,x,s))\,ds, y-x\right)\right\} \approx$$
$$\exp\left\{\frac{1}{2}(Ba(x),a(x))(\tau-t) - (a(x),y-x)\right\} \tag{3.32}$$

This expression leads to the multiplicative functional

$$f_{[t,s)}(x) = \exp\left\{\frac{1}{2}\sum(Ba(x_{j-1}))(t_j-t_{j-1}) - (a(x_{j-1}),x_j-x_{j-1})\right\} =$$
$$\exp\left\{\frac{1}{2}\int_t^s (Ba(x(\theta)),a(x(\theta)))\,d\theta - \int_t^s (a(x(\theta)),dx(\theta))\right\}.$$

Neglecting the small term in (3.32) comes down to replacement of the operator $e^{A_2t}e^{At}$ by an operator close to it. By theorem 2.2 we can justify this change and show it to have no influence on the limit expression.

Since in this example convergence of the functional integral can be derived from positivity of the measure μ, we find that the solution of the Cauchy problem

$$\frac{\partial u}{\partial t} + \frac{1}{2}(B\nabla,\nabla)u + (Ba(x),\nabla)u - \frac{1}{2}(Ba(x),a(x))u = 0,$$
$$u(x,\tau) = \phi(x),$$

is representable as

$$u(x,t) = \int \exp\left\{\frac{1}{2}\int_t^\tau (Ba(x,\theta)),a(x(\theta))\,d\theta - \int_t^\tau (a(x(\theta)),dx(\theta))\right\}\phi(x(\tau))\mu(dx).$$

4. Nonlinear evolution families, and integrals in branching path space

4.1. Branching paths. In this section we show that multiplicative representations of solutions of nonlinear evolution equations lead, as in the linear case, to functional integrals, but in a more complicated space consisting of branching paths (trees).

To do this we first have to develop a proper language for describing such paths.

Consider a finite partially order set Z. An element z_2 is said to be *intermediate* with respect to a pair (or between a pair) z_1, z_3 ($z_i \in Z$, $i = 1,2,3$) if $z_1 < z_2 < z_3$.

A pair of elements $z, \tilde{z} \in Z$ ($z < \tilde{z}$) without intermediates is called an *interval*; notation: $[z,\tilde{z}]$. It goes from z to $\tilde{z}$ (it begins at z and ends at $\tilde{z}$). We denote by $\alpha_-(z)$ the set of elements that are beginnings of intervals ending at z, and by $\alpha_+(z)$ the set of elements that are ends of intervals beginning at z.

Let $k(Z_1)$ be the number of elements in a subset $Z_1 \subset Z$, and put $k_\pm(z) = k(\alpha_\pm(z))$.

We call Z a *frame* if the following conditions hold:

a) there is exactly one element $z_0 \in Z$ (the *root* of the frame) such that the set $\alpha_-(z_0)$ is empty while $k_+(z_0) \neq 1$;

b) for $z \neq z_0$, always $k_-(z) = 1$ and $k_+(z) \neq 1$;

The points of the set $Z' = \{z' \in Z\colon\ k_+(z') = 0\}$ are called *vertices* of the frame, the points of $Z'' = \{z'' \in Z\colon\ k_+(z'') \geq 2\}$ *branching points*, the number $k_+(z'')$ the *multiplicity of a branching point* $z'' \in Z''$, and $k(Z')$ the *order of the frame* Z.

For each $z \in Z \setminus Z'$ we order, in an arbitrary manner, the set $\alpha_+(z)$, and list its points: $\alpha_+(z) = \{z_1, \dots, z_{k_+}\}$. Here, each element $z \in Z$ can be marked by an index composed from the indices of the elements lying between z_0 and z. The set of elements of the frame is then ordered lexicographically; in particular, the vertex set Z' is ordered.

Two frames with the same enumeration of vertices are called *equivalent*. Such frames can be mapped one-to-one onto each other while preserving the partial order. Obviously, re-enumeration of the points of some set $\alpha_+(z)$ leads to an equivalent frame. E.g., there exist, up to equivalence, one frame $Z_1 = \{z_0, z\}$ of order $k(Z_1) = 1$, one frame $Z_2 = \{z_0, z_1, z_{11}, z_{12}\}$ of order $k(Z_2) = 2$, and three frames of order $k(Z) = 3$: $Z_3 = \{z_0, z_1, z_{11}, z_{12}, z_{13}\}$, $Z_3' = \{z_0, z_1, z_{11}, z_{111}, z_{112}, z_{12}\}$, $Z_3'' = \{z_0, z_1, z_{11}, z_{12}, z_{121}, z_{122}\}$.

A finite partially ordered set $\tilde{Z}$ is called an *extension of a frame* Z (an *extended frame*) if it can be obtained from Z by the addition of a finite set of intermediate points $\tilde{z}$, where for each of these $k_+(\tilde{z}) = 1$, $k_-(\tilde{z}) = 1$. Such points will be called *division points*, and $\tilde{Z}$ is called a *division frame*.

An extension $\tilde{Z}$ of a frame Z is called *simple* (a *cut*) if the division points are pairwise incomparable.

If $\tilde{z}$ is a division point of the frame Z, then the set

$$\tilde{z} \cup \{z \in Z : \ \tilde{z} < z\} = Z_{\tilde{z}}$$

of points following $\tilde{z}$ forms a frame, $Z_{\tilde{z}}$, together with the root $\tilde{z}$, is a so-called right subframe of Z.

If $\tilde{z}_j$ $(j = 1, \dots, m)$ is the set of all division points, then

$$Z_0 = Z \setminus \cup_{j=1}^m \{Z_{\tilde{z}_j} \setminus \tilde{z}_j\}$$

also forms a frame, the left subframe of Z with root z_0 and vertex set Z_0' and containing the division points. Thus, each cut gives rise to a representation as a union of subframes:

$$\tilde{Z} = Z_0 \cup \{\cup_{j=1}^m Z_{\tilde{z}_j}\}. \tag{4.1}$$

We say that two cuts are *equivalent* if they give rise to the same representations (4.1). Note that the vertex set of Z is decomposed into the disjoint vertex sets of the frames $Z_{\tilde{z}_j}$ and (possibly) a part of the vertex set of the frame Z_0 (the other part consists of division points).

Let Z be a frame, possibly extended. Consider a monotone map $\theta\colon Z \to [t, \tau]$ having the properties $\theta(z_0) = t$, $\theta(z') = \tau$ $(z' \in Z')$. Let Θ be the set of these maps.

Then to each interval $[z, \tilde{z}]$ we have assigned the numerical interval $[\theta(z), \theta(\tilde{z})]$. The totality of points in these intervals is regarded as the branching interval of $[t, \tau]$, with branching points $\theta(z'')$ $(z'' \in Z'')$; it is called a *tree* and denoted by $\sigma(\theta, Z)$. The set of points of a tree $\sigma(\theta, Z)$ is indexed by a coordinate s. Extension of a frame leads to division of the branching intervals by additional points. In particular, for a frame $\{z_0, z_1\}$ this leads to the familiar interval $[t, \tau]$ and its partition.

To each tree $\sigma(\theta, Z)$ corresponding to a frame Z we assign the point $\tilde{\sigma}(\theta, Z)$ in the multidimensional cube $[t, \tau]^{k(Z'')}$ whose coordinates are the numbers $\theta(z)$, $z \in Z''$ (the images of the branching points), written in lexicographical order.

The set $\Theta(z) = \{\tilde{\sigma}(\theta, Z)\colon \theta \in \Theta\}$ is a subset of this cube; it is bounded by finitely many planes and is the union of a finite number of simplices.

Consider some simple examples.

For $Z = \{z, z_1\}$ there is only one map $\theta\colon [z, z_1] \to [t, \tau]$.

For $Z = \{z_0, z_1, z_{11}, z_{12}\}$ a map θ is determined by its value on one point $\theta(z_1) = s$, where s may run through the interval $[t, \tau]$. The tree $\sigma(\theta, Z)$ consists of the interval $[t, s]$ and the twice-covered interval $[s, \tau]$. The situation is similar for $\{z_0, z_1, z_{11}, z_{12}, z_{13}\}$, except that $[s, \tau]$ is covered thrice.

For $Z = \{z_0, z_1, z_{11}, z_{111}, z_{112}, z_{12}\}$ a map θ is determined by two values, $\theta(z_1) = s_1$ and $\theta(z_{11}) = s_2$, satisfying $t \le s_1 < s_2 \le \tau$. The points of $\tilde{\sigma}(\theta, Z)$, with coordinates (s_1, s_2), fill a triangle in the square $[t, \tau]^2$.

Let Z be a frame and $\sigma(\theta, Z)$ a tree on the interval $[t, \tau]$. A map $x_\theta\colon \sigma(\theta, Z) \to X$ is called a *branching path* in X (of type Z) on $[t, \tau]$. The same term is used for the point set $\{x_\theta(s) = x_\theta(\theta_s)\colon s \in [t, \tau]\}$.

We denote by $\mathcal{M}(x, t, \tau; Z)$ the set of branching paths of type Z on the interval $[t, \tau)$ starting at the point x: $x(t) = x$.

For $Z = Z_1$, $\mathcal{M}(x, t, \tau; Z) = \mathcal{M}(x, t, \tau)$ is the set of ordinary paths.

In each space $\mathcal{M}(x, t, \tau; Z)$ we introduce in a natural manner the algebra of cylinder sets. By a general scheme, it suffices to determine the corresponding system of maps from this space to finite-dimensional spaces.

Let q be a division of Z, $Z_q \supset Z$ the corresponding extended frame, $\tilde{Z}_q = Z_q \setminus Z$ the set of division points, and θ_q the set of images of these points in the tree $\sigma(\theta, Z_q)$.

We consider a pair of maps: the first is

$$S_\theta\colon \mathcal{M}(x, t, \tau; Z) \to \Theta(Z) \times X^{k(Z'')},$$
$$S_\theta x_\theta = \{\tilde{\sigma}(\theta, Z), x(\theta(z'')) : | : z'' \in Z''\}.$$

It assigns to a branching path the point $\tilde{\sigma}(\theta, Z) \in \Theta(Z)$, and describes the set of branching paths of trees and the set of values $x(\theta(Z''))$ of a path at the branching points. The other map,

$$S_{\theta,q}\colon \mathcal{M}(x, t, \tau; Z) \to X^q,$$
$$S_{\theta,q} x_\theta = \{x(\theta(z)) : z \in \tilde{Z}_q\},$$

assigns to a path the set of its values at the division points of a tree.

The algebra of cylinder sets $\mathcal{U}_Z$ is determined by the pair $\{S_\theta, S_q\}$ for all possible θ, q, and is the product of the algebras $\mathcal{U}_Z^1$ and $\mathcal{U}_Z^2$ generated by the pre-images of Borel sets $B^q \subset X^q$, $C'' \subset \Theta(Z)$, $B'' \subset X^{k(Z'')}$ under these maps, $S_\theta^{-1}(C'' \times B'')$ and $S_q^{-1}(B^q)$.

On the algebra of cylinder sets we can define, by the general principles of chapter I, a quasimeasure and the notion of integral with respect to it. We will not describe this general construction, but restrict to integrals connected with evolution families.

4.2. Evolution families in the class of formal power series. Let $\mathcal{L}$ be a topological linear space and $[\mathcal{L}^k]$ the space of continuous k-linear operators on $\mathcal{L}$, equipped with the topology of strong convergence.

A collection

$$A = (a_1, a_2, \ldots, a_k, \ldots) \quad (a_k \in [\mathcal{L}^k])$$

is called a *formal operator* on $\mathcal{L}$.

The expression

$$A(\psi) = \sum_{k=1}^{\infty} a_k \psi^k,$$

where $a_k\psi^k = a_k(\psi, \ldots, \psi)$, is called a *formal power series* in $\mathcal{L}$. This expression is to be taken symbolically, though it determines an element of $\mathcal{L}$ if the series converges.

The set $\mathcal{A}$ of formal operators clearly has the structure of a linear space (with componentwise operations).

Moreover, the operator of composition of formal operators is well defined; it corresponds to substitution of series into series:

$$(A \circ B)(\psi) = (A(B(\psi))).$$

It is readily verified that the coefficients of the formal operator $C = A \circ B$ satisfy the relations

$$c_1 = a_1 b_1, \quad c_2 = a_1 b_2 + a_2(b_1, b_1), \ldots,$$

and in general,

$$c_k = \sum_{j=1}^{k} \sum_{k_1+\cdots+k_j=k} a_j(b_{k_1}, \ldots, b_{k_j}), \tag{4.2}$$

where $a_j(b_{k_1}, \ldots, b_{k_j})(\psi^k) = a_j(b_{k_1}\psi^{k_1}, \ldots, b_{k_j}\psi^{k_j})$.

The definition of this composition operator implies its associativity,

$$(A \circ B) \circ C = A \circ (B \circ C),$$

and left distributivity,

$$(A + B) \circ C = A \circ C + B \circ C.$$

In $\mathcal{A}$ we define a topology by requiring componentwise convergence of formal operators. Below, continuity and differentiability with respect to a parameter are meant in the sense of this topology. A family of formal operators $U(t, \tau)$ ($t \le \tau$, $t, \tau \in \Delta = [a, b] \subset \mathbf{R}$) is called an *evolution family* if

$$U(t, s) \circ U(s, \tau) = U(t, \tau), \quad U(t, t) = I \quad (t < s < \tau).$$

These relations mean that

$$u_1(t,s)u_1(s,\tau) = u_1(t,\tau),$$
$$u_1(t,s)u_2(s,\tau) + u_2(t,s)(u_1(s,\tau)\cdot, u_1(s,\tau)\cdot) = u_2(t,\tau),$$
$$\cdots\cdots\cdots.$$

In particular, the family of linear operators $u_1(t,\tau)$ on $\mathcal{L}$ is an evolution family. The derivative

$$A(s) = \left.\frac{dU(t,s)}{dt}\right|_{t=s} = \lim -\frac{1}{h}\{U(s-h,s) - I\}$$

(if it exists) will be called the *generating operator* of the evolution family $U(t,s)$. Clearly,

$$a_1(s) = \left.\frac{d}{dt}u_1(t,s)\right|_{t=s}$$

is the generating operator of the linear evolution family $u_1(t,s)$.

An evolution family U with generating operator A satisfies the differential equation

$$\frac{\partial U(t,\tau)}{dt} + A(t)\circ U(t,\tau) = 0. \tag{4.3}$$

Written in componentwise manner, we obtain the system

$$\frac{\partial u_1(t,\tau)}{dt} + a_1(t)u_1(t,\tau) = 0, \tag{4.4}$$

$$\frac{\partial u_2(t,\tau)}{dt} + a_1(t)u_2(t,\tau) + a_2(t)(u_1(t,\tau)\cdot, u_1(t,\tau)\cdot) = 0, \tag{4.5}$$
$$\cdots\cdots\cdots.$$

If $\mathcal{L} \subset X$ with X a Banach space, we can similarly construct the space $\mathcal{A}_X$ of formal operators on X. Here we will assume that the multilinear operators in $\mathcal{A}_X$ leave invariant $\mathcal{L}$. Then $\mathcal{A}_X$ is embedded in $\mathcal{A}_{\mathcal{L}}$. The space $\mathcal{A}_X$ is not a metric space, but nevertheless theorems 2.1–2.3 can be generalized to the case under consideration (in terms of componentwise convergence).

By successively solving the system (4.4) we can show that there exists an operator satisfying (4.3). We will not do this, since an explicit construction of the solution is given below.

We introduce the differentiation operation

$$\delta A(B,C) = \left.\frac{d}{d\epsilon}A\circ(B+\epsilon C)\right|_{\epsilon=0}, \quad (A,B,C\in\mathcal{A}).$$

The following properties are easily verified:

1) the formal operator $\delta A(B,C)$ depends linearly on C and A;

2) if $A_1(t)$ and $A_2(t)$ are formal operators which are differentiable with respect to t, then

$$\frac{d}{dt}A_1(t)\circ A_2(t)=\left[\frac{d}{dt}A_1(t)\right]\circ A_2(t)+\delta A_1(t)\left(A_2(t),\frac{dA_2(t)}{dt}\right);$$

3) if $U(t,\tau)$ is an evolution family with continuous generating operator $A(t)$, and B is a formal operator, then

$$\frac{d}{d\tau}U(t,\tau)\circ B=\delta U(t,\tau)(B,A(\tau)\circ B).$$

PROPOSITION 4.1. *An evolution family with continuous generating operator $A(t)$ is uniquely determined by the latter.*

PROOF. Let U and V be a pair of evolution families with the same generating operator $A(t)$. For $t<s<\tau$,

$$\frac{d}{dt}[U(t,s)\circ V(s,\tau)]=$$
$$\frac{d}{dt}U(t,s)\circ V(s,\tau)+\delta U(t,s)\left[V(s,\tau),\frac{dV(s,\tau)}{dt}\right]=$$
$$\delta U(t,s)(V(s,\tau),A(s)\circ V(s,\tau))-\delta U(t,s)(V(s,\tau),A(s)\circ V(s,\tau))=0.$$

Therefore,

$$U(t,\tau)-V(t,\tau)=\int_t^\tau\frac{d}{dt}[U(t,s)\circ V(s,\tau)]\,ds=0.\quad\square$$

The evolution family of formal operators with constant generating operator A will be called an *exponential family*, and it will be denoted by

$$U(t,\tau)=\overset{\circ}{\exp}\{A(\tau-t)\}.$$

By definition we have

$$\overset{\circ}{\exp}t_1A\circ\overset{\circ}{\exp}t_2A=\overset{\circ}{\exp}(t_1+t_2)A.$$

Using analogues of the multiplicative theorems 2.1, 2.2, we can, under certain assumptions about the family of formal operators $A(t)$, prove convergence of the composite multiplicative integral

$$U(t,\tau)=\overset{\frown}{\overset{\circ}{\exp}}\int_t^\tau A(s)\,ds=$$
$$\lim\{\overset{\circ}{\exp}\Delta s_1A(s_1)\circ\cdots\circ\overset{\circ}{\exp}\Delta s_nA(s_n)\}=$$
$$\lim\{[I+\Delta s_1A(s_1)]\circ\cdots\circ[I+s_nA(s_n)]\},$$

and we can show that $A(s)$ is the generating operator.

Here we will not give these considerations, and we will use the symbol of multiplicative integral as a notation for the evolution family with corresponding generating operator.

4.3. Construction of a multiplicative integral. Let Z be an extended frame, and let every point $z \in Z \setminus \{\{z_0\} \cup Z'\}$ (i.e. not being the root of a vertex) be assigned a $k_+(z)$-linear operator A_z, and to every interval $[z, \tilde{z}]$ a linear operator $V_{z\tilde{z}}$.

We consider a $k(Z')$-linear operator on Z, called the *branching composition* of the collection of maps $\{A_z, V_{z\tilde{z}}\colon z, \tilde{z} \in Z\}$ and denoted by

$$S_Z = \prod_Z (V_{z\tilde{z}} A_{\tilde{z}}).$$

This operator is defined by the rule:

a) to an interval $[z, \tilde{z}]$ ending at $\tilde{z}$ we assign the composition $V_{z\tilde{z}} \circ A_{\tilde{z}}$;
b) to a point z at which intervals $[z, z_j]$ $(j = 1, \dots, k_+(z))$ begin, we assing the composition

$$A_z(V_{zz_1}\cdot, \dots, V_{zz_{k_+(z)}}\cdot),$$

the ordering of the operators V_{zz_j} being determined by the ordering of the set $\alpha_+(z)$.

It is readily seen that these rules give a unique operator S_Z. This construction can recurrently be given as follows. Suppose the frame

$$Z = \{z_0, z_1, z_{11}, \dots, z_{1k}, \dots\}, \quad k = k_+(z_1),$$

is divided into the left subframe $Z_0 = \{z_0, z_1\}$ and the right subframes $Z_{ij} = \{z_1, z_{1j}, \dots\}$ with common root z_1 $(j = 1, \dots, k)$. Then

$$S_Z = V_{z_0 z_1} A_{z_1}(S_{Z_{11}}\cdot, \dots, S_{Z_{1k}}\cdot).$$

In particular, for an extended frame $\tilde{Z} = \{z_0, z^{(1)}, \dots, z^{(m)}, z_1\}$ without branching points we obtain the familiar ordered product:

$$S_{\tilde{Z}} = \prod_Z V_{z\tilde{z}} A_{\tilde{z}} = V_{z_0 z^{(1)}} A_{z^{(1)}} V_{z^{(1)} z^{(2)}} \dots A_{z^{(m)}} V_{z^{(m)} z_1}.$$

EXAMPLE 4.1. For $Z_3' = \{z_0, z_1, z_{11}, z_{111}, z_{112}, z_{12}\}$,

$$S_Z(f_1, f_2, f_3) = V_{z_0 z_1} A_{z_1}(V_{z_1 z_{11}} A_{z_1}(V_{z_{11} z_{111}} f_1, V_{z_{11} z_{112}} f_2), V_{z_1 z_{12}} f_3).$$

THEOREM 4.1. *Let $A(t) = (a_1(t), a_2(t), \dots)$ be a family of formal operators on $\mathcal{L}$ depending on a parameter t, and suppose that on $\mathcal{L}$ the family of linear evolution operators $v_1(t, \tau)$ with generating operator $a_1(t)$ exists:*

$$\frac{dv_1(t, \tau)}{dt} + a_1(t) v_1(t, \tau) = 0 \quad (t < \tau; \quad v(\tau, \tau) = id).$$

We assign to each frame Z the $k(Z')$-linear operator on $\mathcal{L}$ defined by

$$U_Z(t, \tau) = \int_{\Theta(Z)} \prod_Z v_1(\theta(z), \theta(\tilde{z})) a_{k_+(\tilde{z})}(\theta(\tilde{z}))\, d\theta, \tag{4.6}$$

where integration is with respect to volume in $\Theta(Z) \subset [t, \tau]^{k(Z'')}$.

Then the family of formal operators $v(t, \tau) = (v_1(t, \tau), \dots, v_k(t, \tau), \dots)$, where

$$v_k(t, \tau) = \sum_{k(Z') = k} U_Z(t, \tau), \tag{4.7}$$

is an evolution family with $A(t)$ as generating operator.

PROOF. Let $t < s < \tau$. The point s lies outside every tree $\sigma(\theta, Z)$, and, neglecting a set of lower dimension, we may assume that division points do not coincide with branching points.

We complete Z by division points $\tilde{z}_j$ so that $\theta(\tilde{z}_j) = s$. Then we obtain a decomposition of the extended frame $\tilde{Z}$ into left and right subframes:

$$Z = Z_0^\theta \cup \{\cup_{j=1}^m Z_{\tilde{z}_j}^\theta\}. \tag{4.8}$$

Consider the multilinear operators

$$S_{Z_0^\theta}(t,s) = \prod_{Z_0} v_1(\theta(z), \theta(\tilde{z}))a_{k_+(\tilde{z})},$$
$$S_{Z_{\tilde{z}_j}^\theta}(s,\tau) = \prod_{Z_{\tilde{z}_j}^\theta} v_1(\theta(z), \theta(\tilde{z}))a_{k_+(\tilde{z})}.$$

The definitions of branching product and the evolution family v_1 imply that

$$S_{Z^\theta}(t,\tau) = S_{Z_0^\theta}(t,s)(S_{Z_{\tilde{z}_1}}(s,\tau)\cdot, \dots, S_{Z_{\tilde{z}_n}}(s,\tau)\cdot). \tag{4.9}$$

We divide the tree space $\Theta(Z)$ into subspaces Θ_k whose elements correspond to equivalent cuts, and hence to the same decomposition (4.7). Distinct sets Θ_k correspond to distinct left subframes $Z_0^\theta \subset Z$, and by taking all these subsets we obtain all possible left subframes of Z, hence all possible decompositions (4.7).

Splitting the integral (4.5) into integrals over the Θ_k, and integrating (4.8), the following representation is readily obtained:

$$U_Z(t,\tau) = \sum_{Z_0} U_{Z_0}(t,s)(U_{Z_1}(s,\tau)\cdot, \dots, U_{Z_{k_+(Z_0)}}(s,\tau)\cdot), \tag{4.10}$$

where summation is over all left subframes $Z_0 \subset Z$. Summing over all frames Z for which $k_+(Z) = k$ we obtain, by (4.6) and (4.9), the formula

$$U_k(t,\tau) = \sum_{j=1}^{k} \sum_{k_1+\dots+k_j=k} U_j(t,s)(U_{k_1}(s,\tau)\cdot, \dots, U_{k_j}(s,\tau)\cdot).$$

By (4.2) this formula implies for formal operators

$$V(t,\tau) = V(t,s) \circ V(s,\tau).$$

To compute the generating operator of the family we set $s = t + \Delta t$ in (4.9), and select terms of the first order with respect to Δt:

$$U_Z(t,\tau) = v_1(t, t+\Delta t)U_Z(t+\Delta t, \tau) +$$
$$\int_t^{t+\Delta t} v_1(t,\alpha)a_{k_+(Z_1)}(\alpha) \times$$
$$(v_1(\alpha, t+\Delta t)U_{Z_1}(t+\Delta t,\tau), \dots, v_1(\alpha, t+\Delta t)U_{Z_{k_+(Z_1)}}(t+\Delta t,\tau)\, d\alpha + o(\Delta t),$$

where $Z_1, \dots, Z_{k_+}$ are the right subframes of Z corresponding to the left subframe $Z_0 = \{z_0, z_1, \tilde{z}_1, \dots, \tilde{z}_m\}$.

Differentiating this relation termwise with respect to t and setting $\Delta t = 0$ we arrive at

$$\frac{d}{dt}U_Z(t,\tau) = -a_1(t)U_Z(t,\tau) - a_{k_+(Z_1)}(t)(U_{Z_1}(t,\tau)\cdot,\dots).$$

Summing again over all frames Z with $k_+(Z) = k$ we obtain

$$\frac{d}{dt}U_k(t,\tau) + \sum_{j=1}^{k}\sum_{k_1+\dots+k_j=k} a_j(t)U_{k_1}(t,\tau)\cdot,\dots,U_{k_j}(t,\tau)\cdot) = 0,$$

which is equivalent to the required equation

$$\frac{d}{dt}V(t,\tau) + A(t)\circ V(t,\tau) = 0. \quad \square$$

4.4. Convergence of formal expansions. We consider in $\mathcal{L}$ the differential equation

$$\frac{du}{dt} + a_1(t)u + B(t,u) = 0, \tag{4.11}$$

where a_1 is a linear operator on $\mathcal{L}$ and

$$B(t,u) = \sum_{k=2}^{\infty} b_k(t)u^k \tag{4.12}$$

is an analytic function in $\mathcal{L}$, represented (in some neighborhood of zero) by a strongly convergent series (4.11), e.g. a polynomial

$$B_N(t,u) = \sum_{k=2}^{N} b_k(t)u^k. \tag{4.13}$$

If the Cauchy problem for (4.10) is well-posed on a subinterval (t,τ) for sufficiently small initial elements $u(\tau) = u_0$, then the solution $u(t)$ may turn out to depend analytically on u_0:

$$u(t) = \sum_{k=1}^{\infty} v_k(t,\tau)u_0^k. \tag{4.14}$$

By virtue of the uniqueness of the evolution family determined by $A = (a_1, b_2, \dots)$ in the class of formal power series, the multilinear functionals $v_k(t,\tau)$ in (4.13) must be defined by (4.6). Thus, in the case under consideration, the formal power series constructed in the previous section, and representing the multiplicative integral, converges in some neighborhood of zero.

Obviously, the converse holds: if the formal power series (4.13) converges in a neighborhood of zero, then it represents the solution of the Cauchy problem for (4.10).

Sometimes these series can be proven to be convergent by estimating the coefficients in (4.13). In fact, let $\mathcal{L}$ be a Banach space and suppose the following conditions hold:

a) the linear evolution family v_1 satisfies the estimate

$$\|v_1(t,\tau)\| \le e^{\alpha_1(\tau-t)};$$

b) the multilinear operators a_k are bounded, and

$$\|a_k\| = \alpha_k \quad (k = 2, 3, \ldots,).$$

Then it is readily shown that

$$\left\| \prod_Z v_1(\theta(z), \theta(\tilde{z})) a_{k_+(\tilde{z})}(\theta(\tilde{z})) \right\| \leq \prod_Z e^{\alpha_1(\theta(\tilde{z}) - \theta(z))} \alpha_k(\theta(\tilde{z})),$$

where the righthand side is the branching product of numbers. Inserting this expression into formulas (4.5) and (4.6) leads to the numerical formal power series with nonnegative coefficients

$$\sum_{k=1}^{\infty} v_k(t, \tau) \psi^k, \tag{4.15}$$

which corresponds to the solution of the scalar Cauchy problem with nonnegative α_k $(k = 2, 3, \ldots)$ and ψ:

$$\frac{du}{dt} + \alpha u + \sum_{k=2}^{\infty} \alpha_k(t) u^k = 0; \quad u(\tau) = \psi. \tag{4.16}$$

Standard considerations from the analytic theory of differential equations show that if (4.15) has an analytic solution in the disc $|\psi| < \rho$ and on the interval $[t, \tau]$, then for $\|u_0\| < \rho$ the formal power series representing the solution of (4.10) converges also on $[t, \tau]$.

If the series occurring in the scalar equation (4.15) has a nonzero radius of convergence, then the above holds a fortiori for sufficiently small $\|u_0\|$ and $|\tau - t|$.

Another approach to the convergence of the expansion under consideration is connected with the multiplicative representations treated in section 2.

For the sake of simplicity, consider (4.10) with a polynomial operator (4.12). Let

$$U(t, s) = v_1(t, s) \left[I + \sum_{k=2}^{N} b_k(s) u^k (t - s) \right],$$

and suppose there is a multiplicative family

$$\tilde{U}(t, s) = \lim_q U_q(t, s).$$

By proposition 2.2, its generating operator is

$$A(t) = a_1(t) + \sum_{k=2}^{\infty} b_k(t).$$

The operator $U_q(t, s)$ is the superposition of a finite number of multilinear maps. By removing the parentheses in this superposition we can establish, by induction, that

$$U_q(t, s) = \sum_Z \sum_{\Theta_q(Z)} \prod_{z \in Z} v(\theta(z), \theta(\tilde{z})) [a_{k_+(\tilde{z})} \Delta t^{k_+}],$$

where summation is over all trees whose branching points belong to the division q. In the limit the individual terms of this sum become the integrals (4.5).

Along these lines we also obtain an independent proof of the fact that the solutions of the problem under consideration can be expanded in formal power series. In certain concrete cases, using additional arguments and estimates, convergence of these series can thus be established.

4.5. Markov quasimeasures and integrals along branching paths. In the construction of Markov quasimeasures in the space of branching paths we will assume that a transition measure

$$P(t_1, x_1, t_2, A) \quad (t \le t_1 < t_2 \le \tau, \quad x_1 \in X, \quad A \in \mathcal{U}),$$

described in section 1, and a collection of 'branching transition measures'

$$Q_j(s; x, A_1 \times \cdots \times A_j) \quad (t \le s \le \tau, \quad x \in X, \quad A_1, \dots, A_j \in \mathcal{U})$$

in the product $X^j = X \times \cdots \times X$ $(j = 2, 3, \dots)$ are given.

From P we construct (see (1.3)) the quasimeasure $\mu_{x,t} = \mu_{x,t}^{(P)}$ in the space of ordinary paths beginning at (t, x).

Now we consider the union σ_z of the intervals of the tree $\sigma(\theta, Z)$ beginning at the branching point $\theta(z)$: $\sigma_z = \{\cup_{j=1}^k [\theta(z), \theta(\tilde{z}_j)]$: $j = 1, \dots, k = k_+(z)$, k being the number of intervals in Z beginning at $z\}$. Let $\mathcal{M}(y, \theta, z)$ be the space of paths $\{x$: $\sigma_z \to X\}$ on σ_z satisfying the condition $x(\theta(z)) = y$. In this space we introduce quasimeasures by setting

$$\mu_{y;\theta(z)}^{(k)} = \int_{X^k} Q_k(\theta(z), y, dy_1, \dots, dy_k) \prod_{j=1}^k \mu_{y_j, \theta(z), \theta(\tilde{z}_j)} \quad (k = k_+(z)).$$

The space of branching paths of type Z can be constructed from those of type $\mathcal{M}(y, \theta, z)$:

$$\mathcal{M}(x, t, \tau; Z) = \cup_{\theta \in \Theta_Z} \prod_{z \in Z \setminus Z'} \cup_{y \in X} \mathcal{M}(y, \theta, z),$$

and in this way we arrive at a quasimeasure in $\mathcal{M}(x, t, \tau; Z)$.

We consider some simple examples.

For $Z = \{z_0, z_1\}$ we put

$$\mu_y^{(Z)} = \mu_{y,t,\tau}^{(P)}.$$

For $Z = \{z_0, z_1, z_{11}, z_{12}\}$ we put

$$\mu_y^{(Z)}(dx_\theta(\cdot)) = \int_t^\tau \mu_{y,t,s}(dx(\cdot))\mu_{x(s),s}^{(2)}(dx_1(\cdot) \times dx_2(\cdot))\, ds$$

(here we present the branching path $x_\theta(t_1)$ as the union of $x(t_1)$ on the interval $[t, s]$ and the pair $(x_1(t_1), x_2(t_2))$ on the twice-covered interval $[s, \tau]$).

The construction of more complicated frames is similar and can be described in a recurrent way (though awkwardly). Instead of giving this description we show that when integrating multiplicative functionals it is not necessary to use the setting of evolution families. The transition measure determines the evolution family of linear operators

$$v_1(t_1, t_2) f(x_1) = \int_X P(t_1, x_1, t_2, dx_2) f(x_2),$$

and the branching transition measures Q_j determine the family of multilinear operators

$$a_j(s,x)(\phi_1,\dots,\phi_j) = \int_{X^j} Q_j(s,x,dx_1 \times \dots \times dx_j)\phi_1(x_1)\dots\phi_j(x_j).$$

Consider the multilinear operator $U_Z(t,\tau)$ introduced in theorem 4.1, and set, by definition,

$$(4.17)\quad \int_{\mathcal{M}(y,t,\tau;Z)} \phi_1(x_1(\tau))\dots\phi_n(x_n(\tau))\mu_y^{(Z)}(dx_\theta(\cdot)) = U_Z(t,\tau)(\phi_1,\dots,\phi_n)(y).$$

It is easily shown that this definition makes sense if $\mu_y^{(Z)}$ is constructed as suggested above.

Consider the more general case of an evolution family v_1 constructed from a multiplicative functional $f_{[t_1,t_2]}$:

$$v_1(t_1,t_2)\phi(x_1) = \int f_{[t_1,t_2]}(x(\cdot))\phi(x(t_2))\mu_{x_1,t_1}(dx).$$

We define the value of this multiplicative functional on a branching path as the product of its values on the individual branches:

$$f_{\sigma(\theta,Z)}(x_\theta(\cdot)) = \prod_{z\in Z} f_{[\theta(z),\theta(\tilde z)]}(x(\cdot)).$$

In this situation formula (4.16) becomes more complicated:

$$U_Z(t,\tau)(\phi_1,\dots,\phi_n)(y) =$$

$$(4.18)\quad \int_{\mathcal{M}(y,t,\tau;Z)} f_{\sigma(\theta,Z)}(x_\theta(\cdot))\phi_1(x_1(\tau))\dots\phi_n(x_n(\tau))\mu_y^{(z)}(dx_\theta(\cdot)).$$

In the space $\mathcal{M}(y,t,\tau) = \cup_Z \mathcal{M}(y,t,\tau;Z)$ of all branching paths on $[t,\tau]$ satisfying the condition $x(t) = y$ we introduce a functional integral, by setting

$$(4.19)\quad \int_{\mathcal{M}(y,t,\tau)} F(x(\cdot))\mu_y(dx(\cdot)) = \sum_Z \int_{\mathcal{M}(y,t,\tau;Z)} F(x_\theta(\dot{)})\mu_y^{(z)}(dx_\theta(\cdot)),$$

provided the series at the righthand side converges absolutely.

In particular, consider the functional

$$F(x_\theta(\dot{)}) = f_{[t,\tau]}(x_\theta(\cdot))\phi(x_\theta(\tau)),$$

determined, for a tree $\sigma(\theta,Z)$ by

$$F(x_\theta(\dot{)})|_{\sigma(\theta,Z)} = f_{\sigma(\theta,Z)}(x_\theta(\cdot)) \prod_{k=1}^{k(Z')} \phi(x_k(\tau)).$$

Formulas (4.17) and (4.18) lead to the relation

$$U(t,\tau)\phi(y) = \sum_Z U(t,\tau)(\phi,\dots,\phi)(y) =$$

$$(4.20)\quad \int_{\mathcal{M}(y,t,\tau)} f_{[t,\tau]}(x_\theta(\cdot)) \prod_{x\in x_\theta(\tau))} \phi(x)\mu_y(dx(\cdot)),$$

representing the nonlinear evolution family $U(t,\tau)$ (provided the integral at the right-hand side exists).

The generating operator of this family is

$$A(t)\phi = a_1(t)\phi + \sum_j a_j(t,\phi,\dots,\phi), \tag{4.21}$$

where $a_1(t)$ is the generating operator of the linear family $v_1(t,\tau)$.

Note that the expression (4.20) makes sense, in any case, if the sum is finite and the multilinear operators a_j are bounded. If the linear evolution family is as suggested in the previous section, then estimates for the rate of convergence of the series (4.19) can be obtained.

As an example, consider the integro-differential equation

$$\frac{\partial u}{\partial t} + l(u) + V(t)u + \sum_{k=2}^{n} \int Q(t,x,dy_1,\dots,dy_k)u(y_1)\dots u(y_k) = 0, \tag{4.22}$$

where $l(u)$ is a linear differential operator.

A particular instance of the equation is the diffusion equation with polynomial nonlinearity:

$$\frac{\partial u}{\partial t} + l(u) + V(t,u) + \sum_{k=2}^{n} a_k(t)u^k(x)0. \tag{4.23}$$

In this case the linear evolution family $v_1(t,\tau)$ is determined by

$$\frac{\partial u}{\partial t} + l(u) = 0,$$

and the multiplicative functional is

$$f_{\sigma(\theta,Z)}(x_\theta(\cdot)) = \exp \int_{\sigma(\theta,\mathcal{Z})} V(x(s))\,ds = \exp \int_t^T V(x_\theta(s))\,ds,$$

where the integral along the tree is understood to be the sum of the integrals along all its intervals.

In this case the functional integral (4.19) is

$$U(t,\tau)\phi(y) = \int_{\mathcal{M}(y,t,\tau)} \exp \int_t^T V(x_\theta(s))\,ds \prod_{x_\theta(s)} \phi(x)\mu_y(dx(\cdot)). \tag{4.24}$$

We make some remarks about its convergence.

a) If $l(u)$ is a second order nonpositive elliptic differential operator, then the transition measure $P(t,x,\tau,dx)$ is nonnegative. Also, the operator v_1 is contractive in the space of continuous functions. If the integral operators occurring in (4.21) are continuous in this space (in particular, this holds for (4.22)), then the integral can be shown to be convergent on a sufficiently small interval, as suggested in the previous section.

b) For elliptic operators $l(u)$ of higher order than two we have to involve spaces of type L_p. Here the measure μ_y is not σ-additive, but in the individual spaces $\mathcal{M}(x,t,\tau;Z)$ integrals of multiplicative functionals make sense. Therefore, for continuous operators with kernels Q as in (4.21), considerations similar to the above hold.

These reasonings are valid in the Hilbert space L_2 for the Schroedinger equation

$$\frac{\partial u}{\partial t} + il(u) + V(t,u) + \sum_{k=2}^{n} a_k(t)u^k(x) = 0$$

with a selfadjoint operator $l(u)$.

The situation for an equation of type (4.22) is more complicated, since the operation $u \to u^k$ is not continuous in L_2. However, here we can apply methods connected with chronological products, mentioned at the end of section 4.

5. Chapter VI: Supplementary remarks and historical comments

The connection between partial differential equations and integration in path space was discovered by Feynman [1] when elaborating a new approach to quantum mechanics. On the other hand, a similar connection arises naturally in the theory of Markov stochastic processes (see chapter VII). Precisely in this way M. Kac [1] has given the first rigorous proof of a formula of type (3.22) (for the Wiener measure μ), now known as the Feynman–Kac formula. It was generalized to arbitrary Markov processes by Dynkin [2].

In Gelfand–Yaglom [1] the problem of generalizing this formula to the wider class of linear evolution equations, including parabolic equations of higher order, was stated. Such a generalization to abstract evolution equations, including various equations and systems of parabolic, hyperbolic and Schroedinger type, was obtained in Dalecky [1,2,3,4,5].

In these papers a method based on a formula of type (2.30) for the chronological product of evolution families was used. Such a formula was also obtained, for other purposes, by Trotter [1] (see also Kato [1]). In particular, (2.30) implies a rigorous proof of the Feynman representation for the Schroedinger equation with a bounded smooth potential.

A proof of this representation for analytic potentials was obtained, using analytic continuation into the complex plane with respect to the parameter in the equation, by Cameron [1] (see chapter II, section 6).

Another approach to constructing Feynman integrals was suggested by Itô [4]. The condition (1.6)–(1.7) of boundedness of the variation of the quasimeasure was obtained by Dalecky [5], and was generalized to complex measures by Feldman [3]. It was pointed out by Cameron [1] and Dalecky [2], independently, that the quasimeasure obtained from the Feynman measure by replacing i by $i-\epsilon$ has unbounded variation.

A Feynman–Kac formula for Schroedinger equations with more general potentials, including singular ones, was proved in Feldman [2] and Nelson [1]. Chronological products of semigroups were also used in these papers (see also Faris [1]). Certain results related to the computation of integrals with respect to quasimeasures were obtained in Krylov [1,2], Ladokhin [1,2], and Maikov [1,2].

There is another version of the Feynman representation (Hamilton's form of the functional integral), which can be obtained from the Feynman–Kac formula in particular cases by using the Fourier transform of the transition measure, and so passing from the configuration space to the phase space. Hamilton's form of the functional integral gives a representation of the symbol of a family of evolution operators in terms of the symbol of the generating operator of this family. It is valid in the quantization procedure and is used to obtain asymptotics for the corresponding Green's function.

A justification of this representation for various classes of symbols was obtained in Alimov [1], Buslaev [1], Alimov–Buslaev [1], Evgrafov [1], Roitburg [1,2], Maslov–Shishmarev [1]. Here we do not aim at giving a review of the large number of recent mathematical papers related to the various aspects of the theory of the Feynman integral (see also the comments to chapters I, II, and Albeverio–Höegh-Krohn [1], also cited there; this paper contains numerous references).

Chronological products of nonlinear evolution families were obtained by Dalecky–Zaplitnaya [1] (see also Zaplitnaya [1]), and were used to represent solutions of nonlinear evolution equations as integrals in the branching path space. This work was influenced by Skorokhod [6], concerned with the theory of branching diffusion. Subsequently multiplicative representations of nonlinear evolution families were independently considered also by Maslov–Chebotarev [1]. The presentation in Dalecky [4] exploits results from Dalecky [17] and Gavrishova–Dalecky [1].

CHAPTER VII

Probabilistic representations of solutions of parabolic equations and systems

In this chapter some of the results from chapters V and VI are, independently of these chapters, considered by methods from the theory of stochastic processes. As a basis we take the theory of Itô stochastic differential equations in Hilbert space, which will be presented in sufficient detail in sections 2, 3.

In section 1 we briefly give the necessary notions from probability theory. Section 4 deals with the Cauchy problem for second order parabolic equations and systems with infinite-dimensional argument. The solutions are represented as averages in path space with respect to the measure determined by the stochastic process satisfying the stochastic equation.

Thus, the differential equation is treated as the inverse Kolmogorov equation for a Markov process. In some respect, this approach proves more effective and leads to stronger results than those obtained in the previous chapters. At the end we consider a probabilistic treatment of quasilinear equations and systems.

1. Wiener process. Stochastic integrals

1.1. Basic notions from the theory of stochastic processes. First of all we briefly consider some well-known facts and notions from the theory of stochastic processes (see Doob [1], Gikhman–Skorokhod [1]) which form the base for further developments.

A measurable space $(\Omega, \mathcal{F})$ with a normalized nonnegative measure P is called a *probability space*, and is denoted by $(\Omega, \mathcal{F}, P)$. Points $\omega \in \Omega$ are called *elementary events*, measurable sets $S \in \mathcal{F}$ *events*, and values $P(S)$ $(S \in \mathcal{F})$ *probabilities* of these events.

Let $(X, \mathcal{B})$ be a measurable space. A measurable map $\xi\colon (\Omega, \mathcal{F}) \to (X, \mathcal{B})$ is called a *random variable* with values in X. The random variable ξ gives rise to a normalized measureμ_ξ: $\mu_\xi(B) = P(\xi^{-1}(B))$ in X, and hence to a new probability space $(X, \mathcal{B}, \mu_\xi)$. The measure μ_ξ is called the *probability distribution* of the random variable ξ.

Let X be the space of functions $x(t)$ defined on a set T and taking values in a space $(Z, \mathcal{U})$. A random variable $\xi\colon \omega \mapsto \xi(t, \omega)$ with values in $(X, \mathcal{B})$, where $\mathcal{B}$ is the σ-algebra containing the algebra of cylinder sets, is called a *random function* on T with values in Z (a *stochastic process* if T is an interval on the real axis). The functions $\xi(t, \omega)$ of the variable t $(\omega \in \Omega)$ are called *realizations* of the random variable ξ. For each $t \in T$ the value $\xi_t(\omega) = \xi(t, \omega)$ is a random variable with values in Z. For each finite collection $\{t_1, \dots, t_n\}$ $(t_k \in T$, $k = 1, \dots, n)$ the random variable $\xi_{(t_1, \dots, t_n)} = (\xi_{t_1}, \dots, \xi_{t_n}) \in Z^n$ determines a distribution $\mu_{t_1, \dots, t_n}$ on $\mathcal{U}^n$. All such distributions form the system of finite-dimensional distributions of the random variable $\xi(t)$. Two random functions ξ and η are said to be *stochastically equivalent* if for every $t \in T$ we have $P\{\omega\colon \xi_t(\omega) = \eta_t(\omega)\} = 1$.

Clearly, the systems of finite-dimensional distributions of stochastically equivalent functions coincide.

The results in chapter I imply that under sufficiently wide conditions on Z a consistent system of finite-dimensional distributions determines a measure μ on the minimal σ-algebra $\mathcal{B}$ in X generated by the cylinder sets. In other words, a consistent system of finite-dimensional distributions determines a random function ξ on T with values in Z ($(X, \mathcal{B}, \mu)$ can be chosen as the probability space). Considering instead of $\mathcal{B}$ the new σ-algebra generated by $\mathcal{B}$ and a collection of sets not in $\mathcal{B}$, assigned measure zero, we obtain a new random function, which is stochastically equivalent to ξ (a *modification* of ξ).

Events $S_1, \ldots, S_n$ are said to be *independent in total* if $P(\prod_{k=1}^n S_k) = \prod_{k=1}^n P(S_k)$; σ-subalgebras $\mathcal{F}_k \subset \mathcal{F}$ $(k = 1, \ldots, n)$ are said to be *independent* if all collections of events $S_k \in \mathcal{F}_k$ $(k = 1, \ldots, n)$ are independent in total.

To each collection of random variables ξ_γ $(\gamma \in \Gamma)$ on $(\Omega, \mathcal{F})$ we can relate a certain σ-algebra $\tilde{\mathcal{F}} \subset \mathcal{F}$, the minimal σ-algebra with respect to which all ξ_γ are measurable. It is well known (see Doob [1]) that a necessary and sufficient condition for a random variable ξ to be $\tilde{\mathcal{F}}$-measurable is that it is a measurable function of the collection ξ_γ. Collections $\{\xi_{\gamma_j}\colon \gamma_j \in \Gamma_j\}$ $(j = 1, \ldots, n)$ of random variables are said to be *independent* if such are the σ-algebras related to these collections. It is easily seen that random variables $\xi_1, \ldots, \xi_n$ are independent in total exactly if the measures determined by them in the product space are products of the distributions of these random variables.

For a random variable ξ with values in a Banach space X the *mean value* is defined:

$$\mathrm{M}\xi = \int_\Omega \xi(\omega) P(d\omega) = \int_X x \mu_\xi(dx) \in X,$$

if only $\mathrm{M}\|\xi\| = \int_\Omega \|\xi(\omega)\| P(d\omega) < \infty$.

Two random variables $\xi(\omega) \in X$, $\eta(\omega) \in Y$ are called *uncorrelated* if $\mathrm{M}\langle \xi, \phi\rangle \langle \eta, \psi\rangle = \mathrm{M}\langle \xi, \phi\rangle \mathrm{M}\langle \eta, \psi\rangle$ $(\phi \in X',\ \psi \in Y')$. For a pair (ξ, η) of random variables with a Gaussian distribution, uncorrelatedness implies independence.

Note that in the computation of mean values of functionals of a random function which is determined by its values at a finite set of points, only the corresponding finite-dimensional distributions are used. Sometimes this makes it possible to extract important information about the properties of the measure μ_ξ in path space from the finite-dimensional distributions.

THEOREM 1.1. *(Kolmogorov.) Let ξ_t $(a \le t \le b)$ be a stochastic process with values in a Banach space X, whose finite-dimensional distributions have the property*

$$M\{\|\xi_{t+\Delta t} - \xi_t\|^\alpha\} \le C|\Delta t|^{1+\epsilon} \quad (\alpha, \epsilon > 0).$$

Then there exists a modification of this process whose distribution is supported in the space $C([a,b], X)$ of continuous functions.

We outline the idea of the proof. The process ξ_t is restricted to the set T_0 of dyadic rational points in T. Then it is proved that almost all functions $\xi(t,\omega)$ are uniformly continuous on T_0. The space of uniformly continuous functions on T_0 is canonically isometrically isomorphic to $C([a,b])$, and this isomorphism gives rise to a measure on $C([a,b])$, which determines the required modification. □

Let $\mathcal{F}_1$ be a σ-subalgebra of the algebra $\mathcal{F}$. Let ξ be a random variable with $\mathrm{M}\|\xi\| < \infty$. In general, it is not $\mathcal{F}_1$-measurable. The Radon–Nikodým theorem implies the existence of an $\mathcal{F}_1$-measurable random variable η for which

$$\int_\Omega \eta(\omega)\phi(\omega)P(d\omega) = \int_\Omega \xi(\omega)\phi(\omega)P(d\omega),$$

where ϕ is an arbitrary bounded $\mathcal{F}_1$-measurable scalar function. This random variable η is called the *conditional mean* of ξ with respect to $\mathcal{F}_1$, and is denoted by $\eta = \mathrm{M}\{\xi/\mathcal{F}_1\} = \mathrm{M}_{\mathcal{F}_1}\xi$. If $\mathcal{F}_1$ is the minimal σ-algebra generated by a family of random variables ξ_α $(\alpha \in A)$, then η is called the *conditional mean* with respect to this family, and is denoted by $\eta = \mathrm{M}\{\xi/\xi_\alpha,\ \alpha \in A\}$. It is a measurable function of ξ_α $(\alpha \in A)$, and its value at $\xi_\alpha = x_\alpha$ is denoted by $\mathrm{M}\{\xi/\xi_\alpha = x_\alpha,\ \alpha \in A\}$. Let

$$j_S(\omega) = \begin{cases} 1, & \omega \in S, \\ 0, & \omega \notin S, \end{cases}$$

be the indicator function of the event S. Its conditional mean is denoted by $\mathrm{M}\{j_S/\mathcal{F}_1\} = P(S/\mathcal{F}_1)$, and is called the *conditional probability* of S with respect to $\mathcal{F}_1$.

Conditional means have the following properties, as can be easily verified:

1) $\mathrm{M}\{\mathrm{M}\{\xi/\mathcal{F}_1\}\} = \mathrm{M}\xi$;
2) $\mathrm{M}\{\xi\eta/\mathcal{F}_1\} = \xi\mathrm{M}\{\eta/\mathcal{F}_1\}$ if ξ is $\mathcal{F}_1$-measurable; in particular, $\mathrm{M}\{\xi/\mathcal{F}_1\} = \xi$;
3) $\mathrm{M}\{\xi/\mathcal{F}_1\} = \mathrm{M}\xi$ if ξ is independent of $\mathcal{F}_1$;
4) $\mathrm{M}\{\mathrm{M}\{\xi/\mathcal{F}_1/\mathcal{F}_2\}\} = \mathrm{M}\{\mathrm{M}\{\xi/\mathcal{F}_2\}/\mathcal{F}_1\} = \mathrm{M}\{\xi/\mathcal{F}_1\}$ for $\mathcal{F}_1 \subset \mathcal{F}_2$;
5) let ξ be measurable with respect to $\mathcal{F}_1$, and let η be independent of $\mathcal{F}_1$. Then $\mathrm{M}\{f(\xi,\eta)/\mathcal{F}_1\} = \mathrm{M}\{f(x,\eta)\}|_{x=\xi}$.

A stochastic process ξ_t $(a \le t \le b)$ with values in a space X is called a *Markov stochastic process* if for any $s < t_1 < t_2 < \cdots < t_n$ the conditional distribution of the random variable $(\xi_{t_1}, \dots, \xi_{t_n})$ with respect to the family of random variables ξ_τ $(\tau \le s)$ coincides with its conditional distribution with respect to ξ_s:

$$P\{\xi_{t_j} \in S_j\ (j = 1, \dots, n)/\xi_\tau,\ \tau \le s\} = P\{\xi_{t_j} \in S_j\ (j = 1, \dots, n)/\xi_s\}.$$

The quantity

$$P(s, x, t, S) = P\{\zeta_t \in S_j/\xi_s = x\}$$

is called the *transition probability* of the process ξ_t. For each pair (s,t) the transition probability is a measurable function with respect to x and a measure with respect to S. From properties of the conditional means and the definition of Markov process, the Chapman–Kolmogorov equation follows:

$$P(s, x, t, S) = \int_X P(s, x, \tau, dy)P(\tau, y, t, S) \quad (s \le \tau \le t).$$

The finite-dimensional distributions of a stochastic process are easily recovered (see chapter VI) from the transition probabilities and initial distribution μ_{ξ_0}.

1.2. Stochastic integrals. First we construct a stochastic process in a Hilbert space X, called the Wiener process, which is important in the sequel. Let B be a nuclear operator on X and μ_{Bt} the Gaussian measure in X with zero mean and correlation operator tB ($t > 0$). We define a *Wiener process* $w(t)$ ($t_0 \le t < \infty$) with values in X as a stochastic process for which:

1) for $t_0 \le t_1 < t_2 \le t_3 < t_4$ the differences $w(t_4) - w(t_3)$ and $w(t_2) - w(t_1)$ are independent;
2) the random variable $w(t+\tau) - w(t)$ has a distribution $\mu_{B\tau}$ (for the sake of being specific we set $w(t_0) = 0$).

It is readily seen that $w(t)$ is a Markov process with transition probability

$$P(t, x, \tau, S) = \mu_{B(\tau - t)}(S - x) = \mu_{B(\tau-t),x}(S).$$

From the formulas in chapter II it follows that

$$\mathrm{M}\|w(t+\tau) - w(t)\|^2 = \tau \operatorname{Tr} B,$$

$$\mathrm{M}\|w(t+\tau) - w(t)\|^4 = \tau^2[(\operatorname{Tr} B)^2 + 2 \operatorname{Tr} B^2]. \tag{1.1}$$

By theorem 1.1, the last equality implies the existence of a Wiener process $w(t)$ almost all realizations of which are continuous functions (with respect to t) with values in X. The space X can be realized as $X = \mathcal{H}_- \supset \mathcal{H}$, where the imbedding $j\colon \mathcal{H} \to \mathcal{H}_-$ is a Hilbert–Schmidt operator, and the Gaussian distribution of the difference $w(t+\tau) - w(t)$ has correlation operator $I\tau$ in $\mathcal{H}$ and is supported in $\mathcal{H}_-$. Such a Wiener process, with almost all trajectories continuous functions in $\mathcal{H}_-$, is called the *standard Wiener process* in $\mathcal{H}_-$. Everywhere below, without any exception, this process will be termed *Wiener process.*

Let $f \in \mathcal{H}$ and $\|f\| = 1$. Since (f, x) is a measurable functional on $\mathcal{H}_-$ (see chapter II, section 3), the scalar stochastic process $w_f(t) = (f, w(t))$ is redaily seen to be a Wiener process in $\mathbf{R}$. Here, for $f, g \in \mathcal{H}$, and $s > t > t_0$, $w(t_0) = 0$:

$$\begin{gathered} \mathrm{M} w_f(t) w_g(s) = \\ \mathrm{M}(f, w(t))(g, w(t)) + \mathrm{M}(f, w(t))(g, w(s) - w(t)) = \\ (f,g)(t - t_0) = (f,g)\min(t - t_0, s - t_0). \end{gathered} \tag{1.2}$$

In particular, for an orthonormal set $\{e_i\} \in \mathcal{H}$ the processes $w_k(t) = w_{e_k}(t)$ are pairwise uncorrelated and (since they are Gaussian) independent in total. If this system is complete in $\mathcal{H}$, then

$$w(t) = \sum_{k=1}^{\infty} e_k w_k(t), \tag{1.3}$$

where the expansion is mean square convergent in the norm of $\mathcal{H}_-$.

We now introduce a *flow of σ-algebras* which are consistent with the Wiener process $w(t)$, i.e. a monotone set of σ-algebras $\mathcal{F}_t$ ($t \ge 0$) such that for $s \le t$ the random variable $w(s)$ is $\mathcal{F}_t$-measurable, and for $\tau > s \geq t$, $w(\tau) - w(s)$ is independent of $\mathcal{F}_t$. For brevity we put $\mathrm{M}_t\xi = \mathrm{M}\{\xi/\mathcal{F}_t\}$.

In the sequel we will often meet the means of functions $f(\xi, w(t) - w(s))$, where ξ is $\mathcal{F}_s$-measurable. The formulas for conditional means given above imply that these can be computed by the formula ($s_1 \leq s$):

$$\mathrm{M}_{s_1} f(\xi, w(t) - w(s)) = \mathrm{M}_{s_1}\mathrm{M}_s f(\xi, w(t) - w(s)) =$$
$$\mathrm{M}_{s_1} \int_{\mathcal{H}_-} f(\xi, y)\mu_{I(t-s)}(dy) = \int_{\mathcal{H}_-} \mathrm{M}_{s_1} f(\xi, z\sqrt{t-s})\mu(dz), \tag{1.4}$$

where μ is a canonical Gaussian measure in $\mathcal{H}_-$.

PROPOSITION 1.1. *Let $w_1(t)$ be a standard Wiener process in the Hilbert space $\mathcal{H}$, and let $\Psi(\omega)$, $\Phi(\omega)$ be $\mathcal{F}_{t_1}$-measurable random variables with values in $\mathcal{L}_2(\mathcal{H}, X)$ and $\mathcal{L}_2(X \times X, X_1)$, respectively, where X and X_1 are Hilbert spaces, $M\sigma_2^2(\Psi) < \infty$, and* $\sup \|\Phi(\omega)\| < \infty$.

Then the random variable $\Psi[w(t_2) - w(t_1)]$, with values in X, is defined and satisfies

$$M_{t_1}\{\Psi[w(t_2) - w(t_1)]\} = 0, \tag{1.5}$$

$$M_{t_1}\|\Psi[w(t_2) - w(t_1)]\|^2 = \sigma_2^2(\Psi)(t_2 - t_1), \tag{1.6}$$

$$M_{t_1}\Phi\{\Psi[w(t_2) - w(t_1)], \Psi[w(t_2) - w(t_1)]\} = (t_2 - t_1)\operatorname{Tr}\Psi^*\Phi\Psi, \tag{1.7}$$

$$M_{t_1}\|\Phi\{\Psi[w(t_2) - w(t_1)], \Psi[w(t_2) - w(t_1)]\}\|^2 =$$
$$(2\sigma_2^2(\Psi^*\Phi\Psi) + (\operatorname{Tr}\Psi^*\Phi\Psi)^2)(t_2 - t_1)^2. \tag{1.8}$$

PROOF. We use the expansion (1.3) and consider

$$\alpha_{n,p} = \Psi \sum_{k=n}^{n+p} e_k[w_k(t_2) - w_k(t_1)] = \sum_k [w_k(t_2) - w_k(t_1)]\Psi e_k. \tag{1.9}$$

Using properties of the conditional mean we obtain

$$\mathrm{M}\{\alpha_{n,p}\|^2 =$$
$$\sum_{j,k=n}^{n+p} \mathrm{M}\|(\Psi e_k, \Psi e_j)\mathrm{M}_{t_1}[w_k(t_2) - w_k(t_1)][w_j(t_2) - w_j(t_1)]\} =$$
$$\sum_{k=n}^{n+p} \mathrm{M}\|\Psi e_k\|^2(t_2 - t_1). \tag{1.10}$$

This formula implies mean square convergence in X of the expansion (1.5), for $\alpha_{1,\infty} = \Psi[w(t_2) - w(t_1)]$. l Formula (1.6) follows from (1.10), if M_{t_1} is taken instead of M. Formulas (1.7), (1.8) follow, by (1.4), from those of chapter II, section 2 for integrals with respect to a Gaussian measure. □

Let X be a Hilbert space. Consider the class $\mathcal{K}(X)$ of stochastic processes $B(t, \omega)$ ($t \in [t_0, \tau]$) with values in the space $\mathcal{L}_2(\mathcal{H}, X)$ of Hilbert–Schmidt operators and satisfying

1) $\beta^2(B) = \int_{t_0}^{\tau} \mathrm{M}\sigma_2^2(B(t))\, dt < \infty$;

2) for every $t \geq t_0$ the quantity $B(t)$ is $\mathcal{F}_t$-measurable.

Let $\mathcal{K}_0(X)$ be the subset of $\mathcal{K}(X)$ consisting of the step functions

$$B(t,\omega) = B_j(\omega) \quad (t_j \leq t < t_{j+1}, \quad j = 0, \dots, n-1),$$

where $t_0 < t < \cdots < t_n = \tau$ is a partition of $[t_0, \tau]$.

For $\mathcal{K}_0(X)$ we define the *Itô stochastic integral* by

$$\mathcal{I}(B) = \int_{t_0}^{\tau} B(t)\,dw(t) = \sum_{j=0}^{n-1} B_j[w(t_{j+1}) - w(t_j)]. \tag{1.11}$$

Proposition 1.1 and the independence of $w(t_{j+1}) - w(t_j)$ on $\mathcal{F}_{t_j}$ imply that this expression makes sense, is a random variable with values in X, and has the properties

$$\mathrm{M}_{t_0}\mathcal{I}(B) = 0,$$

(1.12)

$$|||\mathcal{I}(B)|||^2 = \mathrm{M}_{t_0}\|\mathcal{I}(B)\|_X^2 = \sum_{j=0}^{n-1} \mathrm{M}_{t_0}\sigma_2^2(B_j)(t_{j+1} - t_j) = \int_{t_0}^{\tau} \mathrm{M}_{t_0}\sigma_2^2(B(t))\,dt.$$

Formulas (1.11)–(1.12) show that $\mathcal{I}(B)$ is a linear operator on $\mathcal{K}_0(X)$, with values in the space of random variables $\mathfrak{H} = L_2(\Omega, X)$, which is continuous (even isometric, in the norms $\beta(B)$ and $|||\mathcal{I}(B)|||$). This operator can be extended, in a standard manner, to the class $\mathcal{K}(X)$, in which $\mathcal{K}_0(X)$ is dense in the norm β.

The operator $\mathcal{I}$ thus obtained acts from $\mathcal{K}(X)$ to $\mathfrak{H}$, is also called a *stochastic integral*, and is denoted by

$$\mathcal{I}(B) = \int_t^{\tau} B(t)\,dw(t).$$

It follows from the construction that it has (with probability 1) the properties:

a) $\mathcal{I}(\alpha_1 B_1 + \alpha_2 B_2) = \alpha_1\mathcal{I}(B_1) + \alpha_2\mathcal{I}(B_2)$ for $\mathcal{F}_{t_0}$-measurable and bounded $\alpha_1(\omega)$, $\alpha_2(\omega)$;

b) $\mathrm{M}_{t_0}\mathcal{I}(B) = 0$,

$$\mathrm{M}_{t_0}\|\mathcal{I}(B)\|^2 = \int_{t_0}^{\tau} \sigma_2^2(B(t))\,dt; \tag{1.13}$$

c)

$$\int_{t_0}^{\tau} B(t)\,dw(t) = \sum_{k=1}^{\infty} \int_{t_0}^{\tau} B(t)e_k\,dw_k(t), \tag{1.14}$$

where $w(t) = \sum_{k=1}^{\infty} e_k w_k(t)$ is the expansion (1.3), and (1.14) mean square converges in X.

PROPOSITION 1.2. *Let $B(t,\omega)$ be a random variable with values in $\mathcal{L}(X)$ and suppose that $\sup\|B(t,\omega)\| \leq C$, $A \in \mathcal{L}_2(\mathcal{H}, X)$. Consider a partition $t_0 < t_1 < \cdots <$*

$t_n = \tau$, and let $\Delta t = t_{r+1} - t_r = (\tau - t_0)/n$ and $\Delta_r w = w(t_{r+1}) - w(t_r)$. Then (in the sense of convergence in probability):

$$\lim_{n\to\infty} \sum_{r=0}^{n-1} (B(t_r, \omega) A\Delta_r w, A\Delta_r w) = \int_{t_0}^{\tau} \operatorname{Tr} A^* B(t,\omega) A \, dt.$$

PROOF. Put $b_n = \sum_{r=0}^{n-1} (B(t_r, \omega) A\Delta_r w, A\Delta_r w)$. Then

$$\mathrm{M} b_n = \mathrm{M} \sum_{r=0}^{n-1} \mathrm{M}_{t_r} (B(t_r, \omega) A\Delta_r w, A\Delta_r w) = \mathrm{M} \sum_{r=0}^{n-1} \Delta t \operatorname{Tr} A^* B(t_r) A.$$

So,

$$\lim \mathrm{M} \left(b_n - \int_{t_0}^{\tau} \operatorname{Tr} A^* B(t) A \, dt \right) = 0.$$

It remains to show that

$$\lim_{n\to\infty} \mathrm{M} \left(b_n - \int_{t_0}^{\tau} \operatorname{Tr} A^* B(t) A \, dt \right)^2 = 0, \tag{1.15}$$

since then the proposition follows from Chebyshev's inequality

$$P\left\{ \left| b_n - \int_{t_0}^{\tau} \operatorname{Tr} A^* B(t) A \, dt \right| > \epsilon \right\} \le \frac{1}{\epsilon^2} \mathrm{M} \left[b_n - \int_{t_0}^{\tau} \operatorname{Tr} A^* B(t) A \, dt \right]^2.$$

The relations (1.15) are derived by straightforward computation, using (1.7)–(1.8). □

1.3. Itô's formula. A stochastic integral with variable upper limit is defined by

$$\int_{t_0}^{s} B(t,\omega)\, dw(t,\omega) = \int_{t_0}^{\tau} j_s(t) B(t,\omega)\, dw(t,\omega),$$

where $j_s(t)$ is the indicator function of the interval $[t_0, s]$. Let $a(t,\omega)$ be an $\mathcal{F}_t$-measurable function with values in X such that

$$\int_{t_0}^{\tau} \mathrm{M} \|a(t,\omega)\|^2 \, dt < \infty,$$

and let $\xi_0 \in X$ be an $\mathcal{F}_{t_0}$-measurable random variable. A stochastic process of the type

$$\xi(t,\omega) = \xi_0(\omega) + \int_{t_0}^{t} a(s,\omega)\, ds + \int_{t_0}^{t} B(s,\omega)\, dw(s,\omega) \tag{1.16}$$

is said to have a *stochastic differential*; this is written as

$$d\xi = a(t,\omega)\, dt + B(t,\omega)\, dw(t,\omega).$$

Consider a stochastic process $\eta(t) = f(t, \xi(t))$, where $y = f(t,x)$ $(t \in [t_0, \tau],\ x \in X)$ is a function with values in a Hilbert space Y. The following theorem establishes Itô's formula for the stochastic differential of the stochastic process $\eta(t)$.

THEOREM 1.2. *Suppose that the derivatives* $f_t/(t,x)$, $f'_x(t,x)$, *and* $f''_{xx}(t,x)$ *are bounded and continuous. Then*

$$\eta(t) = f(t_0, \xi_0) +$$

$$\int_{t_0}^{t} \left\{ f'_s(s,\xi(s)) + f'_x(s,\xi(s))a(s) + \frac{1}{2}\operatorname{Tr} B^*(s) f''_{xx}(s,\xi(s))B(s) \right\} ds +$$

$$\int_{t_0}^{t} f'_x(s,\xi(s))B(s)\, dw(s). \tag{1.17}$$

PROOF. The lefthand and righthand sides of (1.17) are linear with respect to f and continuous with respect to ξ in the sense of mean square convergence. On this basis we can make a reduction to a simpler case. It suffices to consider random step functions $a(t,\omega)$, $B(t,\omega)$, and a sufficiently small interval $[t_0,t)$ on which these functions are constant, $B(t,\omega) = B(\omega)$, $a(t,\omega) = a(\omega)$, and $\mathcal{F}_t$-measurable. Further it suffices to take for $f(t,x)$ a scalar function, since applying an arbitrary linear functional reduces the general case to this one. Finally, by means of approximating $f(t,x)$ by functions of the form $\sum \phi_j(t)\psi_j(x)$, the problem reduces to the consideration of functions $f(x)$ which are independent of t.

So, let $\xi(s,\omega) = \xi_0 + a(\omega)(s-t_0) + B[w(s) - w(t_0)]$ $(t_0 \le s < t)$ and $\eta(s) = f(\xi(s))$. Consider a partition $t_0 < s_1 < \cdots < s_n = t$, and let $\xi_r = \xi(s_r)$, $\Delta s_r = s_{r+1} - s_r$, $\Delta_r w = w(s_{r+1}) - w(s_r)$, $\Delta_r \xi = a\Delta s_r + B\Delta_r w$.

Using Taylor's formula we obtain the representation

$$\eta(t) - \eta(t_0) = \sum_{r=0}^{n-1} \{ f(\xi_{r+1}) - f(\xi_r) \} = \alpha_1 + \alpha_2 + \alpha_3,$$

where

$$\alpha_1 = \sum_{r=0}^{n-1} (f'(\xi_r), \Delta_r \xi), \quad \alpha_2 = \frac{1}{2}\sum_{r=0}^{n-1} (f''(\xi_r)\Delta_r\xi, \Delta_r\xi),$$

$$\alpha_3 = \sum_{r=0}^{n-1} \int_0^1 (1-\theta)([f''(\xi_r + \theta\Delta\xi_r) - f''(\xi_r)]\Delta_r\xi, \Delta_r\xi)\, d\theta.$$

It is easily verified that under the present conditions

$$\lim_{n\to\infty} \alpha_1 = \int_{t_0}^{t_1} f'(\xi(s))\, d\xi(s) = \int_{t_0}^{t_1} f'(\xi(s))a(ds) + \int_{t_0}^{t} f'(\xi(s))B\, dw(s),$$

In the same way we obtain from proposition 1.2

$$\lim_{n\to\infty} \alpha_2 = \int_{t_0}^{t_1} \operatorname{Tr} B^* f''(\xi(s))B\, ds.$$

To finish the proof it suffices to show that $\lim_{n\to\infty} \alpha_3 = 0$. Using (1.4) we find

$$\mathrm{M}_{t_0}|\alpha_3| \le$$

$$\sum_{r=0}^{n-1} \int_0^1 (1-\theta) \int_{\mathcal{H}_-} \mathrm{M}_{t_0}|([f''(\xi_r + \theta a\Delta s_r + \theta Bz\sqrt{\Delta s_r}) - f''(\xi_r)] \times$$

$$(a\Delta s_r + Bz\sqrt{\Delta s_r}), a\Delta s_r + Bz\sqrt{\Delta s_r})|\mu_I(dz)\, d\theta \le$$

$$C\sum_{r=0}^{n-1} \Delta s_r \int_0^1 (1-\theta)\, d\theta \int_{\mathcal{H}_-} (1 + \|Bz\|^2)\mathrm{M}_{t_0}\|f''(\xi_0 + a(s_r - t_0) + B(w(s_r) - w(t_0)) +$$

$$\theta a\Delta s_r + \theta Bz\sqrt{\Delta s_r}) - f''(\xi_0 + a(s_r - t_0) + B(w(s_r) - w(t_0)))\|\mu_I(dz) =$$

$$C\sum_{r=0}^{n-1} \Delta s_r \int_0^1 (1-\theta)\, d\theta \int_{\mathcal{H}_-}\int_{\mathcal{H}_-} (1 + \|Bz\|^2)\|f''(\xi_0 + a(s_r - t_0) +$$

$$Bu\sqrt{s_r - t_0} + \theta a\Delta s_r + \theta Bz\sqrt{\Delta s_r}) - f''(\xi_0 + a(s_r - t_0) +$$

$$Bu\sqrt{s_r - t_0})\|\mu_I(dz)\mu_I(du),$$

where $C = C(\omega)$ is $\mathcal{F}_{t_0}$-measurable.

To obtain the required result it now suffices to show that, as $\max|\Delta s_r| \to 0$, the double integral over $\mathcal{H}_- \times \mathcal{H}_-$ at the righthand side tends to zero uniformly with respect to $s_k \in [t_o, t]$. This is done in a standard manner by selecting a compact set in $\mathcal{H}_- \times \mathcal{H}_-$ on which the measure $(1 + \|Bz\|^2)\mu_I(dz)\mu_I(du)$ is supported up to a given, small, $\epsilon > 0$, and by using the uniform continuity on compact sets of the function f''. □

Using Itô's formula it is very simple to estimate moments of stochastic integrals.

PROPOSITION 1.3. *Let $B(t) \in \mathcal{H}(X)$ and*

$$\int_{t_0}^{\tau} M\sigma_2^{2m}(B(t))\, dt < \infty.$$

Then the following estimate holds:

$$M\left\|\int_{t_0}^{\tau} B(t)\, dw(t)\right\|^{2m} \le [m(2m-1)]^m(\tau - t_0)^{m-1}\int_{t_0}^{\tau} M\sigma_2^{2m}(B(t))\, dt. \tag{1.18}$$

PROOF. Apply Itô's formula to the process $\xi(t) = \int_{t_0}^t B(s)\, dw(s)$ and function $f(x) = \|x\|^{2m}$. It is not difficult to estimate that

$$(f''(x)h, h) = 2m\|x\|^{2(m-1)}\|h\|^2 + 4m(m-1)\|x\|^{2(m-2)}(x,h)^2 \le$$

$$2m(2m-1)\|x\|^{2(m-1)}\|h\|^2, \tag{1.19}$$

implying that $\mathrm{Tr}\, B^* f'' B = \sum_{k=1}^{\infty}(f''(x)Be_k, Be_k) \le 2m(2m-1)\|x\|^{2(m-1)}\sigma_2^2(B)$. Itô's formula implies that

$$\mathrm{M}\|\xi(t)\|^{2m} =$$
$$\mathrm{M}\left\{\frac{1}{2}\int_{t_0}^{t} \operatorname{Tr} B^*(s)f''(\xi(s))B(s)\,ds + \int_{t_0}^{t}(B^*(s)f'(\xi(s)), dw(s))\right\} =$$
$$\frac{1}{2}\int_{t_0}^{t} \mathrm{M}\operatorname{Tr} B^*(s)f''(\xi(s))B(s)\,ds.$$

Formula (1.19) implies that f'' is a positive operator, and so $\mathrm{M}\|\xi(t)\|^{2m}$ is monotone increasing with respect to t.

Applying Hölder's inequality, we obtain

$$\mathrm{M}\|\xi(t)\|^{2m} =$$
$$m(2m-1)\int_{t_0}^{\tau} \mathrm{M}\|\xi(t)\|^{2(m-1)}\sigma_2^2(B(t))\,dt \leq$$
$$m(2m-1)\left\{\int_{t_0}^{\tau} \mathrm{M}\|\xi(t)\|^{2m}\,dt\right\}^{(m-1)/m}\left\{\int_{t_0}^{\tau} \mathrm{M}\sigma_2^{2m}(B(t))\,dt\right\}^{1/m} \leq$$
$$m(2m-1)(\tau-t_o)^{(m-1)/m}\{\mathrm{M}\|\xi(t)\|^{2m}\}^{(m-1)/m}\left\{\int_{t_0}^{\tau} \mathrm{M}\sigma_2^{2m}(B(t))\,dt\right\}^{1/m},$$

which yields (1.18) by division and raising to a power. □

COROLLARY 1.1. *For $m = 2$ the estimate (1.18) has the form*

$$M\left\|\int_{t_0}^{\tau} B(t)dw(t)\right\|^4 \leq 36(\tau-t_0)\int_{t_0}^{\tau} M\sigma_2^4(B(t))\,dt.$$

This and theorem 1.1 imply continuity of the stochastic integral $\int_{t_0}^{t} B(t)\,dw(t)$ with respect to the upper limit, provided $\mathrm{M}\sigma_2^4(B(t)) < C < \infty$.

2. Stochastic differential equations

2.1. Equations with bounded operators. Let $\mathcal{H}$ and X be Hilbert spaces, and let $a(t,x)$ and $B(t,x)$ be measurable functions on $[t_0,\tau]\times X$ with values in X and $\mathcal{L}_2(\mathcal{H},X)$, respectively.

An equation of the form

$$d\xi(t) = a(t,\xi(t))\,dt + B(t,\xi(t))\,dw(t) \quad (t_0 \leq t \leq \tau), \tag{2.1}$$

where $w(t)$ is a standard Wiener process in $\mathcal{H}$, is called a *stochastic differential equation.*

A stochastic process $\xi(t,\omega)$ subject to a flow of σ-algebras $\mathcal{F}_t$ consistent with $w(t)$ (i.e. $\xi(t,\omega)$ is $\mathcal{F}_t$-measurable) and having on $[t_0,\tau]$ a stochastic differential (2.1), is called a solution of this equation.

Thus, by definition, the solution $\xi(t)$ of (2.1) must satisfy the relation

$$\xi(t)-\xi(s) = \int_{s}^{t} a(\theta,\xi(\theta))\,d\theta + \int_{s}^{t} B(\theta,\xi(\theta))\,dw(\theta) \quad (t_0 \leq s \leq t \leq \tau); \tag{2.2}$$

in particular, the integrals occurring in this relation must make sense. Note that for $B = 0$ (2.1) becomes the ordinary differential equation

$$\frac{d\xi(t)}{dt} = a(t, \xi(t)),$$

whose solution can be random only if the initial value $\xi_0(t)$ is.

We introduce the Banach space S_p of random functions $\xi(t, \omega)$ ($t \in [t_0, \tau]$, $\omega \in \Omega$) with values in X and consistent with the flow $\mathcal{F}_t$, and the norm

$$\|\xi\|^p = \sup_{[t_0,\tau]} \mathrm{M}\|\xi(t)\|^p.$$

We say that a function $f(t, x)$ ($t \in [t_0, \tau]$, $x \in X$) with values in a Banach space Y satisfies *condition* $R(Y)$, denoted by $f \in R(Y)$, if the following inequalities hold ($t \in [t_0, \tau]$, $x_1, x_2 \in X$):

$$\begin{aligned} &\|f(t, x_1)\|_Y \leq C_1 + C_2\|x_1\|_X, \\ &\|f(t, x_1) - f(t, x_2)\|_Y \leq C_2\|x_2 - x_1\|_X, \end{aligned} \tag{2.3}$$

where C_1, C_2 are constants.

PROPOSITION 2.1. *Let $a \in R(X)$, $B \in R(\mathcal{L}_2(\mathcal{H}, X))$. Then the formula*

$$\eta(t) = G(\xi)(t) = \int_{t_0}^{t} a(\theta, \xi(\theta))\, d\theta + \int_{t_0}^{t} B(\theta, \xi(\theta))\, dw(\theta) \tag{2.4}$$

defines a continuous map in each space S_{2m}.

PROOF. As is clear from (2.4), $\xi(t)$ is subject to the flow $\mathcal{F}_t$, as is $\eta(t)$. From (2.4) and (2.3), using (1.18) we can easily find the estimates

$$\mathrm{M}\|\eta(t)\|^{2m} \leq C\left\{1 + \int_{t_0}^{t} \mathrm{M}\|\xi(\theta)\|^{2m}\, d\theta\right\} \tag{2.5}$$

and

$$\mathrm{M}\|\eta_2(t) - \eta_1(t)\|^{2m} \leq C\int_{t_0}^{t} \mathrm{M}\|\xi_2(\theta) - \xi_1(\theta)\|^{2m}\, d\theta, \tag{2.6}$$

where the constant C depends on m, $\tau - t_0$ and the constants entering in condition R for the coefficients a and B. These estimates prove the proposition. □

THEOREM 2.1. *Let $a \in R(X)$, $B \in R(\mathcal{L}_2(\mathcal{H}, X))$. Then the Cauchy problem for equation (2.1) with initial condition $\xi_0 \in S_{2m}$ has a solution $\xi(t)$ in S_{2m}. This solution is unique up to stochastic equivalence. With probability one it has a continuous modification.*

The stochastic process $\xi(t)$ is Markovian, and its transition probabilities are determined by the formula $P(t, x, s, \Delta) = P\{\xi_{t,x}(s) \in \Delta\}$, where $\xi_{t,x}(s)$ is the solution of

$$\xi_{x,t}(s) = x + \int_{t}^{s} a(\theta, \xi_{t,x}(\theta))\, d\theta + \int_{t}^{s} B(\theta, \xi_{t,x}(\theta))\, dw(\theta). \tag{2.7}$$

PROOF. The Cauchy problem under consideration is equivalent to the equation

$$\xi(t) = \xi_0 + G(\xi)(t) \equiv \tilde{G}(\xi)(t). \tag{2.8}$$

By proposition 2.1, the map $\tilde{G}$ acts in S_{2m} and satisfies the estimate (which follows from (2.6))

$$\mathrm{M}\|\tilde{G}(\xi_2)(t) - \tilde{G}(\xi_1)(t)\|^{2m} \le C \int_{t_0}^{t} \mathrm{M}\|\xi_2(\theta) - \xi_1(\theta)\|^{2m}\, d\theta.$$

Iterating this estimate gives

$$\mathrm{M}\|\tilde{G}^2(\xi_2)(t) - \tilde{G}^2(\xi_1)(t)\|^{2m} \le C \int_{t_0}^{t} \mathrm{M}\|\tilde{G}(\xi_2)(\theta) - \tilde{G}(\xi_1)(\theta)\|^{2m}\, d\theta \le$$
$$C^2 \int_{t_0}^{t} d\theta \int_{t_0}^{\theta} \mathrm{M}\|\xi_2(\theta_1) - \xi_1(\theta_1)\|^{2m}\, d\theta_1 \le \frac{C^2(\tau - t_0)^2}{2!}\|\xi_2 - \xi_1\|^{2m},$$

and by induction

$$\mathrm{M}\|\tilde{G}^r(\xi_2)(t) - \tilde{G}^r(\xi_1)(t)\|^{2m} \le \frac{C^r(\tau - t_0)^r}{r!}\|\xi_2 - \xi_1\|^{2m}.$$

Since for sufficiently large r we have $C^r(\tau - t_0)^r/r! < 1$, the corresponding map $\tilde{G}^r$ is contracting. This implies the existence of a fixed point of $\tilde{G}$, i.e. the unique solution of (2.8) (if processes ξ_1 and ξ_2 with $\|\xi_1 - \xi_2\| = 0$, i.e. stochastically equivalent processes, are identified).

The existence of a continuous modification follows from the estimate

$$\mathrm{M}\|\xi(t) - \xi(s)\|^{2m} \le C_1(t-s)^m(1 + \|\xi\|^{2m}), \tag{2.9}$$

obtained from (2.2) by means of (1.18).

To prove the final assertion we note that $\xi(s)$ and $\xi_{t,\xi(t)}(s)$ coincide almost surely, being solutions of the same stochastic equation. By construction, $\xi_{t,x}(s)$ is independent of $\mathcal{F}_t$ and $\xi(t)$. So it follows from properties of the conditional mean that for any $\mathcal{F}_t$-measurable random variable ψ and bounded measurable function f,

$$\mathrm{M}\psi f(\xi(s)) = \mathrm{M}\mathrm{M}_t\psi f(\xi_{t,\xi(t)}(s)) = \mathrm{M}\psi\{\mathrm{M}_t f(\xi_{t,x}(s))\}|_{x=\xi(t)}.$$

So, $\mathrm{M}_t f(\xi(s))$ depends on $\xi(t)$ only, and coincides with $\mathrm{M}f(\xi_{t,x}(s))$ for $\xi(t) = x$. □

REMARK 2.1. The proof of the existence and uniqueness of the solution can be extended without change to more general integral equations of the form

$$\xi(t) = \xi_0(t) + \int_{t_0}^{t} a(t,s,\xi(s))\, ds + \int_{t_0}^{t} B(t,s,\xi(s)\, dw(s), \tag{2.10}$$

where $\xi_0 \in S_{2m}$, and $a(t,s,x)$ and $B(t,s,x)$ satisfy condition R for $t,s \in [t_0,\tau]$, $x \in X$.

We will now derive some a priori estimates for the solutions of a stochastic equation (2.1). These estimates will be used in the sequel.

PROPOSITION 2.2. *Suppose*

$$\sigma_2^2(B(t,x)) \leq C_1 + C_2\|x\|^2, \quad (x, a(t,x)) \leq C_1 + \beta\|x\|^2$$
$$(t \in [t_0, \tau], \quad x \in X).$$

Then for a solution of (2.1) there are the following estimates:

(2.11)
$$M_{t_0}\|\xi(t)\|^2 \leq e^{(2\beta + C_2)(t-t_0)}\left\{\|\xi_0\|^2 + \frac{3C_1}{2\beta + C_2}\right\} - \frac{3C_1}{2\beta + C_2},$$

(2.12)
$$M\|\xi(t)\|^2 \leq e^{(4\beta+5C_1+6C_2)(t-t_0)}\left\{\|\xi_0\|^4 + \frac{5C_1}{4\beta + 5C_1 + 6C_2}\right\} - \frac{5C_1}{4\beta + 5C_1 + 6C_2}.$$

PROOF. Computing by Itô's formula the stochastic differential of the random variable $\|\xi(t)\|^2$, we arrive at

$$\|\xi(t)\|^2 =$$
$$\|\xi_0\|^2 + \int_{t_0}^t \{2(a(\theta, \xi(\theta)), \xi(\theta)) + \sigma_2^2(B(\theta, xi(\theta)))\}\, d\theta + 2\int_{t_0}^t (\xi(\theta), B(\theta, \xi(\theta)))\, dw(\theta),$$

implying the estimate

$$\mathrm{M}_{t_0}\|\xi(t)\|^2 \leq \|\xi_0\|^2 + 3C_1(t - t_0) + \int_{t_0}^t (2\beta + C_2)\mathrm{M}_{t_0}\|\xi(\theta)\|^2\, d\theta.$$

This estimate leads to the result required by using Gronwall's inequality. The second estimate is obtained in the same way from the similar relation for $\|\xi(t)\|^4$. □

COROLLARY 2.1. *For a $\beta < 0$ with $|\beta|$ sufficiently large we have, on an arbitrary interval $[t_0, \tau]$,*

$$\sup M\|\xi(t)\|^2 < \infty.$$

PROPOSITION 2.3. *Suppose*

$$\sigma_2^2(B(t, x_2) - B(t, x_1)) \leq C_2\|x_2 - x_1\|^2,$$
$$(x_2 - x_1, a(t, x_2) - a(t, x_1)) \leq \beta\|x_2 - x_1\|^2.$$

Then we have for the difference of two solutions of (2.1),

(2.13)
$$M_{t_0}\|\xi_2(t) - \xi_1(t)\|^2 \leq e^{(2\beta+C_2)(t-t_0)}\|\xi_{20} - \xi_{10}\|^2,$$
$$M_{t_0}\|\xi_2(t) - \xi_1(t)\|^4 \leq e^{(4\beta+6C_2)(t-t_0)}\|\xi_{20} - \xi_{10}\|^4,$$

The proof of this proposition is similar to the proof of the previous proposition.

2.2. Equations with unbounded operator. In this section we consider the stochastic equation (2.1) under more general assumptions on its coefficients. Let

$$a(t,x) = a_0(t)x + a_1(t,x), \tag{2.14}$$

where $a_1(t,x) \in R(X)$ and $a_0(t)$ is the unbounded generating operator of a strongly continuous evolution family of linear operators $U(t,s)$ on X:

$$\frac{dU(t,s)x}{dt} = a_0(t)U(t,s)x, \quad U(s,s)x = x \quad (t > s).$$

For simplicity reasons we will regard the domain of definition $D = D_{a_0(t)}$ as being constant. It is of interest to consider two versions:

a) the 'parabolic' version, when

$$\begin{gathered} \|a_0^\alpha(t)U(t,s)a_0^{-\beta}(s)\| \le \frac{C}{(t-s)^{\alpha-\beta}}, \quad \alpha > \beta, \\ \|a_0^\alpha(t)a_0^{-\alpha}(s)\| \le C \quad (t_0 \le s \le t \le \tau); \end{gathered} \tag{2.15}$$

b) the 'hyperbolic' version, with

$$\|a_0(t)U(t,s)a_0^{-1}(s)\| \le C, \quad \|a_0(t)a_0^{-1}(s)\| \le C. \tag{2.15'}$$

Sufficient conditions for the evolution family $U(t,s)$ satisfying estimates of the type (2.15) or (2.15') to exist are well known (see Krein [1]). E.g., in case a) we can take $a_0(t)$ as being selfadjoint, semibounded above, and strongly differentiable in the domain of definition. Here, the fractional powers $a_0^\alpha(t)$ are defined as usual, as functions of the selfadjoint operator. In case b) it is convenient to consider an operator of the form $a_0(t) = ib(t)$, where $b(t)$ is strongly differentiable and selfadjoint. However, these conditions are too restrictive in a number of problems. To avoid unnecessary complications we will not require the fulfillment of other conditions, and we will only use (2.15) or (2.15') (see also chapter VI).

We will consider in some detail the first case, and we will briefly indicate the results obtained for the second case.

First of all, consider the linear inhomogeneous equation

$$d\xi(t) = a_0(t)\xi(t)\,dt + d\eta(t), \quad \xi(t_0) = \xi_0, \tag{2.16}$$

where $\eta(t)$ is a stochastic process with stochastic differential $d\eta = a_1(t)\,dt + B(t)\,dw(t)$. This equation has, for given ξ_0, at most one solution $\xi(t) \in D$ $(t > t_0)$, since the corresponding homogeneous equation has this property. (its unique solution is $\tilde{\xi}(t) = U(t,t_0)\xi_0$).

It follows from the general theory of evolution equations that the solution of (2.16) must have the form

$$\begin{gathered} \xi(t) = U(t,t_0)\xi_0 + \int_{t_0}^t U(t,s)\,d\eta(s) = \\ U(t,t_0)\xi_0 + \int_{t_0}^t U(t,s)a_1(s)\,ds + \int_{t_0}^t U(t,s)B(s)\,dw(s). \end{gathered} \tag{2.17}$$

Below we will give sufficient conditions for the process $\xi(t)$ in (2.17) to have stochastic differential satisfying (2.16).

PROPOSITION 2.4. *Let* $\sigma_2^2(a_0^\beta(s)B(s)) \leq C$, $\|a_0^{\beta_1}(s)a_1(s)\| \leq C$ $(\alpha < \beta + 1/2,\ \alpha < \beta_1 + 1)$. *Then* $\xi(t) \in D_{a_0^\alpha}$ *for* $t > t_0$.

PROOF. It suffices to verify that the integrals in the representation

$$a_0^\alpha(t)\xi(t) =$$
$$a_0^\alpha(t)U(t,t_0)\xi_0 + \int_{t_0}^t a_0^\alpha(t)U(t,s)a_1(s)\,ds + \int_{t_0}^t a_0^\alpha(t)U(t,s)B(s)\,dw(s)$$

exist. It follows from (2.15) that under the present conditions

$$\|a_0^\alpha(t)U(t,s)a_1(s)\| \leq \|a_0^\alpha(t)U(t,s)a_0^{-\beta_1}(s)\| \cdot \|a_0^{\beta_1}(s)a_1(s)\| \leq \frac{C}{(t-s)^{\alpha-\beta_1}};$$
$$\sigma_2^2(a_0^\alpha(t)U(t,s)B(s)) \leq \|a_0^\alpha(t)U(t,s)a_0^{-\beta}(s)\|^2\sigma_2^2(a_0^\beta((s)B(s)) \leq \frac{C}{(t-s)^{2(\alpha-\beta)}},$$

i.e. the integrands have integrable singularities at the point $s = t$. □

PROPOSITION 2.5. *Under the conditions of proposition 2.4 the expression*

$$\alpha(t,\Delta t) = \Delta_t\xi - \Delta_t\eta - [U(t+\Delta t,t) - I]\xi(t),$$

where $\Delta_t\xi = \xi(t+\Delta t) - \xi(t)$, $\Delta_t\eta = \eta(t+\Delta t) - \eta(t)$, *satisfies the estimates*

$$\|M_t\alpha(t,\Delta t)\| \leq C(\Delta t)^{1+\beta_1},$$
$$M\|\alpha(t,\Delta t) - M_t\alpha(t,\Delta t)\|^2 \leq C(\Delta t)^{1+\beta}.$$

PROOF. The representation

$$\xi(t+\Delta t) - U(t+\Delta t,t)\xi(t) + \int_t^{t+\Delta t} U(t+\Delta t,s)\,d\eta(s)$$

follows from (2.17). It yields

$$\alpha(t,\Delta t) = \int_t^{t+\Delta t} [U(t+\Delta t,s) - I]\{a_1(s)\,ds + B(s)\,dw(s)\}.$$

Also,

$$\|\mathrm{M}_t\alpha(t,\Delta t)\| = \left\|\int_t^{t+\Delta t} [U(t+\Delta t,s) - I]\mathrm{M}_t a_1(s)\,ds\right\| \leq C(\Delta t)^{1+\beta_1},$$

since (2.15) implies that $\|[U(t+\Delta t,s) - I]a_0^{-\beta_1}\| = O(\Delta t^{\beta_1})$. We then have

$$\alpha(t,\Delta t) - M_t\alpha(t,\Delta t) =$$
$$\int_t^{t+\Delta t} [U(t+\Delta t,s) - I][a_1(s) - \mathrm{M}_1 a_1(s)]\,ds + \int_t^{t+\Delta t} [U(t+\Delta t,s) - I]B(s)\,dw(s),$$

which implies

$$\begin{gathered}\mathrm{M}\|\alpha(t,\Delta t)-M_t\alpha(t,\Delta t)\|^2\le\\ 4C\Delta t\int_t^{t+\Delta t}\|[U(t+\Delta t,s)-I]a_0^{-\beta_1}(s)\|^2\,ds+\\ 2\int_t^{t+\Delta t}\sigma_2^2([U(t+\Delta t,s)-I]B(s)\,ds\le\\ C_1(\Delta t)^{2(1+\beta_1)}+C_2(\Delta t)^{1+\beta}\le C_3(\Delta t)^{1+\beta}.\quad\square\end{gathered}$$

THEOREM 2.2. *Under the conditions of proposition 2.4, for* $\alpha=1$ $(\beta>1/2,\ \beta_1>0)$ *and every* $t>t_0$ ($t\ge t_0$ *if* $\xi_0\in D_{a_0}$), *the stochastic process* ξ *in (2.17) has a stochastic differential and satisfies (2.16).*

PROOF. Consider the partition $t<t_1<\cdots<t_n=\theta$, $t_{k+1}-t_k=\Delta t_k=(\theta-t)/n$. We show that the quantity $\alpha=\sum_{k=0}^{n-1}\alpha(t_k,\Delta t_k)$ tends to zero in probability as $n\to\infty$. In fact,

$$\mathrm{M}\alpha=\mathrm{M}\sum_{k=0}^{n-1}\mathrm{M}_{t_k}\alpha(t_k,\Delta t_k)=O\left(\frac{1}{n^{\beta_1}}\right)$$

and

$$\begin{gathered}\mathrm{M}\left\|\alpha-\sum_{k=0}^{n-1}\mathrm{M}_{t_k}\alpha(t_k,\Delta t_k)\right\|^2=\\ \mathrm{M}\sum_{k=0}^{n-1}\|\alpha(t_k,\Delta t_k)-\mathrm{M}_{t_k}\alpha(t_k,\Delta t_k)\|^2+2\mathrm{M}\sum_{j<k}\mathrm{M}_{t_k}(\alpha(t_j,\Delta t_j)-\\ \mathrm{M}_{t_j}\alpha(t_j,\Delta t_j),\alpha(t_k,\Delta t_k)-\mathrm{M}_{t_k}\alpha(t_k,\Delta t_k))=\\ \mathrm{M}\sum_{k=0}^{n-1}C(\Delta t_k)^{1+\beta}=O\left(\frac{1}{n^{\beta}}\right).\end{gathered}$$

Therefore we have in the sense of convergence in probability,

$$\xi(\theta)-\xi(t)-[\eta(\theta)-\eta(t)]=\lim_{n\to\infty}\sum_{k=0}^{n-1}[U(t_{k+1},t_k)-I]\xi(t_k),$$

and it is readily verifies that under the present conditions $\xi(t_k)\in D_{a_0^{1+\epsilon}(t_k)}$ $(\epsilon>0)$, and the limit at the righthand side is $\int_t^\theta a_0(s)\xi(s)\,ds$. $\square$

We now turn to the consideration of the stochastic differential equation

$$d\xi(t)=a_0(t)\xi(t)\,dt+a_1(t,\xi(t))\,dt+B(t,\xi(t))\,dw(t),\quad \xi(t_0)=\xi_0. \tag{2.18}$$

A solution of it is a stochastic process $\xi(t)$ whose values belong to D for $t>t_0$ and which has a stochastic differential of the form (2.18).

We will find the solution of (2.18) as an expression of type (2.17):

$$\xi(t)=U(t,t_0)\xi_0+\int_{t_0}^t U(t,s)a_1(s,\xi(s))\,ds+\int_{t_0}^t U(t,s)B(s,\xi(s))\,dw(s). \tag{2.19}$$

By theorem 2.2, under the conditions

$$\sup \|a_0^{\beta_1}(s)a_1(s,x)\| < \infty \quad (\beta_1 > 0), \qquad (2.20)$$
$$\sup \sigma_2^2(a_0^{\beta}(s)B(s,x)) < \infty \quad (\beta > 1/2)$$

the solution (2.19) will also satisfy (2.18), provided it exists.

Equation (2.19) is a particular case of (2.10), and has a solution by remark 2.1 (which is moreover unique up to stochastic equivalence). This solution also satisfies (2.18), under the present conditions, and it is a Markov process, since in proving the Markov property (in theorem 2.1) we only need a special form of stochastic differential, and not properties of the coefficients of the equation.

Hence the following result holds.

THEOREM 2.3. *Suppose the coefficients of equation (2.18) satisfy the conditions (2.15), (2.20), and let* $a_1 \in R(X)$, $B \in R(\mathcal{L}_2(\mathcal{H}, X))$. *Then there is a unique (up to stochastic equivalence), almost surely continuous, stochastic process* $\xi(t)$ *with values in* D_{a_0} *(for* $t > t_0$*) which is a solution of (2.18) for* $t > t_0$ *(and for* $t = t_0$ *if* $\xi_0 \in D_{a_0}$*). This process also satisfies (2.19) for* $t > t_0$*. The process* $\xi(t)$ *is Markovian, and its transition probabilities can be computed as in theorem 2.1.*

REMARK 2.2. Equation (2.19) may have a solution in X also if B is not a Hilbert–Schmidt operator, since the condition $U(t,s)B(s) \in R(L_2(\mathcal{H}, X))$ may hold because of suitable properties of the operator $U(t,s)$, e.g. if $\sigma_2^2(U(t,s))$ has an integrable singularity at $t = s$.

REMARK 2.3. In the 'hyperbolic' case, when (2.15) is replaced by (2.15'), theorem 2.3 remains true for $\beta = \beta_1 = \alpha = 1$ and $\xi_0 \in D_{a_0}$. It is readily verified that all proofs are still valid in this case.

REMARK 2.4. It follows from the proof of theorem 2.1 that solutions of a stochastic equation of type (2.1) can be obtained by the method of successive approximation. This is also true for equation (2.19). It is easily seen that the method of successive approximation for the latter corresponds to the process

$$d\xi^{(n)}(t) - a_0(t)\xi^{(n)}(t)\,dt = a_1(t, \xi^{(n-1)}(t))\,dt + B(t, \xi^{(n-1)}(t))\,dw(t)$$

for (2.18). Thus, the sequence $\xi_n(t)$ obtained from this recursion converges to a solution of (2.18).

2.3. Equations with random coefficients. Consider a stochastic equation (2.1) whose coefficients $a(t,x)$ and $B(t,x)$ are random functions which are consistent with the flow of σ-algebras $\mathcal{F}_t$.

There is a simple case in which all considerations in the proof of theorem 2.1 and concerning existence and uniqueness of a solution remain valid.

PROPOSITION 2.6. *Let* $a \in R(X)$, $B \in R(\mathcal{L}_2(\mathcal{H}, X))$, *almost surely with nonrandom coefficients* C_1, C_2. *If* $\xi_0(t) \in S_{2m}$, *then the equation*

$$\xi(t) = \xi_0(t) + \int_{t_0}^{t} a(s, \xi(s, \xi(s))\,ds + \int_{t_0}^{t} (s, \xi(s))\,dw(s) \qquad (2.21)$$

has a unique solution in S_{2m}.

However, theorem 2.2 cannot be transferred to this case, since the evolution family $U(t,s)$ is random, and thus the coefficients of (2.19) are not consistent with the flow $\mathcal{F}_t$.

In this connection we use another method for studying equations with coefficients of type (2.18).

Let $a_0(t)$ be an, in general unbounded, random operator whose domain of definition D does not depend on t, and which almost surely satisfies the dissipativeness condition

$$(x, [a_0(t) + a_0^*(t)]x) \le 0 \tag{2.22}$$

and the estimate

$$\|[a_0(t) - \lambda I]^{-1}\| \le \frac{C}{1+|\lambda|} \quad (\Re\lambda \ge 0) \tag{2.23}$$

with a nonrandom C.

We say that a function $f(t,x)$ with values in X satisfies the *condition* $G_k(a_0)$ if the following hold almost surely:

$$\begin{gathered} \|a_0^j(t) f(t, a_0^{-j}(t)x)\| \le C_1 + C_2\|x\|, \\ \|a_0^j(t)[f(t, a_0^{-j}(t)x_2) - f(t, a_0^{-j}(t)x_1)]\| \le C\|x_2 - x_1\|, \\ (j = 1, \dots, k). \end{gathered} \tag{2.24}$$

Assume that for each $x \in D$ the stochastic process $a_0(t)x$ has a stochastic differential

$$da_0(t)x = b_0(t)x\,dt + b_1(t)(x, dw(t)), \tag{2.25}$$

with $b_0(t)x \in X$ and $b_1(t)x \in \mathcal{L}_2(\mathcal{H}, X)$ linearly depending on $x \in D$ and satisfying

$$\sup\{\|b_0(t)a_0^{-1}(t)\|, \|a_0(t)b_0(t)a_0^{-2}(t)\|, \|\beta(t)\|, \|\beta_1(t)\|\} \le C \le \infty \tag{2.26}$$

almost surely, with a nonrandom constant C. Here $\beta(t)$, $\beta_1(t)$ are the bilinear operators defined by

$$\beta(t)(x,h) = b_1(a_0^{-1}(t)x, h), \quad \beta_1(t)(x,h) = a_0(t)b_1(a_0^{-2}(t)x, h). \tag{2.27}$$

Note that for fixed $x \in X$ the operators $(\beta(t)x)(h) = \beta(t)(x,h)$ and $(\beta_1(t)x)(h) = \beta_1(t)(x,h)$ are Hilbert–Schmidt operators from $\mathcal{H}$ to X.

Let $a_0^{(n)}(t) = -na_0(t)(a_0(t) - nI)^{-1}$ $(n = 1, 2, \dots)$. The operators $a_0^{(n)}$ are almost surely bounded, commute with $a_0(t)$, satisfy the dissipativity condition (2.22), and approximate $a_0^{(n)}(t)$ almost surely:

$$a_0(t)x = \lim_{n\to\infty} a_0^{(n)}(t)x \quad (x \in D_{a_0}).$$

The equation

(2.28)

$$\xi^{(n)}(t) = \xi(t_0) + \int_{t_0}^t [a_0^{(n)}(s)\xi^{(n)}(s) + a_1(s, \xi^{(n)}(s))]\,ds + \int_{t_0}^t B(s, \xi^{(n)}(s))\,dw(s)$$

satisfies the conditions of proposition 2.6, and hence has a solution $\xi^{(n)}(t)$ $(t > t_0)$.

Below we will state conditions under which the sequence $\xi^{(n)}(t)$ converges to the solution of the equation

$$(2.29)\qquad \xi(t)=\xi(t_0)+\int_{t_0}^{t}[a_0(s)\xi(s)+a_1(s,\xi(s))]\,ds+\int_{t_0}^{t}B(s,\xi(s))\,dw(s).$$

LEMMA 2.1. *Let $a_1(t,x)$ and $B(t,x)$ satisfy the condition $G_2(a_0)$ as elements of the spaces X and $\mathcal{L}_2(\mathcal{H},X)$, respectively. Further, suppose the estimate (2.26) hold. Then $\xi^{(n)}(t)\in D_{a_0^2(t)}$ and*

$$\sup M\|a_0^2(t)\xi^{(n)}(t)\|^2\le C<\infty.$$

PROOF. We first formally compute the stochastic differential of the stochastic process $\eta^{(n)}(t)=a_0^2(t)\xi^{(n)}(t)$. By Itô's formula it can be readily shown that

$$(2.30)\qquad d\eta^{(n)}(t)=a_0^{(n)}(t)\eta^{(n)}(t)\,dt+\gamma(t,\eta^{(n)}(t))\,dt+\Gamma(t,\eta^{(n)}(t))\,dw(t),$$

where

$$\gamma(t,x)=a_0^2(t)a_1(t,a_0^{-2}(t)x)+b_0(t)a_0^{-1}(t)x+a_0(t)b_0(t)a_0^{-2}(t)x+$$
$$\operatorname{Tr}\{\beta(t)\otimes a_0^2(t)B(t,a_0^{-2}(t)x)+\beta_1(t)\otimes a_0^2(t)B(t,a_0^{-2}(t)x)+\beta(t)\otimes\beta_1(t)x\}$$

and $\Gamma(t,x)=a_0^2(t)B(t,a_0^{-2}(t)x)+\beta(t)x+\beta_1(t)x$. Here, $\beta\otimes S$, with $S\in\mathcal{L}_2(\mathcal{H},X)$ denotes the bilinear operator $(\beta\otimes S)(h_1,h_2)=\beta(Sh_1,h_2)$. Under the present conditions, $\operatorname{Tr}(\beta\otimes S)$ exists and belongs to X (see chapter IV).

This relation can be regarded as a stochastic equation with respect to $\eta^{(n)}(t)$, which then has a solution $\eta^{(n)}(t)\in X$. For $\tilde{\xi}^{(n)}(t)=a_0^{-2}(t)\eta^{(n)}(t)$ we can, without difficulty, obtain the initial equation (2.28) (recall that the operator a_0^{-1} is bounded), as well as the initial condition $\xi_0=a_0^{-2}(t_0)\eta^{(n)}(t_0)\in D_{a_0^2(t_0)}$. Uniqueness of the solution implies that $\tilde{\xi}^{(n)}(t)=\xi^{(n)}(t)$, and thus, for $\xi_0\in D_{a_0^2(t_0)}$, also $\xi^{(n)}(t)\in D_{a_0^2(t)}$ and $\eta^{(n)}(t)=a_0^2(t)\xi^{(n)}(t)$.

The lemma now follows from the estimate (2.12) for equation (2.30). □

LEMMA 2.2. *Under the conditions of lemma 2.1, the sequence $\xi^{(n)}(t)$ mean square converges.*

PROOF. We compute the stochastic differential of the difference:

$$d[\xi^{(m)}(t)-\xi^{(n)}(t)]=a_0^{(n)}(t)[\xi^{(m)}(t)-\xi^{(n)}(t)]\,dt+$$
$$[a_1(t,\xi^{(m)}(t))-a_1(t,\xi^{(n)}(t))]\,dt+[B(t,\xi^{(m)}(t))-B(t,\xi^{(n)}(t))]\,dw(t)+$$
$$[a_0^{(m)}(t)-a_0^{(n)}(t)]a_0^{-2}(t)\eta^{(m)}(dt).$$

From Itô's formula we can derive

$$d\|\xi^{(m)}(t)-\xi^{(n)}(t)\|^2 = 2(a_0^{(n)}(t)[\xi^{(m)}(t)-\xi^{(n)}(t)],[\xi^{(m)}(t)-\xi^{(n)}(t)])\,dt +$$
$$2(a_1(t,\xi^{(m)}(t))-a_1(t,\xi^{(n)}(t)),\xi^{(m)}(t)-\xi^{(n)}(t))\,dt +$$
$$2([a_0^{(m)}(t)-a_0^{(n)}(t)]a_0^{-2}(t)\eta^{(m)}(t),\xi^{(m)}(t)-\xi^{(n)}(t))\,dt +$$
$$\sigma_2^2(B(t,\xi^{(m)}(t))-B(t,\xi^{(n)}(t)))\,dt +$$
$$2([B(t,\xi^{(m)}(t))-B(t,\xi^{(n)}(t))]\,dw(t),\xi^{(m)}(t)-\xi^{(n)}(t)).$$

Integrating this relation, averaging the result, and using properties of the coefficients, we can derive

$$\mathrm{M}\|\xi^{(m)}(t)-\xi^{(n)}(t)\|^2 \le$$
$$C\int_{t_0}^{t}\mathrm{M}\|\xi^{(m)}(\theta)-\xi^{(n)}(\theta)\|^2\,d\theta + \int_{t_0}^{t}\mathrm{M}\|[a_0^{(m)}(\theta)-a^{(n)}(\theta)]a_0^{-2}(\theta)\eta^{(m)}(\theta)\|^2\,d\theta.$$

Hence, by Gronwall's inequality we find

$$\mathrm{M}\|\xi^{(m)}(t)-\xi^{(n)}(t)\|^2 \le$$
$$e^{C(t-t_0)}(t-t_0)\sup\mathrm{M}\|\eta^{(n)}(\theta)\|^2\sup\|[a_0^{(m)}(\theta)-a^{(n)}(\theta)]a_0^{-2}(\theta)\|^2,$$

which proves the lemma, since

$$\lim_{m,n\to\infty}\|[a_0^{(m)}(\theta)-a^{(n)}(\theta)]a_0^{-2}(\theta)\| = 0. \quad \square$$

THEOREM 2.4. *Under the conditions of lemma 2.1, equation (2.29) with initial value $\xi(t)\in D_{a_0^2(t_0)}$ has a solution which is unique up to stochastic equivalence. This solution satisfies (2.11), (2.12), and can thus be continuously extended to arbitrary $\xi(t_0)\in\mathcal{H}$.*

PROOF. Consider the Hilbert space of random functions ξ, with scalar product $\langle\xi_1,\xi_2\rangle = \int_t^T \mathrm{M}(\xi_1(t),\xi_2(t))\,dt$. In this space the above constructed sequence converges to a function $\xi(t)$, and the sequence $a_0(t)\xi_n(t)$ is bounded, i.e. weakly compact. If $\eta(t)$ is a limit point of it, then the boundedness of $a_0^{-1}(t)$ implies $a_0^{-1}(t)\eta(t)=\xi(t)$.

Therefore $\xi(t)\in D$ and $\eta(t)=a_0(t)\xi(t)=\lim_{n\to\infty}a_0(A)\xi_n(t)$. Limit transition in (2.28) gives equation (2.29) for $\xi(t)$.

If $\xi(t)$ is an arbitrary solution of the latter equation, then the same estimates as in lemma 2.2 are obtained for the difference $\xi(t)-\xi_n(t)$. This shows that $\xi(t)=\lim\xi_n(t)$ and implies uniqueness. The other assertions are proved as in the case of equations with bounded coefficients. $\square$

REMARK 2.5. These considerations become simpler if the operator $a_0(t)$ is almost surely bounded, though not uniformly bounded by a nonrandom constant. In this case it turns out that we can use the convergence (in a certain sense) of the sequence $\|a_0^{(n)}(t)-a_0^{(m)}(t)\|$ (e.g., $\mathrm{M}\|a_0^{(n)}(t)-a_0^{(m)}(t)\|^2\to 0$) and the uniform boundedness of $\mathrm{M}\|\xi^{(n)}(t)\|^4$.

The conditions of lemma 2.1 are unnecessary in this case. It suffices to assume the dissipativity condition (2.22) and an estimate

$$\mathrm{M}\|a_0(t)\|^4 \le \text{const},$$

with the usual stipulations (2.3) about a_1 and A.

3. Operator multiplicative functionals and the evolution families determined by them

3.1. Basic definitions. Let $\xi(t)$ $(t \in [t_0,\tau])$ be a Markov stochastic process on $(\Omega, \mathcal{F}, P)$ with values in a measurable space $(X, \mathcal{F}_\Delta)$ $(\Delta \subset [t_0,\tau])$, where $\mathcal{F}_\Delta$ is the minimal σ-algebra with respect to which all random variables $\xi(t)$ $(t \in \Delta)$ are measurable, and set $\mathcal{F}_t = \mathcal{F}_{[t_0,t]}$.

Consider a Hilbert space Y. Let $\mathcal{H}(Y)$ be the Hilbert space of random variables on $(\Omega, \mathcal{F})$ with values in Y and norm $|\xi|^2 = \mathrm{M}\|\xi\|_Y^2$. Let $\mathcal{H}_\Delta$ be the subspace of $\mathcal{F}_\Delta$-measurable random variables, and put $\mathcal{H}_t = \mathcal{H}_{[t_0,t]}$.

DEFINITION 3.1. A *multiplicative operator functional* of the stochastic process $\xi(t)$ is a family $T(t,s)$ $(s,t \in [t_0,\tau])$ consisting of bounded linear operators, $T(t,s)\colon \mathcal{H}_s \to \mathcal{H}_t$, satisfying the condition $T(t,s)y \in \mathcal{H}_t$ $(y \in Y)$, the multiplication relation

$$T(t,t) = I, \quad T(t,s) = T(t,\theta)T(\theta,s) \quad (s \le \theta \le t), \tag{3.1}$$

and the estimate

$$\mathrm{M}_s\{\|T(t,s)\eta\|_Y^2\} \le C\|\eta\|_Y^2 \quad (\eta \in \mathcal{H}_s), \tag{3.2}$$

where C is a nonrandom constant.

EXAMPLE 3.1. Let $T(t,s,\omega)$ be a system of random variables on Ω with values in a space $\mathcal{L}(Y)$, satisfying almost surely (3.1), and being uniformly bounded by a nonrandom constant.

If, moreover, $T(t,s,\omega)$ is $\mathcal{F}_{[s,t]}$-measurable, then it determines a multiplicative operator functional, according to the formula

$$(T(t,s)\eta)(\omega) = T(t,s,\omega)\eta(\omega).$$

Such multiplicative operator functionals will be called *homogeneous.* Below it will be shown that not all multiplicative operator functionals are homogeneous.

Consider the space $V(X,Y)$ of bounded measurable functions on X with values in Y.

THEOREM 3.1. *Let T be a multiplicative operator functional of the Markov process $\xi(t)$. The relation*

$$(y, U(s,t)f(x))_Y = M\{(T(t,s)y, f(\xi(t)))_Y / \xi(s) = x\} \quad (y \in Y, \quad x \in X) \tag{3.3}$$

determines an evolution family of bounded operators $U(s,t)$ on $V(X,Y)$:

$$U(t,t) = I, \quad U(s,\tau)U(\tau,t) = U(s,t).$$

PROOF. For $y \in Y$ we have $T(t,s)y \in \mathcal{H}_t$, and thus the latter is a random variable with values in Y. So, the righthand side of (3.3) makes sense. From (3.2) we infer

$$|\mathrm{M}\{(T(t,s)y, f(\xi(t)))_Y/\xi(s)=x\}|^2 \leq \\ \mathrm{M}\{\|T(t,s)y\|^2/\xi(s)=x\}\mathrm{M}\{\|f(\xi(t))\|^2/\xi(s)=x\} \leq \\ \sup_X \|f(x)\|^2 C\|y\|^2, \tag{3.4}$$

which shows that the righthand side of (3.3) is a linear functional on Y (depending on x). It defines an element of Y, which is denoted by $U(s,t)f(x)$. Boundedness of the (obviously linear) operator $U(s,t)$ follows from (3.4).

Consider the relation

$$(y, U(\sigma,t)f(\xi(\sigma)))_Y = \mathrm{M}\{(T(t,\sigma)y, f(\xi(t)))/\mathcal{F}_\sigma\}.$$

It follows from (3.3) and the Markov property of the stochastic process ξ. We can replace y by an $\mathcal{F}_\sigma$-measurable random variable $T(\sigma,s)y_0$:

$$(T(\sigma,s)y_0, (U(\sigma,t)f)(\xi(\sigma))) = \\ \mathrm{M}\{(T(t,\sigma)T(\sigma,s)y_0, f(\xi(t)))/\mathcal{F}_\sigma\} = \\ \mathrm{M}\{(T(t,s)y_0, f(\xi(t)))/\mathcal{F}_\sigma\}.$$

Averaging termwise, under the condition $\xi(s)=x$, we obtain the evolution property

$$(y_0, U(s,t)f(x)) = \mathrm{M}\{(T(t,s)y_0, f(\xi(t)))/\xi(s)=x\} = \\ \mathrm{M}\{(T(\sigma,s)y_0, U(\sigma,t)f(\xi(\sigma)))/\xi(s)=x\} = (y_0, U(s,\sigma)U(\sigma,t)f(x)). \quad \square$$

REMARK 3.1. If T is a homogeneous multiplicative operator functional, then

$$U(s,t)f(x) = \mathrm{M}\{T^*(t,s)f(\xi(t))/\xi(s)=x\}. \tag{3.5}$$

3.2. Ordinary linear equations. Multiplicative functionals of a Markov process naturally raise as solutions of linear differential equations with coefficients depending on this process. We will first consider the simplest version, when the equation is an ordinary, not a stochastic, differential equation.

PROPOSITION 3.1. *Let $\xi(t)$ $(\tau \geq t \geq t_0)$ be a Markov stochastic process in a space X which is almost surely continuous and remains in $X \setminus M$, where M is a subset in X. Let $c(t,x)$ be a function, continuous on $[t_0,\tau] \times [X \setminus M]$, with values in $L(Y)$, and satisfying the condition*

$$([c(t,x)+c^*(t,x)]\phi, \phi)_Y \leq 2\alpha\|\phi\|^2 \quad (\phi \in Y). \tag{3.6}$$

Then a solution of the differential equation

$$\frac{du}{dt} = c(t,\xi(t))u, \quad u(s) = \psi \tag{3.7}$$

determines a homogeneous multiplicative operator functional $T(t,s)$ *of the process* $\xi(t)$ *by*

$$u(t) = T(t,s)\psi.$$

satisfying the condition

$$\|T(t,s)\| \le e^{\alpha(t-s)} \quad (a.s.) \tag{3.8}$$

PROOF. Equation (3.7) almost surely has a continuous coefficient $c(t,\xi(t))$, and thus has, with probability one, a solution which is uniquely determined by the initial condition. The estimate (3.8) follows from (3.6), and the multiplicativity condition is nothing but the natural evolution property, which follows from the uniqueness of the solution. □

Now we consider the case when $c(t,x)$ can be unbounded but when certain conditions implying the well-posedness of the Cauchy problem for (3.7) do hold. Such conditions include, e.g., the following.

PROPOSITION 3.2. *Suppose the operators* $c(t,x)$ $(t \in [t_0,\tau],\ x \in X)$ *have a common domain* D *of definition which is dense in* Y, *and satisfy in it the conditions*

$$\|(c(t,x)-\lambda I)^{-1}\| \le \frac{k}{|\lambda|+1}, \qquad x \in X \setminus M,$$
$$\|[c(t,x)-c(t',x')]c^{-1}(t,x)\| \le \alpha_1\{|t-t'|^\alpha + \|x-x'\|^\beta\}.$$

If, moreover, almost all paths of the process satisfy a Hölder condition

$$\| \subset (t_2) - \xi(t_1)\| \le const|t_2 - t - 1|^\gamma. \tag{3.9}$$

then the Cauchy problem (3.7) is well-posed and defines a homogeneous multiplicative operator functional.

The proof of this assertion can readily be reduced to well-known results in the theory of evolution equations (see Krein [1]), since the operator function $c(t,\xi(t))$ is almost surely Hölder continuous in this case.

REMARK 3.2. It can be shown that almost all solutions of the stochastic differential equation (2.1) satisfy the condition (3.9).

3.3. Stochastic linear equations. Consider an equation

$$d\eta(t) = c(t,\xi(t))\eta(t)\,dt + b(t,\xi(t))(\eta, dw(t)), \tag{3.10}$$

where $c(t,x)$, $b(t,x)$ are functions on $[t_0,\tau] \times X$ (X being a Hilbert space) whose values are, respectively, linear operators on X and bilinear operators from $X \times \mathcal{H}$ to X, and where $\xi(t)$ is a solution of the stochastic differential equation (2.1). We will consider this equation without giving precise conditions on the coefficients, but we assume that there is exactly one solution for each $\eta(t_0)$ that is measurable with respect to $\mathcal{F}_{t_0}$. From the linearity of the equation and the uniqueness of the solution it follows that $\eta(t)$ depends almost surely linearly on $\eta(t_0)$, which is denoted by

$$\eta(t) = T(t,t_0)\eta(t_0). \tag{3.11}$$

Further, the estimate (2.11) for a linear equation leads to (3.2). Finally, the uniqueness implies the multiplicative property of T. This leads to

PROPOSITION 3.3. *If the linear stochastic differential equation (3.10) has a unique solution on each interval $[s,t]$ for each $\mathcal{F}_s$-measurable initial value $\eta(s)$, and if its coefficients satisfy*

$$(c^*(t,x)\phi,\phi) + (c(t,x)\phi,\phi) \leq \beta\|\phi\|^2,$$
$$\sigma_2^2(b(t,x)\phi) \leq \gamma\|\phi\|^2,$$

then its solution can be represented as in (3.11), where $T(t,s)$ is a multiplicative operator functional of the process $\xi(t)$.

We further assume that

$$c(t,x) = c_0(t,x) + c_1(t,x),$$

where $c_0(t,x)$ is an, in general unbounded, operator. To be able to apply theorem 2.4 we have to compute the stochastic differential of the process $c_0(t,\xi(t))z$ ($z \in D_{c_0}$).

Itô's formula implies that if $c_0(t,x)z$ is twice differentiable with respect to x and once with respect to t, then

$$dc_0(t,\xi(t)) =$$
$$\frac{\partial}{\partial t}c_0(t,\xi(t))z\,dt + c_{0_x}(t,\xi(t))(d\xi,z) + \frac{1}{2}\operatorname{Tr} c_{0_{xx}}(t,\xi(t))(B(t,\xi(t))\cdot, B(t,\xi(t))\cdot, z)\,dt,$$

and thus (2.25) holds for $c_0(t,\xi(t))$, where

$$b_0(t)z =$$
$$\frac{\partial}{\partial t}c_0(t,\xi(t))z + c_{0_x}(t,\xi(t))(a(t,\xi(t)),z) + \frac{1}{2}\operatorname{Tr} c_{0_{xx}}(t,\xi(t))(B(t,\xi(t))\cdot, B(t,\xi(t))\cdot, z),$$
$$b_1(t)(z,h) = c_{0_x}(t,\xi(t))(B(t,\xi(t))h, z).$$

Here the conditions (2.26) are readily formulated in terms of uniform boundedness of the first and second order derivatives with respect to x and the first order derivatives with respect to t of the operator functions

$$c_0(t,x)c_0^{-1}(t_0,x_0), \quad c_0(t_0,x_0)c_0(t,x)c_0^{-2}(t_0,x_0). \tag{3.12}$$

The following result can be derived directly from proposition 3.3 and theorem 2.4.

THEOREM 3.2. *Let $\xi(t)$ be a solution of the stochastic differential equation (2.1) in a Hilbert space X. Suppose the operator functions (3.12) are uniformly bounded, together with their first order derivatives with respect to t and their first and second order derivatives with respect to x.*

Suppose further that $c_0(t,x)$ satisfies (2.22) and (2.23), and that $c_1(t,x)y$ and $b(t,x)y$ satisfy condition $G_2(c_0)$ (with respect to y).

Then there is a unique (up to stochastic equivalence) solution of the linear equation

$$\eta(t) = \eta(s) + \int_s^t c_0(\theta, \xi(\theta))\eta(\theta)\, d\theta +$$

$$\int_s^t c_1(\theta, \xi(\theta))\eta(\theta)\, d\theta + \int_s^t b(\theta, \xi(\theta))(\eta(\theta), dw(\theta)), \tag{3.13}$$

and this solution determines a multiplicative operator functional $T(t,s)$ of the process $\xi(t)$:

$$\eta(t) = T(t,s)\eta(s).$$

REMARK 3.3. Since the conditions of theorem 2.4 must hold almost surely, theorem 3.2 remains valid if the coefficients c_0, c_1, and b satisfy all the condition stated not on all of X but only on a set $X \setminus M$ for which M has the property that the stochastic process $\xi(t)$ enters it with probability zero.

REMARK 3.4. If the operator $c_0(t,x)$ is bounded for each t and $x \notin M$, then the conditions of the theorem can be simplified, in accordance with remark 2.5.

We can drop the requirement that $c_0(t,x)$ be smooth, but we must require, e.g., the estimate $\mathrm{M}\|c_0(t,\xi(t))\|^4 < \infty$ to hold and, moreover, replace the condition G_2 by estimates (2.3).

REMARK 3.5. If all coefficients of (3.13) are Hilbert–Schmidt operators, then the solution can be obtained as

$$T(t,s) = I + V(t,s),$$

where V is also a Hilbert–Schmidt operator on Y. In this case the multiplicative operator functional is homogeneous.

REMARK 3.6. If the operator $c_0(t)$ does not depend on x, the conditions of the theorem can be modified, using theorem 2.3. We omit the result which is obtained in this way.

3.4. Smoothness with respect to the initial value of the solution of a stochastic differential equation and of the corresponding multiplicative functional. Consider the pair of equations

$$\xi(t) = x + \int_{t_0}^t a(s, \zeta(s))\, ds + \int_{t_0}^t B(s, \xi(s))\, dw(s), \tag{3.14}$$

$$\eta(t) = \phi = \int_{t_0}^t c(s, \xi(s))\eta(s)\, ds + \int_{t_0}^t b(s, \xi(s))(\eta(s), dw(s)), \tag{3.15}$$

with coefficients $a(s,x)$, $B(s,x)$, $c(s,x)$, $b(s,x)$ that are functionals on $[t_0, \tau] \times X$ with values in X, $\mathcal{L}_2(\mathcal{H}_0, X)$, $\mathcal{L}(Y)$, and $\mathcal{L}(Y, \mathcal{L}_2(\mathcal{H}, Y))$, respectively. Assume that these functions have first order derivatives with respect to x in the corresponding spaces, and consider the equations obtained from (3.14), (3.15) by formal termwise differentiation along $h \in X$,

$$\zeta(t) = h + \int_{t_0}^{t} a_x(s.\xi(s))\zeta(s)\,ds + \int_{t_0}^{t} B_x(s,\xi(s))(\zeta(s), dw(s)), \tag{3.16}$$

$$\chi(t) = \int_{t_0}^{t} c_x(s,\xi(s))(\zeta(s),\eta(s))\,ds + \int_{t_0}^{t} c(s,\xi(s))\chi(s)\,ds +$$

$$\int_{t_0}^{t} b_x(s,\xi(s))(\zeta(s)\eta(s), dw(s)) + \int_{t_0}^{t} b(s,\xi(s))(\chi(s), dw(s)). \tag{3.17}$$

It turns out that the three equations (3.15)–(3.17) can be treated as one linear equation, determining a multiplicative operator functional of the process $\xi(t)$ in the new space $Y + (X \otimes Y)$.

Consider in this space the stochastic process $\tilde{\eta}(t) = (\eta_1(t), \eta_2(t))$ whose components are

$$\eta_1(t) = \eta(t) + \chi(t) \in Y, \quad \eta_2(t) = \zeta(t) \otimes \eta(t) \in X \otimes Y.$$

A bilinear operator $\alpha \in \mathcal{L}(X \times Y, Z)$ determines a linear operator $\hat{\alpha} \in \mathcal{L}((X \otimes Y), Z)$ by $\hat{\alpha}(x \otimes y) = \alpha(x, y)$. Moreover, we introduce the 'tensor sum' operation for operators on $X \otimes Y$. Set, for $\alpha \in \mathcal{L}(X)$, $\beta \in \mathcal{L}(Y)$: $\alpha \boxplus \beta = \alpha \otimes I + I \otimes \beta$, so that $(\alpha \boxplus \beta)(x, y) = \alpha x \otimes y + x \otimes \beta y$. Using this notation we can write

$$\eta_1(t) = \phi + \int_{t_0}^{t} c(s,\xi(s))\eta_1(s)\,ds + \int_{t_0}^{t} b(s,\xi(s))(\eta_1(s), dw(s)) +$$

$$\int_{t_0}^{t} \hat{c}_x(s,\xi(s))\eta_2(s)\,ds + \int_{t_0}^{t} \hat{b}_x(s,\xi(s))(\eta_2(s), dw(s)), \tag{3.18}$$

$$\eta_2(t) = (h \otimes \phi) + \int_{t_0}^{t} (a'(s,\xi(s)) \boxplus c(s,\xi(s)))\eta_2(s)\,ds +$$

$$\int_{t_0}^{t} (B_x(s,\xi(s)) \boxplus b(s,\xi(s)))(\eta_2(s), dw(s)). \tag{3.19}$$

This pair of equations can be written as

$$\tilde{\eta} = \tilde{\eta}_0 + \int_{t_0}^{t} \tilde{c}(s,\xi(s))\tilde{\eta}(s)\,ds + \int_{t_0}^{t} \tilde{b}(s,\xi(s))(\tilde{\eta}(s), dw(s)), \tag{3.20}$$

where $\tilde{c}$ and $\tilde{b}$ are respectively, the linear and bilinear operators defined by

$$\tilde{c}\begin{pmatrix} y_1 \\ z \otimes y_2 \end{pmatrix} = \begin{pmatrix} cy_1 + c'_x(x, y_2) \\ a'z \otimes y_2 + z \otimes cy_2 \end{pmatrix} = \begin{pmatrix} c & \hat{c}' \\ 0 & a' \boxplus c \end{pmatrix}\begin{pmatrix} y_1 \\ x \otimes y_2 \end{pmatrix} \tag{3.21}$$

and

(3.22)
$$\tilde{b}\left(\begin{pmatrix} y_1 \\ z \otimes y_2 \end{pmatrix}, v\right) = \begin{pmatrix} b(y_1, v) + b'(z, y_2, v) \\ B'(z, v) \otimes y_2 + z \otimes b(y_2, v) \end{pmatrix} = \begin{pmatrix} b & \hat{b}' \\ 0 & A' \boxplus b \end{pmatrix} = \left(\begin{pmatrix} y_1 \\ x \otimes y_2 \end{pmatrix}, z\right).$$

These expressions show that formal differentiation of the system (3.14), (3.15) with differentiable coefficients leads to the system (3.14), (3.20) of the same type and with the same properties. Also, it is clear that if the coefficients of the initial system are differentiable up to order k, then the system (3.14), (3.20) has coefficients which are $k-1$ times differentiable. Therefore termwise differentiation can be continued until the kth order derivatives are reached.

We will now show that this differentiation is not only formal.

THEOREM 3.3. *Let, in (3.14), (3.15),*

$$a(t,x) = a_0(t) + a_1(t,x), \quad c(t,x) = c_0(t,x) + c_1(t,x),$$

and suppose that

1) *$a_1(t,x)$, $B(t,x)$, $c_1(t,x)$, and $b(t,x)$ are functions on $[t_0,\tau] \times X$ with values in X, $\mathcal{L}_2(\mathcal{H} \times X)$, $\mathcal{L}(Y)$, and $\mathcal{L}_2(Y \times \mathcal{H}, Y)$, respectively, which are continuously differentiable with respect to $x \in X$ up to order k;*
2) *$a_0(t)$ is a linear operator on X with constant domain of definition D, and satisfying the conditions (2.22), (2.23), and (2.26) for $b_0(t) = a_0'(t)$, $b_1(t) = 0$;*
3) *$c_0(t,x)$ is a linear operator on Y with constant domain of definition D_1, and satisfying the conditions of theorem 3.2;*
4) *the coefficients $a_1(t,x)$, $B(t,x)$ and their derivatives are of class $G_2(a_0)$ (with respect to x), while the coefficients $c_1(t,x)y$, $b(t,x)y$ and their derivatives are of class $G_2(c_0)$ (with respect to y).*

Then the solutions $\xi_{x,t_0}(t)$ and $\eta(t,t_0)\phi$ of equations (3.14)–(3.15) exist, are unique, and have derivatives up to order k with respect to x, in the sense of mean square convergence.

PROOF. Existence of solutions has already been established. Differentiability can be successively proved, and so it suffices to describe the first step only.

First, suppose a_0 and c_0 are bounded. Then the system satisfies the conditions of the contractive mapping theorem, and its solution can be obtained by successive approximation. It is readily checked that termwise differentiation of these successive approximations gives successive approximations for the formally differentiated system which are mean square convergent. This implies differentiability of the limit function, and also the fact that its derivative is a solution of (3.14) (3.20).

If the operators a_0 and c_0 are not bounded, then the solution of (3.14)–(3.15) is obtained by limit transition from solutions constructed in lemma 2.1 for analogous systems with bounded operators $a_0^{(n)}$ and $c_0^{(n)}$. This same limit transition also applies to equation (3.20), leading to the required result. □

REMARK 3.7. Smoothness of the coefficients of the linear equation (3.15) may be violated at points of an exceptional set M which is almost surely avoided by all paths $\xi_x(t)$.

REMARK 3.8. In accordance with remarks 2.5 and 3.4, the conditions are simplified if a_0 and c_0 are almost surely bounded and dissipative.

REMARK 3.9. If $b = 0$, then the multiplicative operator functional is determined by the simpler equation (3.7), and the conditions for its existence (propositions 3.1 and 3,2) are simplified. In this connection it is easy to state simpler conditions for smoothness. We omit these for the sake of brevity.

REMARK 3.10. For $c_0(t)$ independent of x, another version of the theorem is possible (see remark 3.6). In particular, such a situation arises when considering linear equations obtained by termwise differentiation of (3.14), i.e. for $\eta = \xi' x$ when $c_0 = a_0$.

3.5. Invariance of smooth functions under an evolution family. Now we will consider the problem of smoothness of the operator evolution family (3.3) determined by the multiplicative operator functionals determined by solutions of 93.15).

Let $\eta(t) = T(t,s)\eta(s)$ and $u(x,t) = U(t,\tau)\psi(x)$ ($\psi \in V(X,Y)$). Then, by definition,

$$(y, u(x,t)) = \mathrm{M}(T(\tau,t)y, \psi(\xi_{x,t}(\tau))). \tag{3.23}$$

If $\psi(x)$ is a function which is bounded and continuous together with its derivative $\psi'(x)$, then under the conditions of theorem'3.3 the righthand side of (3.23) is differentiable with respect to x and termwise differentiation yields

$$\begin{aligned}(y, u'_x(x,t)h) &= \\ \mathrm{M}\{(T_x(\tau,t)(h,y), \psi(\xi_{x,t}(\tau))) &+ (T(\tau,t)y, \psi'(\xi_{x,t}(\tau))\xi'_{x,t}(\tau)h)\}.\end{aligned} \tag{3.24}$$

Thus $u(x,t)$ is a differentiable function. Moreover, the pair (3.23)–(324) can be written as one relation of type (3.23), with Y replaced by $\tilde{Y} = Y + X \otimes Y$:

$$(\tilde{y}, \tilde{u}(x,t)) = \mathrm{M}(\tilde{T}(\tau,t)\tilde{y}, \tilde{\psi}(\xi_{x,t}(\tau))), \tag{3.25}$$

where $\tilde{u} = \binom{u}{u'}$, $\tilde{y} = \binom{y}{h\otimes y}$, $\tilde{\psi} = \binom{\psi}{\psi'}$, and

$$\tilde{T}(\tau,t)\tilde{y} = \begin{pmatrix} T(\tau,t)y & T_x(\tau,t)(h,y) \\ 0 & \xi'_{x,t}(\tau)h \otimes T(\tau,t)y \end{pmatrix} \tag{3.26}$$

is a multiplicative operator functional determined by (3.20).

Thus, differentiation preserves the form of the equation (3.23), and this enables us to continue the differentiation procedure as far as the coefficient of the equation permit.

This leads to the following result.

THEOREM 3.4. *Under the conditions of theorem 3.3, the evolution family $U(t,x)$ determined by the multiplicative functional determined by the system (3.14)–(3.15) leaves invariant the space $C_k(X,Y)$ of functions that are continuous and bounded together with theier derivatives up to order k (the latter are presumed to be Hilbert-–Schmidt operators).*

REMARK 3.11. Derivatives of solutions of the system (3.14)–(3.15) determine multiplicative operator functionals (3.26), which determine evolution families (3.25) in the space $Y_1 = \tilde{Y}$ and in the similarly constructed spaces $Y_k = \tilde{Y}_{k-1}$ for higher derivatives.

4. The Cauchy problem for second order parabolic systems

4.1. The Cauchy problem for a single linear equation. In this section we will obtain from the above presented general results existence and uniqueness theorems for solutions of the Cauchy problem for linear and quasilinear parabolic systems, as well as representations of these solutions as means on paths of Markov processes. As a consequence we will give conditions for the maximality of a differential operator.

To obtain these results we have to compute the generating operators of the evolution families considered in the previous section.

Recall that the generating operator of an evolution family $U(t,\tau)$ is the operator $A(t) = \lim_{\Delta t \to 0}[U(t, t+\Delta t) - I]/\Delta t$.

LEMMA 4.1. *Let $\alpha(t,x)$ be a continuous function on $[t_0, \tau] \times X$, and let $\xi_{s,x}(t)$ be a solution of equation (2.7). then*

$$\lim_{\Delta s \to 0} \frac{1}{\Delta s} \int_s^{s+\Delta t} \alpha(\theta, \xi_{s,x}(\theta))\, d\theta = \alpha(s,x) \quad a.s.$$

The proof follows from the almost sure continuity of the integrand.

We will first consider the evolution family

$$U(t,\tau)f(x) = \mathrm{M} f(\xi_{t,x}(\tau)) \tag{4.1}$$

in the space $V(X, \mathbf{R})$ of scalar-valued functions $f(x)$ on the Hilbert space X.

The generating operator of the evolution family (4.1) corresponding to the trivial multiplicative functional $T \equiv 1$ is called the *generating operator of the Markov process* $\xi(t)$.

PROPOSITION 4.1. *The restriction of the generating operator of the family (4.1) to functions of class C_2 is, for $x \in D_{a_0}$,*

$$A_t u(x) = \frac{1}{2} \operatorname{Tr} B^*(t,x) u''(x) B(t,x) + (a(t,x), u'(x)) \equiv$$
$$\frac{1}{2}(B(t,x)B^*(t,x)\nabla, \nabla)u(x) + (a(t,x), \nabla)u(x). \tag{4.2}$$

If $u'(x) \in D_{a_0^*}$ and $a_0^* u'(x)$ is continuous, then for all $x \in X$,

$$A_t u(x) = \frac{1}{2} \operatorname{Tr} B^*(t,x) u''(x) B(t,x) + (x, a_0^*(t) u'(x)) + (a_1(t,x), u'(x)). \tag{4.3}$$

PROOF. Let $x \in D_{a_0}$. Then for $\theta \geq s$, $\xi_{s,x}(\theta) \in D_{a_0}$, and Itô's formula and (4.1) imply

$$U(s, s+\Delta s)u(x) - u(x) = \mathrm{M}\{f(\xi_{s,x}(s+\Delta s)) - f(\xi_{s,x}(s))\} =$$
$$\mathrm{M} \int_s^{s+\Delta s} \{(f'(\xi_{s,x}(\theta)), a_0(\theta)\xi_{s,x})\theta) + a_1(\theta, \xi_{s,x}(\theta))) +$$
$$\frac{1}{2} \operatorname{Tr} B^*(\theta, \xi_{s,x}(\theta)) f''(\xi_{s,x}(\theta)) B(\theta, \xi_{s,x}(\theta))\Big\}\, d\theta.$$

By lemma 4.1, the continuity of the coefficients, and the estimate $\mathrm{M}\|a_0(\theta)\xi_{s,x}(\theta)\|^2 \le$ const, we obtain the relation

$$\lim_{\Delta s\to 0} \frac{U(s,s+\Delta s)f(x)-f(x)}{\Delta s} =$$
$$(f'(x), a_0(s)x + a_1(s,x)) + \frac{1}{2}\operatorname{Tr} B^*(s,x)f''(x)B(s,x),$$

which implies (4.2).

If, moreover, the function $a_0^* f(x)$ is continuous, then (4.2) implies (4.3) for all $x \in X$, since the righthand side of this expression is continuous. □

THEOREM 4.1. *Consider the differential equation*

$$\frac{\partial u}{\partial t} + \frac{1}{2}\operatorname{Tr} B^*(t,x)u''B(t,x) + (x, a_0^*(t)u') + (a_1(t,x), u) \equiv$$
$$\frac{\partial u}{\partial t} + \frac{1}{2}(B(t,x)B^*(t,x)\nabla,\nabla)u + (a(t,x),,\nabla)u = 0. \tag{4.4}$$

If the coefficients satisfy the conditions of theorem 3.3 (for $b = c = 0$, $k = 2$), then the Cauchy problem for (4.4) has exactly one solution in the class Δ of functions that are continuous and bounded, together with their derivatives up to order two and together with the function $a_0^*(t)u'(x)$, for every initial value $u(x,\tau) = \phi(x)$ from this class. This solution is representable as

$$u(x,t) = U(t,\tau)\phi(x),$$

where U is the evolution family (4.1) determined by the stochastic equation (2.18).

PROOF. The evolution family (4.1) leaves invariant the class Δ and has generating operator (4.3) in this class. This implies the existence of a solution with the appropriate properties. We prove its uniqueness. Let $u(x,t)$ be some solution of the Cauchy problem under consideration. Consider the auxiliary function $g(s) = u(s,\xi_{t,x}(s))$, and compute its stochastic differential:

$$dg(s) =$$
$$\left\{\frac{\partial u}{\partial s}(s,\xi_{t,x}(s)) + (u'_x(s,\xi_{t,x}(s)), a(s,\xi_{t,x}(s)))+\right.$$
$$\left.\frac{1}{2}\operatorname{Tr} B^*(s,\xi_{t,x}(s)u''_{xx}(s\xi_{t,x}(s))B(s,\xi_{t,x}(s))\right\} +$$
$$(u'_x(s,\xi_{t,x}(s)), B(s,\xi_{t,x}(s))\,dw(s)) =$$
$$(u'_x(s,\xi_{t,x}(s)), B(s,\xi_{t,x}(s))\,dw(s)).$$

This formula implies that

$$\mathrm{M}_t dg(s) = 0,$$

i.e. $\mathrm{M}_t g(s) =$ const, or $g(t) = u(t,x) = \mathrm{M}_t g(t) = \mathrm{M}_t g(\tau) = \mathrm{M}u(\tau,(\xi_{t,x}(\tau))$. □

4.2. The Cauchy problem for a linear system. We turn to the computation of the generating operator $A^{(T)}$ of the general evolution family $U^{(T)}(t,\tau)$ in $V(X,Y)$ determined by (3.2). For $T(t,\tau) \equiv I$ we obtain, as above, the evolution family $U^{(I)}(t,\tau)$ whose generating operator is $A^{(I)}$, and it is the generating operator of the Markov process $\xi(t)$ (on scalarvalued or vectorvalued functions). The difference between these generating operators can be written as

$$(y, [A^{(T)}(s) - A^{(I)}(s)]f(x)) =$$
$$\lim_{\Delta s\to 0} \frac{1}{\Delta s}\{\mathrm{M}_{\xi(s)=x}([T(s+\Delta s,s)y - y], f(x)) +$$
$$\mathrm{M}_{\xi(s)=x}(T(s+\Delta s,s)y - y, f(\xi(s+\Delta s)) - f(x))\}.$$

Suppose the following limits exist:

$$v(t,x)y = \lim_{\Delta s\to 0} \frac{1}{\Delta s}\mathrm{M}_{\xi(s)=x}([T(s+\Delta s,s)y - y], \tag{4.5}$$

and

$$(y, \gamma_s f(x)) = \lim_{\Delta s\to 0} \frac{1}{\Delta s}\mathrm{M}_{\xi(s)=x}(T(s+\Delta s,s)y - y, f(\xi(s+\Delta s)) - f(x)). \tag{4.6}$$

Then we obtain a representation for the generating operator,

$$A^{(T)}(t)f(x) = A^{(I)}(t)f(x) + v^*(t,x)f(x) + \gamma_t f(x), \tag{4.7}$$

which is very suitable for study in particular cases.

First of all, note that in certain cases the last term in (4.7) vanishes. In fact, let

$$\mathrm{M}\|(T(s+\Delta s,s)y - y\|^2 = O(\Delta s^2). \tag{4.8}$$

If the paths of the process $\xi(t)$ are almost surely continuous, then for continuous functions

$$\lim_{\Delta s} \mathrm{M}_{\xi(s)=x}\|f(\xi(s+\Delta s)) - f(x)\|^2 = 0, \tag{4.9}$$

and hence $\gamma_s f(x) = 0$

Condition (4.8) is obviously satisfied if the multiplicative operator functional is determined by the ordinary differential equation (3.7). In this case the operator $v(t,x)$ can also be easily computed. It follows from (4.5) and (3.7) that

$$v(t,x)y = c(t,x)y \quad (y \in D_c).$$

Thus, the following is true:

THEOREM 4.2. *Under the conditions of propositions 3.1 and 3.2 the generating operator of the evolution family (3.3) has the form*

$$A(t)f(x) = A^{(I)}(t)f(x) + c^*(t,x)f(x), \tag{4.10}$$

for functions $f(x)$ for which all terms at the righthand side are defined.

COROLLARY 4.1. *If $\xi(t)$ is a process with generating operator given by (4.3), then the generating operator of the evolution family*

$$U(t,\tau)f(x) = M\left\{\widehat{\exp}\left[\int_t^\tau c(\theta,\xi(\theta))\,d\theta\right] f(\xi(\tau)/\xi(t)=x\right\} \tag{4.11}$$

is given by

$$A(t) = \frac{1}{2}(B(t,x)B^*(t,x)\nabla,\nabla) + (a_1(t,x),\nabla) + (x, a_0^*(t)\nabla) + c^*(t,x). \tag{4.12}$$

Similarly as in the proof of theorem 4.1, sufficient conditions for well-posedness of the Cauchy problem

$$\frac{\partial u}{\partial t} + A(t)u = 0, \quad u(\tau) = u_0$$

can be obtained. Here all considerations remain valid if a_0 and c are bounded.

If these operators are not bounded, an additional reasoning is required (and if both are nonzero, also supplementary conditions). We omit these here.

COROLLARY 4.2. *Consider the differential operator*

$$A^{(I)} = (\nabla, A(x)\nabla) - (A(x)x,\nabla), \quad A = BB^*.$$

It was shown in chapter IV that this operator is symmetric on the Hilbert space $\mathfrak{H}$. If $c = c^*$, then by theorem IV.1.4, well-posedness of the Cauchy problem implies that the operator

$$A^{(I)} + c(x),$$

which is defined on the space Δ of bounded smooth functions (with values in D_c if c is unbounded), is essentially selfadjoint. Here the invariance of Δ under the corresponding evolution family (following from theorem 3.4) plays an important role.

However, theorem 3.4 relates to multiplicative functionals determined by more general equations than (3.7).

Using the specific nature of equation (3.7) we can weaken the restrictions imposed on the dissipative operator function $c(x)$, prescribing only a growth estimate for it (and its derivatives):

$$\|c(x)\| = O(\exp\alpha\|x\|^2).$$

We emphasize that these considerations do not assume invertibility of the coefficient $B(t,x)$.

Finally we consider the case when the multiplicative operator functional is determined by a stochastic equation.

THEOREM 4.3. *The evolution family (3.3) determined by the multiplicative operator functional $T(t,s)$ has, under the conditions of theorem 3.2, generating operator $\tilde{A}_t$ whose restriction to the set $\tilde{\Delta}$ of functions of class C_2 for which $a_0^*f'(x)$ and $c_0^*f(x)$ are defined can be described by the formula*

$$\begin{aligned} A_t f(x) = {}& \frac{1}{2}(B(t,x)B^*(t,x)\nabla,\nabla)f + (x, a_0^*(t)\nabla)f + \\ & (a_1(t,x),\nabla)f + \hat{b}^*(t,x;B^*(t,x)\nabla,f) + c_0^*(t,x)f + c_1(t,x)f. \end{aligned} \tag{4.13}$$

PROOF. It suffices to compute (4.5) and (4.6). The stochastic equation (3.13) readily implies a formula for the multiplicative functional:

$$v(t,x)y = [c_0(t,x) + c_1(t,x)]y \quad (y \in D_{c_0}). \tag{4.14}$$

Using Itô's formula we obtain, up to quantities of higher order of smallness,

$$\begin{gathered}
\mathrm{M}_{\xi(s)=x}(T(s+\Delta s, s)y - y, f(\xi(s+\Delta s)) - f(x)) = \\
\mathrm{M}\left(\int_s^{s+\Delta s} b(\theta, \xi_{s,x}(\theta); T(\theta,s)y, dw(\theta))\,,\right. \\
\left.\int_s^{s+\Delta s} f'(\xi_{s,x}(\theta))B(\theta, \xi_{s,x}(\theta)\, dw(\theta)\right) + o(\Delta s) = \\
\sum_{k,k',j} \mathrm{M} \int_s^{s+\Delta s} (b(\theta, \xi_{s,x}(\theta); T(\theta,s)y, e_k), \phi_j)\, dw_k(\theta) \times \\
\int_s^{s+\Delta s} (f'(\xi_{s,x}(\theta))B(\theta, \xi_{s,x}(\theta))e_{k'}, \phi_j)\, dw_{k'}(\theta) + o(\Delta s),
\end{gathered}$$

where $\{\phi_j\}$ and $\{e_k\}$ are orthonormal bases in the spaces Y and X, respectively. Independence of the scalar Wiener processes w_k, $w_{k'}$ ($k \neq k'$) and properties of stochastic integrals imply by (4.6) that

$$\begin{gathered}
(y, \gamma_s f(x)) = \\
\lim_{\Delta s \to 0} \frac{1}{\Delta s} \sum_k \int_s^{s+\Delta s} \mathrm{M}(b[\theta, \xi_{s,x}(\theta); T(\theta,s)y, e_k], f'(\xi_{s,x}(\theta))B(\theta, \xi_{s,x}(\theta))e_k)\, d\theta = \\
\sum_{k,j} (b(s,x;y,e_k), \phi_j)(\phi_j, f'(x)B(s,x)e_k) = \\
\sum_{k,j} (b(s,x;y,e_k), \phi_j)((f(x), \phi_j)', B(s,x)e_k) = \\
\sum_{k,j} (b(s,x,y,B^*\nabla), \phi_j)(f(x), \phi_j) = \\
(b(s,x;y,B^*\nabla), f(x)) = (\hat{b}(s,x;B^*\nabla, y), f(x)) = \\
(y, \hat{b}^*(s,x;B^*\nabla, f(x)).
\end{gathered}$$

This formula and (4.14), (4.17) imply the required result. □

COROLLARY 4.3. *Theorems 4.3 and 4.4 imply the well-posedness of the Cauchy problem for the system*

$$\begin{gathered}
\frac{\partial u}{\partial t} + A_t u \equiv \\
\frac{\partial u}{\partial t} + \frac{1}{2}(B(t,x)B^*(t,x)\nabla, \nabla)u + (a_1(t,x), \nabla)u + b(t,x;B^*(t,x)\nabla, u) + c_1(t,x)u = 0
\end{gathered}$$

in the class $C_2(X,Y)$, provided all coefficients are bounded operators.

Uniqueness of the solution is proved as in theorem 4.1.

The results can be generalized to the case of unbounded dissipative operators $a_0(t)$ and $c_0(t)$, although an additional reasoning is needed to prove invariance of the class Δ (especially if both a_0 and c_0 are distinct from zero).

COROLLARY 4.4. *If the operator (4.13) is dissipative on the Hilbert space $\mathfrak{H} = L_2(X,Y,\mu)$ (see chapter IV, §1), then well-posedness of the Cauchy problem implies maximal dissipativity of the closure of its restriction to $C_2(X,Y)$.*

4.3. Quasilinear systems. In this section we show how to apply the theory developed above to the study of quasilinear systems

$$\begin{gathered}\frac{\partial u}{\partial t} + \frac{1}{2}(A(t,x,u)\nabla,\nabla)u + \\ (a(t,x,u),\nabla)u + b(t,x;B^*(t,x,u)\nabla,u) + c(t,x,u)u = 0,\\ u(\tau,x) = \phi(x),\end{gathered} \tag{4.15}$$

where B, a, b, c are functions on $[t_0,\tau]\times X\times Y$ with values in $\mathcal{L}_2(\mathcal{H},X)$, X, $\mathcal{L}_2(Y\times\mathcal{H},Y)$, and $\mathcal{L}(Y)$, respectively, and $A = BB^*$.

We assume that the following conditions hold:

A$_1$) restricted growth of the coefficients:

$$\begin{gathered}\|a(t,x,u)\| \le \alpha(1+\|x\|+\|u\|^p) \quad (\alpha>0);\\ \sigma_2^2(B(t,x,u)) \le \alpha(1+\|x\|^2+\|u\|^{2p});\\ ([c(t,x,u)+c^*(t,x,u)]h,h) \le \|h\|^2(\rho_0+\rho_1\|u\|^p);\\ \sigma_2^2(b(t,x,u)(\cdot,h)) \le \|h\|^2\rho_1(1+\|u\|^{2p}) \quad (\rho_1>0);\end{gathered}$$

A$_2$) Lipschitz conditions:

$$\begin{gathered}\|a(t,x_1,u_1)-a(t,x_2,u_2)\| \le \alpha(\|x_2-x_1\|^2+K_{12}\|u_2-u_1\|^2);\\ \sigma_2^2(B(t,x_1,u_1)-B(t,x_2,u_2)) \le \alpha(\|x_2-x_1\|^2+K_{12}\|u_2-u_1\|^2);\\ \sigma_2^2(b(t,x_1,u_1)(\cdot,h)-b(t,x_2,u_2)(\cdot,h)) \le \|h\|\rho_1(\|x_2-x_1\|^2+K_{12}\|u_2-u_1\|^2),\end{gathered}$$

where K_{12} is a constant depending only on $\max\{\|u_1\|,\|u_2\|\}$;

A$_3$) smoothness conditions for the coefficients a, B, b, c, requiring that they have derivatives with respect to x and u up to order three which are continuous with respect to (t,x,u), and satisfy, in the corresponding norms, estimates of the type

$$\|f^{(j)}(t,x,u)\| \le \delta_1+\delta_2\|u\|^{p-j}$$

for the derivatives of orders $j = 1,2,3$.

In the sequel, for functions of type $f(t,x,u)$ and $v(t,x)$ we will use the notation $f_v(t,x) = f(t,x,v(t,x))$. Let $v(t,x)$ be a bounded Lipschitz function:

$$\begin{gathered}\|v(t,x_2)-v(t,x_1)\| \le \mathcal{L}^v(t)\|x_2-x_1\|,\\ \sup_X\|v(t,x)\| = |||v(t)||| < \infty.\end{gathered}$$

Consider the stochastic system

$$\xi(t) = x + \int_s^t a_v(\theta, \xi(\theta))\, d\theta + \int_s^t A_v(\theta, \xi(\theta))\, dw(\theta) \equiv$$

$$x + \mathcal{M}(s, t, v, \xi) \quad (x \in X, \quad t_0 \le s \le t \le \tau), \tag{4.16}$$

$$\eta(t,s)y = y + \int_s^t c_s^*(\theta, \xi(\theta))\eta(\theta, s)y\, d\theta + \int_s^t b_v^*(\theta, \xi(\theta))(dw(\theta), \eta(\theta, s)y)\, d\theta =$$

$$y + N(s, t, v, \xi, \eta)y. \tag{4.17}$$

This system determines, under the given conditions, a Markov stochastic process $\xi_v(t)$ and a multiplicative operator functional $\eta_v(t,s)$, both depending ion the function v. There arise a corresponding evolution family $U_v(t,\tau)$:

$$(y, U_v(t,\tau)\phi(x)) = \mathrm{M}(\eta_0(\tau, t; \xi)y, \phi(\xi_{t,x,v})).$$

For a sufficiently smooth v the function

$$u(t,x) = U_v(t,\tau)\phi(x) \tag{4.18}$$

is a solution of the problem

$$\frac{\partial u}{\partial t} + A_0(u) \equiv$$

(4.19)

$$\frac{\partial u}{\partial t} + \frac{1}{2}(A_v(t,x)\nabla, \nabla)u + (a_v(t,x), \nabla)u + b_v(t, x, B_v^*(t,x)\nabla, u) + c_v(t,x)u = 0,$$

$$u(\tau, x) = \phi(x).$$

Completing the equations (4.16)–(4.18) by the relation

$$v = u, \tag{4.20}$$

we obtain a closed system of equations with respect to u, ξ, η. If the solution of this system exists and is unique, then the Lipschitz function $u(t,x)$ is naturally called a generalized solution of the problem

$$\frac{\partial u}{\partial t} + A_u(u) = 0, \quad u(\tau, x) = \phi(x),$$

which coincides with (4.15).

Clearly, the smooth solution of this problem is generalized in this sense, and, vice versa, a generalized solution having the required derivatives satisfies (4.15).

We will obtain the solution of the problem under consideration by means of the following procedure of successive approximation.

Let $u_0(t,x)$ be a function of class C_2 satisfying

$$u_0(\tau, x) = \phi(x).$$

(E.g., $u_0(t,x) = \phi(x)$.) If the function $u_k(t,x)$ has already been constructed, then the stochastic processes ξ^k, η^k are determined from

$$\xi^k(t) = x + \mathcal{M}(s,t,u_k,\xi^k), \tag{4.21}$$

$$\eta^k(t,s)y = y + N(s,t,u_k,\xi^k,\eta^k)y, \tag{4.22}$$

and we find

$$u_{k+1}(s,x) = U_{u_k}(s,\tau)\phi(x). \tag{4.23}$$

It follows from the properties of solutions of stochastic equations described above that that every function u_{k+1} is a classical solution of the problem

$$\frac{\partial u_{k+1}}{\partial t} + A_{u_k}(u_{k+1}) = 0.$$

In particular, each of them satisfies a Lipschitz condition, and this enables us to continue this procedure indefinitely. To prove its convergence, we need certain estimates, based on the following lemma.

LEMMA 4.2. *Under the conditions A_1 there exist an interval $\Delta = [t_1, \tau]$, depending on $\|\phi\| = \max_x \|\phi(x)\|$, and a function $\psi(t)$ which is bounded on this interval, such that if*

$$\|v(t)\| \le \psi(t),$$

then the same estimate holds for $u(t,x) = u_v\phi$:

$$\|u(t,x)\| \le \psi(t).$$

If $\rho_0 \le -\rho_1(1 + 2\|\phi\|^{2p})$, we can put $t_1 = t_0$, i.e. the estimate is true on the whole interval $[t_0,\tau]$.

PROOF. From proposition 2.2. we readily derive the estimate

$$\mathrm{M}\|\eta(t,s)y\|^2 \le \|y\|^2 \exp\left\{(\rho_0+\rho_1)(t-s) + \int_s^t \|v(\theta)\|^{2p}\,d\theta\right\},$$

which leads to the inequality

$$\|u(s)\|^{2p} \le \exp\left\{p(\rho_0+\rho_1)(\tau-s) + 2p\rho_1\int_s^t \|v(\theta)\|^{2p}\,d\theta\right\}\|\phi\|^{2p}.$$

Consider a function $\psi(t)$ satisfying the functional equation

$$\psi(s) = \psi(t)\exp\left\{p(\rho_0+\rho_1)(\tau-s) + 2p\rho_1\int_s^t \psi(\theta)\,d\theta\right\},$$

$$\psi(\tau) = \|\phi\|^{2p}.$$

This function satisfies also the differential equation

$$-\psi'(s) = p\psi(s)[\rho_0+\rho_1+2\rho_1\psi(s)].$$

If also $\rho_0+\rho_1+2\rho_1\|\phi\|^{2p} > 0$, then

$$\psi(s) = \frac{(\rho_0+\rho_1)\|\phi\|^{2p}}{(\rho_0+\rho_1+2\rho_1\|\phi\|^{2p})e^{-p(\rho_0+\rho_1)(\tau-s)}}$$

is easily seen to be bounded on any interval $[s,\tau]$ where

$$\tau - s < \frac{1}{p(\rho_0+\rho_1)} \ln\left(1+\frac{\rho_0+\rho_1}{2\rho_1\|\phi\|^{2p}}\right)$$

and

$$\tau - s < \frac{1}{p\rho_1\|\phi\|^{2p}} \quad \text{for } \rho_0+\rho_1=0.$$

Conversely, if $\rho_0+\rho_1+2\rho_1\|\phi\|^{2p}\le 0$, then

$$\psi(s)=\frac{(|\rho_0|-\rho_1)\|\phi\|^{2p}}{(|\rho_0|-\rho_1-2\rho_1\|\phi\|^{2p})e^{p(|\rho_0|-\rho_1)(\tau-s)}-2\rho_1\|\phi\|^{2p}}\le\frac{|\rho_0|-\rho_1}{2\rho_1}$$

for any $s\le\tau$.

To end the proof we note that if $|||v(t)|||^{2p}\le\psi(t)$, then also $|||u(t)|||^{2p}\le\psi(t)$. □

COROLLARY 4.5. *The functions $u_k(t,x)$, k=1,2, ... , constructed above satisfy on the interval the uniform estimate*

$$|||u_k(t,x)|||\le C.$$

COROLLARY 4.6. *By differentiation with respect to x of the stochastic system under consideration we obtain similar systems for the derivatives $u'_x, u''_{xx}, u'''_{xxx},\dots$, as was shown in §2. Applying the lemma to these systems, we arrive at*

$$|||u_k^{(j)}(t,x)|||\le C \quad (k=1,2,\dots,\quad j=1,2,3)$$

on, in general smaller, intervals Δ_j:

$$\Delta\supset\Delta_1\supset\Delta_2\supset\Delta_3.$$

However, for fixed $|||\phi|||$ and negative ρ_0 with sufficiently large absolute value these intervals contain the initial interval $[t_0,\tau]$ on which the equations are defined.

THEOREM 4.4. *Suppose the conditions A_1–A_3 hold. There exists a unique generalized solution of the Cauchy problem with initial function of class C_2 on Δ_1. This solution has continuous derivatives up to order two, and hence is a classical solution on the interval Δ_3. The generalized solution $u(t,x)$ is representable as a mean $(y,u(t,x))=M(\eta(\tau,t;\xi)y,\phi(\xi_{t,x}))$, where η is a multiplicative operator functional for the Markov process $\xi(t)$ determined by (4.16)–(4.17) for $v=u$.*

PROOF. By the same reasoning as in the proof of propositions 2.2 and 2.3 we can obtain the estimates

$$\text{(4.24)}\qquad M\|\xi_{s,x,v_2}(t)-\xi_{s,x,v_1}(t)\|^2\le K(t-s)\sup|||v_2(\theta)-v_1(\theta)|||^2,$$

(4.25)

$$M\|\eta_{v_2}(t,s,\xi_{v_2})y-\eta_{v_1}(t,s,\xi_{v_1})y\|^2\le K(t-s)\|y\|^2\sup|||v_2(\theta)-v_1(\theta)|||^2,$$

where K depends on the constants α, ρ_0, ρ_1, the length of the interval Δ_1, and the quantities $\sup_{j=1,2}\{|||v_j(\theta)|||,\mathcal{L}^{v_j}(\theta)\}$.

Using these estimates and taking into account the uniform boundedness of the quantities $|||u_k|||$ and $\mathcal{L}^{u_k}$ on Δ_1, the formula for u_k easily gives the estimates

$$\sup |||u_{k+1}(\theta) - u_k(\theta)||| \leq \tilde{K}_R(\tau - s) \sup_{s\leq\theta\leq\tau} |||u_k(\theta) - u_{k-1}(\theta)|||,$$

where $\tilde{K}_R$ depends only on the radius R of the ball containing the initial function u_0 and on the constants α, ρ_0, ρ_1.

For sufficiently small $(\tau - s) < \epsilon$ we obtain $q = \tilde{K}_R(\tau - s) < 1$. In this case the sequence $u_k(t, x)$ converges uniformly on $[\tau - \epsilon, \tau] \times X$ to some function $u(t, x)$. Formulas (4.24)–(4.25) imply mean square convergence of the sequences: $\xi = \lim \xi_n$, $\eta = \lim \eta_n$.

In the equations (4.21)–(4.23) defining the successive approximations we can pass to the limit, and thus obtain the following equations for u, ξ, η:

$$\xi(t) = x + \mathcal{M}(s, t, u, \xi),$$
$$\eta(t, s)y = y + N(s, t, u, \xi, \eta),$$
$$(y, u(t, x)) = \mathrm{M}(\eta(\tau, t, \xi)y, \phi(\xi_{t,x,u}(t)).$$

We show that $u(t, x)$ is a generalized solution of the problem on the interval $[\tau-\epsilon, \tau]$. Uniqueness of the solution holds, as always under the conditions of the contractive mapping principle.

Let R be chosen such that for fixed $|||\phi|||$ the estimate $|||u_k(\theta)||| \leq R$ holds on all of Δ_1. Then the above described procedure can be repeated on the interval $[\tau - 2\epsilon, \tau - \epsilon]$ with the same $\epsilon > 0$. Therefore, after finitely many steps the generalize solution $u(t, x)$ is extended to the interval Δ_1.

Applying this reasoning to the (sufficiently often) differentiated system, we can show that $u(t, x)$ is twice continuously differentiable on Δ_3, and is thus a classical solution of the problem. □

REMARK 4.1. A more precise analysis shows that under certain assumptions that are restrictive only in the infinite-dimensional case, convergence of the successive approximations holds on all of Δ_1, and together with that of their derivatives up to order two on all of Δ_3.

5. Chapter VII: Supplementary remarks and historical comments

The theory of stochastic differential equations was independently outlined, in different forms, by Gikhman [1,2] and Itô [2,3], and was developed by Skorokhod [3]. The concept of stochastic integral introduced by K. Itô turned out to be a suitable base for further developments. Already in the consideration of equations in finite-dimensional spaces it became clear that the principal part of the theory is almost independent of dimension. This made it possible to generalize it to the infinite-dimensional case. The first papers in this direction were by pupils of I.I. Gikhman and A.V. Skorokhod: Baklan [1,3] and Chantladze [1,2].

For the diffusion processes constructed, certain second order equations in functional derivatives of type (4.4) were obtained as the Kolmogorov equations. Note that Baklan [2] was the first to consider an equation of type (2.16) with unbounded operator

$a_0(t)$ (more exactly, he considered (2.17), not proving its equivalence to (2.16)). An invariant presentation of the theory of stochastic differential equations with bounded operator coefficients in a scale of Hilbert spaces, together with applications to the proof of the well-posedness of the Cauchy problem for equations of type (4.4), was given in Dalecky [8].

It is useful to note that stochastic equations play the same role in the theory of the Cauchy problem for second order parabolic equations as do characteristic equations in the theory of first order partial differential equations. The representation of the solution as a mean over paths of a Markov process has the same meaning as the representation of an evolution family of operators as a functional integral (see chapter VI). However, the possibility of using 'characteristics' permits us to more general results in the case of diffusion equations.

Beginning with M. Kac [1], the method of multiplicative functionals is used to introduce additional terms of the type $c(x)u(x)$ into the equation. The method has been repeatedly generalized by many mathematicians (see also the comments to chapter VI). Prokhorov [1] was the first to introduce a simple multiplicative functional embracing the stochastic integral. To the best of our knowledge, multiplicative operator functionals were first introduced in Dalecky [2,5], and subsequently, in different versions, in Pinsky [1,2], Dalecky–Teterina [1], Teterina [1], Dalecky [13,16]. In Belopolskaya–Dalecky [1] the above described results are transferred to processes in Banach spaces with smooth norm. A number of results in the theory of stochastic equations was given in abstract Wiener spaces (Kuo [1] and Kuo–Piech [1]).

In these papers equations with bounded operator coefficients were considered. Following the above mentioned paper by V.V. Baklan, the case when some coefficients are unbounded was studied by different methods, related to stochastic partial differential equations, and by Bensoussan–Teman [1], Rozovsky [1,2], Krylov–Rozovsky [1], Vishik–Komech [1], Tinfavicius [1], Belopolskaya [1]. Belopolskaya–Nagolkina [1,2] applied the multiplicative method presented in chapter IV to stochastic equations (including those with unbounded operator a_0). Equations with unbounded random operators were considered by Dalecky [16] by means of the methods given in S. Krein [1].

We also note the interesting direction of study of the stochastic partial differential equations related to problems in hydrodynamics. A review of the results in this direction, with additional references, is contained in Vishik–Komech–Fursikov [1].

In the last section we have studied quasilinear parabolic systems by methods from Belopolskaya–Dalecky [2]. These results generalize results by Freidlin [1,2] related to a single equation in finite-dimensional space. In a later paper Belopolskaya–Dalecky [3] it was shown how to consider along this line a parabolic system which is nonlinear with respect to the derivatives.

In Belopolskaya–Dalecky [4] the theory of Itô stochastic equations on infinite-dimensional smooth manifolds is developed, and a short review of the literature on this subject is given.

Bibliography

Albeverio, S., Höegh-Krohn, R.,
[1] *Mathematical theory of Feynman path integrals*, Springer, 1976. (Lecture notes in math. **523**.)
Alimov, A.L.,
[1] *Teoret. Mat. Fiz.* **2: 2** (1972), 182-189. (In Russian.)
[2] *Zap. Nauchn. Sem. LOMI* **42** (1974), 6-11. (In Russian.)
Alimov, A.L., Buslaev, V.S.,
[1] *Vestnik LGU, Mat.* **1** (1972), 5-14. (In Russian.)
Averbukh, V.I.,
[1] *Uspekhi Mat. Nauk* **22: 3** (1967), 233-234. (In Russian.)
Averbukh, V.I., Smolyanov, O.G.,
[1] *Uspekhi Mat. Nauk* **22: 6** (1967), 201-260. (In Russian.)
[2] *Uspekhi Mat. Nauk* **23: 4** (1968), 67-116. (In Russian.)
Averbukh, V.I., Smolyanov, O.G., Fomin, S.V.,
[1] *Tr. Moskov. Mat. Obshch.* **24** (1971), 133-174. (In Russian.)
[2] *Tr. Moskov. Mat. Obshch.* **27** (1972), 248-262. (In Russian.)
Babbitt, D.,
[1] *Bulletin Amer. Math. Soc.* **70: 2** (1964), 254-259.
[2] *Trans. Amer. Math. Soc.* **116: 4** (1965), 66-78.
Badrikian, A.,
[1] *Sem. sur les fonctions aléatoires linéares et les mesures cylindriques*, Springer, 1970. (Lecture notes in math. **139**
Baklan, V.V.,
[1] *Dokl. Akad. Nauk UkrSSR* **10** (1963), 1299-1303. (In Russian.)
[2] *Dokl. Akad. Nauk SSSR* **159: 4** (1964), 707-710. (In Russian.)
[3] *Dokl. Akad. Nauk UkrSSR* **5** (1965), 554-556. (In Russian.)
[4] *Mat. Fiz.R* **7** (1970), 18-25. (In Russian.)
Baklan, V.V., Shatashvili, A.D.,
[1] *Dokl. Akad. Nauk UkrSSR* **9** (1965), 1115-1117. (In Russian.)
Belopolskaya, Ya.I.,
[1] *Abstracts Vilnius Conf.* (1978), 12-16. (In Russian.)
[2] *Veroyatn. Razpredel. v Bezkonechn. Prostr., Kiev* (1978), 5-21. (In Rus-

sian.)
Belopolskaya, Ya.I., Dalecky, Yu.L.,
[1] *Tr. Moskov. Mat. Obshch.* **37** (1978), 107-141. (In Russian.)
[2] *Izv. Vyzov. Mat.* **12** (1978), 5-17. (In Russian.)
[3] *Dokl. Akad. Nauk SSSR* **250: 3** (1980), 521-524. (In Russian.)
[4] *Uspekhi Mat. Nauk* **37: 3** (1982), 95-142. (In Russian.)
Belopolskaya, Ya.I., Nagolkina, Z.I.,
[1] *Dokl. Akad. Nauk UkrSSR* **11** (1977), 966-970. (In Russian.)
[2] *Veroyatn. Razpredel. v Bezkonechn. Prostr., Kiev* (1978), 22-37. (In Russian.)
Bensoussan, A., Temam, R.,
[1] *J. Funct. Anal.* **13: 2** (1973), 195-222.
Bentkus, V.Yu.,
[1] *Litov. Mat. Sb.* **16: 3** (1976), 21-29. (In Russian.)
[2] *Litov. Mat. Sb.* **17: 4** (1977), 5-20. (In Russian.)
Berezansky, Yu.M.,
[1] *Expansions in eigenfunctions of selfadjoint operators*, Amer. Math. Soc., 1968. (Translated from the Russian.)
[2] *Selfadjoint operators in spaces of functions of infinitely many variables*, Amer. Math. Soc., 1986. (Translated from the Russian.)
Berezansky, Yu.M., Gali, I.M.,
[1] *Ukrain. Mat. Zh.* **24: 4** (1972), 247-262. (In Russian.)
Berezansky, Yu.M., Gali, I.M., Zhuk, V.A.,
[1] *Dokl. Akad. Nauk SSSR* **203** (1972), 13-15. (In Russian.)
Berezansky, Yu.M., Samoilenko, V.G.,
[1] *Uspekhi Mat. Nauk* **26: 5** (1981), 3-56. (In Russian.)
Berezansky, Yu.M., Us, G.F.,
[1] *Reports Math. Phys.* **7: 1** (1972), 103-126.
Blecker, P.M., Vishik, M.I.,
[1] *Mat. Sb.* **86: 3** (1971), 446-494. (In Russian.)
Bogdansky, Yu.V.,
[1] *Ukrain. Mat. Zh.* **29: 6** (1977), 781-784. (In Russian.)
Bogolyubov, N.N.,
[1] *Izbr. Trudy*, **2** Kiev, 1970. (In Russian.)
Bokhonov, Yu.E., Dalecky, Yu.L.,
[1] *Ukrain. Mat. Zh.* **31: 6** (1979), 644-649. (In Russian.)
Bourbaki, N.,
[1] *Eléments de mathématique.*, **5. Espaces vectoriels topologiques** Hermann, 1953.
[2] *Eléments de mathématique.*, **6. Intégration** Hermann, 1963.
Buldygin, V.V.,
[1] *Convergence of random elements in topological spaces*, Kiev, 1980. (In Russian.)
Buldygin, V.V., Donchenko, V.S.,

[1] *Preprint Inst. Mat. Akad. Nauk UkrSSR* **76.1** (1976), (In Russian.)
Buslaev, V.S.,
[1] *Probl. Mat. Fiz. Leningrad* **2** (1967), 85-107. (In Russian.)
Cameron, R.H.,
[1] *J. Math. Phys.* **39: 2** (1960), 126-140.
[2] *Proc. Amer. Math. Soc.* **2** (1951), 914-924.
Cameron, R.H., Donsker, M.D.,
[1] *Ann. of Math.* **69** (1959), 15-36.
Cameron, R.H., Martin, W.T.,
[1] *Ann. of Math.* **48: 2** (1947), 385-392.
[2] *Duke Math. J.* **12: 3** (1945), 17-25.
Chanteladze, T.L.,
[1] *Soob. Akad. Nauk GruzSSR* **33: 3** (1964), 529-534.
[2] *Tr. Vychisl. Tsentr. Akad. Nauk GruzSSR* **5: 1** (1965), 106-124.
Chernoff, P.R.,
[1] *J. Funct. Anal.* **2** (1968), 238-242.
[2] *Memoirs Amer. Math. Soc.* **140** (1974),
Chobayan, S.A.,
[1] *Tr. Vychisl. Tsentr. Akad. Nauk GruzSSR* **14: 2** (1975), 80-162. (In Russian.)
Chobayan, S.A., Tarieladze, V.I.,
[1] *J. Mult. Anal.* **7: 1** (1977), 183-203. (In Russian.)
Dalecky, Yu.L.,
[1] *Dokl. Akad. Nauk SSSR* **134: 5** (1960), 1013-1016. (In Russian.)
[2] *Dokl. Akad. Nauk SSSR* **137: 2** (1961), 268-271. (In Russian.)
[3] *Dokl. Akad. Nauk SSSR* **141: 6** (1961), 1290-1293. (In Russian.)
[4] *Izv. Vyzov. Mat.* **3** (1961), 27-48. (In Russian.)
[5] *Uspekhi Mat. Nauk* **17: 5** (1962), 3-115. (In Russian.)
[6] *Dokl. Akad. Nauk SSSR* **166: 5** (1966), 1035-1038. (In Russian.)
[7] *Dokl. Akad. Nauk SSSR* **171: 1** (1966), 21-24. (In Russian.)
[8] *Uspekhi Mat. Nauk* **22: 4** (1967), 3-54. (In Russian.)
[9] *Itogi Nauk. Mat. Anal. 1966* (1967), 83-124. (In Russian.)
[10] *Studia Math.* **38** (1970), 333-340. (In Russian.)
[11] *Proc. Symp. Mechanics of Continuous Media, Tbilisi* (1973), 78-88. (In Russian.)
[12] *Material of the All-Union Sem. on Differential Eqs. with Infinitely Many Independent Variables, and Dynamical Systems with Infinitely Many Degrees of Freedom, Erevan* (1974), 5-49. (In Russian.)
[13] *Uspekhi Mat. Nauk* **30: 2** (1975), 209-210. (In Russian.)
[14] *Dokl. Akad. Nauk SSSR* **227: 4** (1976), 784-787. (In Russian.)
[15] *Limit Theorems for Random Processes, Kiev* (1977), 108-124. (In Russian.)
[16] *Operator multiplicative functionals and their applications*, Springer, 1980. pp. 38-49. (In Russian.) (Lecture notes in control and inform. science **25**, Ed.:

B. Grigelionis.)
[17] *Funkts. Anal. Prilozh.* **14: 4** (1980), 75-76. (In Russian.)
Dalecky, Yu.L., Fomin, S.V.,
[1] *Teor. Veroyatn. Primenen.* **10: 2** (1965), 329-343. (In Russian.)
[2] *Dokl. Akad. Nauk SSSR* **205: 4** (1972), 759-762. (In Russian.)
[3] *Abstracts of Lectures at the Internat. Conf. on Probab. Theory, Vilnius* (1973), 197-200. (In Russian.)
[4] *Tr. Sem. G.I. Petrov.* **4** (1978), 45-64. (In Russian.)
Dalecky, Yu.L., Krylov, V.Yu., Minlos, R.A., Sudakov, V.N.,
[1] *Tr. 4. VseSoyuzn. Mat. Sezd., Leningrad* **V: 2** (1964), 282-292. (In Russian.)
Dalecky, Yu.L., Kukharchuk, N.M.,
[1] *Ukrain. Mat. Zh.* **6** (1965), 114-117. (In Russian.)
Dalecky, Yu.L., Ladokhin, V.I.,
[1] *Teor. Veroyatn. Primenen.* **4** (1974), 845-849. (In Russian.)
[2] *I, Teor. Veroyatn. i Mat. Stat.* **17** (1977), 51-60. *II, Teor. Veroyatn. i Mat. Stat.* **18** (1978), 37-45. (In Russian.)
Dalecky, Yu.L., Samoilenko, Yu.S.,
[1] *Dokl. Akad. Nauk SSSR* **213: 3** (1973), 515-518. (In Russian.)
Dalecky, Yu.L., Stremsky, V.V.,
[1] *Uspekhi Mat. Nauk* **24: 1** (1969), 191-192. (In Russian.)
Dalecky, Yu.L., Teterina, N.I.,
[1] *Uspekhi Mat. Nauk* **27: 2** (1972), 167-168. (In Russian.)
Dalecky, Yu.L., Zaplitnaya, A.T.,
[1] *Ukrain. Mat. Zh.* **5** (1965), 110-114. (In Russian.)
Dieudonne, J.,
[1] *Foundations of modern analysis*, ACad. Press, 1960. (Translated from the French.)
Donsker, M.D.,
[1] *Proc. Conf. Theory and Applic. of Analysis in Function Spaces, Dedham, 1963*, M.I.T., 1964. pp. 17-30.
Doob, J.L.,
[1] *Stochastic processes*, Chapman & Hall, 1953.
Dorfman, I.Ya.,
[1] *Mat. Sb.* **81: 2** (1970), 192-208. (In Russian.)
[2] *Uspekhi Mat. Nauk* **28: 6** (1973), 203-204. (In Russian.)
[3] *Sibirsk. Mat. Zh.* **17: 5** (1976), 1023-1031. (In Russian.)
Dudin, D.N.,
[1] *Uspekhi Mat. Nauk* **27: 2** (1972), 167-170. (In Russian.)
Dudley, R.M.,
[1] *J. Funct. Anal.* **1: 3** (1967), 290-330.
[2] *Ann. Probab.* **1: 1** (1973), 66-103.
Dunford, N., Schwartz, I.T.,
[1] *Linear operators I. General theory*, Interscience, 1958.

Dynkin, E.B.,
[1] *Die Grundlagen der Theorie der Markoffschen Prozesse*, Springer, 1961. (Translated from the Russian.)
[2] *Markov processes*, Springer, 1965. (Translated from the Russian.)
Evgravof, M.A.,
[1] *Dokl. Akad. Nauk SSSR* **191: 5** (1971), 979-982. (In Russian.)
Faris, W.G.,
[1] *Pacific J. Math.* **22** (1967), 147-179.
[2] *J. Funct. Anal.* **1** (1967), 93-108.
Feldman, J.,
[1] *Pacific J. Math.* **8: 4** (1958), 699-708.
[2] *Trans. Amer. Math. Soc.* **108: 2** (1963), 257-264.
[3] *Teor. Veroyatn. Primenen.* **10: 2** (1965), 375-379. (In Rusian.)
Feller, M.N.,
[1] *Dokl. Akad. Nauk UkrSSR* **12** (1965), 1558-1562. (In Russian.)
[2] *Dokl. Akad. Nauk SSSR* **172: 6** (1967), 1282-1285. (In Russian.)
[3] *Dokl. Akad. Nauk SSSR* **205: 1** (1972), 36-39. (In Russian.)
[4] *Dokl. Akad. Nauk SSSR* **221: 5** (1975), 1046-1049. (In Russian.)
[5] *Uspekhi Mat. Nauk* **32: 1** (1980), 69-79. (In Russian.)
[6] *Dokl. Akad. Nauk UkrSSR* **8** (1965), 994-998. (In Russian.)
Feynman, R.P.,
[1] *Reviews Modern Physics* **20: 2** (1948), 367-387.
Feynman, R.P., Hibbs, A.R.,
[1] *Quantum mechanics and path integrals*, New York, 1965.
Foias, C.,
[1] *I, Rad. Sem. Mat. Univ. Radova* **48** (1972), 219-348. *II, Rad. Sem. Mat. Univ. Radova* **49** (1973), 9-123.
Fomin, S.V.,
[1] *Uspekhi Mat. Nauk* **23: 1** (1968), 221-222. (In Russian.)
[2] *Uspekhi Mat. Nauk* **23: 2** (1968), 215-216. (In Russian.)
[3] *Vestnik MGU, Mat. Mekh.* **2** (1970), 57-85. (In Russian.)
[4] *Dokl. Akad. Nauk SSSR* **181: 4** (1968), (In Russian.)
[5] *Nauchn. Dokl. Vys. Shkoly, Sekts. Fiz. Mat. Nauk* **2** (1958), 83-85. (In Russian.)
Freidlin, M.I.,
[1] *Funkts. Anal. Prilozh.* **1: 3** (1967), 74-82. (In Russian.)
[2] *Mat. Sb.* **78: 3** (1969), 332-348. (In Russian.)
Friedman, A.,
[1] *Partial diffrential equations of parabolic type*, Prentice Hall, 1964.
Frolov, N.N.,
[1] *Teor. Veroyatn. Prilozhen.* **3** (1970), 200-210. (In Russian.)
[2] *Dokl. Akad. Nauk SSSR* **203: 1** (1972), 39-42. (In Russian.)
[3] *Mat. Sb.* **90: 3** (1973), 403-414. (In Russian.)
[4] *Mat. Sb.* **92: 3** (1973), 430-445. (In Russian.)

[5] *Funkts. Anal. Prilozhen.* **14: 1** (1980), 85-86. (In Russian.)
Gadzhiev, V.G.,
[1] *Random Processes in Problems of Math. Phys., Kiev* (1979), 57-72. (In Russian.)
Gavrishova, N.I., Dalecky, Yu.L.,
[1] *Boundary Problems in Math. Phys., Kiev* (1981), 19-22. (In Russian.)
Gelfand, I.M.,
[1] *Uspekhi Mat. Nauk* **11: 6** (1956), 3-12. (In Russian.)
Gelfand, I.M., Shilov, G.E.,
[1] *Generalized functions*, **1-3** Acad. Press, 1964-1968. (Translated from the Russian.)
Gelfand, I.M., Vilenkin, N.Ya.,
[1] *Applications of harmonic analysis*, Acad. Press, 1964. (Translated from the Russian.)
Gelfand, I.M., Yaglom, A.M.,
[1] *Uspekhi Mat. Nauk* **11: 1** (1956), 77-114. (In Russian.)
Georgadze, Z.G., Tarieladze, V.I., Chobanyan, S.A.,
[1] *Dokl. Akad. Nauk SSSR* **241: 3** (1978), (In Russian.)
Gestin, G.N.,
[1] *Teor. Funkts., Funkts. Anal. i Prilozhen.* **12** (1970), 69-81. (In Russian.)
[2] *Uspekhi Mat. Nauk* **28: 2** (1976), 170-182. (In Russian.)
[3] *Funkts. Anal. Prilozhen.* **10: 1** (1976), 75-76. (In Russian.)
Getoor, R.K.,
[1] *Pacific J. Math.* **7: 1** (1957), 885-896.
Gikhman, I.I.,
[1] *Uspekhi Mat. Nauk* **2: 3** (1950), 45-69. (In Russian.)
[2] *Uspekhi Mat. Nauk* **2: 4** (1950), 37-63. (In Russian.)
Gikhman, I.I., Skorokhod, A.V.,
[1] *Introduction to the theory of random processes*, Saunders, 1969. (Translated from the Russian.)
[2] *Uspekhi Mat. Nauk* **21: 6** (1966), 83-152. (In Russian.)
[3] *The theory of stochastic processes*, Springer, 1974. (Translated from the Russian.)
Girsanov, I.V., Mityagin, B.S.,
[1] *Nauchn. Dokl. Vysh. Shkoly, Sekts. Fiz. Mat. Nauk* **2** (1959), 5-9. (In Russian.)
Gikhman, I.I., Skorokhod, A.V.,
[1] *Introduction to the theory of linear nonselfadjoint operators*, Amer. Math. Soc., 1969. (Translated from the Russian.)
Goodman, V.,
[1] *Trans. Amer. Math. Soc.* **164** (1972), 411-426.
Gross, L.,
[1] *Proc. Vth Berkeley Symp. Math. Stat. Probab.* **2** (1965), 31-42.
[2] *Trans. Amer. Math. Soc.* **105** (1962), 372-390.

[3] *Memoirs Amer. Math. Soc.* **46** (1963),
[4] *J. Funct. Anal.* **1** (1967), 123-181.
[5] *Trans. Amer. Math. Soc.* **94: 3** (1960), 404-440.

Guseinov, R.V.,
[1] *Vestnik MGU, Mat. Mekh.* **4** (1970), 39-49. (In Russian.)
[2] *Vestnik MGU, Mat. Mekh.* **4: 1** (1969), (In Russian.)

Hajek, I.,
[1] *Chekhoslov. Mat. Zh.* **8 (83)** (1958), 610-618.

Halmos, P.R.,
[1] *Measure theory*, v. Nostrand, 1950.

Hida, T.,
[1] *Nagoya Math. J.* (1969),
[2] *Math. Inst. Nagoya Univ.* (1971), 1-75.

Hopf, E.,
[1] *J. Rat. Mech. Anal.* **1** (1952), 87-123.

Itô, K.,
[1] *Proc. Imper. Acad. Tokyo* **20** (1944), 519-524.
[2] *Proc. Japan Acad.o* **22: 2** (1946), 32-35.
[3] *Memoirs Amer. Math. Soc.* **4** (1951),
[4] *Proc. IVth Berkeley Symp. Math. Stat. Probab.* **2** (1961), 227-238.
[5] *Memoirs College Sci. Univ. Kyoto* **28** (1954), 209-223.
[6] *Japan J. Math.* **22** (1952), 63-86.

Kac, M.,
[1] *Trans. Amer. Math. Soc.* **65** (1949), 1-13.
[2] *Proc. IInd Berkeley Symp. Math. Stat. Probab.* **2** (1951), 189-215.

Kac, M.P.,
[1] *Sibirsk. Mat. Zh.* **13: 5** (1972), 1158-1168. (In Russian.)
[2] *Uspekhi Mat. Nauk* **28: 3** (1973), 183-184. (In Russian.)
[3] *Differentiable measures. Thesis*, MGU, 1976. (In Russian.)

Kakutani, S.,
[1] *Ann. of Math.* **4: 9** (1948), 214-224.

Kato, T.,
[1] *Perturbation theory for linear operators*, Springer, 1966.

Kolmogorov, A.N.,
[1] *Grunbegriffe der Wahrscheinlichkeitstheorie*, Chelsea, 1946. (Translated from the Russian.)
[2] *C.R. Acad. Sci.* **200** (1935), 1717. (In French.)
[3] *Teor. Veroyatn. Primenen.* **4:2** (1959), 237-239. (In Russian.)

Kostyuchenko, A.G., Mityagin, B.S.,
[1] *Tr. Moskov. Mat. Obshch.* **9** (1960), 283-316. (In Russian.)

Kovalchik, I.M.,
[1] *Uspekhi Mat. Nauk* **18: 1** (1963), 97-134. (In Russian.)

Krein, S.G.,
[1] *Linear differential equations in Banach pace*, Amer. Math. Soc., 1971.

(Translated from the Russian.)
Krylov, N.V.,
[1] *Itogi Nauk. Tekhn. Sovrm. Probl. Mat.* **14** (1979), 71-146. (In Russian.)
Krylov, N.V., Rozovsky, B.L.,
[1] *Izv. Akad. Nauk SSSR* **41: 6** (1977), 1329-1347. (In Russian.)
Krylov, V.Yu.,
[1] *Dokl. Akad. Nauk SSSR* **132: 6** (1960), 1254-1257. (In Russian.)
[2] *Dokl. Akad. Nauk SSSR* **163: 2** (1965), 289-292. (In Russian.)
Kuelbs, I.,
[1] *J. Funct. Anal.* **5: 3** (197?), 354-367.
[2] *Pacific J. Math.* **31: 2** (1969), 433-450.
Kuelbs, I., Mandrekar, V.,
[1] *Trans. Amer. Math. Soc.* **149: 1** (1970), 213-231.
Kuo, H.H.,
[1] *Gaussian measures in Banach spaces*, Springer, 1975. (Lecture notes in math. **463**.)
Kuo, H.H., Piech, M.A.,
[1] *Bulletin Amer. Math. Soc.* **79** (1973), 478-482.
Ladokhin, V.I.,
[1] *Uspekhi Mat. Nauk* **19: 1** (1964), 155-159. (In Russian.)
[2] *Izv. Vyzov. Mat.* **4** (1965), 84-90. (In Russian.)
Ladyzhenskaya, O.A, Solonnikov, V.A, Uraltseva, N.N.,
[1] *Linear and quasilinear equations of parabolic type*, Amer. Math. Soc., 1968. (Translated from the Russian.)
Le Cam, L.,
[1] *Univ. Calif. Publ. Stat.* **2: 11** (1957), 207-236.
Lévy, P.,
[1] *Problèmes concrets d'analyse fonctionnelle*, Gauthier-Villars, 1951.
Maikov, E.V.,
[1] *Uspekhi Mat. Nauk* **18: 3** (1963), 243-244. (In Russian.)
[2] *Dokl. Akad. Nauk SSSR* **156: 5** (1964), 1025-1028. (In Russian.)
Maikov, E.V., Fomin, S.V.,
[1] *Dokl. Akad. Nauk* **449: 3** (1963), 525-528. (In Russian.)
Marchenko, A.V.,
[1] *Mat. Sb.* **96: 2** (1975), 276-293. (In Russian.)
Marsden, J.,
[1] *J. Funct. Anal.* **13** (1973), 51-72.
Maslov, V.P.,
[1] *Méthodes opératorielles*, Mir, 1986. (Translated from the Russian.)
Maslov, V.P., Chebotarev, A.M.,
[1] *Uspekhi Mat. Nauk* **34: 5** (1979), 237-238. (In Russian.)
Maslov, V.P., Shishmarev, I.A.,
[1] *Sovr. Probl. Mat.* **8** (1976), 137-197. (In Russian.)
McKean, H.P.,

[1] *Proc. Nat. Acad. Sci. USA* **59: 6** (1966), 1907-1911.
McShane, E.I.,
[1] *Arch. Rat. Mech. Anal.* **18: 5** (1965), 403-421.
Milxaud, R.A.,
[1] *C.R. Acad. Sci.* **271** (1970), 494-496.
Minlos, R.A.,
[1] *Dokl. Akad. Nauk SSSR* **119: 3** (1958), 439-442. (In Russian.)
[2] *Tr. Moskov. Mat. Obshch.* **8** (1959), 497-518. (In Russian.)
Miyamoto, M.,
[1] *Proc. Japan Acad.* **42: 2** (1966), 70-74.
Mourier, E.,
[1] *Eléments aléatoires dans un espace de Banach*, Paris, 1954.
Mushtari, D.H.,
[1] *Teor. Veroyatn. Primenen.* **18: 1** (1973), 66-77. (In Russian.)
Nelson, E.,
[1] *J. Math. Phys.* **5: 3** (1964), 332-343.
Nemirovsky, A.S., Shilov, G.E.,
[1] *Funkts. Anal. Prilozhen.* **3: 3** (1969), 79-86. (In Russian.)
Neveu, J.,
[1] *Bases mathématiques du calcul des probabilités*, Masson, 1964.
Novikov, E.A.,
[1] *Uspekhi Mat. Nauk* **16: 2** (1961), 135-142. (In Russian.)
[2] *Zh. Eksper. Teoret. Fiz.* **47: 5** (1964), 1919-1926. (In Russian.)
Pardoux, E.,
[1] *Equations aux dérivées partielles stochastiques non linéares, Thèse*, Paris, 1975.
Piech, M.A.,
[1] *J. Funct. Anal.* **3** (1969), 85-114.
Pietsch, A.,
[1] *Nukleare lokalkonvexe Räume*, Akad. Verlag, 1965.
Pinsky, M.,
[1] *Bulletin Amer. Math. Soc.* **77** (1971), 377.
[2] *Trans. Amer. Math. Soc.* **167: 5** (1972),
Polishchuk, E.M.,
[1] *Uspekhi Mat. Nauk* **10: 2** (1955), 179-186. (In Russian.)
[2] *Teor. Veroyatn. Primenen.* **6: 4** (1961), 465-469. (In Russian.)
[3] *Ukrain. Mat. Zh.* **15: 1** (1963), 13-24. (In Russian.)
[4] *Uspekhi Mat. Nauk* **19: 2** (1964), 155-162. (In Russian.)
[5] *Sibirsk. Mat. Nauk* **6: 6** (1965), 1322-1331. (In Russian.)
[6] *Mat. Sb.* **72: 2** (1967), 261-292. (In Russian.)
[7] *Differentsialn. Uravn.* **8: 2** (1972), 333-348. (In Russian.)
Prokhorov, Yu.V.,
[1] *Teor. Veroyatn. Primenen.* **1: 2** (1956), 177-238. (In Russian.)
Prokhorov, Yu.V., Sazonov, V.V.,

[1] *Teor. Veroyatn. Primenen.* **6: 1** (1961), 87-93. (In Russian.)
Ramer, R.,
[1] *J. Funct. Anal.* **15: 2** (1974), 166-187.
Reed, M.,
[1] *J. Funct. Anal.* **5: 1** (1970), 94-124.
Roitburd, V.L.,
[1] *Dokl. Akad. Nauk SSSR* **201: 3** (1971), 545-547. (In Russian.)
[2] *Funkts. Anal. Prilozhen.* **10: 2** (1976), 86-87. (In Russian.)
Rozanov, Yu.A.,
[1] *Tr. Mat. Inst. Steklov.* **108** (1966), 1-136. (In Russian.)
Rozovsky, B.L.,
[1] *Mat. Sb.* **96: 2** (1975), 314-341. (In Russian.)
[2] *IInd Internat. Conf. on Probab. Theory, Vilnius* **3** (1977), 196-197. (In Russian.)
Sato, H.,
[1] *Nagoya Math. J.* **36** (1966), 65-81.
Sazonov, V.V.,
[1] *Proc. All-Union Conf. on Probab. Theory and Math. Stat., Vilnius* (1962), 449-454. (In Russian.)
[2] *Dokl. Akad. Nauk SSSR* **142: 5** (1962), 1034-1035. (In Russian.)
[3] *Teor. Veroyatn. Primenen.* **3: 2** (1958), 201-205. (In Russian.)
[4] *Proc. All-Union Conf. on Probab. Theory and Math. Stat., Vilnius* (1962), 455-462. (In Russian.)
Schaeffer, H.H.,
[1] *Topological vector spaces*, Mcmillan, 1966.
Schwartz, L.,
[1] *Radon measures on arbitrary topological spaces and cylindrical measures*, Oxford Univ. Press, 1973.
Schwinger, J.,
[1] *I, Phys. Review* **74: 10** (1948), 1439. *II, Phys. Review* **75: 4** (1949), 651.
Shavgulidze, E.T.,
[1] *Uspekhi Mat. Nauk* **31: 2** (1976), 239. (In Russian.)
Sherstnev, A.N.,
[1] *Izv. Vyzov. Mat.* **6** (1961), 155-158. (In Russian.)
Shilov, G.E.,
[1] *Funkts. Anal. Primenen.* **1: 2** (1967), 81-90. (In Russian.)
[2] *Mat. Issled.* **2: 4** (1968), 166-186. *Mat. Sb.* **74** (1967), 161-168. (In Russian.)
Shilov, G.E., Fahn DucThihn,
[1] *Integral, measure and derivative on linear spaces*, Moscow, 1967. (In Russian.)
Segal,I.E.,
[1] *Trans. Amer. Math. Soc.* **81: 1** (1956), 106-134. *Ann. of Math.* **63: 1**

(1956), 160-175.
[2] *Amer. J. Math.* **76: 3** (1954), 721-732.
[3] *Trans. Amer. Math. Soc.* **88: 1** (1958), 112-41.
Seidman, T.,
[1] *Commun. Pure Appl. Math.* **12: 4** (1959), 611-621. **17: 4** (1964), 493-508.
Simon, B.,
[1] *The $P(\phi)^2$ Euclidean (quantum) field theory*, Princeton Univ. Press, 1974.
Simon, B., Höegh-Krohn, R.,
[1] *J. Funct. Anal.* **9** (1972), 121-180.
Skorokhod, A.V.,
[1] *Teor. Veroyatn. Primenen.* **5: 1** (1960), 45-53. (In Russian.)
[2] *Dokl. Akad. Nauk SSSR* **168: 6** (1966), 1269-1271. (In Russian.)
[3] *Studies in the theory of random processes*, Addison-Wesley, 1965. (Translated from the Russian.)
[4] *Integration in Hilbert space*, Springer, 1974. (Translated from the Russian.)
[5] *Teor. Veroyatn. Primenen.* **20: 2** (1975), 223-238. (In Russian.)
[6] *Teor. Veroyatn. Primenen.* **9: 3** (1964), 492-497. (In Russian.)
Smolyanov, O.G.,
[1] *Dokl. Akad. Nauk SSSR* **170: 3** (1966), 38-41. (In Russian.)
[2] *Mat. Zametki* **25: 2** (1979), 259-269. (In Russian.)
[3] *Dokl. Akad. Nauk SSSR* **221: 6** (1975), 1288-1291. (In Russian.)
[4] *Analysis on topological linear spaces and its applications*, Moscow, 1979. (In Russian.)
Smolyanov, O.G., Fomin, S.V.,
[1] *Uspekhi Mat. Nauk* **31: 4** (1976), 3-56. (In Russian.)
Sudakov, V.N.,
[1] *Dokl. Akad. Nauk SSSR* **127: 3** (1959), 524-525. (In Russian.)
[2] *Dokl. Akad. Nauk SSSR* **185: 1** (1969), 51-53. (In Russian.)
[3] *Tr. Mat. Inst. Steklov.* **CLX: 1** (1976), 1-190. (In Russian.)
Teterina, N.I.,
[1] *Uspekhi Mat. Nauk* **25: 3** (1973), (In Russian.)
Tinfavicius, E.,
[1] *Proc. IInd Vilnius Conf. Probab. Th.* **3** (1977), 231-234. (In Russian.)
Trotter, H.E.,
[1] *Proc. Amer. Math. Soc.* **10: 4** (1959), 545-551.
Uglanov, A.V.,
[1] *Vestnik MGU, Mat. Mekh.* **1** (1971), 52-60. (In Russian.)
[2] *Uspekhi Mat. Nauk* **32: 1** (1977), 207-208. (In Russian.)
[3] *Izv. Akad. Nauk SSSR* **39: 2** (1975), 438-468. (In Russian.)
[4] *Vestnik MGU, Mat. Mekh.* **5** (1972), 14-24. (In Russian.)
[5] *Dokl. Akad. Nauk SSSR* **243: 6** (1978), 1406-1409. (In Russian.)
Umemura, Ja.,
[1] *J. Math. Univ. Kyoto* **4: 3** (1965), 478-492. (In Russian.)

Vakhaniya, N.N.,
[1] *Probability distributions on linear spaces*, North-Holland, 1981. (Translated from the Russian.)
Vershik, A.M.,
[1] *Uspekhi Mat. Nauk* **19: 1** (1964), 210-213. (In Russian.)
[2] *Dokl. Akad. Nauk SSSR* **170: 3** (1966), 497-500. (In Russian.)
[3] *Dokl. Akad. Nauk SSSR* **178: 2** (1968), 278-281. (In Russian.)
Vershik, A.M., Ladyzhenskaya, O.A.,
[1] *Boundary Problems of Mathematical Physics and Related Problems in Function Theory, Leningrad* (1976), 3-24. (In Russian.)
Vershik, A.M., Sudakov, V.N.,
[1] *Zap. Nauchn. Sem. LOMI* **12** (1969), 7-67. (In Russian.)
[2] *Uspekhi Mat. Nauk* **17: 4** (1962), 217-219. (In Russian.)
Viot, M.,
[1] *Solutions faibles d'équations aux dérivées partielles stochastiques non linéaires, Thèse*, Paris, 1976.
Vishik, M.I.,
[1] *Uspekhi Mat. Nauk* **26: 2** (1971), 155-174. (In Russian.)
[2] *Dokl. Akad. Nauk SSR* **208: 4** (1973), 764-767. (In Russian.)
Vishik, M.I., Fursikov, A.V.,
[1] *Sibirsk. Mat. Zh.* **19: 5** (1978), 1005-1031. (In Russian.)
Vishik, M.I., Komech, A.I.,
[1] *Dokl. Akad. Nauk SSSR* **233: 5** (1977), 769-772. (In Russian.)
Vishik, M.I., Komech, A.I., Fursikov, A.V.,
[1] *Uspekhi Mat. Nauk* **34: 5** (1979), 135-210. (In Russian.)
Vishik, M.I., Marchenko, A.V.,
[1] *Mat. Sb.* **90: 3** (1973), 331-371. (In Russian.)
Wiener, N.,
[1] *Nonlinear problems in random theory*, MIT, 1958.
Zaplitnaya, A.T.,
[1] *Dokl. Akad. Nauk UkrSSR* **11** (1965), 1434-1437. (In Russian.)

CHAPTER VIII

Smooth measures

By Yu.L. Dalecky and V.R. Steblovskaya

The contents of this chapter is directly related to the material in chapters II and IV. Generalizing the concept of derivative of a measure in a direction, we introduce the concept of derivative of a measure along a vector field. In this way the class of smooth measures, containing the Gaussian measures, arises. This class is invariant under smooth nonlinear transformations.

Our main purpose is the study of absolute continuity properties of transformed measures, generalizing the considerations in chapter II for Gaussian measures. Our arguments are based on the concept of μ-admissible operator on $L_2(X,\mu)$, which generalizes the concept of admissible shift of a measure.

We consider smoothness conditions for finite-dimensional images of smooth measures. These conditions are effective in the case of quadratic maps.

In the final section we consider concepts generalizing the system of Hermite polynomials and the Fourier–Wiener transformation to the case of smooth nonGaussian measures.

1. Admissible operators

1.1. Definition and properties. Let X be a separable Banach space, $\mathcal{B}$ the Borel σ-algebra, $C(X)$ the Banach space of realvalued continuous bounded functions on X, and $\mathcal{M}(X)$ the Banach space of realvalued bounded measures on the measurable space $(X,\mathcal{B})$. We consider the duality between $C(X)$ and $\mathcal{M}(X)$ given by the natural pairing $\langle f,\mu\rangle = \int_X f(x)\mu(dx)$.

For a linear operator A on $C(X)$ with dense domain D_A we let A' be the dual linear operator on $\mathcal{M}(X)$. satisfying the relation

$$\langle Af,\mu\rangle = \langle f,A'\mu\rangle \quad (f\in D_A);$$

here, $D_{A'}$ is the set of measures $\mu\in\mathcal{M}(X)$ for which the measure $\mu'=A'\mu\in\mathcal{M}(X)$ is defined.

DEFINITION 1.1. Let $\mu\in\mathcal{M}(X)$ be a positive measure. An operator A is said to be *μ-admissible* if the measure $A'\mu$ is absolutely continuous with respect to μ:

$$A'\mu(\Delta) = \int_\Delta \rho(A,\mu,x)\mu(dx) \quad (\Delta\in\mathcal{B}),$$

and the density $\rho(A,\mu,x)$ is μ-square integrable:

$$\int_X |\rho(A,\mu,x)|^2\mu(dx) < \infty.$$

Consider the Hilbert space $\mathcal{H}_\mu = L_2(X,\mu)$. Let A^* be the adjoint of A in $\mathcal{H}_\mu$:

$$\int_X Af(x)g(x)\mu(dx) = \int_X f(x)A^*g(x)\mu(dx).$$

It follows from the definition of μ-admissibility that this property is equivalent to the existence of a function $\phi \in \mathcal{H}_\mu$ satisfying the relation $\int_X Af(x)\mu(dx) = \int_X f(x)\phi(x)\mu(dx)$ for all $f \in D_A \subset C(X) \subset \mathcal{H}_\mu$. In this case $A'\mu = \phi\mu$, and clearly $\rho(A,\mu,x) = \phi(x)$. Thus the following assertion holds.

PROPOSITION 1.1. *An operator A is μ-admissible if and only if $e(x) \equiv 1 \in D_{A^*}$. In this case $\rho(A,\mu,x) = A^*e(x)$.*

It is clear that any operator which is bounded on $\mathcal{H}_\mu$ is μ-admissible. However, it might be nontrivial to verify the boundedness of an operator in concrete situations. In addition, if a certain operator is μ-admissible, then it is important to compute the density $\rho(A,\mu,x)$.

EXAMPLE 1.1. Consider the shift operator on $C(X)$: $S_hf(x) = f(x-h) = f_h(x)$. The corresponding shift operator on measures, $S_h'\mu(\Delta) = \mu(\Delta + h)$ is dual to this operator.

The shift operator is μ-admissible if $S_h'\mu < \mu$ and the density $\rho(S_h,\mu,x) = S_h'\mu(dx)/\mu(dx) \in \mathcal{H}_\mu$. For a canonical Gaussian measure in the space $X = \mathcal{H}_- \supset \mathcal{H}_0 \supset \mathcal{H}_+$ the shift S_h is μ-admissible if h belongs to $\mathcal{H}_0$ (see chapter IV, section 4.3). In this case

$$\rho(S_h,\mu,x) = \exp\left\{-(h,x)_0 - \frac{1}{2}\|h\|_0^2\right\}.$$

EXAMPLE 1.2. Under the same conditions, consider the differentiation operator $\langle h,\nabla\rangle f(x) = \langle h, f'(x)\rangle$, defined on $C_1(X,\mathbf{R})$. μ-admissibility of $\langle h,\nabla\rangle$ means weak differentiability of the measure μ along h and the existence of a μ-square integrable density $\langle h,\nabla\rangle\mu(dx)/\mu(dx)$ (the logarithmic derivative of μ along h). In this case (see chapter IV, section 2.2) μ is differentiable along h also in the sense of convergence in variation.

There is a simple relation between the densities $\rho(\langle h,\nabla\rangle,\mu,x)$ and $\rho(S_{th},\mu,x)$, under the condition that the shifts S_{th} are μ-admissible: $\rho(\langle h,\nabla\rangle,\mu,x) = -(d/dt)\rho(S_{th},\mu,x)|_{t=0}$.

In particular, for a canonical Gaussian measure:

$$\rho(\langle h,\nabla\rangle,\mu,x) = (h,x)_0.$$

The following, most simple, properties of μ-admissible operators are readily verified.

1) Let A_1, A_2 be μ-admissible, $D_{A_1} = D_{A_2}$, $\lambda_1,\lambda_2 \in \mathbf{R}$. Then the operator $A = \lambda_1 A_1 + \lambda_2 A_2$ is μ-admissible, and

$$\rho(A,\mu,x) = \lambda_1\rho(A_1,\mu,x) + \lambda_2\rho(A_2,\mu,x).$$

2) Suppose $A = A_1A_2$ is defined. Let A_1 be μ-admissible, and $A_1^*e(x) \in D_{A_2^*}$. Then A is μ-admissible, and

$$\rho(A,\mu,x) = A_2^*A_1^*e(x).$$

3) Let A_n $(n = 1, 2, \dots)$ be a sequence of μ-admissible operators with common domain D, converging on D to an operator A. Suppose that the corresponding sequence of densities $\rho(A_n, \mu, x)$ converges in $\mathcal{H}_\mu$ to a function $\alpha(x) \in \mathcal{H}_\mu$. Then A is μ-admissible, and $\rho(A, \mu, x) = \alpha(x)$.
4) Let the operators $A(t)$ $(t \in [a, b])$ with constant domain D be continuous with respect to $t \in [a, b]$ on elements of D and μ-admissible. Let the function $\rho(A(t), \mu, x) = A^*(t)e(x)$ be continuous with respect to t as an element of $\mathcal{H}_\mu$. Then the operator $\int_a^b A(t)\,dt$ (convergence in the sense of convergence on the dense set D) is μ-admissible, and

$$\rho\left(\int_a^b A(t)\,dt, \mu, x\right) = \int_a^b \rho(A(t), \mu, x)\,dt.$$

5) Let the operator $A(t)$ be differentiable with respect to $t \in (a, b)$ on D and μ-admissible. Suppose that the density $\rho(A(t), \mu, x)$ is continuously differentiable with respect to t as an element of $\mathcal{H}_\mu$. Then the operator $(d/dt)A(t)$ is μ-admissible for each $t \in (a, b)$, and

$$\rho\left(\frac{d}{dt}A(t), \mu, x\right) = \frac{d}{dt}\rho(A(t), \mu, x).$$

1.2. Admissible operators defined by differential equations. Consider a linear operator A with domain D_A $(\bar{D}_A = C(X))$ and having the following property

For any ϕ in a linear set $D \subset D_A$ $(\bar{D} = C(X))$ the Cauchy problem

$$\frac{\partial u(t, x)}{\partial t} + Au(t, x) = 0, \quad u(0, x) = \phi(x) \quad (t \geq 0) \tag{1.1}$$

has a unique solution $u(t, x) \in D$ $(t \geq 0)$.

Then there is a family of linear operators $U(t)\colon \phi(x) \to u(t, x)$ on D having the semigroup property $U(t + \tau) = U(t)U(\tau)$, $U(0) = I$ $(t, \tau \geq 0)$. The linear set D is invariant under $U(t)$. By definition, the operators $U(t)$ are differentiable with respect to t on the elements of D (for $t \geq 0$), and

$$\frac{d}{dt}U(t) = -AU(t).$$

In this case A is said to be the *generator* (or *generating operator*) of the D-semigroup $U(t)$. Instead of $U(t)$ we will write e^{-tA}, despite the fact that such operators are, in general, unbounded.

This consideration leads to the following result.

THEOREM 1.1. *Let the linear operator A be the generating operator of a D-semigroup e^{-tA} $(t \geq 0)$, $\bar{D} = C(X)$, and let the adjoint A^* be the generating operator of the D^*-semigroup $e^{-tA^*} = (e^{-tA})^*$ $(t \geq 0)$, $\bar{D}^* = C(X)$. If $e(x) \in D^*$, then the family of operators e^{-tA} is μ-admissible for each $t \geq 0$, and the function $v(t, x) = \rho(e^{-tA}, \mu, x)$ is the solution of the Cauchy problem*

$$\frac{\partial v(t, x)}{\partial t} = A^* v(t, x), \quad v(0, x) \equiv 1. \tag{1.2}$$

REMARK 1.1. In the following situation the conditions of theorem 1.1 are fulfilled. Suppose that $-A$ is the generating operator of the semigroup e^{-tA} of bounded operators in $\mathcal{H}_\mu$ (a C_0-semigroup). Due to their boundedness, the operators e^{-At} are μ-admissible. Hence the adjoint operator $-A^*$ generates the C_0-semigroup $e^{-tA^*} = (e^{-At})^*$.

The condition $e(x) \in D^* = D_{A^*}$ is equivalent to the fact that A is μ-admissible. If this condition holds, the density $\rho(e^{-tA}, \mu, x) = e^{-tA^*} \cdot e(x)$ is the solution of the Cauchy problem (1.2). If this condition does not hold, the expression $e^{-tA^*} \cdot e(x)$ could be regarded as a generalized solution of (1.2).

Consider another case. Let $-A$ be the generating operator of a C_0-semigroup $U(t)$ in the space $C(X)$ (or in some space K with $C(X) \subset K \subset \mathcal{H}_\mu$) but not in $\mathcal{H}_\mu$. Let $D^* \subset D_{A^*}$ be the subspace which is invariant under the semigroup $V(t) = U(t)^*$. It is easy to show that $v(t,x) = V(t)e(x)$ is the solution of the Cauchy problem (1.2) if $e(x) \in D^*$.

This situation is often met in concrete examples.

REMARK 1.2. It is possible to enlarge the class of μ-admissible operators, by considering along with e^{-At} other functions of A, e.g. the Laplace transform

$$f(A) = \int_0^\infty g(t)e^{-tA}\,dt \quad \text{where } f(\lambda) = \int_0^\infty g(t)e^{-t\lambda}\,dt.$$

Under conditions guaranteeing the convergence of the corresponding integrals (which we do not state here), we have

$$\rho(f(A), \mu, x) = \int_0^\infty g(t)v(t,x)\,dt,$$

where $v(t,x)$ is the solution of the Cauchy problem (1.2).

It is also possible to use the Fourier transform

$$f(A) = \frac{1}{2\pi}\int_{-\infty}^\infty g(t)e^{itA}\,dt, \quad \text{where } f(\lambda) = \frac{1}{2\pi}\int_{-\infty}^\infty g(t)e^{it\lambda}\,dt.$$

In this case the density is given by

$$\rho(f(A), \mu, x) = \int_{-\infty}^\infty g(t)w(t,x)\,dt,$$

where $w(t,x)$ is the solution of the Cauchy problem

$$\frac{\partial w(t,x)}{\partial t} = -iA^*w(t,x); \quad w(0,x) \equiv 1 \quad (x \in \mathbf{R}). \tag{1.3}$$

EXAMPLE 1.3. By choosing appropriate pairs of functions from the list of Laplace or Fourier transforms it is possible to obtain interesting examples. Suppose that all integral used are convergent and that the initial operators are μ-admissible.

a) Let $g(t) = (\sin at)/a$, $f(\lambda) = 1/(\lambda^2 + a^2)$.

Then the operator $f(A) = (A+a^2 I)^{-1} = (1/a)\int_0^\infty e^{-At}\sin at\,dt$ is μ-admissible. This means that the measure $\mu' = f(A)'\mu$ is absolutely continuous with respect to μ, and

$$\rho((A + a^2 I)^{-1}, \mu, x) = \frac{1}{a}\int_0^\infty \sin at\, v(t,x)\,dt,$$

where $v(t,x)$ is the solution of (1.2).

b) Let $g(t) = J_0(at)$, $f(\lambda) = (\lambda^2 + a^2)^{-1/2}$.
Then $f(A) = (A^2 + a^2 I)^{-1/2} = \int_0^\infty e^{-tA} J_0(at)\,dt$, and

$$\rho((A + a^2 I)^{-1/2}, \mu, x) = \int_0^\infty J_0(at) v(t,x)\,dt.$$

c) Let $g(t) = e^{-t^2/2}$, $f(\lambda) = e^{-\lambda^2/2}$.
Then $f(A) = e^{-A^2/2} = (1/2\pi)\int_{-\infty}^\infty e^{itA - t^2/2}\,dt$, and

$$\rho(e^{-A^2/2}, \mu, x) = \frac{1}{2\pi}\int_{-\infty}^\infty e^{-t^2/2} w(t,x)\,dt,$$

where $w(t,x)$ is the solution of (1.3).

The following situation is more general as compared to theorem 1.1. Suppose that the operators $A(t)$ depend on a parameter $t \in [a,b]$ and have constant domain D_A.

Let for each function ϕ in a linear set $D \subset D_A$ the Cauchy problem

$$\frac{\partial u(t,x)}{\partial t} + A(t)u(t,x) = 0, \quad u(\tau,x) = \phi(x) \quad (0 \le t \le \tau) \tag{1.4}$$

have a unique solution $u(t,\cdot) \in D$. Then the two-parameter family of linear operators $U(t,\tau)\colon \phi \to u(t,\cdot)$ is defined on D, has the evolution property:

$$U(t,s)U(s,\tau) = U(t,\tau), \quad U(t,t) = I \quad (a \le t \le s \le \tau \le b),$$

and D is invariant under $U(t,\tau)$. The family $U(t,\tau)$ is differentiable with respect to t and τ on the elements of D, and satisfies the relations

$$\frac{d}{dt}U(t,\tau) + A(t)U(t,\tau) = 0; \quad \frac{d}{dt}U(t,\tau) = U(t,\tau)A(\tau).$$

In this case A is said to be the generating operator of the D-evolution family $U(t,\tau)$. The following theorem is an analog of theorem 1.1.

THEOREM 1.2. *Let a linear operator $A(t)$ $(a \le t \le b)$ be the generating operator of a D-evolution family $U(t,\tau)$, and let $A^*(t)$ be the generating operator of the D^*-evolution family $V(\tau,t) = U(t,\tau)^*$.*

If $e(x) \in D^*$, then the family $U(t,\tau)$ is μ-admissible, and the density $\rho(U(t,\tau),\mu,x) = v_t(\tau,x)$ is the solution of the Cauchy problem

$$\frac{\partial v_t(\tau,x)}{\partial \tau} = A(\tau)^* v_t(\tau,x), \quad v_t(t,x) = e(x) \quad (a \le t \le \tau \le b). \tag{1.5}$$

REMARK 1.3. Under the conditions of theorem 1.2 the measure $\mu(\tau)' = U(t,\tau)'\mu$, which depends on a parameter τ, is a weak solution of the problem

$$\frac{d}{d\tau}\mu(\tau)' = A(\tau)'\mu(\tau)', \quad \mu(t)' = \mu.$$

2. Admissibility of differential operators

2.1. The logarithmic derivative of a measure along a vector field. Consider the following pair of spaces: a Banach space X and a Hilbert space $\mathcal{H}$ densely embedded in X: $\mathcal{H} \subset X$ (below we will omit the embedding operator). In this case $X^* \subset \mathcal{H}^*$, and by identifying $\mathcal{H} = \mathcal{H}^*$, we obtain the canonical triple of spaces

$$X^* \subset \mathcal{H} \subset X. \tag{2.1}$$

Below we suppose that the operator embedding $\mathcal{H}$ into X is of Hilbert–Schmidt class:

for an orthonormal basis $\{e_k\}_{k=1}^{\infty}$ in $\mathcal{H}$ we have $\sum_{k=1}^{\infty} \|e_k\|_X^2 < \infty$.

In particular, if $X = \mathcal{H}_-$, $X^* = \mathcal{H}_+$ are Hilbert spaces, we obtain the equipped Hilbert space systematically used in this book.

Note that the majority of results can be transferred to the triple (2.1). For example, for a differentiable function $f\colon X \to X^*$ the derivative $f'(x) \in \mathcal{L}(X, X^*)$, i.e. its restriction to $\mathcal{H}$ is of trace class; therefore the operation $\operatorname{div} f(x) = \operatorname{Tr} f'(x)$ is defined. Correspondingly, for a function $f\colon X \to \mathcal{H}$ the restriction to $\mathcal{H}$ of its derivative $f'(x) \in \mathcal{L}(X, \mathcal{H})$ is a Hilbert–Schmidt operator.

Consider the class $\mathfrak{M}_0(X)$ of Borel measures on X having a logarithmic derivative $\rho_\mu(h; x)$ in each direction $h \in X^*$:

$$\int_X \langle h, \nabla\rangle \phi(x)\mu(dx) = -\int_X \phi(x)\rho_\mu(h; x)\mu(dx), \quad \phi \in C_1(X, \mathbf{R}). \tag{2.2}$$

Formula (2.2) readily implies that the logarithmic derivative depends linearly on h (a.e.).

Suppose there is a map $\lambda\colon X \to X$ satisfying the condition

$$\int_X \|\lambda(x)\|_X^2 \mu(dx) < \infty, \tag{2.3}$$

and such that

$$\rho_\mu(h; x) = \langle \lambda(x), h\rangle_X \quad (h \in X^*).$$

Here the brackets $\langle\, ,\, \rangle_X$ denote the pairing between the elements of X and X^*. The function $\lambda(x)$ is called the *logarithmic gradient* of the measure μ.

Obviously, under the condition (2.3) the operators $\langle h, \nabla\rangle$ are μ-admissible, and $\rho(\langle h, \nabla\rangle, \mu, x) = -\rho_\mu(h; x)$.

EXAMPLE 2.1. The Gaussian measure μ on X with characteristic functional $\chi_\mu(\theta) = e^{-\|\theta\|_{\mathcal{H}}^2/2}$ (a canonical Gaussian measure) has the property $\mu_h(dx)/\mu(dx) = \exp\{-\langle x, h\rangle - \|h\|_{\mathcal{H}}^2/2\}$. It follows from example 1.2 that in this case $\lambda(x) = -x$.

We now consider first order differential operators with variable coefficients,

$$\nabla_z = \langle z(x), \nabla \rangle,$$

where $z\colon X \to X^*$ is a vector field of class $C_1(X, X^*)$ which is differentiable along X.

EXAMPLE 2.2. Let $\dim X < \infty$, and let μ be a measure which is absolutely continuous with respect to Lebesgue measure: $\mu(dx) = p(x)\,dx$. If $p(x) > 0$ is a smooth function, then it is easy to check that $\lambda(x) = p'(x)/p(x)$. Using integration by parts we obtain

(2.4)
$$\int_X \langle z(x), \nabla \rangle f(x) g(x) \mu(dx) = -\int_X f(x) \{ \langle \lambda(x), z(x) g(x) \rangle + \operatorname{div}(z(x) g(x)) \} \mu(dx).$$

Thus, the operator $A = \langle z(x), \nabla \rangle$ has adjoint $A^* = -\langle \lambda(x) + \nabla, z(x) \rangle$ in $L_2(X, \mu)$. The measure $\nabla_z \mu = \langle z(x), \nabla \rangle \mu$ is called the *derivative of the measure* μ *along the vector field* $z(x)$, and

$$\rho_\mu(z; x) = \langle \lambda(x) + \nabla, z(x) \rangle \cdot e(x) = \langle \lambda(x), z(x) \rangle + \operatorname{div} z(x)$$

is called the *logarithmic derivative* of μ along $z(x)$.

By limit transition we can extend this result to the infinite-dimensional case.

PROPOSITION 2.1. *Let $\mu \in \mathfrak{M}_0(B)$, and let the vector field $z \in C_1(X, X^*)$ be differentiable along X. Then μ has a logarithmic derivative along z, and*

$$\rho_\mu(z; x) = \langle \lambda(x), z(x) \rangle + \mathit{div}\, z(x) = \langle \lambda(x) + \nabla, z(x) \rangle \cdot e(x), \tag{2.5}$$

where the differential operator $-\langle \lambda(x) + \nabla, z(x) \rangle$ is the adjoint of $\langle z(x), \nabla \rangle$ in $L_2(X, \mu)$.

REMARK 2.1. The conditions of proposition 2.1 are not sufficient for μ-admissibility of the operator $A = \langle z(x), \nabla \rangle$, because the logarithmic derivative $\rho_\mu(z; x) = -\rho(\langle z(x), \nabla \rangle, \mu, x)$ is square integrable only under certain additional conditions, to be considered below.

Denote by $\mathfrak{M}_n^m(X, \mathcal{H}) \subset \mathfrak{M}_0(X)$ the class of positive measures for which their logarithmic gradients $\lambda(x)$ are differentiable along $\mathcal{H}$ up to order n inclusive, and

a) $\lambda'(x)$ is bounded in $\mathcal{H}$, $\lambda^{(k)}(x)$ is a k-linear Hilbert–Schmidt operator on $\mathcal{H}$ $(k > 1)$;

b) $\int_X \|\lambda'(x)\|^m \mu(dx) < \infty$, $\int_X \sigma_2^m(\lambda^{(k)}(x)) \mu(dx) < \infty$ $(k = 2, \dots, n)$, where $\sigma_2(\lambda^{(k)}(x))$ is the Hilbert–Schmidt norm of the k-linear map $\lambda^{(k)}(x)$.

Denote by $W_n^m(X, \mathcal{H}, \mu)$ the Banach space of functions $z\colon X \to \mathcal{H}$ that are differentiable along the space $\mathcal{H}_-$, the completion of $C_n(X, X^*)$ with respect to the norm

$$\sigma_{n,m}(z) = \sum_{j=0}^{n} \left\{ \int_X \sigma_2^m(z^{(j)}(x)) \mu(dx) \right\}^{1/m}. \tag{2.6}$$

THEOREM 2.1. *Let $\mu \in \mathfrak{M}_1^2(X,\mathcal{H})$, $z \in W_1^4(X,\mathcal{H},\mu)$. Then μ has a logarithmic derivative $\rho_\mu \in L_2(X,\mu)$ along the vector field z, satisfying the relations*

$$\int_X |\rho_\mu(z;x)|^2\mu(dx) = \int_X \{\operatorname{Tr} z'(x)^2 - \langle\lambda'(x)z(x), z(x)\rangle\}\mu(dx), \tag{2.7}$$

$$\left\{\int_X |\rho_\mu(z;x)|^2\mu(dx)\right\}^{1/2} \le const\,\sigma_{1,4}(z). \tag{2.8}$$

The differential operator $A = \langle z(x), \nabla\rangle$ is μ-admissible, and

$$\rho(A,\mu,x) = -\rho_\mu(z;x) = \langle\lambda + \nabla, z(x)\rangle. \tag{2.9}$$

PROOF. First, let $z \in C_1(X,X^*)$ be a vector field which is differentiable along $\mathcal{H}$. The existence of the logarithmic derivative follows from proposition 2.1. We estimate the integral

$$\int_X |\rho_\mu(z;x)|^2\mu(dx) = \int_X \rho_\mu(z;x)\langle\nabla, z(x)\rangle\mu(dx) = -\int_X \langle z(x),\nabla\rangle\rho_\mu(z;x)\mu(dx).$$

Using (2.4) it is possible to infer that

$$\begin{gathered}\langle z(x),\nabla\rangle\rho_\mu(z;x) = \langle z(x),\nabla\rangle\{\langle\lambda(x),z(x)\rangle + \operatorname{div} z(x)\} = \\ \langle\nabla_z\lambda(x), z(x)\rangle + \langle\lambda(x),\nabla_z z(x)\rangle + \nabla_z \operatorname{div} z(x) = \\ \rho_\mu(\nabla_z z;x) + \langle\nabla_z\lambda(x), z(x)\rangle - \operatorname{Tr} z'(x)^2.\end{gathered}$$

Note that

$$\int_X \rho_\mu(\nabla_z z;x)\mu(dx) = \int_X \langle\nabla,\nabla_z z\rangle\mu(dx) = -\int_X \langle\nabla_z z,\nabla\rangle\cdot 1\mu(dx) = 0.$$

Thus (2.7) clearly holds. The estimate (2.8) is easily obtained from (2.7):

$$\begin{gathered}\left\{\int_X |\rho_\mu(z;x)|^2\mu(dx)\right\}^{1/2} \le \left\{\int_X \operatorname{Tr} z'(x)^2\mu(dx) + \int_X \|\lambda'(x)\|\,\|z(x)\|^2\mu(dx)\right\}^{1/2} \le \\ \text{const}\left\{\left[\int_X \sigma_2^4(z'(x))\mu(dx)\right]^{1/4} + \left[\int_X \sigma_2^4(z)\mu(dx)\right]^{1/4}\right\} = C\sigma_{1,4}(z).\end{gathered}$$

Now we note that the righthand side of (2.7) makes sense also for a larger class of vector fields, in particular for $z \in W_1^4(X,\mathcal{H},\mu)$. It follows from (2.8) that the linear differential operator $\rho_\mu\colon z \to \langle\lambda(x), z(x)\rangle + \operatorname{div} z(x)$ is a continuous map $C_1(X,X^*) \to L_2(X,\mu)$. Hence it is possible to extend it to $W_1^4(X,\mathcal{H},\mu)$. We will denote this extension by the same symbol $\rho_\mu(z;x)$. After limit transition in (2.7) the theorem is proved. □

REMARK 2.2. For the Gaussian measure with $\lambda(x) = -x$ formula (2.7) becomes (II.3.21) with $A = I$.

In this way it is possible to obtain also estimates of the higher order moments of the logarithmic derivative (see Belopolskaya–Dalecky [1]).

PROPOSITION 2.2. *Let $\mu \in \mathfrak{M}_1^{2^k}(X,\mathcal{H})$, $z \in W_k^{2^{k+1}}(X,\mathcal{H},\mu)$. Then μ has a logarithmic derivative along z, which also satisfies the estimate*

$$\left\{\int_X |\rho_\mu(z;x)|^{2^k}\mu(dx)\right\}^{2^{-k}} \le const\, \sigma_{k,2^{k+1}}(z). \tag{2.10}$$

2.2. Differential operators of higher order. Here we consider the μ-admissibility of differential operators of order $n > 1$,

$$A = \langle z_1(x), \nabla\rangle \dots \langle z_n(x), \nabla\rangle.$$

Let $\mathfrak{M}_n(X,\mathcal{H})$ be the class of measures on X that are n times differentiable along $\mathcal{H}$. Let z_j ($j = 1,\dots,n$) be a set of vectorfields of class $C_n(X, X^*, \mathcal{H})$ such that $z_j(x) \in X^*$ and such that the continuous and bounded nth order derivatives of the z_j along $\mathcal{H}$ exist.

Define the *logarithmic derivative of order n of a measure* $\mu \in \mathfrak{M}_n$ along the set $\{z_j\}_1^n$ by

$$\rho_n(z_1,\dots,z_n;x) = \frac{(\nabla_{z_n}\dots\nabla_{z_1}\mu)(dx)}{\mu(dx)}. \tag{2.11}$$

Using proposition 2.1 it is easy to prove the following.

PROPOSITION 2.3. *Let $\mu \in \mathfrak{M}_n^{4n}(X,\mathcal{H})$, $z_j \in C_n(X,X^*,\mathcal{H})$ ($j = 1,\dots,n$). Then μ has a logarithmic derivative of order n along the set of vector fields $\{z_j\}_1^n$:*

$$\rho_n(z_1,\dots,z_n;x) = \langle\nabla + \lambda(x), z_n(x)\rangle \dots \langle\nabla + \lambda(x), z_1(x)\rangle \cdot 1. \tag{2.12}$$

Here the differential operator $A^* = (-1)^n\langle\nabla + \lambda(x), z_n(x)\rangle \dots \langle\nabla + \lambda(x), z_1(x)\rangle$ is the adjoint of the operator $A = \langle z_1(x),\nabla\rangle \dots \langle z_n(x),\nabla\rangle$ in $L_2(X,\mu)$. The following estimate holds

$$\left\{\int_X |\rho_n(z_1,\dots,z_n;x)|^2\mu(dx)\right\}^{1/2n} \le const \sum_{k=1}^n \sigma_{n,4n}(z_k). \tag{2.13}$$

SKETCH OF THE PROOF. By mathematical induction we can easily verify (2.12). Then we estimate the integral (2.13). First, let $z_1 = \dots = z_n = z \in C_n(X,X^*,\mathcal{H})$. Put $\rho_k(z) = \rho_k(z,\dots,z;x)$ It is possible to express $\rho_k(z)$ in terms of $\rho_1(z)$ and its derivatives. E.g.,

$$\rho_2(z) = \langle\nabla + \lambda(x), z(x)\rangle\rho_1(z) = \rho_1(z)^2 + \langle z(x),\nabla\rangle\rho_1(z),$$
$$\rho_3(z) = \langle\nabla + \lambda(x), z(x)\rangle\rho_2(z) = \rho_1(z)^3 + 3\rho_1(z)\langle z(x),\nabla\rangle\rho_1(z) + \langle z(x),\nabla\rangle^2\rho_1(z),$$

etc. In this way we obtain

$$\left\{\int_X |\rho_n(z)|^2\mu(dx)\right\}^{1/2n} \le \text{const}\sum_{k=1}^n\left\{\int_X |\langle z(x),\nabla\rangle^{k-1}\rho_1(z)|^{2n/k}\mu(dx)\right\}^{1/2n}. \tag{2.14}$$

Using (2.12) it is possible to show that

$$|\langle z(x), \nabla\rangle^{k-1} \rho_1(z)| \leq$$
$$\text{const}\{|\rho_1(\langle z(x), \nabla\rangle^{k-1} z)| + |\langle z, \langle z, \nabla\rangle^{k-1}\rangle \lambda| + |\langle\langle z, \nabla\rangle z, \nabla\rangle^{k-2} \lambda\rangle| + \cdots +$$
$$|\langle\langle z, \nabla\rangle^{k-2} z, \langle z, \nabla\rangle \lambda\rangle| + |\langle z, \nabla\rangle^{k-2} \operatorname{Tr} z'^2|\}. \tag{2.15}$$

Estimating each term at the righthand side of (2.15) and using (2.8) we obtain

$$\left\{\int_X |\rho_n(z)|^2 \mu(dx)\right\}^{1/2n} \leq \text{const} \sigma_{n,4n}(z) \quad (n = 2^p, \quad p \in \mathbf{N}). \tag{2.16}$$

The estimate (2.13) can be similarly obtained, but requires a more tedious computation. □

REMARK 2.3. The righthand sides of (2.13) and (2.16) make sense for the larger class of vector fields $z_k \in W_n^{4n}(X, \mathcal{H}, \mu)$. After limit transition we can generalize the theorem to this class.

3. Absolute continuity of smooth measures

3.1. The shift of a measure along a vector field. Let X be a separable Banach space, $\mu \in \mathfrak{M}(X)$, $z(t,x)$ a vector field depending on a parameter $t \in [a;b]$ and of class $C_1(X, X)$. Denote by $S(\sigma, t)$ the integral flow of z:

$$S(\sigma, t)\colon X \to X, \quad S(\sigma, t) \circ S(t, \tau) = S(\sigma, \tau),$$
$$\frac{dS(\sigma, t)x}{d\sigma} = z(\sigma, S(\sigma, t)x), \quad S(t,t)x = x.$$

Consider the differential operator $A(t) = \langle z(t,x), \nabla\rangle$ related to z. Its domain D_A is the entire space $C_1(X, \mathbf{R})$. The operator $A(t)$ is the generating operator of a D_A-evolution family $U(t,\tau)$. The operators $U(t,\tau)$, which are bounded in $C(X, \mathbf{R})$, are determined by the Cauchy problem

$$\frac{\partial u(t,x)}{\partial t} + A(t)u(t,x) = 0, \quad u(\tau, x) = \phi(x) \in D_A; \quad U(t,\tau)\colon \phi \to u(t, \cdot). \tag{3.1}$$

They realize shifts of functions along the trajectories of z. In fact, solving (3.1) by the method of characteristics, we obtain

$$u(t,x) = U(t,\tau)\phi(x) = \phi(S(\tau, t)x). \tag{3.2}$$

Clearly, the set $D_A = C_1(X, \mathbf{R})$ is invariant under $U(t,\tau)$. Consider the dual operator $U(t,\tau)'$ on $\mathfrak{M}(X)$:

$$\int_X U(t,\tau)\phi(x)\mu(dx) = \int_X \phi(S(\tau, t)x)\mu(dx) =$$
$$\int_X \phi(x)(\mu \circ S(t,\tau))(dx) = \int_X \phi(x)[U(t,\tau)'\mu](dx).$$

Thus, the dual operator $U(t,\tau)$ realizes a corresponding shift of measures:

$$U(t,\tau)'\mu = \mu_{\tau,t} = \mu \circ S(t,\tau); \quad \mu_{t,t} = \mu. \tag{3.3}$$

Suppose that the measure μ has a logarithmic derivative $\rho_\mu(z;x)$ along z which is μ-square integrable. Then the differential operator $A(t) = \langle z(t;x), \nabla\rangle$ is μ-admissible for any $t \in [a,b]$. The adjoint operator $A(\tau)^*$ is given by

$$A(\tau)^* = -\langle \nabla + \lambda(x), z(\tau;x)\rangle = -\langle \nabla, z(\tau;x)\rangle - \langle \lambda(x), z(\tau;x)\rangle,$$

where $\lambda(x)$ is the logarithmic gradient of μ. The differential equation generated by $A(\tau)^*$ is (see theorem 1.2):

$$\frac{\partial v(\tau,x)}{\partial \tau} + \langle z(\tau,x), \nabla\rangle v(\tau,x) = -\rho_\mu(z(\tau,\cdot),x)v(\tau,x). \tag{3.4}$$

Suppose, for the moment, that $z(\tau,x) = 0$ if $\|x\| \geq R$ for some R. Then the coefficients of (3.4) are bounded. Therefore the evolution family $V(\tau,t) = U(t,\tau)^*$ corresponding to this equation leaves invariant the domain $D^* = D_A = C_1(X,\mathbf{R})$. So the operators $U(t,\tau)$ are μ-admissible for any $t,\tau \in [a,b]$. This means that the measure (3.3) is absolutely continuous with respect to μ for any $t,\tau \in [a,b]$. The density $\rho(U(t,\tau),\mu,x) = \mu_{\tau,t}(dx)/\mu(dx)$ is the solution of equation (3.4) with initial function $v(t,x) = e(x) \equiv 1$.

The characteristic system of (3.4) is:

$$\frac{dx}{d\sigma} = z(\sigma,x); \quad \frac{dv}{d\sigma} = -\rho_\mu(z(\sigma,\cdot);x)v(\sigma,x).$$

Substituting the solution $S(\sigma,\tau)x$ of the first equation into the second equation and solving it with initial condition $v(t,x) = 1$, we arrive at a formula for the density:

$$\frac{\mu_{\tau,t}(dx)}{\mu(dx)} = v_t(\tau,x) = \exp\left\{-\int_t^\tau \rho_\mu(z(\sigma,\cdot);S(\sigma,\tau)x)\,d\sigma\right\}. \tag{3.5}$$

Now we note that the density $v_t(\tau,x)$ at a point x does not depend on the behavior of the vector field z outside some neighborhood of this point. Therefore we can drop the requirement that $z = 0$ ($\|x\| \geq R$). Thus the following theorem holds.

THEOREM 3.1. *Suppose the measure μ has a logarithmic derivative along the vector field z, satisfying the condition*

$$\int_X |\rho_\mu(z;x)|^2\mu(dx) < \infty.$$

Then the measure $\mu_{\tau,t} = \mu \circ S(t,\tau)$ is absolutely continuous with respect to μ for any $t,\tau \in [a,b]$, and the density $\mu_{\tau,t}(dx)/\mu(dx)$ is given by (3.5).

EXAMPLE 3.1. Let μ be a canonical Gaussian measure, $z(x) \equiv h$ a constant direction. Then $T_{-t}x = x - th$ is the corresponding shift operator. The density of the 'shifted' measure μ_t with respect to μ is given by

$$\frac{\mu_t(dx)}{\mu(dx)} = \exp\left\{\int_0^t \langle h, x - \sigma h\rangle\,d\sigma\right\} = \exp\left\{t\langle h,x\rangle - \frac{t^2}{2}\|h\|^2\right\}$$

(see theorem II.4.2).

3.2. Equivalence of solutions of the forward Kolmogorov equation. Here we consider the second order differential operator

$$\mathcal{A}(t,x) = \langle B(t,x)\nabla, \nabla\rangle + \langle a(t,x), \nabla\rangle$$

in a Hilbert space X; $a(t,x)$ is a vector field in X depending on a parameter t. The values of $B(t,x)$ are symmetric trace class operators on X. Suppose that $B = (AA^*)/2$, with $A(t,x)\colon \mathcal{H} \to X$ a Hilbert–Schmidt operator and $\mathcal{H}$ a Hilbert space. In this case it is possible to regard the differential equation

$$\frac{\partial u(t,x)}{\partial t} + \mathcal{A}(t,x)u(t,x) = 0; \quad u(\tau,x) = \phi(x) \quad (t \le \tau) \tag{3.6}$$

as a backward Kolmogorov equation for the Markov stochastic process $\xi(t)$ determined by the stochastic equation

$$d\xi(\sigma) = a(\sigma,\xi(\sigma))\,d\sigma + A(\sigma,\xi(\sigma))\,dw(\sigma) \tag{3.7}$$

(see chapter VII). Let $\xi_{t,x}(\sigma)$ be the solution of this equation satisfying the condition $\xi_{t,x}(t) = x$.

Then, under the conditions of chapter VII, the solution of the Cauchy problem (3.6) is given by the evolution family

$$U(t,\tau)\phi(x) = \mathrm{M}\phi(\xi_{t,x}(\tau)).$$

The dual evolution family in the space $\mathfrak{M}(X)$ of measures is:

$$U(t,\tau)'\colon \mu \longrightarrow \mu_{\tau,t} = \mathrm{M}\mu \circ S^{-1}(\tau,t),$$

where $S(\tau,t)\colon \xi(t) \to \xi(\tau)$ is the random evolution map determined by the stochastic equation (3.7). The measure $\mu_{\tau,t}$ is the generalized solution of the forward Kolmogorov equation

$$\frac{\partial \mu_{\tau,t}}{\partial t} = [\mathcal{A}(\tau,x)]'\mu_{\tau,t}$$

or

$$\frac{\partial \mu_{\tau,t}}{\partial \tau} = \{\langle \nabla\otimes\nabla, B(\tau,x)\rangle - \langle \nabla, a(\tau,x)\rangle\}\mu_{\tau,t}. \tag{3.8}$$

There arise a natural question: is it possible to differentiate this equation with respect to μ, or, in other words, does the density $v_t(\tau,x) = \mu_{\tau,t}(dx)/\mu(dx)$, $\mu_{t,t} = \mu$, exist? Reasoning similarly as in the previous case, we choose the space $C_2(X,\mathbf{R}) \subset C(X)$ as the domain of $\mathcal{A}(t,x)$. Let $\mu \in \mathfrak{M}_1(X)$. The adjoint of $\mathcal{A}$ has the form

$$\mathcal{A}(\tau,x)^* = \langle (\nabla + \lambda(x)) \otimes (\nabla + \lambda(x)), B(\tau,x)\rangle - \langle \nabla + \lambda(x), a(\tau,x)\rangle,$$

where $\lambda(x)$ is the logarithmic gradient of μ. When the coefficients $A^*(\tau,x)\lambda(x)$ and $\langle \lambda(x), a(\tau,x)\rangle$, are bounded, $\mathcal{A}(\tau,x)^*$ acts also on $C(X)$, with domain $C_2(X,\mathbf{R})$.

In this way it is possible to obtain the following result.

THEOREM 3.2. *Let $\mu \in \mathfrak{M}_1(X)$ and let the Cauchy problem*

(3.9)
$$\frac{\partial v(\tau, x)}{\partial \tau} = \{\langle(\nabla + \lambda(x)) \otimes (\nabla + \lambda(x)), B(\tau, x)\rangle - \langle \nabla + \lambda(x), a(\tau, x)\rangle\} v(\tau, x),$$
$$v(t, x) = 1,$$

have a μ-square integrable solution $v_t(\tau, x)$. Then the solution $\mu_{\tau,t}$ of (3.7) is absolutely continuous with respect to μ, and

$$\frac{\mu_{\tau,t}(dx)}{\mu(dx)} = v_t(\tau, x).$$

REMARK 3.1. It is possible to analyze the solvability of the Cauchy problem (3.9) by methods of the theory of stochastic equations, using the Feynman–Kac formula (see section VII.4.1). For solvability of (3.9) we need the following conditions: $\|\lambda'(x)\| \leq$ const, $\|\lambda(x)\| \leq \alpha\|x\| + \beta$ ($\alpha, \beta \in \mathbf{R}$). It is possible to drop the conditions that $A^* * \lambda$, $\langle a, \lambda\rangle$ are bounded, by limit transition, if $a(\tau, x)$, $A(\tau, x)$ are bounded in the corresponding norms.

3.3. Absolute continuity of measures under nonlinear transformations. As an application of the above theory we consider the problem of μ-admissibility of nonlinear transformations of the space $X \supset \mathcal{H} \supset X^*$. Consider a map

$$f\colon x \to x + F(x), \quad F\colon X \to X^*. \tag{3.10}$$

Deform it to a map depending on a parameter t: $f(t, x) = x + tF(x) = S(t) \circ x$. This map is generated by the vector field $z(t, x) = S(t)' S^{-1}(t)] \circ x = F(S^{-1}(t) \circ x)$.

If $\lambda(x)$ denotes the logarithmic gradient of μ, then

$$\rho_\mu(z(\sigma, \cdot) S(\sigma) \circ x) =$$
$$\langle \lambda(S(\sigma) \circ x), F(S^{-1}(\sigma) \cdot S(\sigma) \cdot x)\rangle + \operatorname{div} z(\sigma, S(\sigma) \circ x) =$$
$$\langle \lambda(x + \sigma F(x)), F(x)\rangle + \operatorname{Tr} F'(x)(I + \sigma F'(x))^{-1}.$$

Noting that

$$\int_0^1 \operatorname{Tr} F'(x)(I + \sigma F'(x))^{-1}\, d\sigma = \ln \det(I + F'(x)),$$

theorem 3.1 gives the following result.

THEOREM 3.3. *Let $\mu \in \mathfrak{M}_1^2(X, X^*)$ and $\mu^f = \mu \circ f^{(-1)}$, where $f(x) = x + F(x)$ is a transformation of X such that*

$$F\colon X \to X^*, \quad \|F'(x)\|_H < \infty, \quad \|[I + \tau F'(x)]^{-1}\|_H \leq C, \quad 0 \leq \tau \leq 1.$$

Then μ and μ^f are equivalent, and

$$\frac{\mu^f(dx)}{\mu(dx)} = \det(I + F'(x)) \exp\left\langle \int_0^1 \lambda(y + \sigma F(y))\, d\sigma, F(y)\right\rangle. \tag{3.11}$$

REMARK 3.2. For a canonical Gaussian measure $\lambda(x) = -x$, and therefore (3.11) leads to the results of section II.4.3.

REMARK 3.3. If $F\colon X \to \mathcal{H}$, then the restriction of $F'(x)$ to $\mathcal{H}$ is Hilbert–Schmidt. In this case (3.11) becomes

$$\frac{\mu^f(dx)}{\mu(dx)} = \tilde{\det}(I + F'(x)) \exp\left\{\int_0^1 \beta(\sigma, F, x)\, d\sigma\right\}, \tag{3.12}$$

where the expression β is defined on Hilbert–Schmidt operators $F'(x)$ and for symmetric trace class operators has the form

$$\beta(\sigma, F, x) = \langle \lambda(x + \sigma F(x)), F(x)\rangle + \operatorname{Tr} F'(x).$$

4. Maps of spaces and admissible operators

4.1. Compatible pairs of operators. Let X and Y be Banach spaces, $\mu \in \mathfrak{M}(X)$, $f\colon X \to Y$ a measurable map, and $\mu^f = \mu \circ f^{(-1)}$ the f-image of μ. Two operators A on $L_2(X,\mu)$ and B on $L_2(Y,\mu^f)$ are said to be *f-compatible* if for any function $\phi \in D_B$ the composite $\phi \circ f$ belongs to D_A, and

$$(B\phi) \circ (f(x)) = A(\phi \circ f)(x). \tag{4.1}$$

Clearly, if A_1, B_1 and A_2, B_2 are f-compatible pairs, then $A_1 \cdot A_2, B_1 \cdot B_2$ are f-compatible.

EXAMPLE 4.1. Let $f\colon X \to Y$ be a differentiable map, and let z and ξ be vector fields on X and Y, respectively. The pair z, ξ is said to be *f-compatible* if

$$\xi(f(x)) = f'(x) \cdot z(x). \tag{4.2}$$

Consider the differential operators $A = \langle z(x), \nabla\rangle$, $B = \langle \xi(y), \nabla\rangle$ corresponding to the vector fields z and ξ. It follows from (4.2) that the operators A, B are f-compatible.

Consider now differential operators of order n,

$$A = \langle z_n(x), \nabla\rangle \dots \langle z_1(x), \nabla\rangle, \quad B = \langle \xi_n(x), \nabla\rangle \dots \langle \xi_1(x), \nabla\rangle. \tag{4.3}$$

Suppose all pairs $z_1, \xi_1; \dots; z_n, \xi_n$ are f-compatible. Then the operators (4.3) are f-compatible.

THEOREM 4.1. *Suppose the operators A on $L_2(X,\mu)$ and B on $L_2(Y,\mu^f)$ are f-compatible, and let A be μ-admissible. Then B is μ^f-admissible, and the following relations hold:*

$$B'\mu^f = (A'\mu)^f, \tag{4.4}$$

$$\rho(B, \mu^f, y) = M\{\rho(A, \mu, \eta)|_{f(\eta)=y}\}, \tag{4.5}$$

where η is the random variable taking values in X and with probability distribution μ, and

$$\|\rho(B, \mu^f, y)\|_{L_2(Y,\mu^f)} \le \|\rho(A, \mu, x)\|_{L_2(X,\mu)}. \tag{4.6}$$

If f is invertible, then

$$\rho(B, \mu^f, y) = \rho(A, \mu, f^{(-1)}(y)). \tag{4.7}$$

PROOF. For any $\phi \in D_B$,

$$\left|\int_Y (B\phi)(y)\mu^f(dy)\right| = \left|\int_X A(\phi \circ f)(x)\mu(dx)\right| =$$
$$\left|\int_X (\phi \circ f)(x)\rho(A,\mu,x)\mu(dx)\right| \leq$$
$$\left\{\int_Y |\phi(y)|^2\mu^f(dy)\right\}^{1/2}\left\{\int_X |\rho(A,\mu,x)|^2\mu(dx)\right\}^{1/2} =$$
$$\|\rho(A,\mu,x)\|_{L_2(X,\mu)}\|\phi\|_{L_2(Y,\mu^f)}.$$

These estimates imply that $\int_Y (B\phi)(y)\mu^f(dy)$ is a continuous linear functional on $L_2(Y,\mu^f)$. Therefore there is a function $\beta \in L_2(Y,\mu^f)$ such that

$$\int_Y (B\phi)(y)\mu^f(dy) = \int_Y \phi(y)\beta(y)\mu^f(dy).$$

Clearly, $\beta(y) = \rho(B,\mu^f,y)$, and the estimate (4.6) holds. The relation (4.4) follows from a simple computation. From this computation we also obtain:

$$\int_Y (B\phi)(y)\mu^f(dy) = \int_X (\phi \circ f)(x)(A'\mu)(dx) =$$
$$\int_X (\phi \circ f)(x)\rho(A,\mu,x)\mu(dx) = \mathrm{M}\phi(f(\eta))\rho(A,\mu,\eta) =$$
$$\mathrm{M}\phi(f(\eta))\mathrm{M}\{\rho(A,\mu,\eta)|_{f(\eta)}\} =$$
$$\int_X (\phi \circ f)(x)\mathrm{M}\{\rho(A,\mu,\eta)|_{f(\eta)=f(x)}\}\mu(dx) = \int_Y \phi(y)\mathrm{M}\{\rho(A,\mu,\eta)|_{f(\eta)=y}\}\mu^f(dy).$$

Formula (4.5) follows from this; for invertible f we obtain (4.7). □

COROLLARY 4.1. *Let z be a vector field on X such that the differential operator $\langle z(x),\nabla\rangle$ is μ-admissible. Then the logarithmic derivative $\rho_\mu(z;x) = -\rho(\langle z(x),\nabla\rangle,\mu,x)$ exists, and*

$$\int_X |\rho_\mu(z;x)|^2\mu(dx) < \infty.$$

Let a vector field ξ on Y and a vector field z be f-compatible. Then the differential operator $\langle \xi(y),\nabla\rangle$ is μ^f-admissible, the measure μ^f has a μ^f-square integrable logarithmic derivative along ξ, and the following relations hold:

$$\langle \nabla,\xi\rangle\mu^f = [\langle \nabla,z\rangle\mu]^f, \tag{4.8}$$
$$\rho_{\mu^f}(\xi;y) = M\{\rho_\mu(z;\eta)|_{f(\eta)=y}\}, \tag{4.9}$$
$$\|\rho_{\mu^f}(\xi;y)\|_{L_2(Y,\mu^f)} \leq \|\rho_\mu(z;x)\|_{L_2(X,\mu)}. \tag{4.10}$$

If f is invertible, then

$$\rho_{\mu^f}(\xi;y) = \rho_\mu(z;f^{(-1)}(y)). \tag{4.11}$$

Let now $z_1,\xi_1;\dots;z_n,\xi_n$ be a set of pairs of f-compatible vector fields. Suppose the differential operator $A = \langle z_n(x),\nabla\rangle\dots\langle z_1(x),\nabla\rangle$ be μ-admissible. Then the operator $B = \langle \xi_n(y),\nabla\rangle\dots\langle \xi_1(y),\nabla\rangle$ is μ^f-admissible, and the corresponding logarithmic *derivative of order n satisfies the estimate*

$$\int_Y |\rho_n^f(\xi_1, \ldots, \xi_n; y)|^2 \mu^f(dy) \le \int_X |\rho_n(z_1, \ldots, z_n; x)|^2 \mu(dx). \tag{4.12}$$

4.2. Smoothness of finite-dimensional images of measures. Let $\mathcal{H}_- \supset \mathcal{H}_0 \supset \mathcal{H}_+$ be an equipped Hilbert space, $\mu \in \mathfrak{M}(\mathcal{H}_-)$, $Y = \mathbf{R}^n$, $f\colon \mathcal{H}_- \to \mathbf{R}^n$ a map which is differentiable along $\mathcal{H}_0$, $\mu^f = \mu \circ f^{(-1)}$. We use theorem 4.3 for analyzing the smoothness of the measure μ^f.

REMARK 4.1. Let μ be a measure on $\mathbf{R}^n$, $p(x)$ its density with respect to Lebesgue measure, $\{e_j\}_{j=1}^n$ some orthonormal basis in $\mathbf{R}^n$, $x = \sum_{j=1}^n x_j e_j$. If $p(x) > 0$ is sufficiently smooth, then

$$\rho_n(e_1, \ldots, e_n; x) = \frac{1}{p(x)} \frac{\partial^n p(x)}{\partial x_1 \ldots \partial x_n}. \tag{4.13}$$

Conversely, if $\rho_n(e_1, \ldots, e_n; x)$ is an integrable function, then (4.13) implies that $p(x)$ is continuous. More precisely (see Averbukh–Smolyanov–Fomin [1]), if $\rho_m(e_{k_1}, \ldots, e_{k_m}; x)$ $(k_1, \ldots, k_m = 1, \ldots, n)$ exist and are integrable, then μ is absolutely continuous with respect to Lebesgue measure, and the density $p(x)$ is of class $C_{m-n}(\mathbf{R}^n)$. In particular, the existence of integrable logarithmic derivatives of arbitrary order is equivalent to the existence of a density $p(x) \in C_\infty(\mathbf{R}^n)$.

First, let $n = 1$.

Consider the constant vector field $\xi \equiv 1$ on $\mathbf{R}^1$. Construct a vector field $z(x)$ f-compatible with ξ:

$$(f'(x), z(x)) \equiv 1.$$

Clearly,

$$z(x) = \frac{f'(x)}{\|f'(x)\|^2} \tag{4.14}$$

is the solution of this equation.

Then we compute the norm $\sigma_{n,4n}(z)$. For $h_i \in \mathcal{H}$ $(i = 1, \ldots, n)$ we obtain

$$z^{(i)}(x)(h_1, \ldots, h_i) =$$

$$\sum_{k=0}^{i} \sum_{(j_1,\ldots,j_i)} f^{(i-k+1)}(x)(h_{j_1}, \ldots, h_{j_{i-k}}) \left(\frac{1}{\|f'(x)\|^2}\right)^{(k)} (h_{j_{i-k+1}}, \ldots, h_{j_i}),$$

and for some orthonormal basis $\{e_r\}_{r=1}^\infty$ in $\mathcal{H}$:

$$\sigma_2^2(z^{(i)}(x)) =$$

$$\sum_{r_1,\ldots,r_i=1}^{\infty} \left\| \sum_{k=0}^{i} \sum_{(r_1,\ldots,r_i)} f^{(i-k+1)}(x)(e_{r_1}, \ldots, e_{r_{i-k}}) \left(\frac{1}{\|f'(x)\|^2}\right)^{(k)} (e_{r_{i-k+1}}, \ldots, e_{r_i}) \right\|^2 \le$$

$$\text{const} \sum_{k=0}^{i} \sigma_2^2 \left[\left(\frac{1}{\|f'(x)\|^2}\right)^{(k)} \right] \sigma_2^2[f^{(i-k+1)}(x)].$$

Integrating the last inequality, we find

$$\left\{\int_{\mathcal{H}_-} \sigma_2^{4n}(z^{(i)}(x))\mu(dx)\right\}^{1/4n} \leq$$

$$\text{const}\left\{\int_X \sum_{k=0}^{i} \sigma_2^{4n}\left[\left(\frac{1}{\|f'(x)\|^2}\right)^{(k)}\right] \sigma_2^{4n}[f^{(i-k+1)}(x)]\mu(dx)\right\}^{1/4n} \leq$$

$$\text{const}\sigma_{i,32n}(f'(x))\left\{\int_{\mathcal{H}_-} \sum_{k=1}^{i+1} \frac{1}{\|f'(x)\|^{16nk}}\mu(dx)\right\}^{1/8n},$$

where

$$\sigma_{n,p}(f') = \sum_{j=0}^{n}\left\{\int_{\mathcal{H}_-} \sigma_2^p(f^{(j+1)}(x))\mu(dx)\right\}^{1/p}. \tag{4.15}$$

Let f have the property that

$$\sigma_{n,32n}(f') < \infty. \tag{4.16}$$

Then

$$\sigma_{n,4n}(z) = \sum_{i=0}^{n}\left\{\int_{\mathcal{H}_-} \sigma_2^{4n}(z^{(i)}(x))\mu(dx)\right\}^{1/4n} \leq$$

$$\text{const}\sum_{i=0}^{n}\sum_{k=1}^{i+1}\left\{\int_{\mathcal{H}_-} \frac{\mu(dx)}{\|f'(x)\|^{16nk}}\right\}^{1/8n} \leq$$

$$\text{const}\left\{\int_{\mathcal{H}_-} \frac{\mu(dx)}{\|f'(x)\|^{16n(n+1)}}\right\}^{1/8n}.$$

Taking into account remark 4.1, we arrive at the following result.

THEOREM 4.2. *Let $\mu \in \mathfrak{M}_n^{4n}(\mathcal{H}_-, \mathcal{H}_0)$ ($n = 2^p$, $p \in \mathbf{N}$), and let $f \in C_{n+1}(\mathcal{H}_-, \mathbf{R})$ have the property (4.16). If*

$$\int_{\mathcal{H}_-} \frac{\mu(dx)}{\|f'(x)\|^{16n(n+1)}} < \infty, \tag{4.17}$$

then the measure μ^f on $\mathbf{R}^1$ has a density with respect to Lebesgue measure, of class $C_{n-1}(\mathbf{R})$.

REMARK 4.2. The conditions (4.16) are fulfilled if μ has finite moments $\int_{\mathcal{H}_-} \|x\|^q\mu(dx) < \infty$ ($q = 32n$) and the derivatives of f satisfy the estimates $\sigma_2^q(f^{(j)}(x)) \leq C_{1q}\|x\|^q + C_{2q}$.

If these conditions hold for arbitrary n, j, q, and $\int_{\mathcal{H}_-}(\mu(dx))/\|f'(x)\|^{2q} < \infty$ ($q = 1, 2, \ldots$), then the density $p(x)$ of μ is infinitely differentiable.

We now consider m-dimensional images of smooth measures.

THEOREM 4.3. *Let* $\mu \in \mathfrak{M}_n^{4n}(\mathcal{H}_-, \mathcal{H}_0)$; *let the function* $f = (f_1, \dots, f_m) \in C_{n+1}(\mathcal{H}_-, \mathbf{R}^m)$ *be differentiable along* $\mathcal{H}$ *and have the property* $\sigma_{n,32n}(f_k') < \infty$ $(k = 1, \dots, m)$. *If for any* $h \in \mathbf{R}^m$ $(\|h\| = 1)$,

$$\int_{\mathcal{H}_-} \frac{\mu(dx)}{\left\|\sum_{j=1}^m h_j f_j'(x)\right\|^{16k(k+1)}} \leq C_k \quad (k = 1, \dots, n), \tag{4.18}$$

then μ^f *has a density with respect to Lebesgue measure, of class* $C_{k-1}(\mathbf{R}^m)$.

PROOF. Denote by $\chi(\theta) = \int_{\mathbf{R}^m} e^{i(\theta,y)} \mu^f(dy)$ the characteristic functional of μ^f. Suppose that for some k we have an estimate

$$\|\theta\|^k \chi(\theta) \leq C_k. \tag{4.19}$$

In this case it is well-known that the density $p^f(dy) = \mu^f(dy)/dy$ (the Fourier transform of $\chi(\theta)$) is a continuously differentiable function, up to order $k-1$ inclusive.

For a fixed $h \in \mathbf{R}^m$ we define a map $g_h \colon \mathbf{R}^m \to \mathbf{R}^1$, $g_h(y) = (h, y)$, and denote by γ_h the composite map $\gamma_h(x) = (g_h \circ f)(x) = (h, f(x))$. The map γ_h gives rise to a measure $\mu^{\gamma_h} = \mu \circ \gamma_h^{(-1)} = \mu^f \circ g_h^{(-1)}$ on $\mathbf{R}$, and smoothness conditions for this measure imply smoothness conditions for the measure μ^f.

Let, for fixed $\theta \in \mathbf{R}^m$, $h = \theta/\|\theta\|$. Then

$$\|\theta\|^k \chi(\theta) = \int_{\mathbf{R}^m} \|\theta\|^k e^{i(\theta,y)} \mu^f(dy) =$$
$$\int_{\mathbf{R}^m} \|\theta\|^k e^{i(h,y)\|\theta\|} \mu^f(dy) = \int_{-\infty}^{\infty} \|\theta\|^k e^{i\alpha\|\theta\|} \mu^{\gamma_h}(d\alpha) =$$
$$i^{-k} \int_{-\infty}^{\infty} \frac{\partial^k}{\partial \alpha^k} e^{i\alpha\|\theta\|} \mu^{\gamma_h}(d\alpha) = i^{-k} \int_{-\infty}^{\infty} e^{i\alpha\|\theta\|} \rho_k^{\gamma_h}(\underbrace{1, \dots, 1}_{k}; \alpha) \mu^{\gamma_h}(d\alpha).$$

Denote by v a vector field in $\mathcal{H}_-$ which is γ_h-compatible with the vector field $z(\alpha) \equiv 1$ in $\mathbf{R}$. Using the estimates of theorem 4.1, we obtain

$$\|\theta\|^k |\chi(\theta)| \leq \left\{ \int_{-\infty}^{\infty} |\rho_k^{\gamma_h}(1, \dots, 1; \alpha)|^2 \mu^{\gamma_h}(d\alpha) \right\}^{1/2} \leq$$
$$\left\{ \int_X |\rho_k(v \dots, v; x)|^2 \mu(dx) \right\}^{1/2} \leq \text{const}\, \sigma_{k,4k}^k(v(x)) \leq$$
$$\text{const} \left\{ \int_X \frac{\mu(dx)}{\|\gamma'(x)\|^{16k(k+1)}} \right\}^{1/8} \leq \text{const} \left\{ \int_X \frac{\mu(dx)}{\|(h, f'x))\|^{16k(k+1)}} \right\}^{1/8}. \quad \square$$

REMARK 4.3. If the measure μ and the function f satisfy the conditions of theorem 4.3 for any n and

$$\int_X \frac{\mu(dx)}{\|f'(x)\|^{2q}} < \infty \quad (q = 1, 2, \dots), \tag{4.20}$$

then the measure μ^f has an infinitely differentiable density with respect to Lebesgue measure.

4.3. The quadratic map of smooth measures. Here we consider the case $f(x) = (Ax, x)/2$, where $A = A^* \in L(\mathcal{H}_-, \mathcal{H}_+)$. Note that the restriction of A to $\mathcal{H}$ is of trace class. Satisfaction of (4.16) is guaranteed by the existence of the corresponding moment of the measure μ:

$$(\sigma_{n-1,p}(f'))^p = \left\{\sum_{j=1}^{n}\left[\int_{\mathcal{H}_-} \sigma_2^p(f^{(j)}(x))\mu(dx)\right]^{1/p}\right\}^p \leq$$
$$\int_{\mathcal{H}_-} \|Ax\|^p \mu(dx) \leq \|A\|^p \int_{\mathcal{H}_-} \|x\|^p \mu(dx).$$

THEOREM 4.4. *Let a measure $\mu \in \mathfrak{M}_n^{4n}(\mathcal{H}_-, \mathcal{H}_0)$ have a finite moment of order $p = 32n$, $f(x) = (Ax, x)/2$, where the operator $A = A^* \in L(\mathcal{H}_-, \mathcal{H}_+)$ has a nontrivial invariant subspace L_l ($\dim L_l = l \geq 2q+1$; $q = 8n(n+1)$): $\|Ax\| \geq const\|x\|$ ($x \in L_l$). Then the measure μ^f on $\mathbf{R}^1$ has a density $p(x)$ of class $C_{n-1}(\mathbf{R})$.*

PROOF. In our case condition (4.17) has the form

$$\int_X \frac{\mu(dx)}{\|Ax\|^{2q}} < \infty \quad (q = 8n(n+1)). \tag{4.21}$$

Form the spectral decomposition of the operator $A = \sum_{k=1}^{\infty} \lambda_k P_k$, where the $\lambda_k \neq 0$ are the eigenvalues of A and P_k are the corresponding spectral projectors ($\dim P_k = 1$).

Choose a constant $\epsilon > 0$, and split the spectrum $\sigma(A)$ into two parts, $\sigma(A) = \sigma_1(\epsilon) \cup \sigma_2(\epsilon)$, where $\sigma_1(\epsilon) = \{\lambda \in \sigma(A)\colon |\lambda| > \epsilon\} = \{\lambda_1, \dots, \lambda_{l(\epsilon)}\}$, $\sigma_2(\epsilon) = \{\lambda \in \sigma(A)\colon |\lambda| \leq \epsilon\}$. One obtains the estimate:

$$\|Ax\|^2 = \sum_{\lambda_k \in \sigma_1(\epsilon)} |\lambda_k|^2 \|P_k x\|^2 + \sum_{\lambda_k \in \sigma_2(\epsilon)} |\lambda_k|^2 \|P_k x\|^2 \geq$$
$$\epsilon^2 \sum_{k=1}^{l} \|P_k x\|^2 \quad (l = \operatorname{card} \sigma_1(\epsilon) < \infty). \tag{4.22}$$

Using (4.22) we find

$$\int_X \frac{\mu(dx)}{\|Ax\|^{2q}} \leq \frac{1}{\epsilon^{2q}} \int_{L_l} \frac{p_l(x)(dx)}{\|x\|_{L_l}^{2q}}; \quad p_l(x) = \frac{(P^{(l)}\mu)(dx)}{dx},$$

where $P^{(l)} = \sum_{k=1}^{l} P_k$, $L_l = P^{(l)}\mathcal{H}_-$ is an invariant subspace of A.

Split L_l into two parts, $L_l = S_{0,1} \cup S_{0,1}^c$, where $S_{0,1} = \{x \in L_l\colon \|x\|_{L_l} \leq 1\}$. Then

$$\int_X \frac{\mu(dx)}{\|Ax\|^{2q}} \leq \frac{1}{\epsilon^{2q}} \left\{\int_{S_{0,1}} \frac{p_l(x)(dx)}{\|x\|_{L_l}^{2q}} + \int_{S_{0,1}^c} \frac{p_l(x)(dx)}{\|x\|_{L_l}^{2q}}\right\} \leq$$
$$\frac{1}{\epsilon^{2q}} \left\{\int_0^1 \int_{\|x\|=r} p(x) r^{l-2q-1}\, ds\, dr + \int_{S_{0,1}^c} p_l(x)\, dx\right\}.$$

Clearly, if $l \geq 2q + 1$ then (4.21) is true. □

COROLLARY 4.2. *If A is an infinite-dimensional operator, then (4.21) holds for any q.*

Suppose μ has finite moments of all orders, and let the logarithmic gradient of μ be infinitely differentiable. Suppose all powers of the derivatives $\lambda^{(k)}(x)$ $(k \geq 1)$ are integrable. Then the measure μ^f has an infinitely differentiable density $p^f(x)$.

Theorem 4.4 admits the following generalization.

THEOREM 4.5. *Let μ be a measure satisfying the conditions of the previous corollary. Consider the map*

$$f = (f_1, \dots, f_m) \colon X \to \mathbf{R}^m;$$

$$f_i(x) = \frac{1}{2}(A_i x, x), \quad A_i = A_i^* \in L(\mathcal{H}_-, \mathcal{H}_+) \quad (i = 1, \dots, m).$$

If for each $h \in \mathbf{R}^m$, $\|h\| \neq 0$, the operator $A_h = \sum_{j=1}^m h_j A_j$ is infinite dimensional, then the measure $\mu^f = \mu \circ f^{(-1)}$ on $\mathbf{R}^m$ has an infinitely differentiable density.

The proof of this theorem is a refined version of that of theorem 4.4: start with the reasoning concerning the spectrum of A_h for each h ($\|h\| \neq 0$), and then use the compactness of the unit ball in $\mathbf{R}^m$.

REMARK 4.4. It is possible to state theorems 4.4 and 4.5 for a larger class of quadratic maps; in particular, for measurable quadratic functionals $\tilde{f}(x) = [(\tilde{A}x, x) - \operatorname{Tr} \tilde{A}]/2$, where $\tilde{A} \in L(\mathcal{H}_-, \mathcal{H}_0)$ (see section II.3).

Note that the conditions of these theorems are fulfilled for a Gaussian measure.

5. Biorthogonal systems in $L_2(X, \mu)$

Let μ be a Borel measure on $X \supset \mathcal{H} \supset X^*$. Suppose it has finite moments of arbitrary order:

$$M(\phi_1, \dots, \phi_n) = \int_X \langle x, \phi_1 \rangle \dots \langle x, \phi_n \rangle \mu(dx) \quad (\phi_j \in X^*, \quad j = 1, \dots, n, \quad n = 1, 2, \dots).$$

In this case the characteristic functional $\chi(\theta) = \int_X e^{i\langle \theta, x \rangle} \mu(dx)$ is infinitely differentiable along X^*.

Consider the system of polynomials

$$P(\phi_1, \dots, \phi_n; x) = \left\langle \phi_1, \frac{1}{i} \nabla_\theta \right\rangle \dots \left\langle \phi_n, \frac{1}{i} \nabla_\theta \right\rangle \frac{e^{i\langle \theta, x \rangle}}{\chi(\theta)} \bigg|_{\theta=0}. \tag{5.1}$$

It is possible to show by direct computation that these polynomials have properties similar to those of the classical Hermite polynomials.

PROPOSITION 5.1. *The following relations hold:*

$$\left\langle \psi, \frac{1}{i} \nabla_x \right\rangle P(\phi_1, \dots, \phi_n; x) = \sum_{j=1}^n \langle \psi, \phi_j \rangle P(\phi_1, \dots, \phi_{j-1}, \phi_{j+1}, \dots, \phi_n; x),$$

$$P(\phi_1, \dots, \phi_n; x + y) =$$

$$\sum_{k+l+r=n} \sum_{(j_1, \dots, j_n)} i^k M(\phi_{j_1}, \dots, \phi_{j_k}) P(\phi_{j_{k+1}}, \dots, \phi_{j_{k+l}}; x) P(\phi_{j_{k+l+1}}, \dots, \phi_{j_n}; y).$$

Suppose μ has a logarithmic gradient $\lambda(x)$ which is infinitely differentiable along X^* and which satisfies the estimate

$$\int_X |\lambda^{(k)}(x)(\phi_1,\dots,\phi_k)|^m\,\mu(dx) < \infty \quad (k,m = 1,2,\dots;\phi_j \in X^*).$$

Consider the system of polynomials

$$R(\phi_1,\dots,\phi_n;x) = \langle\phi_1,x\rangle\dots\langle\phi_n,x\rangle \quad (\phi_j \in X^*, \quad x \in X),$$

and the corresponding system of differential operators

$$R(\phi_1,\dots,\phi_n;\nabla) = \langle\phi_1,\nabla\rangle\dots\langle\phi_n,\nabla\rangle.$$

It was shown in section 2 that

$$R(\phi_1,\dots,\phi_n;\nabla)^* = (-1)^n Q(\phi_1,\dots,\phi_n;\nabla), \tag{5.2}$$

where $Q(\phi_1,\dots,\phi_n;\nabla) = \langle\phi_1,\lambda(x)+\nabla\rangle\dots\langle\phi_n,\lambda(x)+\nabla\rangle$.

THEOREM 5.1. *Let $e_j \in X^* \subset \mathcal{H}$ $(j=1,2,\dots)$ be an orthonormal basis in $\mathcal{H}$. The system of functions*

$$\tilde{Q}(e_{j_1},\dots,e_{j_n};x) = Q(e_{j_1},\dots,e_{j_n};\nabla)\cdot 1 \quad (n, j_k = 1,2,\dots)$$

is biorthogonal with respect to the system of polynomials $P(e_{j_1},\dots,e_{j_n};x)$:

$$\int_X P(e_{j_1},\dots,e_{j_m};x)\tilde{Q}(e_{r_1},\dots,e_{r_n};x)\mu(dx) = \delta_{nm}\sum_{(r_1,\dots,r_n)}\delta_{j_1r_1}\dots\delta_{j_nr_n}.$$

PROOF. Using the properties of the operators $Q(\phi_1,\dots,\phi_n;\nabla)$ we find:

$$\int_X P(\phi_1,\dots,\phi_m;x)R(\psi_1,\dots,\psi_n;\nabla)^*\cdot 1\mu(dx) =$$

$$\int_X \langle\psi_1,\nabla_x\rangle\dots\langle\psi_n,\nabla_x\rangle\left\langle\phi_1,\frac{1}{i}\nabla_\theta\right\rangle\dots\left\langle\phi_m,\frac{1}{i}\nabla_\theta\right\rangle\frac{e^{i\langle\theta,x\rangle}}{\chi(\theta)}\mu(dx)\bigg|_{0-0} =$$

$$i^{n-m}\langle\phi_1,\nabla_\theta\rangle\dots\langle\phi_m,\nabla_\theta\rangle\langle\theta,\psi_1\rangle\dots\langle\theta,\psi_n\rangle|_{\theta=0} =$$

$$\delta_{nm}\sum_{(r_1,\dots,r_n)}\langle\phi_1,\psi_{r_1}\rangle\dots\langle\phi_n,\psi_{r_n}\rangle.$$

For $m > n$ this expression is equal to zero, because the order of the derivative exceeds the degree of the polynomial. For $m < n$ it is also equal to zero, since $\theta = 0$. □

COROLLARY 5.1. *If a function $f(x)$ $(x \in X)$ is representable as the sum of a series converging in $L_2(X,\mu)$,*

$$f(x) = \sum_{n=0}^{\infty}\sum_{(r_1,\dots,r_n)}\alpha(r_1,\dots,r_n)\tilde{Q}(e_{r_1},\dots,e_{r_n};x),$$

then

$$\alpha(r_1,\dots,r_n) = \frac{1}{n!}\int_X f(x)P(e_{r_1},\dots,e_{r_n};x)\mu(dx).$$

The generating function of the polynomials (5.1) gives rise to the integral transformation

(5.3)
$$\Phi\colon f(x) \to \hat{f}(\theta) = \frac{1}{\chi(\theta)} \int_X e^{i\langle\theta,x\rangle} f(x)\mu(dx) = \frac{\int_X e^{i\langle\theta,x\rangle} f(x)\mu(dx)}{\int_X e^{i\langle\theta,x\rangle}\mu(dx)} \quad (x \in X, \quad \theta \in X^*).$$

Note that for a canonical Gaussian measure both systems of functions $\{P_n\}$ and $\{\tilde{Q}_n\}$ coincide with the Hermite polynomials, and the transformation (5.3) is the Fourier–Wiener transformation (see section II.5).

Everywhere below we assume that for some $\epsilon > 0$ the measure μ satisfies the condition

$$\int_X e^{\epsilon\|x\|_X} \mu(dx) < \infty. \tag{5.4}$$

In this case, if $f \in L_2(X, \mu)$, then $\hat{f}(\theta)$ is an analytic function at $\theta = 0 \in X^*$:

$$\hat{f}(\theta) = \sum_{n=0}^{\infty} \sum_{k_1,\dots,k_n=1}^{\infty} c(k_1, \dots, k_n) R(e_{k_1}, \dots, e_{k_n}; \theta),$$

where

$$c(k_1, \dots, k_n) = R(e_{k_1}, \dots, e_{k_n}; \nabla)\hat{f}(\theta)|_{\theta=0} =$$
$$\int_X f(x) R(e_{k_1}, \dots, e_{k_n}; \nabla) \frac{e^{i\langle\theta,x\rangle}}{\chi(\theta)}\bigg|_{\theta=0} \mu(dx) =$$
$$\int_X f(x) P(e_{k_1}, \dots, e_{k_n}; x)\mu(dx).$$

We then verify the relation

$$\Phi\colon \tilde{Q}(\phi_1, \dots, \phi_n; x) \mapsto i^n R(\phi_1, \dots, \phi_n; \theta). \tag{5.5}$$

In fact,

$$\int_X \frac{e^{i\langle\theta,x\rangle}}{\chi(\theta)} \tilde{Q}(\phi_1, \dots, \phi_n; x)\mu(dx) = \int_X \frac{e^{i\langle\theta,x\rangle}}{\chi(\theta)} Q(\phi_1, \dots, \phi_n; \nabla) \cdot 1\mu(dx) =$$
$$\int_X R(\phi_1, \dots, \phi_n; \nabla_x) \frac{e^{i\langle\theta,x\rangle}}{\chi(\theta)} \mu(dx) = \frac{R(\phi_1, \dots, \phi_n; \theta)}{\chi(\theta)} \int_X e^{i\langle\theta,x\rangle}\mu(dx) = R(\phi_1, \dots, \phi_n; i\theta).$$

Comparison of (5.3) and (5.5) leads to the following result, describing the inverse of the integral transformation Φ.

THEOREM 5.2. *If a function*

$$F(\theta) = \sum_{n=0}^{\infty} \sum_{(k_1,\dots,k_n)} c(k_1, \dots, k_n) R(e_{k_1}, \dots, e_{k_n}; \theta)$$

which is analytic at $\theta = 0$ is the Φ-transform of a function $f \in L_2(X, \mu)$, then f is representable by the following series (provided this series converges):

$$f(x) = \sum_{n=0}^{\infty} \sum_{(k_1,\dots,k_n)} c(k_1, \dots, k_n)\tilde{Q}(e_{k_1}, \dots, e_{k_n}; x) =$$

$$\sum_{n=0}^{\infty} \sum_{(k_1,\dots,k_n)} R\left(e_{k_1}, \dots, e_{k_n}; \frac{1}{i}\nabla\right) \hat{f}(\theta)\Big|_{\theta=0} \tilde{Q}(e_{k_1}, \dots, e_{k_n}; x). \tag{5.6}$$

Formula (5.6) holds identically at least in case $\hat{f}(\theta)$ is a polynomial; in this case the series is finite. In a more general case an analysis of the convergence is required. This analysis is based on estimating the functions $\tilde{Q}(\phi_1, \dots, \phi_n; x)$.

It is possible to obtain such estimates by the methods developed in section 2.

6. Chapter VIII: Supplementary remarks and historical comments

The notion of logarithmic derivative of a measure along a vector field was used in Dalecky [1] in connection with the development of an infinite-dimensional analog of the Malliavin calculus. The further development of this notion and some formulas are given in Dalecky–Marjanin [1] (including the case of smooth infinite-dimensional manifolds) and in Norin [1]. Yu. Kondratjev has obtained close results, and considered differential operators with infinitely many independent variables generated by Dirichlet forms (see Berezansky–Kondratjev [1]).

A detailed account of the results in this area is given in Belopolskaya–Dalecky [1]. Formulas for the densities of measures shifted along the trajectories of a vector field were obtained in Dalecky [2] under the (superfluous) condition of analyticity, and in Dalecky–Sochadze [1] (under general conditions). In the latter, formula (3.11) was obtained also. On the other hand, this formula follows from earlier results in Skorokhod [1], based on other reasonings. See also Bell [1], which considers special cases.

The approach presented, based on the theory of μ-admissible operators, as well as corresponding definitions and properties of these operators, belongs to the authors (see Dalecky–Steblovskaya [1]).

The results of section 4, concerning absolute continuity of quadratic images of smooth measures, were obtained in Steblovskaya [1]. The mathematicians Uglanov [1], Smorodina [1], and Bogachev [1] have obtained results close to these, but by different methods. In connection with this, see the recent extensive review Bogachev–Smolyanov [1].

The polynomials (5.1) and the functions $\tilde{Q}$ biorthogonal to them (and reducing to the Hermite polynomials for Gaussian measures) were introduced in Dalecky [3]. We do not know whether they were known earlier in the literature, not even in the finite-dimensional case. Note that it is possible to interpret the system $\{\tilde{Q}\}$ as a generalization to the nonGaussian case of multiple stochastic Itô integrals.

Bibliography

Averbukh, V.I., Smolyanov, O.G., Fomin, S.V.,
[1] Extended functions and differential equations in linear spaces, *Proc. Moscow Math. Soc.* **24** (1971), 1133-174. (Translated from the Russian.)

Bell, D.,
[1] A quasi-invariance theorem for measures on Banach spaces, *Trans. Amer. Math. Soc.* **290** (1985), 851-855.

Belopolskaya, Ya.I., Dalecky, Yu.L.,
[1] *Stochastic equations and differential geometry*, Kluwer, 1989. (Translated from the Russian.)

Berezansky, Yu. M., Kondratyev, Yu.G.,
[1] *Spectral methods in infinite-dimensional analysis*, Nauk. Dumka, Kiev, 1980. (In Russian.)

Bogachev, V.I.,
[1] Distributions of smooth functionals of random processes and infinite-dimensional oscillating integrals, *Proc. Vth Internat. Vilnius Conf. Probab. Theory and Math. Stat.* **3** (1989), 76-77. (In Russian.)

Bogachev, V.I., Smolyanov, O.G.,
[1] Analytic properties of infinite-dimensional distributions, *Uspekhi Mat. Nauk* **45: 3** (1990), 3-83. (In Russian.)

Dalecky, Yu.L.,
[1] Stochastic differential geometry, *Russian Math. Surveys* **38: 3** (1983), 97-125. (Translated from the Russian.)
[2] Absolute continuity of measures generated by diffusion processes on smooth manifolds, *Proc. IIIth Internat. Vilnius Conf. Probab. Theory and Math. Stat.* **1** (1981), 167-170. (In Russian.)
[3] Biorthogonal analogy of Hermite polynomials and inversion of Fourier transformation with respect to a non-gaussian measure, *Funct. Anal. Appl.* **Forthcoming** (1991), (Translated from the Russian.)

Dalecky, Yu.L., Maryanin, B.D.,
[1] Smooth measures on infinite-dimensional manifolds, *Soviet Math. Dokl.* **32: 3** (1985), 863-866. (Translated from the Russian.)

Dalecky, Yu.L., Sochadze, G.A.,

[1] Absolute continuity of smooth measures, *Funct. Anal. Appl.* **22: 2** (1988), 77-78. (Translated from the Russian.)

Dalecky, Yu.L., Steblovskaya, V.R.,

[1] Smooth measures on Banach spaces and their smooth mappings, *Proc. Vth Internat. Vilnius Conf. Probab. Theory and Math. Stat.* **1** (1990), 245-257. (In English.)

Norin, N.V.,

[1] Stochastic integrals and differentiabl emeasures, *Teor. Veroyatnost. Prilozh.* **32: 1** (1987), 114-124. (In Russian.)

Skorokhod, A.V.,

[1] *Integration in Hilbert space*, Springer, 1974. (Translated from the Russian.)

Smorodina, N.V.,

[1] Smoothness conditions for densities of distributions of functionals of random processes, *Proc. IVth Internat. Vilnius Conf. Probab. Theory and Math. Stat.* **3** (1985), 143-145. (In Russian.)

Steblovskaya, V.R.,

[1] Smoothness of finite-dimensional images of measures, *Ukrainian Mat. Zh.* **41: 2** (1989), 261-265. (In Russian.)

Uglanov, A.V.,

[1] On the division of extended functions of an infinite number of variables on polynolmials, *Soviet Math. Dokl.* **264: 5** (1982), 1096-1099. (Translated from the Russian.)

Supplement to chapters IV–V

Cauchy problem for the simplest parabolic equation with essentially infinite-dimensional elliptic operator

By Yu.V. Bogdansky and Yu.L. Dalecky

This supplement partially completes the deficiency noted in the additional remarks and historical comments to chapter IV.

We introduce the notion of an essentially infinite-dimensional differential operator, of which the Laplace–Lévy operator is a special case.

The Cauchy problem for the evolution equation with such (and even more general) operators is investigated. The characteristics of solutions of Cauchy problems in the case of essentially infinite-dimensional operators are studied

1. Essentially infinite-dimensional elliptic operators

Let $\mathcal{H}$ be a separable infinite-dimensional real Hilbert space, and $B_s(\mathcal{H})$ the Banach space of selfadjoint bounded operators on $\mathcal{H}$ (with operator norm). Let j be a nonnegative linear functional on $B_s(\mathcal{H})$. The differential expression L is defined on the space $C^2(\mathcal{H})$ of twice continuously Fréchet-differentiable functions by the formula

$$(Lu)(x) = \frac{1}{2}j(u''(x)), \tag{1.1}$$

where, in general, j_x depends on $x \in \mathcal{H}$.

EXAMPLE 1.1. Let $\mathbf{A}(x)u = \operatorname{Tr} A(x)u''(x)/2$, where $A(x) \geq 0$ is a nonnegative nuclear operator. Following Averbukh–Smolyanov Fomin [1] we will call this differential operator *regular*.

EXAMPLE 1.2. Let $Lu = \lim_{n\to\infty} \mathcal{A}_n(x)u = \lim_{n\to\infty} \operatorname{Tr} A_n(x)u''(x)/2$, where $A_n(x)$ is a sequence of nonnegative nuclear operators for which $\|\mathbf{A}_n(x)\| \to 0$ and the sequence $\operatorname{Tr} A_n(x)$ has a finite limit for all $x \in \mathcal{H}$. For cylindrical functions u, $Lu = 0$. In general, L is defined not on all of $C^2(\mathcal{H})$, but it can be extended to $C^2(\mathcal{H})$ (by a well-known theorem of Krein).

A special case of such operators is the *Laplace–Lévy operator*

$$Lu = \frac{1}{2}\lim_{n\to\infty}\frac{1}{n}\operatorname{Tr} P_n u'',$$

where $\{P_n\}$ $(\dim \Im P_n = n)$ is a sequence of orthogonal projectors on increasing subspaces of $\mathcal{H}$.

DEFINITION 1.1. A linear functional $j\colon B_S(\mathcal{H}) \to \mathbf{R}$ is called *essentially infinite-dimensional* if all finite-dimensional (hence compact) operators belong to its kernel. The corresponding differential operator L is also called essentially infinite-dimensional.

The differential operator in example 1.2 is essentially infinite-dimensional.

PROPOSITION 1.1. *Let $j \in B_S^*(\mathcal{H}$ and $j \geq 0$. Then there exists a unique decomposition $j = j_1 + j_2$, where $j_1, j_2 \geq 0$, $j_1(C) = \operatorname{Tr} AC$ $(A \geq 0$, nuclear), and j_2 is essentially infinite-dimensional.*

PROOF. Let $A \geq 0$ be the unique nuclear operator such that $j(K) = \operatorname{Tr} AK$ for all compact operators $K \in B_s(\mathcal{H})$. It suffices to get convinced of the fact that the functional $j_2(C) = j(C) - \operatorname{Tr} AC$ is nonnegative.

Fix $C \geq 0$ and $\epsilon > 0$. By the Weyl–von Neumann theorem there is a $C_\epsilon \in B_S(\mathcal{H})$ such that $\|C - C_\epsilon\| < \epsilon$, $C - C_\epsilon$ compact, and C_ϵ having a complete orthonormal system of eigenvectors, $\mathcal{F} = \{f_1, f_2, \dots\}$: $C_\epsilon x = \sum_{k=1}^\infty \lambda_k(x, f_k) f_k$ (the λ_k are corresponding eigenvalues).

The subspace $\mathcal{M}_\mathcal{F} \subset B_S(\mathcal{H})$ of operators that leave invariant $\mathcal{F}$ is isomorphic to ℓ_∞: $\Lambda\colon \mathcal{M}_\mathcal{F} \ni C \to (\lambda_1, \lambda_2, \dots) \in \ell_\infty$. Let $j_\mathcal{F} = j \circ \Lambda^{-1}$, $\alpha_\mathcal{F} = j_\mathcal{F} \circ \Lambda^{-1}$. Then $\alpha_\mathcal{F}\colon \Lambda(C) \to \sum_{k=1}^\infty a_k \lambda_k$, where $a_k = (Af_k, f_k) \geq 0$, $\sum_{k=1}^\infty a_k < \infty$, and $j_\mathcal{F}(\vec{\lambda}) = \alpha_\mathcal{F}(\vec{\lambda})$ for $\vec{\lambda} = (\lambda_1, \dots, \lambda_m, 0, 0, \dots)$. Hence $j_\mathcal{F}(\vec{\lambda}) \geq \alpha_\mathcal{F}(\vec{\lambda})$ for all $\vec{\lambda} \geq 0$. Thus, $j_2(C) = j_2(C_\epsilon) \geq -\epsilon j_2(I)$, and since ϵ is arbitrary, $j_2(C) \geq 0$. □

THEOREM 1.1. *Let j be an essentially infinite-dimensional nonnegative linear functional on $B_S(\mathcal{H})$. For any separable subset X in $B_S(\mathcal{H})$ there is a sequence of nonnegative nuclear operators A_m such that $\|A_m\| \to 0$, $\operatorname{Tr} A_m = \|j\|$, and $j(C) = \lim_{m\to\infty} \operatorname{Tr} A_m C$ for all $C \in X$. Moreover, for any compact set $X \subset B_S(\mathcal{H})$ the convergence of $\operatorname{Tr} A_m(\cdot)$ to $j(\cdot)$ is uniform on X.*

PROOF. It suffices to prove the theorem for a normalized functional j, $\|j\| = 1$. Let $C \in B_S(\mathcal{H})$ and let C_ϵ be the operator with complete orthonormal system of eigenvectors such that $C - C_\epsilon$ is compact. Let $\{\lambda_n\}$ be the sequence of eigenvalues of C_ϵ, taking into account multiplicity. Since variation of C_ϵ by a finte-dimensional operator does not change the value of $j(C_\epsilon)$, we conclude: $\liminf \lambda_n \leq j(C) = j(C_\epsilon) \leq \limsup \lambda_n$. Therefore $j(C) = \lim_{m\to\infty}(\lambda_{n_1} + \dots + \lambda_{n_m})/m$ for some subsequence $\{\lambda_{n_m}\}$, and $j(C) = \lim_{m\to\infty} \operatorname{Tr} P_m C/m$, where P_m is the othogonal projector on the subspace spanned by the eigenvectors $\{f_{n_1}, \dots, f_{n_m}\}$ of C_ϵ.

Let $\sigma_+(C) = \sup_{\omega\in\Omega} \omega(C)$, where Ω is the set of all essentially infinite-dimensional nonnegative normalized functionals on $B_S(\mathcal{H})$. Then (with previous notation), $\sigma_+(C) = \limsup\{\lambda_n\}$, $j(C) \leq \sigma_+(C)$.

Fix a finite set $\{C_1, \dots, C_n\} \subset B_S(H)$. For any $\vec{\alpha} \in \mathbf{R}^n$ we put $g(\vec{\alpha}) = j(\sum_{k=1}^n \alpha^k C_k)$; g is a linear functional on $\mathbf{R}^n$.

Consider the function $h_+(\vec{\alpha}) = \sigma_+(\sum_{k=1}^n \alpha^k C_k)$. Since

$$\sup_{\omega\in\Omega} \omega(\beta B_1 + (1-\beta)B_2) \le \beta \sup_{\omega\in\Omega} \omega(B_1) + (1-\beta) \sup_{\omega\in\Omega} \omega(B_2)$$

$(0 \le \beta \le 1)$, h_+ is a convex function on $\mathbf{R}^n$.

Since $g(\vec{\alpha}) \le h_+(\vec{\alpha})$ for all $\vec{\alpha} \in \mathbf{R}^n$, and $g(\vec{0}) = h_+(\vec{0}) = 0$, the graph of g is the support plane of $\operatorname{epi} h_+$ at $(\vec{0}, 0)$.

But then $g(\vec{\alpha}) = (\vec{\alpha}, \vec{\gamma})$, where $\vec{\gamma}$ is the subgradient of h_+ at $\vec{0}$. Since the differential $\partial h_+(\vec{0})$ is the closed convex hull of the set of limit points of sequences of the type $\{\nabla h_+(\vec{\alpha}_k)\}$, where $\vec{\alpha}_k \to \vec{0}$, and h_+ is differentiable at the point $\vec{\alpha}$, we can, with arbitrarily high precision, replace $\vec{\gamma}$ by a linear combination $\sum_{k=1}^p \beta_k \vec{\gamma}_k$. Here $\beta_k \ge 0$, $\sum_{k=1}^p \beta_k = 1$, and γ_k are the gradients of the function h_+ at the points of differentiability $\vec{\alpha}_k$ $(k = 1, \dots, p)$.

If $j_k \in \Omega$ is such that $j_k(\sum_{m=1}^n \alpha_k^m C_m) = h_+(\vec{\alpha}_k)$, then, since the support plane to the set $\operatorname{epi} h_+$ at $(\vec{\alpha}_k, h_+(\vec{\alpha}_k))$ is unique, the function $g_k(\vec{\alpha}) = j_k(\sum_{m=1}^n \alpha^m C_m)$ has the form $g_k(\vec{\alpha}) = (\vec{\alpha}, \vec{\gamma}_k)$.

If

$$j_k(C) = \lim_{m\to\infty} \operatorname{Tr} A_m^{(k)} C \tag{1.2}$$

is the expression for the value of j_k on the operator $C = \sum_{m=1}^n \alpha_k^m C_m$ obtained above, then by choosing a subsequence, if necessary, we can state that $\lim_{m\to\infty} \operatorname{Tr} A_m^{(k)} C_l$ exists for $l = 1, \dots, n$. Since the functional at the righthand side of (1.2) belongs to Ω (see example 1.2), the equality (1.2) holds for all $C = \sum_{m=1}^n \alpha^m C_m$, again by the uniqueness of the support plane.

Thus, for any $\epsilon > 0$ it is possible to construct functionals $j_k \in \Omega$ of the form $\lim \operatorname{Tr} A_m(\cdot)$ such that $|g(\vec{e}_l) - \sum_{k=1}^p \beta_k j_k(C_l)| < \epsilon$ $(l = 1, \dots, n)$, where $\vec{e}_1, \dots, \vec{e}_n$ is a canonical basis in $\mathbf{R}^n$. In addition, the functional $\omega = \sum_{k=1}^r \beta_k j_k$ again has the form $\lim \operatorname{Tr} A_m(\cdot)$.

By this procedure, for $\epsilon = 1/m$ we can construct a nonnegative nuclear operator A_m such that $\|A_m\| < \epsilon$, $\operatorname{Tr} A_m = 1$, and $|j(C_l) - \operatorname{Tr} A_m C_l| < 2\epsilon$. Hence we have the representation

$$j(C_l) = \lim_{m\to\infty} \operatorname{Tr} A_m C_l \quad (l = 1, \dots, n). \tag{1.3}$$

Let $\{C_k\}_{k=1}^\infty$ be a countable everywhere dense set in $X \subset B_S(\mathcal{H})$. By the diagonal method we can select a sequence $\{A_m\}$ such that (1.3) holds for $C_1, C_2, \dots$; since the sequence of linear functionals $\operatorname{Tr} A_m(\cdot)$ is uniformly bounded, (1.3) holds for all $C \in X$.

A standard argument involving ϵ-nets for compact susets X now completes the proof. □

It is useful to note that the A_m are finite dimensional.

2. Properties of essentially infinite-dimensional elliptic operators, and solutions of the corresponding Cauchy problem

PROPOSITION 2.1. *Let L be an essentially infinite-dimensional differential operator. Then:*

$$L(uv) = uLv + vLu, \quad L(\phi \circ u) = \phi'(u)Lu, \tag{2.1}$$

for $u, v \in C^2(\mathcal{H})$, $\phi \in C^2(\mathbf{R})$.

PROOF. $(uv)''(x) = u(x)v''(x) + v(x)u''(x) + u'(x) \otimes v'(x) + v'(x) \otimes u'(x)$ (using the natural isomorphism between $\mathcal{H}$ and $\mathcal{H}^*$). Since the operator $u'(x) \otimes v'(x) + v'(x) \otimes u'(x)$ has finite rank, $j((uv)''(x)) = u(x)j(v''(x)) + v(x)j(u''(x))$. The second assertion follows from the formula $(\phi \circ u)''(x) = \phi'(u(x))u''(x) + \phi''(u(x))u'(x) \otimes u'(x)$. □

For functions $u(x,t)$ defined on $\mathcal{H} \times \mathbf{R}$ we can consider the differential operator $L_x u(x,t) = j(x,t)(u''_x(x,t))/2$ (with respect to x). Suppose that $j(x,t) \geq 0$ is continuous with respect to (x,t).

THEOREM 2.1. *Let G be a bounded domain in $\mathcal{H}$, and let the function u be defined and continuous in the 'closed' cylinder $\bar{Z} = \bar{G} \times [0,T) \subset \mathcal{H} \times \mathbf{R}$, $u(x,t) \in C^2(G)$ for all $t \in (0,T)$, and $u(x,t) \in C^1((0,T))$ for all $x \in G$. Moreover, $u''_{xx}, u'_t \in C(G \times (0,\tau))$. Suppose that the inequality*

$$\frac{\partial u}{\partial t} \leq L_x u \tag{2.2}$$

holds everywhere in $Z = G \times [0,T)$. Then the inequality $u(x,t) \leq \sup_{\partial Z} u$ is true for all $(x,t) \in Z$, where $\partial Z = \bar{G} \times \{0\} \cup \partial G \times [0,T)$.

We will first prove some lemmas.

Suppose that $u \in C^2(G)$. Then for all $x \in G$ we have

$$u(x+h) = u(x) + (u'_x, h) + \frac{1}{2}(u''_{xx}h, h) + \omega(x,h),$$

where $\omega(x,h)/\|h\|^2 \to 0$ as $\|h\| \to 0$. (Here $u'_x = u'(x)$, $u''_{xx} = u''(x)$.)

Suppose that $\rho > 0$ and $x \in G$, and consider the set

$$\mathcal{A}(x,\rho) = \{h \in \mathcal{H} : \ x + h \in G, \ |\omega(x,h)| < \rho\|h\|^2\},$$

and the numbers

$$r_S(u,x,\rho) = \sup\{a : \ S_a \subset \mathcal{A}(x,\rho)\} > 0,$$
$$r_B(u,x,\rho) = \sup\{a : \ B_a \subset \mathcal{A}(x,\rho)\} > 0.$$

(Here S_a and B_a are the sphere and ball with centre 0 and radius a, respectively.)

LEMMA 2.1. *Suppose that $u \in C^2(G)$. Then for any $x \in G$ and $\rho > 0$ there are $\epsilon, \delta > 0$ such that for all $z \in G$ which belong to the ϵ-neighborhood $B(x,\epsilon)$ of x the inequality $r_S(u,z,\rho) \geq \delta$ holds.*

PROOF. Fix $x \in G$. There is an $\epsilon > 0$ such that

$$\epsilon \leq \min\left\{\frac{1}{3}r_B\left(u,x,\frac{\rho}{5}\right);\ \frac{\rho r_B(u,x,\rho/5)}{6\|u''_x\|}\right\};$$

$$\|u'_{x+\tau} - u'_x\| < \frac{\rho}{4}\|\tau\|;\quad \|u''_{x+\tau} - u''_x\| < \frac{\rho}{4}\quad \text{for } \|\tau\| < \epsilon.$$

Put $\delta = (2/3)r_B(u,x,\rho/5)$. Then for $\|h\| = \delta$ we find $\|\tau + h\| < r_B(u,x,\rho/5) = 3\delta/2$, and

$$|\omega(x+\tau,h)| = \left|u(x+\tau+h) - u(x+\tau) - (u'_{x+\tau},h) - \frac{1}{2}(u''_{x+\tau}h,h)\right| =$$

$$\left|\omega(x,\tau+h) - \omega(x,\tau) - (u'_{x+\tau} - u'_x, h) + (u''_x\tau,h) - \frac{1}{2}([u''_{x+\tau} - u''_x]h,h)\right| <$$

$$\rho\delta^2 = \rho\|h\|^2. \quad \square$$

LEMMA 2.2. *Let j be an essentially infinite-dimensional nonnegative functional. Let A be a bounded selfadjoint linear operator such that $j(A) > 0$. Then there is an infinite-dimensional subspace $V \subset \mathcal{H}$ such tat $(Ah,h) \geq 0$ for all $h \in V$.*

PROOF. Let $A = \int_\alpha^\beta t\,dE_t$ be the spectral decomposition of A. Suppose that $(E_\beta - E_0)\mathcal{H}$ is finite-dimensional. Then $\int_\alpha^0 t\,dE_t \leq 0$, and $j(\int_\alpha^0 t\,dE_t) = j(A) > 0$, contradicting the nonnegativity of j. Hence $V = (E_\beta - E_0)\mathcal{H}$ is infinite-dimensional, and for all $h \in V$ the inequality $(Ah,h) = \int_0^\beta t\,d(E_th,h) \geq 0$ holds. $\square$

PROOF OF THE THEOREM. First we assume that everywhere in Z the inequality $\partial u/\partial t < L_x u$ holds.

Take $\rho > 0$ and consider $u_\rho(x,t) = u(x,t) + \rho\|x\|^2$. Then for $h \in \mathcal{H}$ such that $x + h \in G$ we have

$$u_\rho(x+h,t) = u_\rho(x,t) +$$
$$(u'_x(x,t),h) + \frac{1}{2}(u''_x(x,t)h,h) + 2\rho(x,h) + \rho\|h\|^2 + \omega_t(x,h). \tag{2.3}$$

Take $y_1 = (x_1,t_1) \in Z$; by induction we construct a sequence $y_n = (x_n,t_n) \in Z$. Suppose that y_n has already been constructed. Then two cases are possible:

A) $L_x u(x_n,t_n) > 0$.
Now, by lemma 2.2, there are an infinite-dimensional subspace V_n in $\mathcal{H}$ such that V_n is orthogonal to the vectors $x_n, u'_x(x_n,t_n)$, as well as $h_1,\dots,h_{n-1}$ (whose meaning will be explained below) such that $(u''_x(x_n,t_n)h,h) \geq 0$ for all $h \in V_n$. Put $v_n(x) = u(x,t_n)$. Take $h_n \in V_n$ with norm $\|h_n\| > r_S(v_n,x_n,\rho)/2$ and such that $|\omega_{t_n}(x_n,h_n)| < \rho\|h_n\|^2$.
Put $x_{n+1} = x_n + h_n$, $t_{n+1} = t_n$, $h'_n = 0$. Then (2.3) implies that $u_\rho(x_{n+1},t_{n+1}) > u_\rho(x_n,t_n)$. This step will be called the 'horizontal step'.

B) $L_x u(x_n, t_n) \le 0$.
Then $(\partial u/\partial t)(y_n) < 0$. Put $t_{n+1} = 0$ if $(\partial u/\partial t)(x_n, t) < 0$ for all $t \in (0, t_n)$; otherwise $t_{n+1} = \sup\{t < t_n\colon\ (\partial u/\partial t)(x_n, t) = 0\}$. In both cases we put $x_{n+1} = x_n$, $h'_n = t_n - t_{n+1}$, $h_n = 0$. We have $u_\rho(x_{n+1}, t_{n+1}) > u_\rho(x_n, t_n)$. This step will be called the 'vertical step'.

Two cases are possible: either $y_n \in \partial Z$ or the sequence $y_n \in Z$ is infinite.

Suppose that the sequence $y_n \in Z$ is infinite. Then it converges in the Hilbert space $\mathcal{H} \oplus \mathbf{R}$, since the number series $\sum_{n=1}^\infty \|h_n\|^2 + \sum_{n=1}^\infty h'_n$ converges. We prove that $y_0 = \lim y_n \in \partial Z$.

Assume that $y_0 \in Z$. Again, two cases are possible:

a) $L_x u(y_0) > 0$.
Since $L_x u$ is continuous, there is a neighborhood W of y_0 in which $L_x u(y) > 0$. Take N such that $y_n \in W$ for $n > N$. Then $t_n = t_0$ for $n > N$; applying lemma 2.1 to the function $v(x) = u(x, t_0)$ we arrive at a contradiction with the condition $\|h_n\| \to 0$ as $n \to \infty$.

b) $L_x u(y_0) \le 0$.
Then in some neighborhood W of y_0 the inequality $(\partial u/\partial t)(y) < 0$ holds, and by the construction of the sequence y_n all steps from an N onwards must be horizontal. Then we reason as in a).

So, either $y_n \in \partial Z$ or $y_n \to y_0 \in \partial Z$. We have $u_\rho(x_1, t_1) \le \sup_{\partial Z} u_\rho$.

Since ρ is arbitrary we are finished.

To prove the theorem in the case of nonstrict inequality it suffices to consider the function $u_\epsilon(x,t) = u(x,t) - \epsilon t$, and to refer to the fact that $\epsilon > 0$ is arbitrary. □

COROLLARY 2.1. *Let G be a bounded domain in $\mathcal{H}$, let u be a function defined and continuous on $\bar{G}$, $u \in C^2(G)$, let $j(x) > 0$, and $Lu(x) \ge 0$ everywhere in G. Then $u(x) \le \sup_{\partial G} u$ for all $x \in G$.*

PROOF. In the case of strict inequality $Lu(x) > 0$ in G we consider the function $v(x,t) = u(x)$. The assertion then follows from theorem 2.1 (all steps in the proof are horizontal). The general case reduces to this one by replacing $u(x)$ by $u_\epsilon(x) = u(x) + \epsilon\|x\|^2$ and referring to the fact that $\epsilon > 0$ is arbitrary. □

COROLLARY 2.2. *Suppose that, under the conditions of the previous corollary, $Lu(x) = 0$ everywhere in G. Then $\inf_{\partial G} u \le u(x) \le \sup_{\partial G} u$ for all $x \in G$.*

Let Ω be the algebra of continuous functions on $\mathcal{H} \times [0,T)$ which are twice continuously differentiable with respect to x and continuously differentiable with respect to t, in $\mathcal{H} \times (0,T)$ (and $u'_x, u'_t \in C(\mathcal{H} \times (0,T))$).

THEOREM 2.2. *Suppose there exist nonnegative continuous functions $a(t)$ and $b(t)$ such that*

$$b(t) \le \|j(x,t)\| \le a(t) \tag{2.4}$$

for $(x,t) \in \mathcal{H} \times (0,T)$. Let $u \in \Omega$ be a solution of the equation

$$\frac{\partial u}{\partial t} = L_x u. \tag{2.5}$$

Then for every finite-dimensional subspace $V \subset \mathcal{H}$, any $\epsilon > 0$, and $(x_0, t_0) \in \mathcal{H} \times (0, T)$ the following inequalities hold:

$$\inf_{x \in X_\epsilon} u(x, 0) \leq u(x_0, t_0) \leq \sup_{x \in X_\epsilon} u(x, 0),$$

where X_ϵ is the ϵ-neighborhood of a finite-codimensional annuls

$$X = \left\{ x \in x_0 + V : \int_0^{t_0} b(\tau)\, d\tau \leq \|x - x_0\|^2 \leq \int_0^{t_0} a(\tau)\, d\tau \right\}.$$

PROOF. Without loss of generality we may assume that $u(x,t)$ is bounded on $\mathcal{H} \times (0,T)$. In fact, by proposition 2.1, along with u the function $\arctan u$ is a solution of equation (2.5).

Suppose $|u(x,t)| \leq C$ in $\mathcal{H} \times (0,T)$.

Consider an arbitrary $t_1 \in (t_0, T)$. Let $M_\delta = \{x\colon \|x - x_0\|^2 < \int_0^{t_0} a(\tau)\, d\tau + \delta\}$, $Z_\delta = M_\delta \times (0, t_1)$, $\bar{Z}_\delta = \bar{M}_\delta \times [0, t_1)$, $\partial Z_\delta = \bar{M}_\delta \times \{0\} \cup \partial M_\delta \times [0, t_1)$.

Consider the following functions F_1 and F_2 on $\mathbf{R}$:

$$F_1(s) = \begin{cases} 0 & \text{for } s \leq \int_0^{t_0} a(\tau)\, d\tau, \\ 2C \left(\frac{s - \int_0^{t_0} a(\tau)\, d\tau}{\delta} \right)^3 & \text{for } s > \int_0^{t_0} a(\tau)\, d\tau; \end{cases}$$

$$F_2(s) = \begin{cases} 0 & \text{for } s \geq \int_0^{t_0} b(\tau)\, d\tau, \\ 2C \left(\frac{s - \int_0^{t_0} b(\tau)\, d\tau}{\delta} \right)^3 & \text{for } s < \int_0^{t_0} b(\tau)\, d\tau. \end{cases}$$

Then $F_1, F_2 \in C^2(\mathbf{R})$, $F_1 \geq 0$, $F_2 \leq 0$, $F_1' \geq 0$, $F_2' \geq 0$ on $\mathbf{R}$.

Let P be the orthogonal projector onto V. Then $j_{x,t}(P) = \|j_{x,t}\|$ for all $(x,t) \in \mathcal{H} \times (0,T)$.

Consider the following function on $\bar{Z}_\delta$:

$$w(x,t) = F_1\left(\|x - x_0\|^2 + \int_0^t a(\tau)\, d\tau \right) + F_1\left(\|P(x - x_0)\|^2 + \int_0^t a(\tau)\, d\tau \right) - F_2\left(\|x - x_0\|^2 + \int_0^t b(\tau)\, d\tau \right) - F_2\left(\|P(x - x_0)\|^2 + \int_0^t b(\tau)\, d\tau \right).$$

By proposition 2.1 we have

$$\frac{\partial w}{\partial t} \geq L_x w \quad \text{everywhere in } Z_\delta.$$

On the other hand, $w(x,t) \geq 0$ in $\bar{Z}_\delta$, $w(x_0, t_0) = 0$, and $w(x,t) \geq 2C$ on $\partial Z_\delta \setminus (X'_\delta \times \{0\})$, where

$$X'_\delta = \left\{ x : \int_0^{t_0} b(\tau)\, d\tau - \delta < \|x - x_0\|^2 < \int_0^{t_0} a(\tau)\, d\tau + \delta \right\} \cap \left\{ x : \int_0^{t_0} b(\tau)\, d\tau - \delta < \|P(x - x_0)\|^2 < \int_0^{t_0} a(\tau)\, d\tau + \delta \right\}.$$

Let $v_1(x,t) = u(x,t) - w(x,t)$. Then v_1 satisfies the inequality $\partial v_1/\partial t \leq L_x v_1$ everywhere in Z_δ, and theorem 2.1 can be applied to it. Hence,

$$u(x_0,t_0) = v_1(x_0,t_0) \leq \sup_{\partial Z_\delta} v_1 \leq \sup_{X'_\delta} u(x,0),$$

since for $(x,t) \in \partial Z_\delta \setminus X'_\delta$ we have $v_1(x,t) < -C$.

By analogy, considering $v_2 = u + w$, we find $u(x_0,t_0) \geq \inf_{X'_\delta} u(x,0)$.

A simple geometrical argument shows that $X'_{\epsilon^2/3} \subset X_\epsilon$. □

REMARK 2.1. If $u(x,0)$ is uniformly continuous on bounded sets in $\mathcal{H}$, then we can strengthen the assertion of the theorem by putting $\epsilon = 0$.

COROLLARY 2.3. *If $\|j(x,t)\| = a(t)$, with a a continuous function on* **R**, *then under the conditions of theorem 2.1 and remark 2.1 we obtain*

$$\inf_Q u(x,o) \leq u(x_0,t_0) \leq \sup_Q u(x,0),$$

where Q is a large finite-codimensional subsphere of the sphere $S(x_0, \sqrt{\int_0^{t_0} a(\tau)\,d\tau})$.

REMARK 2.2. If we replace (2.4) in theorem 2.2 by the condition $b(t) \leq j(x,t)(B) \leq a(t)$, where B is a linear homeomorphism of $\mathcal{H}$, then it is easy to obtain a similar estimate, in which the annulus is replaced by an ellipsoidal annulus:

$$X = \left\{x \in x_0 + V : \int_0^{t_0} b(\tau)\,d\tau \leq (B(x-x_0), x-x_0) \leq \int_0^{t_0} a(\tau)\,d\tau\right\}.$$

(here V is a finite-codimensional invariant subspace of B.)

COROLLARY 2.4. *Suppose that, under the conditions of theorem 2.2, $T = \infty$, $\int_0^\infty b(\tau)\,d\tau$ is divergent, and the function $u(x,0)$ vanishes outside a ball $B(x_0,R) \subset \mathcal{H}$. Then there is a $\tilde{t}$ such that $u(x,t) = 0$ for all $x \in \mathcal{H}$, $t \geq \tilde{t}$.*

PROOF. Let $\tilde{t}$ be chosen such that $\int_0^{\tilde{t}} b(\tau)\,d\tau > R^2$. Then for all $x \in \mathcal{H}$, $t > \tilde{t}$ there is a finite-codimensional annulus with inner radius $\sqrt{\int_0^t b(\tau)\,d\tau}$ and not intersecting the ball $B(x_0,R)$. □

REMARK 2.3. This assertion can be formulated as follows: if $u_1(x) = u_2(x)$ outside some ball, then for $t \geq \tilde{t}$ the solutions of the Cauchy problem with these initial conditions coincide.

COROLLARY 2.5. *If $\|j(x,t)\| \leq a(t)$, where $a \in C(\mathbf{R})$, then the Cauchy problem for (2.5) has at most one solution in the class Ω.*

3. Existence of solutions of the Cauchy problem

Let $\mathfrak{A}$ be the set of functions $u \in C^2(\mathcal{H})$ satisfying

1) for any function $u \in \mathfrak{A}$ and any $R > 0$ there are a compact set $\mathfrak{N}_R = \mathfrak{N}_{u,R} \subset B_S(\mathcal{H})$ and a number $d = d_{u,R} > 0$ such that for all $x \in \{x\colon \|x\| \le R\}$ we have $u''(x) = S(x) + \gamma(x)$, with $S(x) \in \mathfrak{N}_R$, $\gamma(x) \in \Gamma_d$ (where Γ_d is the set of selfadjoint Hilbert–Schmidt operators with norm $\sigma_2(\gamma(x)) \le d$);
2) the map $\gamma\colon \mathcal{H} \to \Gamma$ (Γ the space of selfadjoint Hilbert–Schmidt operators with norm σ_2) is continuous;
3) $u''(x)$ is uniformly continuous on the balls of $\mathcal{H}$.

It is easily verified that $\mathfrak{A}$ is an algebra under the usual operations of addition and multiplication of functions.

Let $B_1, \dots, B_n \in B_S(\mathcal{H})$, let T be a compact linear operator on $\mathcal{H}$, let $f \in C^2(\mathbf{R}^n \oplus \mathcal{H})$ be uniformly continuous on the balls in $\mathbf{R}^n \oplus \mathcal{H}$. Then the function $u(x) = f((B_1x, x), \dots, (B_nx, x); Tx)$ belongs to $\mathfrak{A}$.

REMARK 3.1. Because the closed absolutely convex hull of a compact subset in a Banach space is compact, we may take $\mathfrak{N}_R$ to be a closed absolutely convex compact set. Then $\mathfrak{N}_R + \Gamma_d$ is a weakly closed absolutely convex set.

Let $\tilde{\mathfrak{A}}$ be the subalgebra of $\mathfrak{A}$ consisting of the functions u for which $|u(x)| \le C$, $\|u'(x)\| \le C$, $\|u''(x)\| \le C$, and $\sigma_2(\gamma(x)) \le C$ everywhere on $\mathcal{H}$, for a certain constant $C = C(u) > 0$. E.g., all finite functions in $\mathfrak{A}$ satisfy this condition.

Let j be an essentially infinite-dimensional nonnegative functional on $B_S(\mathcal{H})$. Then for any $u \in \mathfrak{A}$ we can write $j(u''(x)) = \lim_{m\to\infty} \operatorname{Tr} A_m u''(x)$, by theorem 1.1. Moreover, the convergence is uniform on balls in $\mathcal{H}$.

THEOREM 3.1. *The Cauchy problem*

$$\frac{\partial u}{\partial t} = \frac{1}{2} \operatorname{Tr} A u''_x + \frac{1}{2} j(u''_x); \quad u(x,0) = \phi(x) \tag{3.1}$$

($A \ge 0$ nuclear, $j \ge 0$ essentially infinite-dimensional) has a solution on $(0,\infty)$ for any $\phi \in \tilde{\mathfrak{A}}$. This solution can be represented as

$$\begin{gathered} u(x,t) = \lim_{n\to\infty} \int_{\mathcal{H}} \phi(x-y)\mu_{t(A+A_n)}(dy) = \\ \int_{\mathcal{H}} \left(\lim_{n\to\infty} \int_{\mathcal{H}} \phi(x-y-z)\mu_{tA_n}(dz) \right) \mu_{tA}(dy), \end{gathered} \tag{3.2}$$

where A_n is a sequence of nonnegative nuclear operators such that $\operatorname{Tr} A_n = \|j\|$, $\|A_n\| \to 0$ as $n \to \infty$.

LEMMA 3.1. *Let $\phi \in \tilde{\mathfrak{A}}$. Then for any $T > 0$, $p > 0$, and $r > 0$ there are a compact set $\mathfrak{N} \subset B_S(\mathcal{H})$ and a $d > 0$ such that for the function $v(x,t) = \int_{\mathcal{H}} \phi(x-y)\mu_{tB}(dy)$ we have $v''_x(x,t) \in \mathfrak{N} + \Gamma_d$ for all $t \in (0,T)$, $x \in \{x\colon \|x\| \le r\}$ and $B \ge 0$ such that $\operatorname{Tr} B \le p$.*

PROOF. The function v is of class C^2 with respect to x, and $v''_x(x,t) = \int_{\mathcal{H}} \phi''(x-y)\mu_{tB}(dy)$.

Choose $\epsilon > 0$ and $R > r$ such that $C \cdot T \cdot p/(R-r)^2 < \epsilon$. Then for $\|x\| \leq r$, $t \in (0,T)$, $\operatorname{Tr} B \leq p$:

$$\left\| \int_{\|y\|>R} S(x-y)\mu_{tB}(dy) \right\| < \frac{T \cdot \operatorname{Tr} B}{(R-r)^2} \cdot C < \epsilon.$$

Since $\int_{\|y\|<R} S(x-y)\mu_{tB}(dy) \in \mathfrak{N}_{\phi,R+r}$ ($\mathfrak{N}_{\phi,R+r}$ is a weakly closed and absolutely convex set) we may conclude that $\mathfrak{N} = \{\int_{\mathcal{H}} S(x-y)\mu_{tB}(dy)\colon t \in (0,T),\ \|x\| \leq r,\ \operatorname{Tr} B \leq \rho\}$ has a compact ϵ-net for any $\epsilon > 0$, hence $\mathfrak{N}$ is compact.

We can similarly verify that $\int_{\mathcal{H}} \gamma(x-y)\mu_{tB}(dy) \in \Gamma_d$. □

LEMMA 3.2. *Let f be a bounded continuous function on $\mathcal{H}$ which is uniformly continuous on the balls in $\mathcal{H}$, and let $\operatorname{Tr} B_n \leq p$. Then for each $r > 0$, $T_2 > T_1 > 0$ the sequence of functions $\{\int_{\mathcal{H}} f(x-y)\mu_{tB_n}(dy)\}$ is equicontinuous on $\{x\colon \|x\| \leq r\} \times (T_1, T_2)$.*

The proof of this lemma is easy, but a little awkward, so we omit it.

PROOF OF THE THEOREM. Consider the sequences

$$v_n(x,t) = \int_{\mathcal{H}} \phi(x-y)\mu_{t(A+A_n)}(dy),$$

$$(v_n)'_x(x,t) = \int_{\mathcal{H}} \phi'(x-y)\mu_{t(A+A_n)}(dy),$$

$$(v_n)''_x(x,t) = \int_{\mathcal{H}} \phi''(x-y)\mu_{t(A+A_n)}(dy),$$

$$\frac{\partial}{\partial t} v_n(x,t) = \frac{1}{2}\int_{\mathcal{H}} \operatorname{Tr}(A+A_n)\phi''(x-y)\mu_{t(A+A_n)}(dy)$$

(the last equality is extended to any function $\phi \in \tilde{\mathfrak{A}}$, using cylindrical functions and limit transition).

For each point $(x,t) \in \mathcal{H} \times (0,\infty)$ we can select a convergent subsequence $v_{n_k}(x,t)$ such that the sequences $(v_{n_k})'_x(x,t)$ and $(v_{n_k})''_x(x,t)$ converge weakly (this can be done by Alaoglu's theorem).

By Cantor's diagonal method we can choose a sebsequence $v_{n_m}(x,t)$ satisfying this condition for all (x,t) in a countable everywhere dense subset of the halfspace $\mathcal{H} \times (0,\infty)$. Without loss of generality we assume that this is true for the original sequence $v_n(x,t)$.

By lemma 3.2 the sequences $v_n(x,t)$, $((v_n)'_x(x,t),h)$, $((v_n)''_x(x,t)h,h)$ are equicontinuous on the sets $\{x\colon \|x\| < r\} \times (T_1,T_2) \times \{h\colon \|h\| < \delta\}$. This implies their pointwise convergence to continuous functions on $\mathcal{H} \times (0,\infty) \times \mathcal{H}$.

Let $u(x,t) = \lim v_n(x,t)$, $\xi(x,t) = \lim (v_n)'_x(x,t)$, $V(x,t) = \lim(v_n)''_x(x,t)$.

By limit transition on both sides of the equation

$$v_n(x+h,t) - v_n(x,t) - (v'_n(x,t),h) = \frac{1}{2}\int_0^1 (v''_n(x+sh)h,h)\,ds,$$

we obtain

$$u(x+h,t) - u(x,t) - (\xi(x,t),h) = \frac{1}{2}\int_0^1 (V(x+sh)h,h)\,ds.$$

This implies $u \in C^2(\mathcal{H})$, $u'_x(x,t) = \xi(x,t)$, $u''_x(x,t) = V(x,t)$. By the weak closedness of $\mathfrak{N} + \Gamma_d$ we have $u''_x(x,t) \in \mathfrak{N} + \Gamma_d$ (for $\|x\| < r < \infty$, $t < T < \infty$).

On the other hand, for all x the sequence $(\partial/\partial t)v_n(x,t)$ is equicontinuous with respect to t in the intervals (T_1, T_2) $(0 < T_1 < T_2 < \infty)$. By a similar argument we find that $u(x,t)$ is a continuously differentiable function with respect to $t \in (0,\infty)$, and

$$\frac{\partial}{\partial t}u(x,t) = \lim_{n\to\infty}\frac{\partial}{\partial t}v_n(x,t).$$

Since for fixed $x \in \mathcal{H}$, $t \in (0,\infty)$ we have $\int_{\mathcal{H}} \phi''(x-y)\mu_{t(A+A_n)}(dy) = S_n + \gamma_n$, where $S_n \in \mathfrak{N}$, $\gamma_n \in \Gamma_d$, by passing to a subsequence we may assume that S_n converges in norm to $S_0 \in \mathfrak{N}$. Therefore $\lim_{n\to\infty} \operatorname{Tr}(A + A_n)S_n = \lim_{n\to\infty} \operatorname{Tr}(A + A_n)S_0$.

Also, γ_n weakly converges to $\gamma_0 \in \Gamma_d$, so $\lim_{n\to\infty} \operatorname{Tr}(A + A_n)\gamma_n = \operatorname{Tr} A\gamma_0$. Hence

$$\lim_{n\to\infty}\frac{\partial}{\partial t}v_n(x,t) = \frac{1}{2}(\operatorname{Tr} Au''_x + j(u''_x)).$$

Let us verify the equality $\lim_{t\to 0+} u(x,t) = \phi(x)$, with uniform convergence on $\mathcal{H}$.

Take $\epsilon > 0$. Then there is a $\delta > 0$ such that for $\|y\| < \delta$, $x \in \mathcal{H}$: $|\phi(x-y) - \phi(x)| < \epsilon/3$. Take t_δ to satisfy

$$\sup|\phi(x)| \cdot \frac{t_\delta(\operatorname{Tr} A + \|j\|)}{\delta^2} < \frac{\epsilon}{3}.$$

Then for each n and $t \in (0, t_\delta)$ we have

$$\left|\int_{\mathcal{H}} \phi(x-y)\mu_{t(A+A_n)}(dy) - \phi(x)\right| \le$$

$$\left|\int_{\|y\|>\delta} \phi(x-y)\mu_{t(A+A_n)}(dy)\right| + \left|\int_{\|y\|>\delta} \phi(x)\mu_{t(A+A_n)}(dy)\right| +$$

$$\int_{\|y\|<\delta} |\phi(x-y) - \phi(x)|\mu_{t(A+A_n)}(dy) < \epsilon.$$

□

REMARK 3.2. It is not difficult to see that $u_t(x) = u(x,t) \in \tilde{\mathfrak{A}}$ for all $t > 0$, and $|u(x,t)| \le \sup_{\mathcal{H}} |\phi(y)|$ for all $x \in \mathcal{H}$, $t \in (0,\infty)$.

REMARK 3.3. Theorem 3.1 can be generalized, in fact, the following is valid.

THEOREM 3.1'. *Suppose $\phi \in \mathfrak{A}$, and there are $C, a > 0$ such that $\|\phi''(x)\| \le Ce^{a\|x\|^2}$ and $d_{\phi,R} \le Ce^{aR^2}$. then a solution of the Cauchy problem (3.1) on the interval $(0, 1/2a\|A\|)$ exists; it can be presented in the form (3.2). Moreover, for fixed t the function $u_t(x) = u(x,t)$ satisfies the conditions imposed on the function ϕ of this theorem.*

Let Ω be the class of functions that are bounded and continuous in $\mathcal{H} \times [0,\infty)$, twice continuously differentiable with respect to x and continuously differentiable with respect to t, in $\mathcal{H} \times (0,\infty)$, and let for $R > 0$, $0 < T < \infty$ there be a compact set $\mathfrak{N} = \mathfrak{N}_{u,R,T}$ and a $d = d_{u,R,T} > 0$ such that $u''_x(x,t) \in \mathfrak{N} + \Gamma_d$ for $\|x\| < R$, $t \in (0,T)$.

It is clear that the function u in theorem 3.1' belongs to Ω.

THEOREM 3.2. *The Cauchy problem (3.1) has at most one solution in* Ω.

PROOF. Let $u \in \Omega$ be a solution of (3.1), $|u(x,t)| \leq C$ for $(x,t) \in \mathcal{H} \times [0,\infty)$, $u(x,0) \equiv 0$. We show that $u(x,t) \equiv 0$ everywhere in $\mathcal{H} \times [0,\infty)$.

Consider the function $v_\epsilon(x,t) = u(x,t) - \epsilon(\|x\|^2 + 2t\|\jmath\|)$. Then $(\partial/\partial t)v_\epsilon = (\jmath(v_\epsilon)/2) - \epsilon\|\jmath\|$ in $\mathcal{H} \times (0,\infty)$. For $R > \sqrt{C/\epsilon}$, on the boundary $\partial W_{R,T}$ of the cylinder $W_{R,T} = \{x\colon \|x\| < R\} \times (0,T)$ the inequality $v_\epsilon(x,t) < 0$ holds. Moreover, $(v_\epsilon)''_x \in \mathfrak{N} + \Gamma_d$ in $W_{R,T}$ for some compact set $\mathfrak{N} \subset B_S(\mathcal{H})$ and number $d > 0$.

By theorem 1.1 there is a finite-dimensional nonnegative operator A for which $(\partial/\partial t)v_\epsilon < (\operatorname{Tr} A(v_\epsilon)''_x)/2$ in $W_{R,T}$. Let P be a finite-dimensional orthoprojector for which $PAP = A$, and let $(x_0,t_0) \in W_{R,T}$. Then the inequality $(\partial/\partial t)v_\epsilon^P < (\operatorname{Tr} PAP(v_\epsilon^P)''_x)/2$ holds in the finite-dimensional cylinder $W_P = ((x_0 + PH) \cap \{x\colon \|x\| < R\}) \times (0,T)$ for the function $v_\epsilon^P = v_\epsilon|_{W_P}$. Hence, by the finite-dimensional maximum principle, the inequality $v_\epsilon(x_0,t_0) = v_\epsilon^P(x_0,t_0) \leq \sup_{\partial W} v_\epsilon(x,t) \leq 0$ is valid.

Since $R > \sqrt{C/\epsilon}$ is arbitrary, this proves that $u(x,t) \leq \epsilon(\|x\|^2 + 2t\|\jmath\|)$ everywhere in $\mathcal{H} \times (0,\infty)$, and since $\epsilon > 0$ is arbitrary, we have $u(x,t) \leq 0$. Replacing u by $-u$ we arrive at the reverse inequality. □

REMARK 3.4. Theorem 3.2 can be generalized to the class of unbounded functions satisfying an estimate $|u(x,t)| \leq Ce^{a\|x\|^2}$ for some $C, a > 0$ and the conditions of the class Ω in the cylinders $\{x\colon \|x\| < R\} \times (0,T)$.

REMARK 3.5. If the differential expression at the righthand side of (3.1) is the Laplace–Lévy operator $Lu(x) = (1/2)\lim_{n\to\infty}(1/n)\operatorname{Tr} P_n u''(x)$, then the solution (3.2) of (3.1) has the form

$$u(x,t) = \lim_{n\to\infty}\left(\frac{n}{2\pi}\right)^{n/2}\int_{P_nH}\phi(x-y)e^{-n\|y\|^2/2}\,dy,$$

where P_n is a sequence of orthogonal projectors to increasing subspaces of $\mathcal{H}$.

4. Supplementary remarks and historical comments

The Laplace–Lévy operator appeared for the first time in Lévy [1]. The properties of this operator, as well as various problems for equations involving this operator, were studied in Shilov [1], Polishchuk [1,2,3,4], Feller [1,2,3], and others.

Another definition of operators of Laplace–Lévy type (not requiring a fixed ordered sequence of orthoprojectors) was proposed by I.Ya. Dorfman, and also in Nemirovsky–Shilov [1]. Various problems for equations with modified operators were studied in Dorfman [1,2] and Sokolovsky [1].

Another generalization of the Laplace–Lévy operator can be observed in Sikiryavy [1,2].

The problem of studying the more general operators of type $\lim_{m\to\infty} \operatorname{Tr} A_m u''(\cdot)$ (see example 1.2) was formulated by Yu.L. Dalecky. The presentation given here is based on results in Bogdansky [1,2,3], and emerged from fruitful cooperation by the authors.

Bibliography

Averbukh, V.I., Smolyanov, O.G., Fomin, S.V.,

[1] Generalized functions and differential equations in linear spaces, *Tr. Moskov. Mat. Obshch.* **27** (1972), 247-262. (In Russian.)

Bogdansky, Yu.V.,

[1] The Cauchy problem for parabolic equations with essentially infinite-dimensional elliptic operators, *Ukrainian Mat. Zh.* **29** (1977), 781-784. (In Russian.)

[2] The maximum principle for a nonregular elliptic differential equation in a countably-dimensional Hilbert space, *Ukrainian Mat. Zh.* **40** (1988), 21-25. (In Russian.)

[3] The Cauchy problem for the heat equation with a nonregular elliptic operator, *Ukrainian Mat. Zh.* **41** (1989), 584-590. (In Russian.)

Dorfman, I.Ya.,

[1] On means and the Laplacian of functions on Hilbert space, *Mat. Sb.* **81: 2** (1970), 192-208. (In Russian.)

[2] On the heat equation on Hilbert space, *Vestnik M.G.U.* **4** (1971), 46-51. (In Russian.)

Feller, M.N.,

[1] On the Poisson equation in the space $L_2(C)$, *Dokl. Akad. Nauk SSSR* **4** (1966), 426-429. (In Russian.)

[2] On infinite-dimensional operators, *Dokl. Akad. Nauk SSSR* **205** (1972), 36-39. (In Russian.)

[3] Infinite-dimensional elliptic equations and operators of Levy type, *Uspekhi Mat. Nauk* **41: 4** (1986), 97-140. (In Russian.)

Lévy, P.,

[1] *Les problèmes concrets d'analyse fonctionnelle*, Gauthier-Villars, 1951.

Naroditsky, V.A.,

[1] On operators of Laplace–Lévy type, *Ukrainian Mat. Zh.* **29: 3** (1977), 513-517. (In Russian.)

Nemirovsky, A.S.,

[1] On the general theory of Laplace–Lévy operators, *Funktsion. Anal.* **8: 3** (1974), 79-80. (In Russian.)

[2] Laplace–Lévy operators on some class of functions, *Tr. MMO* **32** (1975), 227-250. (In Russian.)

Nemirovsky, A.S., Sholiv, G.E.,

[1] On the axiomatic description of the Laplace operator for functions on Hilbert space, *Funktsion. Anal.* **3: 3** (1969), 79-85. (In Russian.)

Polishchuk, E.M.,

[1] On the functional Laplacian and equations of parabolic type, *Uspekhi Mat. Nauk* **19: 2** (1964), 155-162. (In Russian.)
[2] Linear equations in functional Laplacians, *Uspekhi Mat. Nauk* **19: 2** (1964), 163-170. (In Russian.)
[3] On functional analogs of the heat equation, *Sibirsk. Mat. Zh.* **6: 6** (1965), 1322-1331. (In Russian.)
[4] *Continual means and boundary value problems in function spaces*, Akad. Verlag, 1988.

Shilov, G.E.,
[1] On some questions in analysis on Hilbert space, *I., Funktsion. Anal.* **1: 2** (1967), 81-90. *II., Mat. Issled. Kishin. Izd. ANMoldavSSR* **2: 4** (1968), 166-186. *III., Mat. Sb.* **74: 1** (1967), 161-168. (In Russian.)

Sikiryavy, V.Ya.,
[1] The operator of quasidifferentiation and the boundary value problems related with it, *Tr. Moskov. Mat. Obshch.* **27** (1972), 195-246. (In Russian.)
[2] The invariant Laplace operator as the operator of pseudospherical differentiation, *Vestnik MGU* **3** (1972), 66-73. (In Russian.)

Sokolovsky, V.B.,
[1] New relations between certain problems for equations with the Laplace–Lévy operator and problems in mathematical physics, *Izv. Vyzov. Mat.* **11** (1980), 82-84. (In Russian.)

Supplement to chapter VII

Stochastic regularization of the ill-posed abstract parabolic problem

By Yu.L. Dalecky and N.Yu. Goncharuk

Consider a linear differential equation

$$\frac{d\xi(t)}{dt} = \alpha A\xi(t), \quad t \in \mathbf{R}, \quad \xi(t_0) = u_0, \tag{0.1}$$

in a Banach space B, where A is the unbounded generating operator of the analytic semigroup e^{tA} in B, $t \in \mathbf{R}$, $u_0 \in B$.

It is well known that if $\alpha(t - t_0) > 0$, then the solution of the Cauchy problem for equation (0.1) exists and is given by

$$\xi(t) = e^{\alpha(t-t_0)A}u_0,$$

while if $\alpha(t - t_0) < 0$, then the corresponding Cauchy problem is ill-posed.

Let $W(t)$ be the standard scalar Wiener process. Adding th white noise $\dot{W}(t)$ with coefficient $\beta \in \mathbf{R}$ to the coefficient α, we obtain the equation

$$\frac{d\xi(t)}{dt} = (\alpha + \beta\dot{W}(t))A\xi(t), \quad \xi(t_0) = u_0,$$

which has a meaning as a stochastic differential equation in B:

$$d\xi(t) = \alpha A\xi(t)\,dt + +\beta A\xi(t)\,dW(t)), \quad t \in \mathbf{R}, \tag{0.2}$$

$$\xi(t_0) = u_0. \tag{0.3}$$

It will be shown below that the Cauchy problem (0.2)–(0.3) is well-posed for $\beta \neq 0$ and $\alpha(t - t_0)$ of arbitrary sign. Moreover, its solution is given by the formula

$$\xi(t) = V_\beta(t, t_0)u_o = \exp\left\{\alpha\Delta tA - \frac{1}{2}\beta^2|\Delta t|A^2 + \beta\Delta WA\right\}u_0, \quad t \in \mathbf{R}, \tag{0.4}$$

where $\Delta t = t - t_0$, $\Delta W = W(t) - W(t_0)$. This formula may be obtained from Itô's formula if the operator exponential makes sense. In this case it is possible to speak of the *stochastic regularization* of the ill-posed Cauchy problem.

As an example we can consider the Cauchy problem for the partial differential equation with a second order elliptic operator A,

$$A = \sum_{k,j=1}^{n} \frac{\partial}{\partial x_k} a_{kj} \frac{\partial}{\partial x_j} + \sum_{k=1}^{n} b_k(x) \frac{\partial}{\partial x_k}.$$

Here $x = (x_1, \dots, x_n) \in \mathbf{R}^n$ (ξ and u_0 are assumed to be scalar functions).

We consider in detail the example of the operator exponential (0.4).

Let $-A$ be a sectorial operator on B with the following properties:

1) there exist $a \in \mathbf{R}$ and $\phi \in (0, \pi/4)$ such that $S_{a,\phi} = \{\mu \in \mathbf{C}\colon \phi \leq |\arg(\mu - a)| \leq \pi,\ \mu \neq a\} \subset \rho_{-A}$, where ρ_{-A} is the resolvent set of the operator $-A$, $\rho_{-A} \neq \emptyset$;
2) There exists an $M \geq 1$ such that for $\mu \in S_{a,\phi}$,

$$\|(\mu I + A)_{-1}\| \leq \frac{M}{|\mu - a|}, \tag{0.5}$$

with I the identity operator on B.

To simplify the exposition we set $a = 0$.

It is well known (see Henry [1]) that A generates the analytic semigroup $e^{tA} = (1/2\pi i) \int_\gamma e^{t\lambda} (\lambda I - A)^{-1}\, d\lambda$, $t \geq 0$, in B, with γ a piecewise smooth curve in ρ_A consisting of two rays $\{re^{-i\theta}\colon 1 \leq r < \infty\}$, $\{re^{i\theta}\colon 1 \leq r < \infty\}$, and the arc $\{e^{i\delta}\colon -\theta \leq \delta \leq \theta\}$, $3\pi/4 < \theta < \pi$, of the unit circle.

Let D_A be the domain of A, $\langle \cdot, \cdot \rangle$ a pairing in (B, B^*), and $U_c(A)$ the set of entire vectors of A. By definition (see Goodman [1]), $U_c(A)$ is the class $C^\infty(A) = \bigcap_{n=1}^{\infty} D_{A^n}$ of infinitely differentiable vectors of A, defined by the condition

$$U_c(A) = \{f \in C^\infty(A) : \forall \alpha > 0 \exists c(\alpha) > 0 : \|A^n f\| \leq c(\alpha)(\alpha n)^n\}.$$

THEOREM 0.1. *For $\beta \neq 0$ the random function (0.4), with $V_\beta(t, t_0)$ defined by the formulas*

$$V_\beta(t, t) = I, \quad t \in \mathbf{R}, \tag{0.6}$$

$$V_\beta(t, t_0) = \frac{1}{2\pi i} \int_\gamma e^{\alpha \Delta t \lambda - \beta^2 \lambda^2 |\Delta t|/2 + \beta \lambda \Delta W} (\lambda I - A)^{-1}\, d\lambda, \quad t \neq t_0, \tag{0.7}$$

has a stochastic differential, and satisfies equation (0.2) for $t \neq t_0$ when $u_0 \in B$, and for $t \in \mathbf{R}$ when $u_0 \in D_A$. In addition, the family of operators $V_\beta(t, t_0)$ has almost surely the properties:

a) *$V_\beta(t, \tau) = V_\beta(t, s) V_\beta(s, \tau)$, $t \geq s \geq \tau$ or $t \leq s \leq \tau$; $V_\beta(t, \tau) V_\beta(\tau, t) = e^{-\beta^2 |t - \tau| A^2}$ is nonrandom;*
b) *$\|V_\beta(t, t_0)\| < \infty$;*
c) *for any $x \in D_A$, $\lim_{t \to t_0} \|V_\beta(t, t_0)x - x\| = 0$;*
d) *$V_\beta(t, t_0)$ is the unique evolution family for the Cauchy problem (0.2)–(v3);*

e) *for any* $x \in U_c(A)$,

$$\lim_{\beta\to 0} \|V_\beta(t,t_0)x - e^{\alpha\Delta t A}x\| = 0; \tag{0.8}$$

moreover, for $\alpha\Delta t > 0$, *equation (8) is valid for any* $x \in B$;

f) *for* $\alpha\Delta t > 0$ *and any* $x \in B$, $f \in B^*$,

$$M\langle V_\beta(t,t_0)x, f\rangle = \langle e^{\alpha\Delta t A}x, f\rangle.$$

We reduce the proof of this theorem to the proof of a number of propositions.

PROPOSITION 0.1. *For* $\beta \neq 0$ *the integral (v7) converges absolutely almost surely; moreover, for almost every trajectory of* $W(t)$ *the following estimates hold: for any* $\epsilon > 0$ *and any* Δt *such that* $0 < |\Delta t| < T$ *there are positive constants* $a_1, \dots, a_7, b_1, b_2, b_3$, *depending on* M, θ, α, T, *and* ϵ *only, such that*

a) *for* $\alpha\Delta t < 0$,

$$\|V_\beta(t,t_0)\| \le \frac{a_1}{|\beta|\sqrt{|\Delta t|}} \exp\left\{a_2\left(\frac{\sqrt{|\Delta t|}}{\beta} + |\Delta t|^{-\epsilon}\right)^2\right\} + b(\beta, |\Delta t|); \tag{0.9}$$

b) *for* $\alpha\Delta t > 0$,

$$\|V_\beta(t,t_0)\| \le \frac{a_3}{\sqrt{|\Delta t|}} \exp\{a_4(-|\Delta t| + |\Delta t|^{-2\epsilon})\} + b(\beta, |\Delta t|); \tag{0.10}$$

c) *for* $\alpha\Delta t > 0$ *there exists a* $\beta_0 > 0$ *such that for any* β, $0 < |\beta| < \beta_0$,

(0.11)
$$\|V_\beta(t,t_0)\| \le \frac{a_5}{\sqrt{|\Delta t|}} \exp\{a_6((\beta^2 - a_7)|\Delta t| + |\beta|\cdot|\Delta t|^{(1-2\epsilon)/2})\} + b(\beta, |\Delta t|),$$

where

$$b(\beta, |\Delta t|) = b_1 \exp\{b_2((\beta^2 + b_3)|\Delta t| + |\beta|\cdot|\Delta t|^{(1-2\epsilon)/2})\}.$$

PROOF. Computing the line integral in (0.7) and estimating its norm using the boundedness condition for the resolvent of A along γ following from (0.5), we find

$$\|V_\beta(t,t_0)\| \le \frac{M}{\pi}\left[\int_1^\infty \exp\left\{(\alpha\Delta t + \beta\Delta W)r\cos\theta - \frac{1}{2}\beta^2 r^2 |\Delta t|\cos 2\theta\right\}\frac{dr}{r} + \right.$$

$$\left.\int_0^\theta \exp\left\{(\alpha\Delta t + \beta\Delta W)\cos\delta - \frac{1}{2}\beta^2|\Delta t|\cos 2\delta\right\} d\delta\right]. \tag{0.12}$$

Denote the first and second integral in (0.12) by I_1 and I_2, respectively. It is clear that

$$I_2 \le \pi \exp\left\{|\alpha\Delta t| + |\beta|\cdot|\Delta W| + \frac{1}{2}\beta^2|\Delta t|\right\}. \tag{0.13}$$

For I_1 we find

$$I_1 < \left(\frac{4\pi}{\beta^2|\Delta t|\cos 2\theta}\right)^{1/2} \exp\left\{\frac{\cos^2\theta}{2\cos 2\theta}\left(\left|\frac{\alpha}{\beta}\right|\sqrt{|\Delta t|} + \frac{|\Delta W|}{\sqrt{|\Delta t|}}\right)^2\right\}. \tag{0.14}$$

For $\alpha\Delta t > 0$ we can improve the estimate (0.14). In fact, for the maximum of the function $\phi_1(r) = \exp\{\beta\Delta W r\cos\theta - (1/2)\beta^2 r^2|\Delta t|\cos 2\theta\}$ on $[1,\infty)$ we have

$$\max_{r\geq 1}\phi_1(r) \leq \max_{r\in\mathbf{R}}\phi_1(r) = \exp\left\{\frac{\cos^2\theta}{2\cos 2\theta}\cdot\frac{\Delta W^2}{|\Delta t|}\right\}.$$

Then

$$\begin{aligned} I_1 = \int_1^\infty \phi_1(r) e^{\alpha\Delta t r\cos\theta}\frac{dr}{r} &\leq \\ \max_{r\geq 1}\phi_1(r)\int_1^\infty e^{\alpha\Delta t r\cos\theta}\frac{dr}{r} &\leq \\ -\frac{1}{\alpha\Delta t\cos\theta}\exp\left\{\alpha\Delta t\cos\theta + \frac{\cos^2\theta}{2\cos 2\theta}\cdot\frac{\Delta W^2}{|\Delta t|}\right\}&. \end{aligned} \tag{0.15}$$

Analyzing the function $\phi_2(r) = \exp\{(\alpha\Delta t r\cos\theta)/2\}\phi_1(r)$, we find that it attains a global maximum on $(-\infty,\infty)$ at the point $r_0 = \cos\theta(\alpha\Delta t + 2\beta\Delta W)/(2\beta^2|\Delta t|\cos 2\theta)$. Note that $r_0 < 1$ if $\alpha\Delta t > 0$ and β is sufficiently small, and $\phi_2(r)$ has no other extrema on $(-\infty,\infty)$. Thus, for $\alpha\Delta t > 0$ there is a $\beta_0 > 0$ such that for any β, $0 < |\beta| < \beta_0$,

$$\max_{r\geq 1}\phi_2(r) = \phi_2(1) = \exp\left\{\left(\frac{1}{2}\alpha\Delta t + \beta\Delta W\right)\cos\theta - \frac{1}{2}\beta^2|\Delta t|\cos 2\theta\right\}$$

and

$$\begin{aligned} I_1 = \int_1^\infty \phi_2(r) e^{(\alpha\Delta t r\cos\theta)/2}\frac{dr}{r} &\leq \\ \phi_2(1)\int_1^\infty e^{(\alpha\Delta t r\cos\theta)/2}\frac{dr}{r} &\leq \\ \frac{2}{\alpha\Delta t\cos\theta}\exp\left\{\alpha\Delta t\cos\theta - \frac{1}{2}\beta^2|\Delta t|\cos 2\theta - |\beta|\cdot|\Delta W|\cos\theta\right\}&. \end{aligned} \tag{0.16}$$

The estimates (0.9)–(0.11) now follow from inequalities (0.12)–(0.16) and Hölder's condition for the increment of a Wiener process with any exponent less than 1/2 (almost surely), i.e. the fact that for any $\epsilon > 0$ there is for almost every trajectory of $W(t)$ a $c_{\epsilon,t}$ such that

$$|\Delta W| < c_{\epsilon,t}|\Delta t|^{(1-2\epsilon)/2}. \quad \square$$

PROPOSITION 0.2. *For $\beta \neq 0$ the family (0.7) has almost surely the evolution property for $t \geq s \geq \tau$ or $t \leq s \leq \tau$, i.e.*

$$V_\beta(t,\tau) = V_\beta(t,s)V_\beta(s,\tau),$$

and $V_\beta(t,\tau)V_\beta(\tau,t) = \exp\{-\beta^2|t-\tau|A^2\}$ is nonrandom.

PROOF. Since $V_\beta(t,t_0) = f(A,t,t_0) = (1/2\pi i)\int_\gamma f(\lambda,t,t_0)(\lambda I - A)^{-1}\,d\lambda$ with

$$f(\lambda,t,t_0) = \exp\left\{\alpha\lambda(t-t_0) - \frac{1}{2}\beta^2\lambda^2|t-t_0| + \beta\lambda(W(t)-W(t_0))\right\}, \quad \lambda\in\gamma,$$

then the proposition follows immediately from the condition $f(\lambda,t,\tau) = f(\lambda,t,s)\times f(\lambda,s,\tau)$ for $t\ge s\ge\tau$ or $t\le s\le\tau$ and the equation $f(\lambda,t,\tau)f(\lambda,\tau,t) = \exp\{-\beta^2\lambda^2|\tau-t|\}$. □

PROPOSITION 0.3. *For $\beta\neq 0$ the family of operators $V_\beta(t,t_0)$ defined by (0.6)–(0.7) leaves invariant the set D_A, almost surely, and the solution of the Cauchy problem (0.2)–(0.3) exists and is given by (0.4) for $t\neq t_0$ when $u_0\in B$, and for $t\in\mathbf{R}$ when $u_0\in D_A$.*

PROOF. The first assertion of the proposition follows from the fact that the family of operators $V_\beta(t,t_0)$ commutes with A on D_A. Consider the scalar random process $\eta_\lambda(t) = \alpha\Delta t - \beta^2|\Delta t|\lambda/2 + \beta\Delta W$, $\lambda\in\gamma$, and the scalar function

$$\phi(\eta_\lambda(t)) = \exp\{\lambda\eta_\lambda(t)\} = e^{\alpha\lambda\Delta t-\beta^2\lambda^2|\Delta t|/2+\beta\lambda\Delta W}, \lambda\in\gamma. \tag{0.17}$$

The integral $(1/2\pi i)\int_\gamma \lambda\phi(\eta_\lambda(t))(\lambda I - A)^{-1}\,d\lambda$, which defineds for every $t\neq t_0$ and $\beta\neq 0$ the operatorvalued function $AV_\beta(t,t_0)$, converges absolutely almost surely, and for its norm estimates of the type (0.9)–(0.11) hold for almost every trajectory of $W(t)$.

By Itô's formula for the differential of a scalar random process we have

$$\begin{gathered} d\phi(\eta_\lambda(t)) = \\ \left(\alpha - \frac{1}{2}\lambda\beta^2\operatorname{sgn}\Delta t\right)\phi'_x(\eta_\lambda(t))\,dt + \frac{1}{2}\beta^2\operatorname{sgn}\Delta t\phi''_{xx}(\eta_\lambda(t))\,dt + \beta\phi'_x(\eta_\lambda(t))\,dW(t) = \\ \left(\alpha - \frac{1}{2}\lambda\beta^2\operatorname{sgn}\Delta t\right)\lambda\phi(\eta_\lambda(t))\,dt + \frac{1}{2}\beta^2\operatorname{sgn}\Delta t\lambda^2\phi(\eta_\lambda(t))\,dt + \beta\lambda\phi(\eta_\lambda(t))\,dW(t) = \\ \lambda\phi(\eta_\lambda(t))(\alpha\,dt + \beta\,dW(t)), \end{gathered}$$

or, by the definition of stochastic differential,

$$\phi(\eta_\lambda(t)) - \phi(\eta_\lambda(\tau)) = \alpha\lambda\int_\tau^t \phi(\eta_\lambda(s))\,ds + \beta\lambda\int_\tau^t \phi(\eta_\lambda(s))\,dW(s). \tag{0.18}$$

Computing the increment $\xi(t)-\xi(\tau)$ by formula (0.18) and the representation $\xi(t) = (1/2\pi i)\int_\gamma \phi(\eta_\lambda(t))(\lambda I - A)^{-1}u_0\,d\lambda$, $u_0\in B$, $t\neq t_0$, which follows from (0.4) and (0.7), we obtain

$$\begin{gathered} \xi(t)-\xi(\tau) = \frac{1}{2\pi i}\int_\gamma [\phi(\eta_\lambda(t)) - \phi(\eta_\lambda(\tau))](\lambda I - A)^{-1}u_0\,d\lambda = \\ \frac{1}{2\pi i}\int_\gamma \alpha\lambda\left[\int_\tau^t \phi(\eta_\lambda(s))\,ds\right](\lambda I - A)^{-1}u_0\,d\lambda + \\ \frac{1}{2\pi i}\int_\gamma \beta\lambda\left[\int_\tau^t \phi(\eta_\lambda(s))\,dW(s)\right](\lambda I - A)^{-1}u_0\,d\lambda. \end{gathered}$$

Changing the order of integration gives

$$\xi(t) - \xi(\tau) =$$
$$\alpha \int_\tau^t \frac{1}{2\pi i} \int_\gamma \lambda \phi(\eta_\lambda(s))(\lambda I - A)^{-1} u_0 \, d\lambda \, ds +$$
$$\beta \int_\tau^t \frac{1}{2\pi i} \int_\gamma \lambda \phi(\eta_\lambda(s))(\lambda I - A)^{-1} u_0 \, d\lambda \, dW(s). \tag{0.19}$$

But according to the definition of $\xi(s)$, and since $AV_\beta(s, t_0)$ is almost surely bounded for $s \neq t_0$, we have

$$\frac{1}{2\pi i} \int_\gamma \lambda \phi(\eta_\lambda(s))(\lambda I - A)^{-1} u_0 \, d\lambda = AV_\beta(s, t_0) u_o = A\xi(s), \quad s \neq t_0. \tag{0.20}$$

Substituting (0.20) into (0.19) we obtain

$$\xi(t) - \xi(\tau) = \alpha \int_\tau^t A\xi(s)\, ds + \beta \int_\tau^t A\xi(s)\, dW(s).$$

Hence $\xi(t)$ has a stochastic differential and satisfies equation (2) for $t \in \mathbf{R}$ when $u_0 \in B$. The last assertion can be similarly proved when $t \neq t_0$ when $u_0 \in D_A$. □

PROPOSITION 0.4. *For any $x \in D_A$ and $\beta \neq 0$,*

$$\lim_{t \to t_0} \|V_\beta(t, t_0)x - x\| = 0 \quad a.s..$$

PROOF. Let $x \in D_A$ and let $\phi(\eta_\lambda(t))$ be the random function defined by (0.17). Then for $\beta \neq 0$,

$$V_\beta(t, t_0)x - x = \frac{1}{2\pi i} \int_\gamma \phi(\eta_\lambda(t))[(\lambda I - A)^{-1} - (\lambda I)^{-1}] x \, d\lambda =$$
$$\frac{1}{2\pi i} \int_\gamma \phi(\eta_\lambda(t))(\lambda I - A)^{-1} Ax \frac{d\lambda}{\lambda},$$

and

$$\|V_\beta(t, t_0)x - x\| \le \left\| \frac{1}{2\pi i} \int_\gamma \phi(\eta_\lambda(t))(\lambda I - A)^{-1} \frac{d\lambda}{\lambda} \right\| \cdot \|Ax\|. \tag{0.21}$$

Changing variables $\mu = \sqrt{|\Delta t|}\lambda$ in the integral (0.21), we find

$$J_1 = \frac{1}{2\pi i} \int_\gamma \phi(\eta_\lambda(t))(\lambda I - A)^{-1} \frac{d\lambda}{\lambda} =$$
$$\frac{1}{2\pi i} \int_{\gamma'} \exp\left\{ (\alpha \Delta t + \beta \Delta W) \frac{\mu}{\sqrt{|\Delta t|}} - \frac{\beta^2 \mu^2}{2} \right\} \left(\frac{\mu}{\sqrt{|\Delta t|}} I - A \right)^{-1} \frac{d\mu}{\mu}, \tag{0.22}$$
$$\gamma' = \left\{ \mu \in \mathbf{C} : \frac{\mu}{\sqrt{|\Delta t|}} \in \gamma \right\}.$$

If we note that by Cauchy's theorem for $|\Delta t| < 1$ integration along γ' in (0.22) can be replaced by integration along γ, and using the boundedness of the operator $((\mu/\sqrt{|\Delta t|})I - A)^{-1}$ when $\mu \in \gamma'$, we obtain

$$\|J_1\| \leq$$

$$\frac{M}{\pi}\sqrt{|\Delta t|}\left[\int_1^{\infty} \exp\left\{(\alpha\Delta t + \beta\Delta W)\frac{r\cos\theta}{\sqrt{|\Delta t|}} - \frac{1}{2}\beta^2 r^2 \cos 2\theta\right\}\frac{dr}{r^2} + \right.$$

$$\left.\int_0^{\theta} \exp\left\{(\alpha\Delta t + \beta\Delta W)\frac{\cos\delta}{\sqrt{|\Delta t|}} - \frac{1}{2}\beta^2 \cos 2\delta\right\} d\delta\right] <$$

$$\frac{M}{\pi}\sqrt{|\Delta t|}\left[\left(\frac{4\pi}{\beta^2\cos 2\theta}\right)^2 \exp\left\{\frac{\cos^2\theta}{2\cos 2\theta}\left(\left|\frac{\alpha}{\beta}\right|\sqrt{|\Delta t|} + \frac{|\Delta W|}{\sqrt{|\Delta t|}}\right)^2\right\} + \right.$$

$$\left.\pi\exp\left\{|\alpha|\sqrt{|\Delta t|} + \frac{1}{2}\beta^2 + \frac{|\Delta W|}{\sqrt{|\Delta t|}}\right\}\right].$$

The law of the iterated logarithm for the increment of a Wiener process (see, e.g., Skorokhod [1]) implies that

$$\lim_{\Delta t\to 0}\sqrt{|\Delta t|}\exp\left\{\delta_1\left(\frac{|\Delta W|^2}{|\Delta t|} + |\Delta W|\right)\right\} = 0 \quad \text{(a.s.)},$$

and

$$\lim_{\Delta t\to 0}\sqrt{|\Delta t|}\exp\left\{\delta_2\frac{|\Delta W|}{\sqrt{|\Delta t|}}\right\} = 0 \quad \text{(a.s.)},$$

where δ_1 and δ_2 are positive constants.

Thus, $\lim_{\Delta t\to 0}\|J_1\| = 0$ (almost surely) and

$$\lim_{\Delta t\to 0}\|V_\beta(t,t_0)x - x\| = \lim_{\Delta t\to 0}\|J_1\|\cdot\|Ax\| = 0 \quad \text{(a.s.)}. \quad \square$$

PROPOSITION 0.5. *The evolution family $V_\beta(t,t_0)$ defined for $\beta \neq 0$ is almost surely the unique evolution family of the Cauchy problem(0.2)-(0.3).*

PROOF. Suppose there is another evolution family, $U(t,t_0)$, of the Cauchy problem (0.2)-(0.3). Then for any $y \in D_A$ the random function $U(t,t_0)y$ has a stochastic differential:

$$dU(t,t_0)y = \alpha AU(t,t_0)y\,dt + \beta AU(t,t_0)y\,dW(t),$$

leaves invariant the set D_A (almost surely), and $U(t,t_0) \in \mathcal{B}(B)$ (almost surely) (where $\mathcal{B}(B)$ is the Banach space of all bounded linear operators on B) has a stochastic differential

$$dU(t,t_0) = \alpha AU(t,t_0)\,dt + \beta AU(t,t_0)\,dW(t). \tag{0.23}$$

Similar as in proposition 0.3 we can prove that for $\beta \neq 0$, $t \neq s$, $A^2 V_\beta(t,s) \in \mathcal{B}(B)$ (almost surely), and $V_\beta(t,s)$ has a stochastic differential:

$$dV_\beta(t,s) = (\beta^2 A^2 - \alpha A) V_\beta(t,s)\,ds - \beta A V_\beta(t,s)\,dW(s). \tag{0.24}$$

Consider the random function $\xi(s) = V_\beta(t,s)U(s,t_0)x$, $x \in D_A$, $\beta \neq 0$, $t \neq s$, in B. By Itô's formula, (0.23)–(0.24) imply

$$d\xi(s) =$$
$$((\beta^2 A^2 - \alpha A)V_\beta(t,s)U(s,t_0)x + V_\beta(t,s)\alpha A U(s,t_0)x - \beta A V_\beta(t,s)\beta A U(s,t_0)x)\,ds +$$
$$(-\beta A V_\beta(t,s)U(s,t_0)x + V_\beta(t,s)\beta A U(s,t_0)x)\,dW(s), \quad s \neq t.$$

Taking into account the fact that $V_\beta(t,s)$ commutes with A on D_A, we obtain $d\xi(s) = 0$ (almost surely), $s \neq t$. Proposition 0.4 implies that $\xi(s)$ is an (almost surely) continuous function at $s = t$ for $\beta \neq 0$, $x \in D_A$. Hence for any $s \in \mathbf{R}$ and $x \in D_A$,

$$d\xi(s) = 0 \quad \text{(a.s.)}.$$

Thus, $\xi(s)$ does not depend on s (almost surely). But $\xi(t) = V_\beta(t,t)U(t,t_0)x = U(t,t_0)x$ and $\xi(t_0) = V_\beta(t,t_0)U(t_0,t_0)x = V_\beta(t,t_0)x$ for any $x \in D_A$ (almost surely). The condition of boundedness of $V_\beta(t,t_0)$ for $\beta \neq 0$ and almost every trajectory of $W(t)$, and the fact that D_A is dense in B imply that $U(t,t_0) = V_\beta(t,t_0)$ (almost surely). □

REMARK 0.1. Suppose the spectrum of A consists of countably many spectral sets σ_k ($k = 1, 2, \dots$), and the resolvent $(\lambda I - A)^{-1}$ of A satisfies conditions guaranteeing the uniform boundedness of the projectors

$$P_k = \frac{1}{2\pi i} \int_{\gamma_k} (\lambda I - A)^{-1}\,d\lambda,$$

with γ_k a curve in ρ_A bounding a domain containing σ_k. In this case the solution (0.4) of the Cauchy problem (0.2)–(0.3) is unique (almost surely). Indeed, applying the operator P_k ($k = 1, 2, \dots$) to both parts of equation (0.2) and taking into account that all P_k commute with A on D_A, we obtain

$$\begin{gathered} d\xi_k(t) = \alpha A_k \xi_k(t)\,dt + \beta A_k \xi_k(t)\,dW(t), \quad t \in \mathbf{R}, \\ \xi_k(t_0) = P_k u_0, \end{gathered} \tag{0.25}$$

with $\xi_k(t) = P_k \xi(t)$ and $A_k = (1/2\pi i) \int_{\gamma_k} \lambda(\lambda I - A)^{-1}\,d\lambda$ being a bounded operator on B, $\xi(t) = \sum_{k=1}^{\infty} \xi_k(t)$. Then the (almost sure) uniqueness of the solution of the Cauchy problem (0.25) (see theorem VII.2.1) implies the (almost sure) uniqueness of the solution of the initial Cauchy problem.

PROPOSITION 0.6. *For any $x \in U_c(A)$, and any $x \in B$ when $\alpha \Delta t > 0$,*

$$\lim_{\beta \to 0} \| V_\beta(t,t_0)x - e^{\alpha \Delta t A} x \| = 0 \quad (a.s.).$$

PROOF. For $\alpha\Delta t > 0$ the family of operators $e^{\alpha\Delta tA}$ is defined on B, and is given by

$$e^{\alpha\Delta tA} = \frac{1}{2\pi i}\int_\gamma e^{\alpha\Delta t\lambda}(\lambda I - A)^{-1}\,d\lambda. \tag{0.26}$$

For $\alpha\Delta t < 0$ the family $e^{\alpha\Delta tA}$ is defined, as the inverse of the family $e^{-\alpha\Delta tA}$ on the set $U_c(A)$ which is dense in B (see Gorbachuk–Knyazyuk [1]).

Let $x \in U_c(A) \subset D_{Ae^{\alpha\Delta tA}}$, $\beta \neq 0$. Then

$$\|V_\beta(t,t_0)x - e^{\alpha\Delta tA}x\| =$$

$$\left\|\frac{1}{2\pi i}\int_\gamma e^{-\beta^2\lambda^2|\Delta t|/2+\beta\lambda\Delta W}[(\lambda I - A)^{-1} - (\lambda I)^{-1}]e^{\alpha\Delta tA}x\,d\lambda\right\| \leq$$

$$\left\|\frac{1}{2\pi i}\int_\gamma e^{-\beta^2\lambda^2|\Delta t|/2+\beta\lambda\Delta W}(\lambda I - A)^{-1}\frac{d\lambda}{\lambda}\right\| \cdot \|Ae^{\alpha\Delta tA}x\|. \tag{0.27}$$

Changing variables $\mu = |\beta|\lambda$ in (0.27), using the boundedness condition for the operator $(\mu I/|\beta| - A)^{-1}$ for $\mu \in \gamma$, $|\beta| < 1$, and Hölder's condition for the increment of a Wiener process, we obtain that for almost every trajectory of $W(t)$, any $\epsilon > 0$, and any Δt such that $0 < |\Delta t| < T$, there is a constant $c > 0$, depending on M, θ, α, T, and ϵ only, such that

$$\|V_\beta(t,t_0)x - e^{\alpha\Delta tA}x\| \leq c|\beta| \cdot \|Ae^{\alpha\Delta tA}x\|.$$

Thus, for any $x \in U_c(A)$, $\lim_{\beta\to 0}\|V_\beta(t,t_0)x - e^{\alpha\Delta tA}x\| = 0$ (almost surely).

The second part of the proposition follows from the facts that $U_c(A)$ is dense in B, and that for $\alpha\Delta t > 0$ there is a $\beta_0 > 0$ such that for any β, $0 < |\beta| < \beta_0$, the estimate (0.11) for the norm of $V_\beta(t,t_0)$ holds. □

REMARK 0.2. For $\alpha\Delta t < 0$ and $\beta = 0$ the operator (0.7) is unbounded, and is defined, as the limit of $V_\beta(t,t_0)$ as β tends to 0, on the set $U_c(A)$, which is dense in B:

$$e^{\alpha\Delta tA}x = \lim_{\beta\to 0} V_\beta(t,t_0)x, \quad x \in U_c(A).$$

PROPOSITION 0.7. *For $\alpha\Delta t > 0$ and any $x \in B$, $f \in B^*$,*

$$M\langle V_\beta(t,t_0)x, f\rangle = \langle e^{\alpha\Delta tA}x, f\rangle.$$

PROOF. Let $\alpha\Delta t > 0$. Then the family $e^{\alpha\Delta tA}$ can be defined by formula (0.26). Computing the expectation of the random variable $\langle V_\beta(t,t_0)x, f\rangle$, we find

$$\mathrm{M}\langle V_\beta(t,t_0)x, f\rangle = \mathrm{M}\left\langle\frac{1}{2\pi i}\int_\gamma \phi(\eta_\lambda(t))(\lambda I - A)^{-1}x\,d\lambda, f\right\rangle =$$

$$\mathrm{M}\frac{1}{2\pi i}\int_\gamma \phi(\eta_\lambda(t))\langle(\lambda I - A)^{-1}x, f\rangle\,d\lambda =$$

$$\int_{-\infty}^{\infty}\frac{e^{-\tau^2/(2|\Delta t|)}}{\sqrt{2\pi|\Delta t|}}\left[\frac{1}{2\pi i}\int_\gamma e^{\alpha\lambda\Delta t+\beta\lambda\tau-\beta^2\lambda^2|\Delta t|/2}\langle(\lambda I - A)^{-1}x, f\rangle\,d\lambda\right]d\tau.$$

Changing the order of integration we find

$$\mathrm{M}\langle V_\beta(t,t_0)x,f\rangle =$$

$$\frac{1}{2\pi i}\int_\gamma e^{\alpha\lambda\Delta t-\beta^2\lambda^2|\Delta t|/2}\left(\int_{-\infty}^{\infty}\frac{e^{\beta\lambda\tau-\tau^2/(2|\Delta t|)}}{\sqrt{2\pi|\Delta t|}}\,d\tau\right)\langle(\lambda I-A)^{-1}x,f\rangle\,d\lambda =$$

$$\left\langle\frac{1}{2\pi i}\int_\gamma e^{\alpha\lambda\Delta t}(\lambda I-A)^{-1}x\,d\lambda,f\right\rangle = \langle e^{\alpha\Delta tA}x,f\rangle. \quad \square$$

1. Supplementary remarks and historical comments

Stochastic parabolic partial differential equations with coefficients containing white noise were considered in Gikhman–Mestechkina [1], but they did not pay attention to the possibility of regularization of the ill-posed problem in this way. There is another approach to ill-posed problems via semigroup theory, as expounded in Krein [1].

The contents of this supplement is based on the authors' lecture (with minor corrections) at the Vth international Vilnius conference on probability theory and mathematical statistics (see Dalecky–Goncharuk [1]). Another version of the operator exponential (0.4) in a Hilbert space, based on the spectral decomposition of a selfadjoint operator A, was considered also in this lecture. In this case it turns out to be possible to consider the stochastic variant of the formula for differentiation of functions of selfadjoint operators with respect to a parameter (see Dalecky–Krein [1]).

Bibliography

Dalecky, Yu.L., Goncharuk, N.Yu.,
[1] Random operator functions defined by stochastic differential equations, *Proc. Vth Vilnius Conf. Probab. Theory and Math. Stat.* **1** (1990), pp. 433–445.

Dalecky, Yu.L., Krein, S.G.,
[1] Integration and differentiation of Hermitian operator functions and applications to perturbation theory, *Proc. Sem. Functional Anal., Voroneg* (1956), 81-105. (In Russian.)

Gikhman, I.I., Mestechkina, T.M.,
[1] The Cauchy problem for the parabolic equation with coefficients of the type of 'white noise', *Teor. Sluch. Prots.* **15** (1987), 19-28. (In Russian.)

Goodman, R.,
[1] Analytic and entire vectors for representations of Lie groups, *Trans. Amer. Math. Soc.* **143** (1969), 55-76.

Gorbachuk, V.I., Knyazyuk, A.V.,
[1] Boundary values of solutions of operator differential equations, *Uspekhi Mat. Nauk.* **44: 3** (1989), 55-91. (In Russian.)

Henry, D.,
[1] *Geometric theory of semilinear parabolic equations*, Springer, 1981. (Lecture notes in math. **840**.)

Krein, S.G.,
[1] *Linear differential equations in Banach space*, Amer. Math. Soc., 1971. (Translated from the Russian.)
Skorokhod, A.V.,
[1] *Random processes with independent increments*, Nauka, Moscow, 1986. (In Russian.)

Index

MIX
Papier aus verantwortungsvollen Quellen
Paper from responsible sources
FSC® C105338

If you have any concerns about our products,
you can contact us on
ProductSafety@springernature.com

In case Publisher is established outside the EU,
the EU authorized representative is:
Springer Nature Customer Service Center GmbH
Europaplatz 3, 69115 Heidelberg, Germany

Printed by Libri Plureos GmbH
in Hamburg, Germany